5TH EDITION
GLOBAL SHIFT

5TH EDITION

GLOBAL SHIFT

MAPPING THE CHANGING CONTOURS
OF THE WORLD ECONOMY

PETER DICKEN

THE GUILFORD PRESS
New York · London

First published in the United States of America by
The Guilford Press
A Division of Guilford Publications, Inc.
72 Spring Street, New York, NY 10012
www.guilford.com

Printed in the United States of America

This book is printed on acid-free paper.

Last digit is print number: 9   8   7   6   5   4   3   2   1

Library of Congress Cataloging-in-Publication Data

Dicken, Peter.
   Global shift: mapping the changing contours of the world economy / by Peter Dicken.—5th ed.
      p. cm.
Includes bibliographical references and index.
ISBN-13: 978-1-59385-436-2 ISBN-10: 1-59385-436-6 (pbk.; alk. paper)
ISBN-13: 978-1-59385-437-9 ISBN-10: 1-59385-437-4 (cloth; alk. paper)
   1. Industries — History — 20th century.   2. International economic relations.
3. International business enterprises.   4. Economic policy.
5. Globalization — Economic aspects.   6. Technological innovations —
Economic aspects. I. Title.
HD2321.D53 2007
338.09'051—dc22   2006030433

This one is for Harry, who arrived just too late last time

Praise for Global Shift

FIFTH EDITION

"A comprehensive, balanced, thorough, interdisciplinary review of one of the critical issues of our time. A 'must' for anyone interested in globalization."

—Stephen J. Kobrin, The Wharton School, University of Pennsylvania

"Impressive in the extent of empirical research, *Global Shift* successfully captures the historical continuities and basic changes marking the world economy. Peter Dicken's new edition is a vividly written guide to globalizing processes."

—James H. Mittelman, School of International Service, American University

"*Global Shift, Fifth Edition,* remains the benchmark for studies of the geography of globalization. In accessible prose, Dicken presents tightly argued propositions about the emerging economic landscape. The fields of international business, economic geography, international relations, and economic sociology can profitably use the book to communicate the fundamentals of globalization. Clear, effective, and engaging case studies are ideal for classroom use. There is no other text with such a broad reach or appeal for anyone interested in understanding the contemporary international economy."

—Amy K. Glasmeier, Department of Geography, The Pennsylvania State University

"*Global Shift* just keeps on getting better. There is no other source that gives you the full story on globalization in such a fluent and authoritative way. This book is not just recommended, but essential."

—Nigel Thrift, Vice-Chancellor, University of Warwick, UK

"With this edition of *Global Shift*, Dicken confirms his mastery as one of the preeminent authorities in the study of globalization. This careful and penetrating analysis of the complexities of a unifying world should prove a seminal text for students, scholars, and policymakers. If you wish to explore beyond 'flatland,' I can't recommend a better source."

—William E. Halal, Department of Management Science, George Washington University

"The fifth edition of *Global Shift* remains at the top of the ever more crowded field of globalization texts. Peter Dicken is a master of weaving together new theoretical arguments, visually compelling charts and graphs, and insightful industry case studies. If you had to use just one book to convey globalization's promise and perils, this is the book I would recommend."

—Gary Gereffi, Department of Sociology and Center on Globalization, Governance and Competitiveness, Duke University

FOURTH EDITION

"Dicken identifies states and transnational corporations as the two key actors in the multiple processes of restructuring and institutionalization that we usually call the global economy. In so doing, he has written a political economy of globalization and produced a far more comprehensive account than is typically the case in books about the global economy, most of which tend to confine the analysis to firms and markets."

—Saskia Sassen, Ralph Lewis Professor of Sociology, The University of Chicago

"In these uncertain times, it is reassuring to have Peter Dicken as our guide to the world economy. No other commentator has his eye equally attuned to both the big picture of global corporations and capital flows, and the fascinating stories of local places, people, and industries. In this new edition of *Global Shift*, Dicken shows us once again why he has become one of the most respected social scientists studying the world of global business and economy."

—Meric Gertler, Department of Geography, University of Toronto, Canada

"...the book presents not only a thorough and balanced description and analysis of globalization, but also a nuanced explanation of the globalization–antiglobalization debates and provocative examination of the distributional consequences of globalization. I will certainly continue to use *Global Shift* in my graduate seminar. In fact, I am contemplating using it in my introductory economic geography course as well."

—Robin Leichenko in *Economic Geography*

"...a solid 640-page text on the phenomena of globalization in the modern age....provides detailed case studies of crucial global industries, more than 200 updated figures and tables, and well serves to broaden and illustrate the critical points toward understanding the world's economic future. This is an ideal text for classroom instruction and recommended to the attention of non-specialist general readers with an interest in understanding the complexities of global economics."

—*Library Bookwatch*

THIRD EDITION

"*Global Shift* has become a landmark and a classic. It remains a popular text whose strength lies in its clear presentation and analysis of empirical data and in its focus on the production chain. This alone makes it a welcome corrective for the many speculative works on globalization based, as Dicken says, more 'on rhetoric and hype than on reality'."

—Paula Cerni in *Review of Radical Political Economics*

"By far the best and most readable account of the past three decades of economic globalization. Replete with maps, graphs, and tables, the book offers the clearest and most complete exposition of the scale and depth of the transformation currently affecting all societies."

—John O'Loughlin in *Lingua Franca*

"A first-rate and eminently readable work, with a unique blend of empirical and conceptual material and an analytical depth rarely achieved in textbooks. The third edition of *Global Shift* continues to be one of the most useful, interesting, and readable texts in the field of economic geography. I thoroughly recommend it both to students of geography and to readers in other disciplines who are interested in seeing what contemporary economic geography is really all about."

—John Holmes, Department of Geography, Queen's University, Kingston, Ontario, Canada

Summary of Contents

Contents

List of Abbreviations

AFTA	ASEAN Free Trade Agreement
APEC	Asia–Pacific Economic Cooperation Forum
ASEAN	Association of South East Asian Nations
CAFTA	Central American Free Trade Agreement
CSO	Civil society organization
CUSFTA	Canada–United States Free Trade Agreement
ECB	European Central Bank
EMU	European Monetary Union
EOI	Export-oriented industrialization
EPZ	Export processing zone
EU	European Union
FDI	Foreign direct investment
GATS	General Agreement on Trade in Services
GATT	General Agreement on Tariffs and Trade
GCC	Global commodity chain
GCSO	Global civil society organization
GDP	Gross domestic product
GNP	Gross national product
GPN	Global production network
GVC	Global value chain
ILO	International Labour Organization
IMF	International Monetary Fund
ISI	Import-substituting industrialization
JIT	Just-in-time
LDC	Less developed country
MAI	Multilateral Agreement on Investment
MFA	Multi-Fibre Arrangement
NAC	Newly Agriculturalizing Country
NAFTA	North American Free Trade Agreement
NIE	Newly Industrializing Economy
OECD	Organization for Economic Co-operation and Development

PTA   Preferential trading arrangement
RIA   Regional integration agreement
TNC   Transnational corporation
TPN   Transnational production network
TRIPS   Trade-related intellectual property rights
TRIMS   Trade-related investment measures
UNCTAD   United Nations Conference on Trade and Development
WTO   World Trade Organization

Preface to the Fifth Edition

As I write this preface to the fifth edition of *Global Shift*, it is exactly 20 years since the first edition was published. During that time, immense changes − significant global shifts − have occurred in the contours of the world economic map. For example, in 1986 the EU (then the EC) had only 12 members compared with today's 25; the NAFTA did not exist. The United States was obsessed by the economic threat posed by Japan (as was, to a lesser extent, the EC). The Soviet Union still existed, as did the economic bloc of countries in Eastern Europe it had created in 1949. Yugoslavia was heralded as a newly industrializing economy, along with what had recently come to be called the four East Asian 'tigers' (Hong Kong, Korea, Singapore, Taiwan). China had only just begun to open its economy to the outside world and was still a very small presence on the world's trade map. Certainly, China was not seen as the future economic giant it has since become. Nobody foresaw that Japan's phenomenal growth would grind to a halt in the late 1980s.

In the early 1980s, there was very little debate about what we now call 'globalization'. Certainly there was nothing like the deluge of books, articles and TV programmes or the street demonstrations that have become the norm today. A count of books and academic papers with 'global' or 'globalization' in the title shows only 13 between 1980 and 1984, and only a further 78 between 1985 and 1989. Then came the take-off − 600 between 1992 and 1996 − and the pace has not slackened since: the rate is now several hundred items every year. If anything the obsession with globalization has increased. So where does this fifth edition of *Global Shift* fit? What does the book set out to do? How does it do it? What is different about this edition? Who is the book aimed at?

The basic aims of this book are to provide a clear path through the dense thickets of what are large, often conflicting, often confusing debates and arguments about globalization: to show how the global economy works and what its effects are. It tries to separate the reality from the hype: to provide a balanced − but emphatically not an uncritical − perspective on a topic often richer in rhetoric than reality. It focuses on the longer-term, underlying processes of global economic change within which 'events of the moment' can be better understood. In particular, it aims to:

- present a comprehensive account of the changing contours of the global economic map, as well as of the globalization/anti-globalization debates
- provide a clear guide to how the global economy is continually being transformed, especially through the interactions between transnational corporations and states (and other interest groups in civil society), set within a rapidly changing technological environment, in which digital technologies are increasingly dominant
- demonstrate how these transformations create highly differentiated outcomes which impact very unevenly upon the lives and livelihoods of people and communities across the world and upon the environment
- explore ways in which such processes can be shaped to improve, rather than to diminish, people's lives, especially those most disadvantaged by the huge tides of change sweeping across them.

As in previous editions, the book is organized into four closely related, but distinct, parts. *Part One* focuses on the *shifting contours* of the global economy, both the debates about globalization as a phenomenon and the actual empirical changes that are occurring at different geographical scales on the global economic map. *Part Two* explores the closely connected *processes* creating these changes: the 'gales of creative destruction' set in motion by new technologies; the increasingly complex and extensive transnational production networks created and controlled by transnational corporations; the actions of states in their roles as containers of distinctive institutions and practices, as competitors, and as collaborators with other states; and the uneasy relationships between TNCs and states, as each tries to exercise bargaining power over the other. *Part Three* shows how these processes are worked out differently in different kinds of economic activity through *detailed case studies* of the clothing, automobile, semiconductor, agro-food, finance and logistics industries. Finally, *Part Four* is concerned with *winning and losing* in the global economy: the impact of TNCs on home and host countries; the economic and social problems facing developed and developing countries in the face of globalizing processes; the development of 'resistances' to globalization; and how the world might be made a better place for all.

What is new about the fifth edition? As in the case of its predecessors, it is emphatically not a mere cosmetic exercise. I don't believe in such things. I have always set out to produce the most up-to-date and comprehensive account of economic-geographic globalization. Hence, all the empirical data have been fully updated using the latest available sources as of early 2006. At the same time, all of the illustrations have been drawn or redrawn specially for this book. There are now 212 figures, many of them completely new for this edition, as well as 32 tables. Every chapter has been completely revised and extensively rewritten not just to take into account new empirical developments but also to incorporate new ideas on the shaping and reshaping of production, distribution and consumption in the global economy. In particular, I have introduced an entirely new chapter on the agro-food industries, one of the most sensitive and significant industries of all. I have substantially expanded the discussion of the role of global civil society organizations, problems of governance (including the tensions over trade regulation within the WTO and labour and environmental issues), and of alternative economic systems.

Global Shift is both a *cross-disciplinary* and a *multilevel* book. It deliberately spans, and draws from, a wide range of academic disciplines, including business and management, development studies, economics, economic geography, political science and sociology, amongst others. At the same time the book is designed for use at different levels. On the one hand, my aim has been to make the book accessible to readers without prior specialist knowledge by ensuring that all key terms are clearly defined, by avoiding excessive jargon, and by making extensive use of graphics. On the other hand, for the specialist reader, each chapter contains a set of end-of-chapter notes that connect to the extremely extensive and up-to-date research bibliography. Through such means, the book should be useful to graduate students and researchers, as well as to policy makers and to people in business. Certainly my experience of the reception of previous editions suggests that this is the case.

Without a rich network of friends and colleagues from all round the world, a book like this simply could not exist. To all of them, I offer my sincere thanks. However, several people deserve special mention. First of all, I want to thank Nick Scarle, Senior Cartographer at the University of Manchester. Nick has produced all the illustrations for every edition. Always superb, they have got better and better. Everybody who uses this book praises the graphics; they add so much. I can even forgive him for being a Yorkshireman.

I want to express my deep gratitude to all those who contributed to *Remaking the Global Economy* (Sage, 2003), especially Jamie Peck and Henry Yeung who conceived and edited the collection of essays. I will always remain deeply honoured and humbled by that wonderful gesture. Neil Coe and Roger Lee very generously devoted much time to making detailed comments on drafts of various chapters of this new edition and on its overall shape and content (despite their both supporting the wrong football teams). Roger, especially, has provided the most remarkable support and inspiration over many years and, above all, with Lesley, friendship and sharing of mutual interests and pleasures. Henry Yeung, as always, deserves a very special thank you, as does Liu Weidong, for introducing me to China and for translating the book into Chinese; Anders Malmberg, who suggested the subtitle (or was it Anna?) as well as many other ways of improving the book; and Martin Hess and Philip Kelly, who have contributed in more ways than they realize. Kris Olds and Ray Hudson made detailed evaluations of the various editions that made me think about some ways forward. Both Ray and Nigel Thrift commented very helpfully on the draft proposal for this edition. None of these, of course, bears any responsibility for all the weaknesses that remain but they can certainly claim credit for any strengths and I am enormously grateful to all of them. I am also very fortunate to remain part of the community of geographers at the University of Manchester and I much appreciate the friendship of colleagues there, especially Kevin Ward and Noel Castree (who do support the right football team).

Robert Rojek is the most caring, encouraging and stimulating editor, and Vanessa Harwood has again lavished her great skill and care on creating a visually stimulating book. To them, and to their colleagues at Sage Publications, notably Janey Walker, my very sincere thanks. Thanks, too, to Seymour Weingarten and the staff at The Guilford Press in New York.

Finally – as always – it all comes back to the people who matter to me most of all: my (now dispersed) family. My two young grandsons, Jack and Harry, have already learned to give me a bad time, encouraged by their father, mother and uncle. Thanks a lot, Michael, Sally and Chris! Above all, though, it is Valerie who makes everything worthwhile and who does so with so much love and tolerance. This is for you (again).

5TH EDITION
GLOBAL SHIFT

PART ONE
THE SHIFTING CONTOURS OF THE GLOBAL ECONOMY

One
Questioning 'Globalization'

Globalization is a fact ... Not just in finance, but in communication, in technology, increasingly in culture, in recreation. In the world of the Internet, information technology and TV, there will be globalization. And in trade, the problem is not there's too much of it; on the contrary there's too little of it ... The issue is not how to stop globalization. The issue is how we use the power of community to combine it with justice ... the alternative to globalization is isolation.

Tony Blair, UK Prime Minister, 2001

Globalization – as announced, promised and asserted to be inevitable in the 1970s, '80s and much of the '90s – has now petered out. Bits and pieces continue. Other bits have collapsed or are collapsing. Some are blocked. And a flood of other forces have come into play, dragging us in a multitude of directions ... The desire of people to organize their lives around the reality of where they live is central to the return of nationalism.

John Rawlston Saul, 2005

What in the world is going on?

'Globalization' is one of the most used, but also one of the most *mis*used and one of the most *con*fused, words around today. It's virtually everywhere. Hardly a day goes by without its being invoked by politicians, by academics, by business or labour union leaders, by journalists. In fact, there is an interesting geography of the awareness of, and attitudes towards, globalization.[1] In the last 25 years or so it has entered the popular imagination in a big way, although it is a concept whose roots go back at least to the nineteenth century notably in the ideas of Karl Marx. The current explosion of interest in 'globalization' reflects the pervasive feeling that something fundamental is happening in the world; that there are lots of 'big issues' that are somehow interconnected under the broad umbrella term 'globalization'. In the words of one contemporary commentator,

> We live in an era in which everything has changed and most things are still changing. The ice has melted on the familiar landscape of the second half of the 20th century. Power in all its forms is shifting rapidly and unpredictably. You might even say that we are at the beginning of history.[2]

Such feelings of uncertainty are intensified by an increased awareness that what is happening in one part of the world is deeply – and often very immediately – affected by events happening in other parts of the world. Part of this is simply the result of the revolution in electronic communications that has transformed the speed with which information spreads. But part of it is also to do with the fact that many of the things we use in our daily lives are derived more and more from an increasingly complex geography of production, distribution and consumption, whose scale has become, if not totally global, at least vastly more extensive, and whose choreography has become increasingly intricate. Many products, indeed, have such a complex geography – with parts being made in different countries and then assembled somewhere else – that labels of origin rarely have meaning any more. To the individual citizen the most obvious indicators of change are those which impinge most directly on her/his daily activities – making a living, acquiring the necessities of life, providing for their children to sustain their future.

In the industrialized countries, there is fear that the dual (and connected) forces of technological change and global shifts in the location of economic activities are adversely transforming employment prospects. The current waves of concern about the outsourcing and offshoring of jobs in the IT service industries (notably, though not exclusively, to India), or the more general fear that many manufacturing jobs are being sucked into a newly emergent China, suddenly growing at breakneck speed, are only the most recent examples of such fears. But the problems of the industrialized countries pale into insignificance when set against those of the poorest countries in what used to be called the 'Third World'. Although there are indeed losers in the developed and affluent countries their magnitude is totally dwarfed by the poverty and deprivation of much of Africa and of many parts of South Asia and of Latin America. The development gap continues to widen: the disparity between rich and poor continues to grow.

It is, of course, totally naïve to explain all such problems in terms of a single causal mechanism called 'globalization',

> a term … [too often] used by a lot of woolly thinkers who lump together all sorts of superficially converging trends … and call it globalization without trying to distinguish what is important from what is trivial, either in causes or in consequences.[3]

'Globalization', in fact, has become a convenient 'catch-all' term, used by many to bundle together virtually all the 'goods' and 'bads' facing contemporary societies.

Not surprisingly, then, it generates heated and polarized argument across the entire political and ideological spectrum. Most dramatic of all, since the turn of the millennium, has been the proliferation of global protest movements: the explosion of street demonstrations at major international political meetings. These have involved a remarkable *mélange* of pressure groups, ranging from long-established civil society organizations (CSOs) to totally new groups with either very specific, or very general, foci for their protest, together with anarchist and revolutionary elements with a broad anti-capitalist agenda.

First coming to real prominence at the meeting of the World Trade Organization in Seattle in November 1999, the global protest movements immediately became a permanent feature of every subsequent international meeting of governmental organizations. In some cases, they have manifested themselves in such globally generated events as the 'Live8' concerts in 2005, organized to coincide with the meeting of the world's largest economies (G8) and the 'Make Poverty History' campaign. At the other political extreme, the leaders of big business have their own 'tribal' meetings, most notably the World Economic Forum held every year at Davos in Switzerland.

Conflicting perspectives on 'globalization'

Because 'globalization' is such a highly contested term,[4] it is important to be clear about the position taken in this book. Its primary focus is the global *economy*. There are, of course, other forms of 'globalization' – political, cultural and social – and these are often difficult to separate. Indeed, the 'economy' itself is not some kind of isolated entity. Not only is it deeply embedded in social, cultural and political processes and institutions but also these are, themselves, often substantially imbued with economic values. This is especially so in the kind of capitalist market economy that now prevails throughout most of the world.

'Hyper-globalists' to the right and to the left
Probably the largest body of opinion – and one that spans the entire politico-ideological spectrum – consists of what might be called the *hyper-globalists,* who argue that we live in a borderless world in which the 'national' is no longer relevant. In such a world, 'globalization' is the new economic (as well as political and cultural) order. It is a world where nation-states are no longer significant actors or meaningful economic units and in which consumer tastes and cultures are homogenized and satisfied through the provision of standardized global products created by global corporations with no allegiance to place or community. Thus, the 'global' is claimed to be the *natural* order, an inevitable state of affairs, in which time–space has been compressed, the 'end of geography' has arrived and everywhere is becoming the same. Such a hyper-globalist view is shown in Figure 1.1 as an inexorable process of increasing geographical spread and increasing functional integration between economic activities.

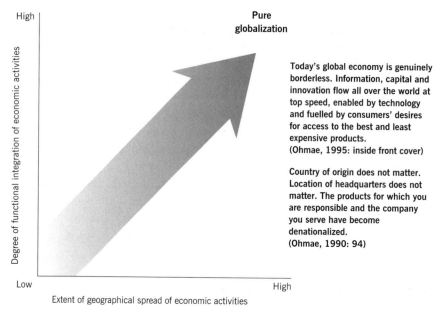

High

Pure
globalization

Degree of functional integration of economic activities

Today's global economy is genuinely
borderless. Information, capital and
innovation flow all over the world at
top speed, enabled by technology
and fuelled by consumers' desires
for access to the best and least
expensive products.
(Ohmae, 1995: inside front cover)

Country of origin does not matter.
Location of headquarters does not
matter. The products for which you
are responsible and the company
you serve have become
denationalized.
(Ohmae, 1990: 94)

Low

High

Extent of geographical spread of economic activities

Figure 1.1   The hyper-globalist view

This hyper-globalist view of the world is a myth. It does not – and is unlikely
to – exist. Nevertheless, its rhetoric retains a powerful influence on politicians,
business leaders and many other interest groups. It is a world-view shared by many
writers, on both the right and the left. Where they differ is in their evaluation of the
situation and in their policy positions. To the 'neo-liberals' on the right – the *pro*-
globalizers – globalization is a political-economic project,[5] one which, it is argued,
will bring the greatest benefit for the greatest number. Simply let free markets
(whether in trade or finance) rule and all will be well. The 'rising tide' of globalization will 'lift all boats'; human material well-being will be enhanced. Although the
neo-liberal pro-globalizers recognize that such a state of perfection hasn't yet been
achieved, the major problem, in their view, is that there is too little, rather than too
much, globalization.[6]

To the hyper-globalizers of the left – the *anti*-globalizers – the problem is
globalization itself. The very operation of those market forces claimed to be
beneficent by the right are regarded as the crux of the problem: a malign and
destructive force. Markets, it is argued, inevitably create inequalities. By extension,
the globalization of markets increases the scale and extent of such inequalities.
Unregulated markets inevitably lead to a reduction in well-being for all but
a small minority in the world, as well as creating massive environmental problems.
Markets, therefore, *must* be regulated in the wider interest. To some anti-
globalists, in fact, the only logical solution is a complete rejection of globalization
processes and a return to the 'local'.

'Sceptical internationalists'

Although the notion of a global*ized* economic world has become widely accepted, some adopt a more *sceptical* position, arguing that the 'newness' of the current situation has been grossly exaggerated. The world economy, it is claimed, was actually more open and more integrated in the half century prior to World War I (1870–1913) than it is today.[7] The empirical evidence used to justify this position is *quantitative* and *aggregative*, based on national states as statistical units. Such data reveal a world in which trade, investment and, especially, population migration flowed in increasingly large volumes between countries. Those levels of international trade and investment were not reached again (after the world depression of the 1930s and the Second World War) until the later decades of the twentieth century. Indeed, international population migration has not returned to those earlier levels, at least in terms of the proportion of the world population involved in cross-border movement. On the basis of such quantitative evidence Hirst and Thompson assert that 'we do not have a fully globalized economy, we do have an international economy'.[8]

'Grounding globalization'

Such national-level quantitative data need to be taken seriously. But they are only part of the story. They do not tell us what kinds of *qualitative* changes have been occurring in the global economy. Most important have been the changes in both the *where* and the *how* of the material production, distribution and consumption of goods and services (including, in particular, finance). Old geographies of production, distribution and consumption are continuously being disrupted; new geographies of production, distribution and consumption are continuously being created. There has been a huge transformation in the *nature* and the *degree* of interconnection in the world economy and, especially, in the *speed* with which such connectivity occurs, involving both a *stretching* and an *intensification* of economic relationships.[9]

International economic integration before 1914 – and even until only about four decades ago – was essentially *shallow* integration, manifested largely through arm's-length trade in goods and services between independent firms and through international movements of portfolio capital and relatively simple direct investment. Today, we live in a world in which *deep integration*, organized primarily within and between geographically extensive and complex *transnational production networks*, and through a diversity of mechanisms, is increasingly the norm.

Such qualitative changes are simply not captured in aggregative trade or investment data of the kind used by the sceptics. For example, in the case of international trade, what matters are not so much changes in volume – although these are important – as changes in *composition*. There has been a huge increase in both *intra-industry* and *intra-firm* trade, both of which are clear indicators of more functionally fragmented and geographically dispersed production processes.[10] There have been, too, dramatic changes in the operation of financial markets, with 'money' moving round the world at unprecedented speeds, generating enormous repercussions for national and local economies.

The position taken in this book is that although there are undoubtedly global-*izing* forces at work, we do not have a fully global*ized* world economy as depicted in Figure 1.1. Indeed, what we have is 'not a single, unified phenomenon, but a *syndrome* of processes and activities'.[11]

> Globalization is a … supercomplex series of multicentric, multiscalar, multi-temporal, multiform and multicausal processes.[12]

Globalizing processes, therefore, are reflected in, and influenced by, multiple geographies, rather than a single global geography: the 'local and the global inter-mesh, running into one another in all manner of ways'.[13] In fact, as Figure 1.2 shows, several tendencies can be identified, reflecting different combinations of geographical spread and functional integration or interconnection:

- *localizing* processes: geographically concentrated economic activities with varying degrees of functional integration
- *internationalizing* processes: simple geographical spread of economic activities across national boundaries with low levels of functional integration
- *globalizing* processes: both extensive geographical spread and also a high degree of functional integration
- *regionalizing* processes: the operation of 'globalizing' processes at a more geographically limited (but supranational) scale, ranging from the highly integrated and expanding European Union to much smaller regional economic agreements.

Unravelling the complexity of the new geo-economy: economies as networks

Even allowing for the hype of much of the globalization debate, there is no doubt that we *are* witnessing the emergence of a new geo-economy that is *qualitatively different* from the past.[14] To understand what is going on, however, we need to adopt an approach that is firmly *grounded* in the uneven geographical reality of globalizing structures, processes and outcomes and not submerged in the hype and inflated language that characterizes so much of the globalization debates. The question is: how can we begin to unravel its dynamic, kaleidoscopic complexity?

Figure 1.3 represents one point of entry. It seeks to provide both a structural perspective on globalization processes and outcomes and also a sense of how the

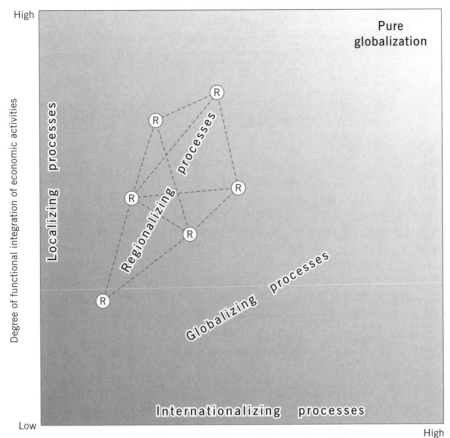

Figure 1.2   Processes and scales of global economic transformation

key 'actors' behave. In particular, it emphasizes the complex ways in which they are interconnected and governed through highly unequal power relationships. Of course, a simplified diagram like Figure 1.3 attempts the impossible: to capture and represent the multidimensionality of the global economy in just two dimensions. Inevitably, it both grossly oversimplifies, and also distorts, relationships and causalities. We need to bear in mind that it is an *idealized* representation of a world that is infinitely more complex.

We can choose to cut into the highly interconnected system of Figure 1.3 at many different points, according to our specific interest. In this book, I choose to focus primarily (though not exclusively) on the 'central slice' of Figure 1.3: the major actors in the global economy and the webs of *networked relationships* between them.

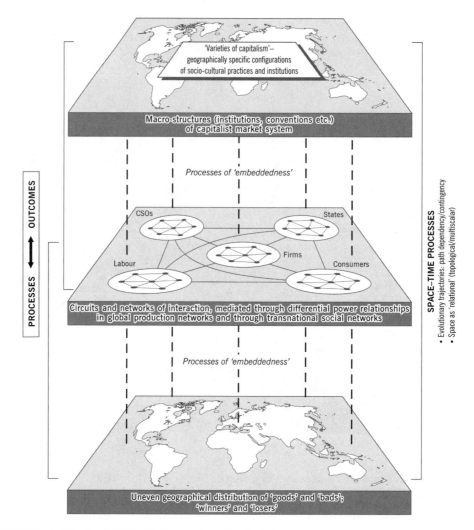

Figure 1.3   A simplified framework of the global economy

Source: based on Dicken 2004: Figure 2

> Economic processes must be conceptualized in terms of a complex circuitry
> with a multiplicity of linkages and feedback loops rather than just 'simple'
> circuits or, even worse, linear flows.[15]

Here lies the key: to think of economic processes (production, distribution,
consumption) in terms of *connections* of activities, linked through *flows* of both
material and non-material phenomena (like services) into circuits and networks.
Such circuits and networks constitute *relational structures* and *processes* in which the
power relationships between the key 'actors' – firms, states, individuals, social groups –
are uneven.

> The critical point about networks is that they involve relational thinking. What links people together across time and space? How are things and people connected and embedded economically, politically, and culturally? In what ways do goods and information and capital flow and why are they channelled down particular vertices and nodes? ... Thinking in terms of networks forces us to theorize socioeconomic processes as intertwined and mutually constitutive.[16]

Networks are always in flux; they are always in a process of becoming. They also, of course, do not exist in isolation. In particular, they are embedded within the broader macro-structures of the global economy as well as grounded in the prevailing geographical structures of the real world. In that sense, what is being adopted here is a *situated network* approach.

Macro-structures of the global economy

The macro-structures of the global economy are essentially the institutions, conventions and rules of the capitalist market system. These are not naturally given but socially constructed – in their present form predominantly as a neo-liberal political-economic ideology. The rules and conventions of the capitalist market economy relate to, for example, private property, profit-making, resource allocation on the basis of market signals, and the consequent commodification of production inputs (including labour).

The International Monetary Fund (IMF), the World Trade Organization (WTO) and the World Bank, together with the various 'G' meetings (such as the G8), are the most obvious manifestations of global institutions (see Chapter 19), although there is, of course, a myriad of other, more specific, bodies. These institutions are only a part of the broader socio-cultural matrix of practices, rules and conventions that shape how the world works. Such institutions and conventions continue to be manifested in *specific configurations* in specific places (notably within national states, but not only at that scale). In other words, they, too, are *territorially embedded*. There are *varieties of capitalism*, not one single universal form.

Actor-centred networks

Within this geographically differentiated macro-structural framework, it is primarily the actions of, and especially the interactions between, the five actor-centred networks shown in the central section of Figure 1.3 that shape the changing geographical configuration of the global economy at different spatial scales. Despite the suggestion sometimes made that we should not 'privilege' one set of actors over others, there is no doubt that some actors *are* more significant than others. It is for this reason that, throughout this book, I place particular emphasis on transnational corporations (TNCs) and states. More generally, however, the entire system is one of *asymmetric power relations*.

Significant variables in determining relative power are, first, control over key assets (such as capital, technology, knowledge, labour skills, natural resources, consumer markets) and, second, the spatial and territorial range and flexibility of each of the actors. The two are not unconnected. Ability to control access to specific assets is a major bargaining strength. Where such assets are available virtually everywhere, then the power gradient is shallow or even non-existent. But where assets are 'localized', whether geographically, organizationally, or even personally, then the power gradient may be very steep. However, actors able to tap into localized assets across geographical space have a significant advantage over those without such spatial flexibility.

This leads to a more general observation. Each of the major sets of actors shown in the central part of Figure 1.3 is involved in *both* cooperation and collaboration on the one hand *and* conflict and competition on the other. Such apparently paradoxical behaviour warns us against assuming that relationships between certain actors are always of one kind: for example, that those between TNCs, or between TNCs and states, or between TNCs and labour are always conflictual or competitive. Or, conversely, that relationships between groups of workers or labour organizations are always cooperative (in the name of class solidarity). Not so. These various actor networks are imbued with an ever-changing mixture of both conflict and collaboration.

So, for example, TNCs in the same industry are fierce competitors but also, invariably, enmeshed in a complex web of collaborative relationships (see Chapter 5). States compete in cut-throat fashion with other states to entice internationally mobile investment by TNCs (see Chapter 8) or to find ways to keep out certain types of imports whilst, at the same time, increasingly engaging in preferential trading arrangements, including bilateral and multilateral agreements, often within broader regional groupings (see Chapters 6 and 7). Labour unions in one country engage in competition with labour unions in other countries in the scramble for new, or to protect existing, jobs whilst, at the same time, unions strive to create international alliances with unions in other countries, especially those involved in the geographically dispersed operations of major TNCs. Civil society organizations (CSOs), likewise, are not immune from these conflicting actions. In the context of the anti-globalization protests, for example, CSOs have developed collaborations across national boundaries but, at the same time, the goals and values of individual CSOs are not always compatible, to say the least.

Mapping outcomes of globalization

Globalization is, then, a complex syndrome of processes, in which actor networks and macro-structures interconnect in extremely intricate and dynamic ways. We need to map, and to analyse, the concrete outcomes of these processes. We do this in general terms in Chapter 2 and, more specifically, for individual industries (clothing, automobiles, semiconductors, agro-food, financial services, and the logistics and distribution industries) in the case study chapters of Part Three. Such processes not only are geographically grounded and embedded (in the sense of deriving some of their characteristics and resources from place-specific contexts) but also generate geographically specific, highly uneven, concrete outcomes (Figure 1. 3). Of course, the geography of outcomes at one point in time constitutes the context within which subsequent processes operate. The whole process is circuitous and highly path dependent.

The processes of production, distribution and consumption may generate either 'good' or 'bad' outcomes. They produce 'goods' in the form of employment opportunities, incomes, and access to an increasing variety of consumer products, services and cultural artefacts. They produce 'bads' in the form of unemployment, poverty, resource depletion, environmental pollution and cultural damage. The extent to which the 'goods' exceed the 'bads' is a contested issue, as is the issue of who are the 'winners' and who are the 'losers', because such goods and bads are, themselves, highly unequally distributed both geographically and socially. These are the issues we address in the chapters of Part Four.

Production circuits; production networks

The conventional unit of analysis of the global economy is the country. Virtually all the statistical data on production, trade, investment and the like are aggregated into national 'boxes'. Indeed, the word 'statistics' originally denoted facts collected about the 'state'. However, such a level of statistical aggregation is less and less useful in light of the changes occurring in the organization of economic activity. Unfortunately, we have to rely heavily on national-level data to explore the changing maps of production, trade and investment. But, because national boundaries no longer 'contain' production processes in the way they once did, we need to find ways of getting both below and above the national scale – to break out of the constraints of the 'national boxes' – in order to understand what is really going on in the world. One way is to think in terms of production circuits and networks. These cut through, and across, all geographical scales, including the bounded territory of the state.

Production chains or production circuits?

It has become common to conceive of the production of any good or service as a production *chain*, that is, as a transactionally linked sequence of functions in which each stage adds value to the process of production of goods or services. In the global context, the idea of the *global commodity chain* (GCC) or, more recently, the *global value chain* (GVC) has been extensively developed by Gary Gereffi and his colleagues.[17] As its name suggests, a production chain is essentially linear. It represents the sequence of operations required to produce and distribute a good or service (services, like any other item of consumption, have to be 'produced'). But, as we have seen, economic processes are circuitous, rather than linear. There is a circularity involved in connecting the major stages of the production process, such circularity consisting of the feedback loops connecting consumption – a fundamental, though often neglected, component of the production process – with the processes of production and distribution.

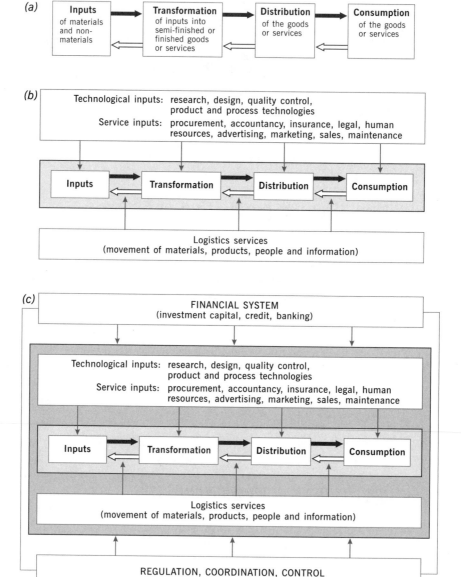

Figure 1.4   **Basic components of a production circuit**

Figure 1.4a shows a 'stripped-down' version of a hypothetical production circuit. At the core is a set of four basic operations, connected by a series of trans-actions between one element and the next. Inputs are transformed into products which are distributed and consumed. But note that the processes are two-way:

- flows of materials, semi-finished goods and final products in one direction
- flows of information (the demands of customers – tastes, preferences etc.) and money (payments for goods and services) in the other direction.

But there is much more to it than this, as Figures 1.4b and 1.4c show. Each indi-vidual element in the production circuit depends upon:

- technological inputs
- service inputs
- logistical (movement) systems
- financial systems
- coordination and control systems.

Hence, each of the individual elements in a production circuit depends upon many other kinds of input, both those directly related to production and also those related to circulation. In particular,

> service activities not only provide linkages between the segments of produc-tion within a [production circuit] and linkages between overlapping [production circuits], but they also bind together the spheres of production and circulation. Services have come to play a critical role … because they not only provide geographical and transactional connections, but they *integrate* and *coordinate* the atomized and globalized production process.[18]

Financial systems are especially important. The decisions of financiers have an extra-ordinarily powerful role not only in 'lubricating' production circuits but also in shaping them through their evaluative decisions on what (and where) to invest in order to gain the highest (and sometimes the quickest) return.

Production networks

Individual production circuits are, themselves, enmeshed in broader *production networks* of inter- and intra-firm relationships.[19] Such networks are, in reality, extremely complex structures with intricate links – horizontal, vertical, diagonal – forming multidimensional, multilayered lattices of economic activity. They vary considerably both within, and between, different industries, as we shall see in the case study chap-ters of Part Three. The complex nature of production networks will be discussed in detail in Chapter 5. At this stage, we simply need to note that three dimensions of production networks are especially important:

- *governance* – how they are coordinated and regulated
- *spatiality* – how they are configured geographically
- *territorial embeddedness* – the extent to which they are connected into particular bounded political, institutional and social settings.

Governance of production networks

In market economies, production networks are coordinated and regulated primarily by *business firms,* through the multifarious forms of intra- and inter-organizational relationships that constitute an economic system. As Figure 1.5 shows, economies are made up of many different types of business organization – transnational and domestic, large and small, public and private – in varying combinations and interrelationships. The firms in each of the segments shown in Figure 1.5 operate over widely varying geographical ranges and perform rather different roles in the economic system.

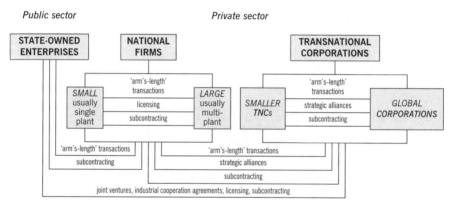

Figure 1.5   Types of firms in an economy

A major theme of this book is that the *transnational corporation* plays the key role in coordinating production networks and, therefore, in shaping the new geo-economy (see Chapters 4 and 5). A broad definition of the TNC – one that goes beyond the conventional definition based upon levels of ownership of internationally based assets – is used here to capture the diversity and complexity of transnational networks.

> *A transnational corporation is a firm that has the power to coordinate and control operations in more than one country, even if it does not own them.*

In fact, TNCs generally do own such assets but they are also, as we will see in Chapter 5, typically involved in intricate and multiple spiders' webs of collaborative relationships with other legally independent firms across the globe.

The nature of the coordination process within a TNC's production network depends, in part, on where the firm draws the boundary between those functions it *internalizes* (i.e. performs 'in-house') and those it *externalizes* (i.e. outsources to other firms). Theoretically, at one extreme, the whole TNC production network may be internalized within the firm as a *vertically integrated* system crossing national boundaries. In this case, the links consist of a series of *internalized transactions*, organized 'hierarchically' through the firm's internal organizational structure. At the other extreme, each function may be performed by separate firms. In this case, the links consist of a series of *externalized transactions*, organized either through 'the market' or in collaboration with other firms in a kind of 'virtual' network.

This dichotomy – between externalized, market-governed transactions and internalized, hierarchically governed transactions – grossly simplifies the richness and diversity of the governance mechanisms in the contemporary economy. In fact, there is a *spectrum* of different forms of coordination, consisting of networks of interrelationships within and between firms. Such networks increasingly consist of a mix of intra-firm and inter-firm structures. These networks are dynamic and in a continuous state of flux; the boundary between internalization and externalization is continually shifting. Precisely how they are coordinated depends, to a considerable degree, on the precise nature of the production, distribution and consumption processes involved. We look at this in some detail in Chapter 5 and in the industry cases of Part Three. Suffice it to note at this stage that the governance of production networks reflects the particular configuration of *power* within them. In some cases, one dominant actor calls all the shots; in other cases, power may be more widely dispersed with a greater degree of collaboration involved.

Spatiality of production networks

Every production network has *spatiality*: the particular geographical configuration and extent of its component elements and the links between them. At the most basic level, production within a network may be organized along a spectrum from geographically concentrated to geographically dispersed. But such terms are relative and beg the important question of geographical scale. In one sense, we may simply think of geographical scale as a continuum and, therefore, of production networks as being 'more or less long and more or less connected'.[20] Today, virtually all production networks have become geographically more extensive: they are increasingly long and increasingly more connected.

We are witnessing, therefore, the emergence of *global production networks* (GPNs) or, as some would prefer to call them, *transnational production networks* (TPNs). One, although not the only, reason for such increased geographical spread has been the 'revolutions' in transportation and communications technologies. Such transformations of time–space relationships have led some to claim 'the end of geography' or 'the death of distance'.[21] These two phrases resonate, either explicitly or implicitly, throughout much of the globalization literature. But although developments in transportation and communications have, indeed, 'shrunk the world', they have not done so in the simplistic way often envisaged (see Chapter 3).

Global/transnational production networks span global, regional, national and local scales. Thinking of the world in terms of such discrete spatial scales is helpful in many ways. But it can too easily imply that each scale is a self-contained 'box'. For example, it has become very common to separate the global scale from the local (or the national) scale and to imply that the processes operating at these two scales are separate and distinct: in particular that the 'global' determines what happens at the local scale. That is not the case, as we saw earlier. The spatiality of production networks is far more complex than is often claimed.

Territorial embeddedness of production networks

Capital, it is often argued, has become 'hyper-mobile', freed from the 'tyranny of distance' and no longer tied to 'place'. In other words, economic activity is becoming 'deterritorialized' or 'disembedded'. The sociologist Manuel Castells argues that the forces of globalization, especially those driven by the new information technologies, are replacing this 'space of places' with a 'space of flows'.[22] Anything can be located anywhere and, if that does not work out, can be moved somewhere else with ease.

Seductive as such ideas might seem, they are highly misleading. The world is *both* a 'space of places' *and* a 'place of flows'. Production networks don't just float freely in a spaceless/placeless world. Although transportation and communications technologies have, indeed, been revolutionized, both geographical distance and, especially, place are fundamental. Every component in a production network – every firm, every economic function – is, quite literally, 'grounded' in specific locations. Such grounding is both physical (in the form of the built environment) and also less tangible (in the form of localized social relationships and in distinctive institutions and cultural practices). Hence, the precise nature and articulation of firm-centred production networks are deeply influenced by the concrete socio-political, institutional and cultural contexts within which they are embedded, produced and reproduced.[23]

The *national state* continues to be the most important bounded territorial form in which production networks are embedded. All the elements in a production network are regulated within some kind of political structure, whose basic unit is the national state, but which also includes such supranational institutions as the International Monetary Fund and the World Trade Organization, regional economic groupings such as the European Union or the North American Free Trade Agreement, and 'local' states at the subnational scale.

All *transnational* production networks, by definition, have to operate within *multiscalar* regulatory systems. They are, therefore, subject to a multiplicity of geographically variable political, social and cultural influences. On the one hand, TNCs attempt to take advantage of national differences in regulatory regimes whilst, on the other hand, states attempt to minimize such 'regulatory arbitrage'. The result is a very complex situation in which firms and states are engaged in various kinds of power play: a triangular nexus of interactions comprising firm–firm, state–state and firm–state relationships (Figure 1.6).[24] In other words, the new geo-economy is essentially being structured and restructured not simply by the actions of either firms or states alone but also by complex, dynamic interactions between the two sets of institutions. Of course, relationships with states are not the only ones involved. As we saw in Figure 1.3, TNCs are continuously engaged in relationships (sometimes conflictual, sometimes collaborative) with other major actors – labour, consumers, CSOs – all of which have strong territorial bases.

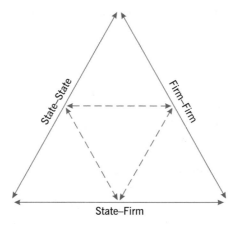

Figure 1.6   The triangular nexus of relationships between firms and states

Source: based on Stopford and Strange, 1991: Figure 1.6

Consumption as a driving force

Production circuits and production networks involve more than just 'production': they are driven, ultimately, by the necessity, the willingness and the ability of customers to acquire and consume the products themselves, and to continue doing so (see Figure 1.4). And yet, despite its central importance, consumption rarely figures in the script. It is important to redress this imbalance. Each of the case study chapters of Part Three shows how the nature of consumption varies according to the specific sector involved. Here, we need simply to emphasize some basic aspects of consumption processes.

First, we need to distinguish between the consumption of 'producer' goods or services (sometimes called intermediate products because they are purchased by firms within a production circuit for further transformation) and 'consumer' goods or services ('final demand' goods: those purchased by individuals and households).[25] Consumption is very much more than the economic process of 'demand'. Obviously, it is greatly influenced by levels of income. But it is also a complex set of *social and cultural* processes, in which all kinds of personal motivations are involved. People buy (or aspire to buy) particular goods for a bewildering variety of reasons, ranging from the satisfaction of basic *needs* to ensure survival (food, shelter, clothing) through to ever more sophisticated *wants* (discretionary goods, such as fashionable clothing, particular kinds of car, exotic or organic foods and the like). The social psychologist Abraham Maslow saw these needs/wants as forming a hierarchy, with 'lower' needs being satisfied before those of 'higher' needs (or wants).[26] But that is a little too simplistic.

Consumption may be driven by the desire to acquire particular kinds of products (even specific varieties or brands) either because they are regarded as desirable in themselves or because they send out social messages signifying the particular lifestyles, attitudes, social positions or self-evaluations of the consumer. 'Positional' goods have become increasingly important. However, they lose their value as more and more people have access to them. New positional goods have to be sought. [27]

> 'The material object being sold is never enough' … Commodities meet both the functional and symbolic needs of consumers. Even commodities providing for the most mundane necessities of daily life must be imbued with symbolic qualities and culturally endowed meanings.[28]

It is, of course, precisely these symbolic qualities of consumption that the advertising, retailing and media industries attempt to manipulate.

How far consumption is, or can be, manipulated in such ways is open to question. Some argue that consumption (and consumers) is becoming increasingly more important in the global economy than production (and producers). In Miller's view, the consumer has become the 'global dictator' and he describes consumption as the 'vanguard of history'.[29] Some support for such a view is provided by the increasing significance of retailers in many production circuits. Gereffi, for example, has argued that *buyer-driven* (rather than 'producer-driven') production networks are becoming increasingly important in 'those industries in which large retailers, brand-named merchandisers, and trading companies play the pivotal role in setting up decentralized production networks in a variety of exporting countries'.[30]

The bewildering proliferation of choice within many product areas is a direct reflection of producers' perceived need to meet the increasingly fragmented demands of consumers. The days when Henry Ford could dictate to his potential customers by telling them that they could have any colour Model T, as long as it was black, are long gone. Of course, in many cases the variety on offer is more apparent than real (heavily advertised 'newness' often being little more than superficial modification). But, in some cases, there is no doubt that consumer demands directly drive production circuits. It is also clear that the emergence of the Internet (Chapter 3) is transforming the abilities of consumers to make informed choices:

> Consumers select what they want from a far greater variety of sources – especially with a few clicks of a computer mouse. Thanks to the internet, the consumer is finally seizing power … consumer power has profound

implications for companies, because it is changing the way the world shops. Many firms already claim to be 'customer-driven' or 'consumer-centric'. Now their claims will be tested as never before … it is also intensifying competition. Today, window-shopping takes place online. People can compare products, prices and reputations.[31]

It is extremely important, therefore, to integrate consumption into our analysis of the global economy. However, whether or not the consumer or the producer is the dominant player is not really the point. What matters is that we must see the entire circuit of production, distribution and consumption as a *dynamically interconnected whole* (as in Figure 1.4). In some circumstances, power will lie at different positions within the circuit and different circuits may have different configurations of power. The question is an empirical one.

Even in a globalizing world, economic activities are geographically localized

The view of the 'hyper-globalizers' is that increasing geographical dispersal at a global scale is now the norm. But, if that is the case, why do *geographical concentrations* of economic activity not only still exist but also represent the normal state of affairs? Why do 'sticky places' continue to exist in 'slippery space'?[32] When we look at the geo-economic map we find tendencies of *both* concentration *and* dispersal – but with a very strong propensity for economic activities to agglomerate into *localized geographical clusters*.[33]

The bases of geographical clusters

Figure 1.7 identifies two types of geographical cluster: *generalized* and *specialized*. Both are based on the notion of *externalities*, the positive 'spillovers' created when activities in a particular place are connected with one another, either directly (through specific transactions) or indirectly. Both are based on the idea that the 'whole' (the cluster) is greater than the sum of the parts because of the benefits that spatial proximity provides.

- *Generalized clusters* simply reflect the fact that human activities tend to agglomerate to form urban areas. Hence, such benefits have traditionally been labelled *urbanization economies*. General clustering of activities creates the basis for sharing the costs of a whole range of services. Larger aggregate demand in, say, a large city encourages the emergence and growth of a variety of infrastructural, economic, social and cultural facilities that cannot be provided where their customers are geographically dispersed. The larger the city, quite obviously, the greater the variety of available facilities and vice versa.

- *Specialized clusters*, on the other hand, reflect the tendency for firms in the same, or closely related, industries to locate in the same places to form what are sometimes termed 'industrial districts' or 'industrial spaces'. Such benefits have been called *localization economies*. The bases of specialized clusters arise from the geographical proximity of firms performing different – but *linked* – functions in particular production networks.

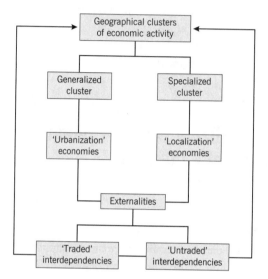

Figure 1.7   The bases of geographical clusters

Clusters generate two types of interdependency:

- *Traded interdependencies* are direct transactions between firms in the cluster (e.g. the supply of specialized inputs of intermediate products and services). In such circumstances, spatial proximity is a means of reducing transaction costs either through minimizing transportation costs or by reducing some of the uncertainties of customer–supplier relationships.
- *Untraded interdependencies* are the less tangible benefits, ranging from the development of an appropriate pool of labour, to particular kinds of institutions (such as universities, business associations, government institutions and the like), to broader socio-cultural phenomena. In particular, geographical agglomeration or clustering facilitates three important processes: face-to-face contact; social and cultural interaction; and enhancement of knowledge and innovation.

Why do clusters develop in the first place?

But why do clusters form in the first place? Why do they arise in one place rather than another? And how do they develop over time? These are difficult questions to answer. The reasons for the origins of specific geographical clusters are highly contingent and often shrouded in the mists of history.

> Within broad limits the power of attraction today of a center has its origin mainly in the historical accident that something once started there, and not in a number of other places where it could equally well or better have started, and that the start met with success.[34]

Once established, a cluster tends to grow through a process of *cumulative, self-reinforcing development* involving:

- attraction of linked activities
- stimulation of entrepreneurship and innovation
- deepening and widening of the local labour market
- economic diversification
- enrichment of the 'industrial atmosphere'
- 'thickening' of local institutions
- intensification of the socio-cultural milieu
- enhanced physical infrastructures.

The cumulative nature of these processes of localized economic development suggests that the process is *path dependent*. In other words, an economy becomes 'locked into' a pattern that is strongly influenced by its particular history. This may be either a source of continued strength or, if it embodies too much organizational or technological rigidity, a source of weakness. However, even for 'successful' regions, such path dependency does not imply the absolute inevitability of continued success. Rigidity of local practices may reduce the capacity to adapt to external changes.

Networks of networks

The global economy, therefore, can be envisaged as the linking together of two sets of networks:

- the *organizational* (in the form of production circuits and networks)
- the *geographical* (in the form of localized clusters of economic activity).

The major advantage of adopting such a grounded network approach to understanding the global economy is that it helps us to appreciate the interconnectedness of economic activities across different geographical scales and within and across territorially bounded spaces. The production of any commodity, whether it is a manufactured product or a service, involves an intricate articulation of individual activities and transactions across space and time. Such production networks – the nexus of interconnected functions and operations through which goods and services are produced and distributed – have become both organizationally and geographically more complex. Global and regional production networks not only integrate firms (and parts of firms) into structures which blur traditional organizational boundaries (for example, through the development of diverse forms of equity and non-equity relationships) but also integrate national and local economies (or parts of such economies) in ways which have enormous implications for their economic development and well-being. At the same time, the specific characteristics of national and local economies influence and 'refract' the operation and form of larger-scale processes. In that sense, 'geography matters' a lot.

The process is especially complex because, while states and local economies are essentially territorially specific, production networks themselves are not.[35] Production networks 'slice through' boundaries in highly differentiated ways, influenced in part by regulatory and non-regulatory barriers and by local socio-cultural conditions, to create structures which are 'discontinuously territorial'. This has major implications for the relative bargaining powers of the actors involved. The geo-economy, therefore, can be pictured as a geographically uneven, highly complex and dynamic web of production networks, economic spaces and places connected together through threads of flows.

Figure 1.8 captures the major dimensions of these relationships. Individual production networks can be regarded as vertically organized structures configured across increasingly extensive geographical scales. Cutting across these vertical structures are the territorially defined political-economic systems which, again, are manifested at different geographical scales. It is at the points of intersection of these dimensions in 'real' geographical space where specific outcomes occur, where the problems of existing within a globalizing economy – whether as a business firm, a government, a local community or an individual – have to be resolved.

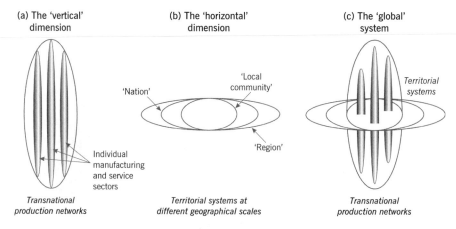

Figure 1.8   Interconnecting dimensions in a globalizing economy

Source: based, in part, on Humbert, 1994: Figure 1

The geo-economy and the environment

Economy and nature

In our earlier description of the production process as a circuit, in which inputs are transformed into products which are then consumed (Figure 1.4), we ignored the fact that the inputs have to come from somewhere and that the consumption of products is not the end of the story. Ultimately, the 'somewhere' from which all inputs of materials and energy derive, and to which what is left over after production, distribution and consumption have taken place must go, is the *natural environment*.[36] In the final analysis, '*all* production depends on, and is grounded in, the natural environment'.[37] Although the primary purpose of the production process shown in Figure 1.4 is the production of 'goods' for consumption (driven, in a capitalist market system, of course, by profits), the process itself — in a way unintentionally — also produces 'bads' in the form of environmental degradation. There are, in other words, unintended external effects (*negative externalities* or *spillovers*) involved in all economic activities.

Three aspects of such environmental damage are especially important:[38]

- over-use of non-renewable and renewable resources (including exploitation of fossil fuels, depletion of water resources, clearance of forests)
- over-burdening of natural environmental 'sinks' (for example, the increasing concentration of greenhouse gases in the earth's atmosphere and of heavy metals in the soil)
- destruction of increasing numbers of ecosystems to create space for urban and industrial development.

Production as a system of materials flows and balances

The bases of such environmental damage can be understood most easily if we produce a diagram parallel to that of the production circuit discussed earlier. Figure 1.9 depicts the production circuit of Figure 1.4 in terms of *materials flows* and *materials balances*.[39] The key point of the process is that what goes in has to come out again, albeit transformed, but without being reduced. In Figure 1.9, the materials used in the production process are

> dispersed and chemically transformed. In particular, they enter in a state of low entropy (as 'useful' materials) and leave in a state of high entropy (as 'useless' materials, such as low temperature heat emissions, mixed municipal wastes, etc.) ... No material recycling processes can therefore ever be 100 per cent efficient.[40]

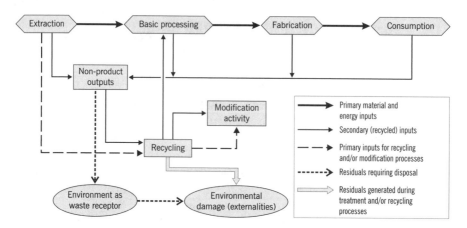

Figure 1.9   Production circuits and the environment

Source: based on Turner et al., 1994: Box 1.2

In effect, economic systems in general, and production circuits in particular, place demands on the natural environment in two ways:

- in terms of *inputs* to the process of production which are derived from the natural environment as *resources*
- in terms of *outputs* to the natural environment in the form of pollution of various kinds.

Let's look briefly at each of these in turn.

The resources issue

Natural resources are not 'naturally' resources. An element or a material occurring in nature is only a 'resource' if it is defined as such by potential users. In other words, there must be an effective demand, there must be an appropriate technology with which to exploit it, and there must be some means of ensuring 'property rights' over its use: 'if any of these conditions ceases to hold, resources could "unbecome"'.[41] Resources, once defined as such, are of two broad kinds:

- *Renewable resources* – those which, over time and with good management, can be replenished. Nevertheless, most so-called renewable resources may become exhausted if they are not managed in a sustainable manner.
- *Non-renewable resources* – those which are fixed in overall quantity, at least under known technological conditions. The more we use today, the less will be available for tomorrow.

One of the major current debates, therefore, is the extent to which non-renewable resources, in particular, are becoming increasingly scarce through excessive exploitation and, therefore, facing imminent exhaustion. Here, as in all areas of the environmental debate, views become polarized. On the one hand there is the 'Malthusian' view that resource exhaustion is inevitable; the only question is the time-scale over which such exhaustion will occur. On the other hand, there is the view that new technologies of exploration leading to discoveries of new reserves, better means of exploitation leading to more efficient use of the resource (including recycling), and the development of appropriate substitutes will put off the dreadful day. So far, at least, all of these have happened. The dire predictions of imminent resource exhaustion have not been borne out (yet). Nevertheless, there is a real danger of resource exhaustion in specific areas and of continuing localized environmental damage. There is the additional geopolitical complication that access to a localized resource (like oil, for example) may be restricted from time to time by the states within whose territory it is located.

Unintended effects of production

As Figure 1.9 shows, after all efforts are made to recycle the unused energy and materials involved in production, there will still be 'things' left over in the form of residual waste and environmental damage. This is simply because the fundamental laws of thermodynamics cannot be overruled:

> the total mass of inputs to a transformation process is equal to the total mass of outputs. If inputs do not emerge as desired products, they must therefore appear as unwanted by-products or wastes.[42]

Such negative externalities are of various kinds and of varying spatial extent. For example, the negative externalities from a factory or from an airport are, at one level, geographically localized. The impact is greatest at the location of the facility itself and its immediate neighbourhood but then declines with increasing distance away from that location. On the other hand, the smoke pollution from the factory or the effect of aircraft fuel combustion may have much more extensive geographical effects, particularly on the atmosphere. The problem is that many adverse environmental effects cannot be contained within geographical boundaries.

By far the most contentious aspect of negative environmental externalities relates to potential *atmospheric damage*, that is, damage to the gaseous membrane that sustains all life on earth. The processes of material transformation involve the use of massive quantities of energy, especially of fossil fuels whose combustion products are the major source of damage to the earth's atmosphere. The problems arise because some of the key gaseous components of the earth's atmosphere – notably carbon dioxide, methane and ozone – are becoming excessively concentrated. The issue is one of balance. Without these, and other, gases the earth would have a surface temperature like that of the planet Mars; that is, it would be uninhabitable. The earth's surface remains habitable precisely because of their presence in the atmosphere. In combination, they act like a 'greenhouse', preventing both excessive solar heating and excessive cooling. But it is a very delicate balance.

Most scientists believe that this balance is dangerously disturbed by human action. In the case of the main greenhouse gas, carbon dioxide, for example, there is clear evidence of a significant acceleration after around 1800.[43] Before that time, carbon dioxide levels in the atmosphere remained fairly stable at between 270 and 290 parts per million (ppm). Since then, the following escalation in levels has occurred:

- 1900: 295 ppm
- 1950: between 310 and 315 ppm
- 1995: 360 ppm
- 2003: 376 ppm.

These progressive increases were closely associated with the processes of indus-trialization and urbanization, through the burning of fossil fuels and through deforestation. At the same time, levels of methane have increased dramatically, partly through the same fossil fuel usage and partly through agricultural prac-tices like rice irrigation, livestock production and garbage decomposition. As a result, the atmosphere has become more efficient at trapping heat from the sun.[44]

So, the overwhelming scientific consensus is that human-induced *global warm-ing* is occurring, although there are some scientists – and some politicians, notably in the George W. Bush US administration (2000–2008) – who dispute its nature, its rate and its causes. However, if current trends continue they will generate poten-tially enormous social and economic damage in many parts of the world, notably: major dislocation of climatic zones around the world, causing flooding of many coastal and low-lying areas, altering agricultural economies; changing patterns of disease; and increasing volatility of atmospheric systems (for example, an increase in severe storms). But how much, and how fast, such warming will be – and the precise role of human activity – is unclear.

The other aspect of atmospheric – or, rather, stratospheric – damage concerns the earth's *ozone layer*. Ozone is formed in the stratosphere through the chemical reaction of oxygen and sunlight. At this level, ozone is vital to the sustainability of human life on earth because it absorbs almost all the ultra-violet radiation from the sun, which would otherwise make human life impossible. Any damage to this vital protective shield poses a serious problem. Just such thinning of – or even holes in – the ozone layer (beyond natural occurrences) began to be identified in certain parts of the world in the early 1970s. One of the major effects of ozone depletion is an increase in the incidence of skin cancer. A primary cause was believed to be chlorofluorocarbons (CFCs), which had become extensively used in refrigeration and aerosols. Although CFCs are now heavily restricted, the fact that these chemicals are immensely stable means that the amount already in the stratosphere will still be affecting the ozone layer until about 2087.[45]

Hence, the environmental problems that are inherent in all aspects of production, distribution and consumption raise serious questions about the future sustainability of economy and society as we know them. They raise big issues relating to the future of the world's economic and trading system and, indeed, to most aspects of contemporary economic life. As such, they have come to form a major element in the globalization debates and in the anti-globalization movements. This raises a major problem of global governance, which we will address in the final chapter of this book.

Conclusion

The purpose of this chapter has been to interrogate the concept of 'globalization' and to refute the popular view that it is some kind of all-embracing, inexorable, irreversible, homogenizing force. Rather, the world in which we live is constituted through, and transformed by, a complex of interrelated processes rather than by some single force called 'globalization'. The processes that are transforming the geo-economy are highly uneven in their operation and in their effects. Without doubt the world *is* a qualitatively different place from that of only 60 or 70 years ago, although it is not so much more open as increasingly *interconnected* in rather different ways.

One way of understanding the nature of this change is to think in terms of production circuits and networks, configured at a multiplicity of geographical scales, from the local through to the global. These are the structures through which different parts of the world are connected together through flows of material and non-material phenomena in a system of differential power relationships, in which consumers, as well as producers, can exert significant influence. Production networks are intrinsically geographical in terms of their differentiated spatial configurations and their territorial embeddedness in specific places. In particular, economic activities tend to cluster or agglomerate in particular kinds of location. Such clusters, once formed, have a strong tendency to develop in path dependent ways, which influence – although they do not totally determine – future geographies. Seeing production circuits as systems of materials flows and balances, subject to the inexorable laws of thermodynamics, helps us to appreciate the ultimate dependence of production on the natural environment, both as a source of materials inputs in the form of renewable and non-renewable resources and as a receptor of the waste products of production.

NOTES

1 ILO (2004b: 12–23).
2 Stephens (2005). The reference to the 'beginning of history' seems to be a deliberate contrast to Fukuyama's (1992) prediction of the 'end of history' following the collapse of the Soviet system.
3 Strange (1995: 293).
4 Held et al. (1999: 1–28) provide a useful discussion of some of the major strands in the globalization debates. See also Cameron and Palan (2004).
5 See Tickell and Peck (2003).
6 For an example of this position, see Friedman (1999; 2005). More nuanced writers within this general framework include Bhagwati (2004) and Wolf (2004).
7 Hirst and Thompson (1992; 1999), Kozul-Wright (1995).
8 Hirst and Thompson (1992: 394).
9 Held et al. (1999: 15).
10 See Feenstra (1998), Gereffi (2005).
11 Mittelman (2000: 4).
12 Jessop (2002: 113–14).
13 Thrift (1990: 181). See also Amin (2002), Brenner (1998), Swyngedouw (2000).
14 The material in this section is based on Dicken (2004).
15 Hudson (2004: 462).
16 Mitchell (2000: 392).
17 See, for example, Gereffi and Korzeniewicz (1994), Gereffi et al. (2005).
18 Rabach and Kim (1994: 123)
19 Recent literature on global, or transnational, production networks includes Coe et al. (2004), Dicken (2005), Ernst and Kim (2002), Henderson et al. (2002).
20 Latour quoted in Thrift (1996: 5).
21 See, for example, O'Brien (1992), Cairncross (1997).
22 Castells (1996).
23 Granovetter (1985) pioneered the concept of 'embeddedness' within the field of economic sociology. It has become a ubiquitous (though contested) term since then. See Hess (2004) for a recent discussion of the concept in a spatial/territorial context.
24 Stopford and Strange (1991).
25 This distinction is made for analytical convenience. In fact, the boundary between them is generally very blurred.
26 Maslow (1970).
27 Hirsch (1977).
28 Hudson (2005: 65).
29 Miller (1995: 1).
30 Gereffi (1994: 97).
31 *The Economist* (2 April 2005: 9).
32 Markusen (1996).
33 'Clustering' has recently become a hot topic in policy debates in different parts of the world (Krugman, 1998; Porter, 1990; 1998; 2000). However, the concept has a long history: see, for example, Amin and Thrift (1992), Bathelt et al. (2004), Dicken and

Lloyd (1990), Malmberg (1999), Markusen (1996), Martin and Sunley (2003), Scott (1998), Storper (1997).

34 Myrdal (1958: 26).

35 Dicken and Malmberg (2001).

36 See Cairncross (1992), Dauvergne (2005), Hudson (2001: Chapter 9), McNeill (2000), Turner et al. (1994).

37 Hudson (2001: 300).

38 Simonis and Brühl (2002: 98).

39 Turner et al. (1994: 15–23).

40 Turner et al. (1994: 17).

41 Hudson (2001: 301).

42 Hudson (2001: 288).

43 McNeill (2000: 109).

44 McNeill (2000: 109).

45 McNeill (2000: 114).

Two
Global Shift: The Changing Global Economic Map

What's new? The imprint of past geographies

In Chapter 1, I argued that old geographies of production, distribution and consumption are continuously being disrupted and that new geographies are continuously being created. In that sense, the global economic map is always in a state of 'becoming'; it is never finished. But the new does not simply obliterate the old. On the contrary, there are complex processes of path dependency at work. What already exists constitutes the preconditions on which the new develops. Today's global economic map, therefore, is the outcome of a long period of evolution during which the structures and relationships of one historical period help to shape the structures and relationships of subsequent periods. In that sense, we cannot fully understand the present without at least some understanding of the past. Indeed, traces of earlier economic maps – earlier patterns of geographical specialization or divisions of labour – continue to influence what is happening today, although there are debates amongst economic historians over when we can first identify a 'world' or a 'global' economy. To some, this appeared during what has been called the 'long sixteenth century' (1450 to 1640).[1] To others, the key period was the second half of the nineteenth century.[2] Whatever,

> by 1914, there was hardly a village or town anywhere on the globe whose prices were not influenced by distant foreign markets, whose infrastructure was not financed by foreign capital, whose engineering, manufacturing, and even business skills were not imported from abroad, or whose labour markets were not influenced by the absence of those who had emigrated or by the presence of strangers who had immigrated. The economic connections were intimate.[3]

Over a period of 300 years or so, therefore, a *global division of labour* developed, and intensified with industrialization, in which the newly industrializing economies

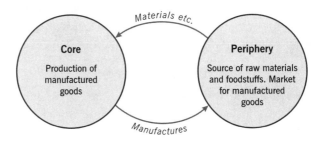

Figure 2.1   A simple geographical division of labour: core and periphery in the global economy

of the West (led by the 'Atlantic' economies, notably Britain, some Western European countries, and later the United States) became increasingly dominant in a *core–periphery* configuration (Figure 2.1). Of course, over time, this structure became more complex in detail, and also changed in its geographical composition. Some core economies experienced a progressive decline to semi-peripheral status during the eighteenth century and new economies emerged, especially in the late nineteenth and early twentieth centuries. Figure 2.2 shows some of these dramatic changes, notably the steep decline of Asia and the emergence to unrivalled dominance of the United States, measured in terms of shares of global gross domestic product. (Gross domestic product, or GDP, is the total value of goods and services produced by a country.)

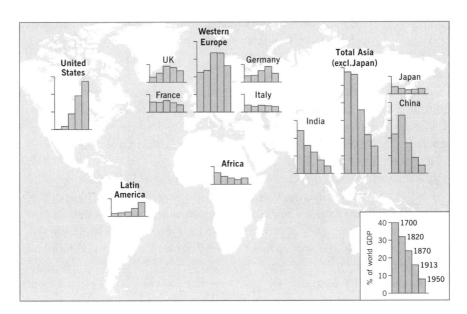

Figure 2.2   Global shifts in GDP, 1700–1950

Source: calculated from Maddison, 2001: Table B-20

The broad contours of this core–periphery global economic map persisted until the outbreak of the Second World War in 1939. Manufacturing production remained strongly concentrated in the core: 71 per cent of world manufacturing production was concentrated in just four countries and almost 90 per cent in only 11 countries. Japan produced only 3.5 per cent of the world total. The group of core industrial countries sold two-thirds of its manufactured exports to the periphery and absorbed four-fifths of the periphery's primary products.[4] This long-established global division of labour was shattered by the Second World War. Most of the world's industrial capacity (outside North America) was destroyed and had to be rebuilt. At the same time, new technologies were created and many existing industrial technologies were refined and improved in the process of waging war. Hence, the world economic system that emerged after 1945 was, in many ways, a new beginning. It reflected both the new political realities of the post-war period – particularly the sharp division between East and West – and also the harsh economic and social experiences of the 1930s.

The major political division of the world after 1945 was essentially that between the capitalist West (the United States and its allies) and the communist East (the Soviet Union and its allies). Outside these two major power blocs was the so-called 'Third World', a highly heterogeneous – but generally impoverished – group of nations, many of them still at that time under colonial domination. The Third World was far from immune from the East–West confrontation. Both major powers made strenuous efforts to extend their spheres of influence, with considerable implications for the subsequent pattern of global economic change.

The Soviet bloc drew clear boundaries around itself and its Eastern European satellites and created its own economic system, quite separate from the capitalist market economies of the West, at least initially. In the West the kind of economic order built after 1945 reflected the economic and political domination of the United States. Alone of all the major industrial nations, the United States emerged from the war strengthened, rather than weakened. It had both the economic and technological capacity, and also the political power, to lead the way in building a new order. As Figure 2.2 shows, by 1950 the United States accounted for no less than 27 per cent of global GDP.

It is from this historical baseline, therefore, that recent global shifts in economic activity will be examined in this chapter. Today's world is far more complex than it was even a few decades ago. There has been a truly fundamental transformation of the world economy: a new geo-economic map has come into being, although one which, of course, bears many traces of the contours of the old. Since 1950, two highly significant political events have occurred with huge implications: the emergence of China into the global market economy, although still under Communist Party control, starting in 1979; the collapse of the prevailing political systems in the Soviet Union and its Eastern European satellites in 1989. More broadly, we are seeing the *re*-emergence of Asia as one of the world's most dynamic economic regions.[5] In 1700, Asia's share of global GDP had been 62 per cent compared with the West's 23 per cent. By 1950 those positions had been almost exactly reversed: the combined GDP of Western economies was almost 60 per cent; that of Asia (including Japan) was a mere 19 per cent. Much of this was due to the relative economic decline of China and India. In 1700, their combined share of global GDP was almost 50 per cent; by 1950, it was only 9 per cent.

Roller-coasters and interconnections

Two particularly important features have characterized the global economy since 1950: the increased volatility of aggregate economic growth; the growing interconnectedness between different parts of the world, as reflected in the differential growth rates of production, trade and foreign direct investment.

Aggregate trends in global economic activity

> The world economy performed better in the last half century than at any time in the past. World GDP increased six-fold from 1950 to 1998 with an average growth of 3.9 per cent a year compared with 1.6 from 1820 to 1950, and 0.3 per cent from 1500 to 1820.[6]

However, even within this broad surge of economic growth there were some pretty large interruptions to the upward curve. In fact, the path of economic change is best seen as being like a roller-coaster. Sometimes the ride is relatively gentle, with just minor ups and downs; at other times, the ride is truly stomach-wrenching, with steep upward surges separated by vertiginous descents to what seem like bottomless depths (Figure 2.3).

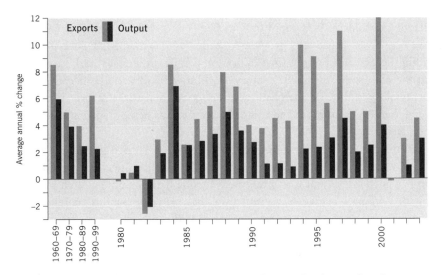

Figure 2.3   The roller-coaster of world merchandise production and trade

Source: calculated from GATT/WTO *International Trade Statistics*, various issues

The years immediately following the Second World War were ones of basic reconstruction of war-damaged economies throughout the world. At the time it was felt that growth rates would then slacken in the 1960s. This did not happen. Instead, rates of economic growth reached unprecedented levels. Indeed, the period between the early 1950s and the early 1970s became known as the 'golden age'. In fact, it was only partly golden: it was more golden in some places than others, and for some people than others.[7] But then, in the early 1970s, the sky fell in. The long boom suddenly went 'bust', the 'golden age' of growth became tarnished. As Figure 2.3 shows, growth rates declined dramatically. Throughout the 1980s and 1990s annual rates of growth became very volatile indeed. The roller-coaster had come back with a vengeance.

Rates of growth have become extremely variable, ranging from the negative growth rates of 1982 through to two years (1984 and 1988) when growth of world merchandise trade reached the levels of the 1960s once again. But then, in the early 1990s, recession occurred again. In 1994 and 1995, strong growth reappeared, especially in exports. A similarly volatile pattern characterized the last years of the century. There was spectacular growth in world trade in 1997, followed by far slower growth in 1998 and 1999 (partly related to the East Asian financial crisis and to its contagious effects on other parts of the world). Then, once again, there was spectacular acceleration in world trade in 2000, followed by a spectacular bursting of the growth bubble, a problem certainly exacerbated (though not caused) by the 9/11 terrorist attacks on New York City and by the crisis in the IT (dotcom) sector of the so-called 'new' economy.

Growing interconnectedness within the global economy

One major characteristic of global economic growth, therefore, is its inherent volatility: periods of very rapid growth being interspersed with periods of very slow – or even negative – growth. A second is the increasing *interconnectedness* within the global economy. One indication of this is the fact that, between 1950 and the end of the twentieth century, world merchandise trade increased almost twentyfold while world merchandise production increased just over sixfold. The overall trend is clear: more and more production was being traded across national boundaries.

But there is another dimension to this process of increased interconnectedness: the growth of TNC activities, as measured by foreign direct investment (FDI) data.[8] 'Direct' investment is an investment by one firm in another, with the intention of gaining a degree of control over that firm's operations. 'Foreign' direct investment is simply direct investment across national boundaries, that is, when a firm from one country buys a controlling investment in a firm in another country, or where a firm sets up a branch or subsidiary in another country. It differs from 'portfolio' investment, the situation in which firms purchase stocks/shares in other companies purely for financial reasons. Unlike direct investment, portfolio investments are not made to gain control. FDI is only one measure – albeit a very important one – of TNC activity. It does not capture the increasingly diverse ways in which firms organize their production networks, for example, through various kinds of collaborative ventures and alliances, or through their coordination and control of production network transactions. We will look at these issues in Chapter 5.

Although there was very considerable growth and spread of foreign direct investment during the first half of the twentieth century, that was as nothing compared with its spectacular acceleration and spread after the end of the Second World War. The post-war surge of FDI was an integral part of the 1960s 'golden age' of economic growth. Figure 2.4 shows that during the 1970s and into the first half of the 1980s the trend lines of both FDI and exports ran more or less in parallel. Then from 1985 to 1990 the rate of growth of FDI and of exports and GDP diverged rapidly. Between 1986 and 1990 FDI outflows grew at an average annual rate of 25 per cent and cumulative FDI stocks at a rate of 18 per cent a year compared with a growth rate of world exports of 12.7 per cent. FDI during the 1980s grew more than four times faster than world GNP. The recession of the early 1990s reduced the FDI growth rates significantly, but by the mid 1990s the upward trend had resumed. The situation changed again dramatically in 2001–3, when FDI flows went into steep reverse and FDI stocks grew more slowly. However, the strong upward trend resumed in 2004.

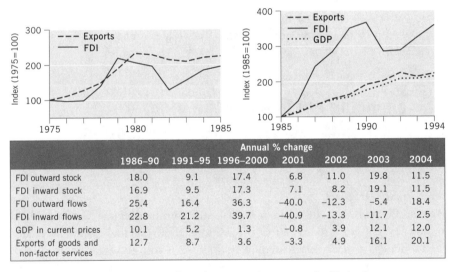

	Annual % change						
	1986–90	1991–95	1996–2000	2001	2002	2003	2004
FDI outward stock	18.0	9.1	17.4	6.8	11.0	19.8	11.5
FDI inward stock	16.9	9.5	17.3	7.1	8.2	19.1	11.5
FDI outward flows	25.4	16.4	36.3	–40.0	–12.3	–5.4	18.4
FDI inward flows	22.8	21.2	39.7	–40.9	–13.3	–11.7	2.5
GDP in current prices	10.1	5.2	1.3	–0.8	3.9	12.1	12.0
Exports of goods and non-factor services	12.7	8.7	3.6	–3.3	4.9	16.1	20.1

Figure 2.4   Growth of foreign direct investment compared with trade and production

Source: calculated from UNCTAD *World Investment Report*, various issues

This divergence in growth trends between FDI and trade is extremely significant. The fact that, especially after the mid 1980s, FDI grew much faster than trade suggests that the primary mechanism of interconnectedness within the global economy has shifted from trade to FDI. Of course, these trends in the growth of FDI, trade and production are not independent of one another. The common element is the transnational corporation. According to UNCTAD, the number of TNCs has grown exponentially over the past three decades. Today there are around 70,000 parent company TNCs controlling around 700,000 foreign affiliates.[9] Of course, this is only a very small proportion of the total number of business firms in the world, the vast majority of which are small and entirely domestically oriented.

TNCs account for around two-thirds of world exports of goods and services, of which a significant share is *intra-firm trade*. In other words, it is trade within the boundaries of the firm – although across national boundaries – in the form of transactions between *different parts of the same firm*. Unfortunately, there are no comprehensive and reliable statistics on intra-firm trade. The 'ballpark' figure is that approximately one-third of total world trade is intra-firm although, again, that could well be a substantial underestimate. Unlike the kind of trade assumed in international trade theory – and in the trade statistics collected by national and international agencies – intra-firm trade does not take place on an 'arm's-length' basis. It is, therefore, subject not to external market prices but to the internal decisions of TNCs. It has become especially important as the production networks of TNCs have become more complex and, in particular, as production circuits have become more 'disintegrated'. Such 'disintegration of production itself leads to more trade, as intermediate inputs cross borders several times during the manufacturing process'.[10] These are processes we will examine in some detail in Chapter 5.

The changing contours of the global economic map: global shifts in production, trade and direct investment

The macro-scale

Around 20 years ago, the Japanese management writer Kenichi Ohmae coined the term *global triad*[11] to argue that the world economy is now essentially organized around a tripolar, macro-regional structure, whose three pillars are North America, Europe and East Asia. It is a view that has received very wide acceptance, especially within business circles. Certainly, if we look at the statistical data on production, trade and foreign direct investment the case looks pretty convincing. Together, these three macro-regions contain 86 per cent of both total world GDP and total world merchandise exports (Figure 2.5) and are the focus of the vast majority of the world's foreign direct investment (Table 2.1).

The 'triad' appears to sit astride the global economy like a modern three-legged Colossus, constituting the world's 'mega-markets' and 'sucking in' more and more of the world's production, trade and direct investment. Whether it is 'real' or more of a statistical artefact is subject to debate.[12] However, if the 'triad' does represent a functional reality (actual or potential) then it poses major problems for those parts of the world – notably the least developed countries – which are not integrated into the system. In fact, although developing countries, as a group, have increased their share of global exports and of inward foreign direct investment, their share remains very limited, as Table 2.2 shows. Most of the world's FDI moves between developed countries as complex patterns of cross-investment. Developing countries' share of global GDP remains virtually static. Such figures must be seen in the context of the fact that developing countries contain the overwhelming majority of the world's population. More than that, there is a very high level of concentration of economic activities among developing countries (see the lower section of Table 2.2): a handful of countries accounts for the majority of the total.

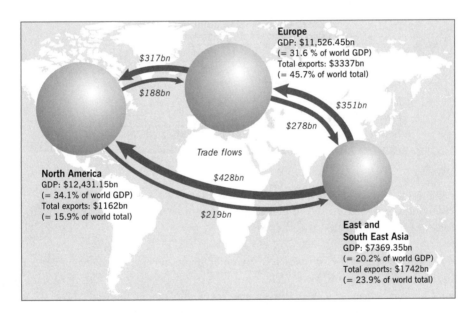

Figure 2.5   A global triad: concentration of world GDP and exports

Source: calculated from WTO, 2004, *World Trade Report*: Tables III.1, III.3; World Bank, 2005a: Table 4.2

The national scale

Such broad-brush perspectives on the shape of the global economic map are quite useful but, of course, they obscure the finer-grained texture of what is actually

Table 2.1   A global triad: concentration of world foreign direct investment, 2004 (% of total)

	FDI stocks		FDI flows	
	Inward	*Outward*	*Inward*	*Outward*
Europe	50.5	63.0	35.5	46.4
North America	22.0	24.7	18.3	38.2
East and South East Asia	13.8	11.0	21.4	13.4
Total	86.3	98.7	75.2	98.0

Source: calculated from data in UNCTAD, 2005

Table 2.2   (a) Developing countries' share of world production, trade and foreign direct investment, 1990 and 2003–4 (%)

	1990	2003–4
Share of world GDP	18.4	19.5
Share of world exports	19.0	26.3
Share of world inward FDI stocks	20.6	25.0
Share of world inward FDI flows	15.0	35.6

(b) Share of developing countries' total by leading 10 and leading 5 developing countries (%)

	Leading 10	Leading 5
GDP	64	53
Exports	77	59
Inward FDI stocks	65	54
Inward FDI flows	75	62

Source: calculated from data in World Bank, 2005a; UNCTAD, *World Investment Report* 1994; 2005

happening. To see this, we need to refocus our lens, first, to the national scale. The changing configuration of the global economic map at this scale is shown in the series of figures collected together on pages 51 to 61:

- *Shifting geographies of production* in manufacturing, services and agriculture are illustrated in Figures 2.6 to 2.11.
- *Changing patterns of trade* in the same sectors are shown in Figures 2.12 to 2.21.
- *Changing patterns of foreign direct investment* are shown in Figures 2.22 to 2.24.

These show not only what is going on outside the 'triad' but also, and just as importantly, that there is considerable variation between countries *within* each triad bloc, as well as within both developed and developing country groups. The maps and tables are largely self-explanatory and do not need individual discussion, although they do merit close attention. All I will do here is to summarize some of the major trends.[13]

Continuing geographical concentration within the global economy

Each of the three major categories of production – manufacturing, services and agriculture – has a very uneven and highly concentrated global geography. Around four-fifths of global manufacturing and services production, and almost two-thirds of world agricultural production, are concentrated in just 15 countries. Between one-fifth and one-quarter of world trade in goods, services and agriculture is accounted for by the leading two countries in each sector. The picture is similar in the case of foreign direct investment: almost 90 per cent of outward FDI stock originates from 15 countries; the leading two source countries – the United States and the United Kingdom – account for one-third of the world total. More than half of all the inward FDI in developing countries is concentrated in just five host countries; almost one-third is concentrated in China and Hong Kong alone.

The newly *industrializing* economies (NIEs), that have become so important in global manufacturing, have their agro-food counterparts: the newly *agriculturalizing* countries (NACs).[14] As in the case of manufacturing, a small number of developing countries – Argentina, Brazil, China, Kenya, Mexico – account for a disproportionate share of high-value food production and exports. For example,

> Brazil is to agriculture what India is to business offshoring and China to manufacturing: a powerhouse whose size and efficiency few competitors can match. Despite facing one of the highest agricultural tariffs in the western hemisphere ... the country is the world's largest or second largest exporter of sugar, soybeans, orange juice, coffee, tobacco and beef and is rapidly building a strong position in products such as cotton, chicken and pork. Brazil has the largest agricultural trade surplus in the world.[15]

The United States still dominates the global economy – though a little less so

The United States has been the pre-eminent force in the global economy for almost 100 years, having superseded the original industrial leader, the United Kingdom, early in the twentieth century. The United States accounts for one-quarter of the world's manufacturing production, one-third of services production, and 15 per cent of agricultural production. It is also the world's biggest foreign direct investor, generating one-fifth of the world total, the largest exporter of commercial services and agricultural products, and the second largest exporter of manufactured goods. Between 1980–1990 and 1990–2003, the United States' GDP grew at an annual average rate of 3.6 and 3.3 per cent respectively, slightly above the world growth rate of 3.3 and 2.8 per cent.

But, although still the world's leading economic power, its dominance has been much reduced as other competitors have emerged. This is most apparent in the trade data, although it should be noted that trade is a smaller proportion of GDP in the United States than in all its major competitors, apart from Japan.

Nevertheless, the United States' share of world merchandise exports has fallen from 17 per cent in 1963 to less than 10 per cent. At the same time, its share of merchandise imports has surged from less than 9 per cent to almost 17 per cent. Although US merchandise exports have grown at around 5 per cent a year, imports have grown at between 8 and 9 per cent a year. As a result, the United States has an enormous merchandise trade deficit of $580 billion, a figure hardly dented by the small surplus in commercial services trade. In effect, the United States has become the 'importer of last resort' for the global economy. Given the political sensitivity involved, this poses serious potential problems for the world trade system as a whole.

There have been very substantial changes in the United States' position as a source of, and destination for, foreign direct investment. In 1960, the United States generated almost 50 per cent of all the world's FDI, compared with 21 per cent today. The biggest change, however, has been in the country's position as a host for FDI. Although the United States has attracted FDI for many decades, such inward investment was always a tiny fraction of the country's outward direct investment. Even in 1975, its outward investment was four-and-a-half times greater than its inward investment. Since then, however, the United States has become significantly more important as an FDI destination as firms from Europe, Japan and, more recently, some East Asian NIEs have reoriented the geographical focus of their overseas direct investments. Inward and outward FDI are now much closer than in the past. Europe remains the most important destination for the United States' FDI: around half of the total is located there, with the largest regional share being in the UK. But, as might be expected, Asia has become increasingly important as an FDI destination, as has Mexico.

Europe is still a major player – but its performance is uneven

Europe, as a region, is the world's biggest trading area and the primary focus of foreign direct investment. However, despite being the most politically integrated region in the world (see Chapter 6), the European economy is actually very diverse, experiencing uneven rates of economic growth over the past two decades. Germany is by far the biggest economy in global terms: it is the third largest manufacturing producer (after the United States and Japan), the largest merchandise exporter, the third largest commercial services exporter, and the third most important source of foreign direct investment. But its growth in GDP has been below the world average for a long period, and it still faces especially difficult problems in integrating the former East Germany into the economy as a whole.

Europe's second biggest economy, the United Kingdom, has experienced the greatest long-term relative decline in so far as it once dominated the world. However, it is still the world's second biggest source of FDI and second biggest exporter of commercial services. Indeed, its GDP growth rate has been very close to the world average over the past 20 years, significantly better than either

Germany or France, Europe's third largest economy. In general, the fastest-growing European countries have been the more 'peripheral' economies of Ireland and Spain, along with Finland and Norway.

There are considerable differences in trade performance between individual European countries. Whereas the UK, Spain and Sweden have large merchandise trade deficits, Germany, the Netherlands and Ireland have surpluses, while France and Italy are more or less in balance. In contrast, in commercial services the UK has a big trade surplus, France and Spain modest surpluses, while Germany has a substantial deficit. Overall, Europe constitutes the world's major trading region. More than two-thirds of European trade is *intra-regional*. North America is Europe's most important export destination.

Europe remains a major magnet for inward investment as well as the leading source of outward FDI. For all the major European countries (excluding the United Kingdom), more than half of their FDI outflows are to other European countries. In most cases, this regional orientation has actually increased. For example, in 1985 49 per cent of Italy's outward FDI was in Europe. By the end of the 1990s, this share had grown to 69 per cent. The comparable figures for Germany were 44 and 60 per cent; for the Netherlands 40 and 55 per cent; for France 58 and 60 per cent. For the United Kingdom, in contrast, the figures were only 28 and 38 per cent.

'Back to the future' – four tigers, a dragon, and the resurgence of Asia

Without any doubt, the most significant global shift in the geography of the world economy during the past 40 years has been the resurgence of Asia – especially East Asia. As noted earlier, this is, in fact, a real 'back to the future' event, although a complex one that is too often overhyped and oversimplified in the media. Observers in the early 1980s began to write about the twenty-first century being 'the Pacific Century', and of the future belonging to Asia, rather than to the North Atlantic economies, as had been the case for 200 years.[16] But commentators are fickle. The unexpected East Asian financial crisis of 1997 brought out the doomsayers (see Chapter 18). Ten years on, the position looks more like it did before 1997: the Asia boosters are out again in force, although now their focus is rather different. The future, it seems, is China – and, possibly, India.

Empirically, the resurgence of Asia during the past 40 years can be seen as consisting of four major processes:

- The rise of Japan after World War II.
- The rapid growth of what came to be called the 'four tigers': the newly industrializing economies of Hong Kong, Korea, Singapore and Taiwan. This was followed by the emergence of a 'second tier' of East Asian developing economies (the 'tiger cubs' in journalese), primarily Indonesia, Malaysia and Thailand.

- The (re-)emergence of China – the 'dragon' – as a major participant in the global market economy.
- The potential economic dynamism of India.

Japan's post-war economic growth as a manufacturing power was truly spectacular. In the early 1960s it ranked fifth in the world economy; by 1980 it had risen to second place. During the 1960s, Japan's rate of manufacturing growth averaged 13.6 per cent per year: two-and-a-half times greater than the United States and four times greater than the United Kingdom. The Japanese economy continued to grow at very high rates throughout the 1970s and most of the 1980s. Japan's share of world FDI grew from less than 1 per cent in 1960 to almost 12 per cent in 1990. As a result, 'Japan Inc.' came to be seen as the biggest threat facing both the United States and Europe, as a deluge of polemical, protectionist literature (especially in the United States) at the time demonstrated.

In the late 1980s, however, Japan's rapid growth rate fell as dramatically as it had increased in the 1960s, with the collapse of the so-called 'bubble economy' in the late 1980s. Between 1990 and 2003, Japanese GDP grew at an annual average rate of only 1.2 per cent and its manufacturing sector by a mere 0.7 per cent. Merchandise exports, which had grown at almost 9 per cent a year between 1980 and 1990, grew at less than 3 per cent a year between 1990 and 2003. The United States' fear of the Japanese threat receded; the 'bash Japan' literature virtually disappeared. Nevertheless, Japan remains the world's second largest economy after the United States and today there are signs of significant economic recovery.[17] Japan's decline has been much exaggerated.

At the same time as Japan was surging up the ranks of industrialized countries, a small group of East Asian developing countries also appeared on the scene as foci of manufacturing growth, especially in labour-intensive industries. The 'pioneers' were the so-called four 'tiger' economies of Hong Kong, Korea, Singapore and Taiwan. In terms of manufacturing production, for example:

- Korea's manufacturing sector grew at annual average rates of 18 per cent during the 1960s, 16 per cent during the 1970s, 13 per cent during the 1980s, and 7 per cent during the 1990s (to 2003).
- Taiwan's manufacturing sector grew at rates of 16, 14, 8 and 6 per cent respectively during the same periods.

Subsequently, Malaysia, Thailand and Indonesia also displayed extremely high rates of manufacturing growth.

The relative importance of these East Asian economies is especially marked in the sphere of exports. Indeed, in the global reorganization of manufacturing trade the increased importance of East Asia as an exporter of manufactures is unique in its magnitude. Seven East Asian NIEs (Korea, Hong Kong, Singapore, Taiwan,

Indonesia, Malaysia, Thailand) increased their collective share of total world manufactured exports from a mere 1.5 per cent in 1963 to almost 20 per cent in 1999 (and remember that this period includes the East Asian financial crisis of 1997–8, which had a devastating effect on most of the East Asian economies).

So, it is especially in their role as exporters that the East Asian economies are most significant. In some cases the transformation has been nothing short of spectacular. For example, in 1980, less than 20 per cent of Malaysia's exports were of manufactures; by 1998 the figure was 79 per cent. Indonesia provides an even more striking experience: in 1980 a mere 2 per cent of the country's exports were of manufactures; in 1998 almost half were in that category. Others show a similar transformation.

The most recent – and potentially the biggest – development within East Asia is, without question, the (re-) emergence of China. As a consequence, 'China bashing' is replacing the 'Japan bashing' of an earlier period. China has rather suddenly become a hugely significant presence in the global economy.

> The entry of China's massive labour force into the global economy may prove to be the most profound change for 50, and perhaps even for 100 years ... China's growth rate is not exceptional compared with previous or current emerging economies in Asia, but China is having a more dramatic effect on the world economy because of two factors: not only does it have a huge, cheap workforce, but its economy is also unusually open to trade. As a result, China's development is not just a powerful driver of global growth; its impact on other economies is also far more pervasive ... China's growing influence stretches much deeper than its exports of cheap goods: it is revolutionising the relative prices of labour, capital, goods and assets in a way that has never happened so quickly before.[18]

Between 1980 and 2003, China's growth rate (of GDP as a whole and of manufacturing) was the highest in the world: annual average rates of around 10 per cent. Its average annual rate of growth of merchandise exports was 13 per cent in the 1980s and 14 per cent between 1990 and 2003. As a result, China is now the world's fourth largest manufacturing producer, the second largest agricultural producer, the fourth largest exporter of merchandise (soon to overtake Japan into third place) and the world's third largest importer.

China's growth, together with the continued resilience of the Korean and Taiwanese economies, and the apparent recovery of Japan, has made *North East* Asia the most dynamic part of the world. This creates potential problems for some of the smaller NIEs of *South East* Asia, notably Malaysia, Thailand, the Philippines, Indonesia and even, possibly, Singapore. These are issues we will address in Chapter 18. Recently, however, attention has been drawn to the other very large Asian country (in population terms): India. Indeed, some commentators envisage a world economy that will increasingly be dominated by 'Chindia', defined by one writer as 'where the world's workshop meets its office',[19] an allusion to China's growth as a manufacturer and to India's growth in IT services. But beware the hype.

India has recently shown spectacular growth in one specific type of economic activity: the outsourcing of IT services (software, data processing, call centres and the like). As such it has attracted huge publicity and a growing view that India could be 'the next China', given the size of its population and other advantages. That may be so. But, at present, the evidence is slender. India's GDP growth rate between 1980 and 2003, though well above the world average at between 5 and 6 per cent, was half that of China during the same period. India is the world's 13th largest manufacturing economy; China is the fourth largest. India is not in the top 15 merchandise exporters; China is the fourth largest. Of course, it might be argued that India's strength lies in services rather than in manufacturing. Certainly it is true that the share of services in India's GDP is much higher than China's: 51 per cent compared with 33 per cent. Conversely, India has only 27 per cent of its GDP in manufacturing, compared with China's 52 per cent. Despite this, China generated almost twice the commercial services exports as India in 2003. Unlike all the other fast-growing East Asian NIEs, India does not have a strong export base in manufactures. China's merchandise exports are eight times larger than India's. Indeed, if India is ranked along with the nine leading NIEs of East Asia in terms of merchandise exports, it would be placed ninth. None of this is to suggest that India does not have the *potential* to become a really major economic power but, at present, the evidence is rather thin.

A significant aspect of East Asian development has been the increasing tendency for Korean, Taiwanese, Hong Kong, Singaporean and, most recently, Chinese firms to expand overseas through direct investment in addition to trade. The most publicized case recently was the takeover of IBM's PC business by the Chinese firm Lenovo. Overall, these five East Asian economies account for two-thirds of all FDI from developing countries.

Emergence of the 'transitional economies' of Eastern Europe and the former Soviet Union

Since 1989, there has been a further significant development in the changing geography of the global economy. The political collapse of the Soviet-led group of countries, and, indeed, of the Soviet Union itself, produced a group of so-called 'transitional economies': former command economies now in various stages of transition to a capitalist market economy. The process of transition, from a centrally planned economic system, with a heavy emphasis on basic manufacturing industries, to a capitalist market system, has been painful in many cases. The kinds of industries favoured in the centrally planned system are less viable in the context of today's highly competitive global economy, as are the kinds of industrial organization themselves. In 1985, for example, the USSR accounted for almost 10 per cent of world manufacturing output; by the mid 1990s, the share of the Russian Federation was around 1 per cent.

There are big differences within the group of transitional economies, both in terms of their scale and in terms of their potential for growth within a market, as opposed to a centrally planned, system. The four most significant transitional economies are the Russian Federation, Poland, the Czech Republic and Hungary, although their combined share of global GDP is a mere 2 per cent. Together, these four account for most of the manufacturing production and exports of the transitional economies. The latter three became members of the European Union in 2005 and this undoubtedly changes their potential prospects for economic development. Certainly their growth between 1990 and 2003 was far stronger than that of the Russian Federation (whose average growth rate was negative). Poland and Slovenia grew at above the world rate; Hungary and Slovakia at just below the world rate.

However, these economies achieved much more impressive export performance during the 1990s. Poland, Hungary and the Czech Republic each had double-digit export growth while the Russian Federation and Slovenia grew at around 7–8 per cent per year. Such export growth figures compare very favourably with those of the East Asian economies during the same period. Much of this growth is underpinned by inward FDI which has grown substantially since the early 1990s, especially in the Czech Republic, Hungary, Poland, Slovakia and the Russian Federation.

Latin America – unfulfilled potential

Latin American countries are among the most resource-rich in the world. Several also have a long history of industrialization. Some, like Brazil and Mexico, in population terms are very large indeed. And yet, most of the Latin American economies have not figured very prominently in the redrawing of the global economic map. Certainly, their modest economic performance contrasts markedly with that of East Asia. Within Latin America itself, there is a clear contrast between relatively faster-growing economies, like Chile and Mexico, on the one hand, and relatively slower-growing economies like Argentina and Brazil, on the other. None of these countries 'punches its weight' as exporters; over the past 20 years, their average export growth has been significantly lower than that of the East Asian economies. During the 1990s, the major exception was Mexico (which does not really regard itself as 'Latin American' anyway). Mexico's export growth rate of over 14 per cent undoubtedly reflected its increasing integration with the United States through the North American Free Trade Agreement (see Chapters 6 and 7).

Persistent peripheries

Alongside the areas of strong, though differential, economic growth in the global economy – the peaks, as it were – are those parts of the world whose economic growth remains very limited. These are the 'persistent peripheries'. All of the maps shown on pages 51 to 61 tell more or less the same story: most of the continent of Africa, parts of Asia, parts of Latin America constitute the 'troughs' of the global economic map. Sub-Saharan Africa, as is so often noted, is the largest single area of 'economic peripherality'. These are the parts of the world enmeshed in the deepest poverty and deprivation and whose existence poses one of the biggest social challenges of the twenty-first century. We will return to this issue in Part Four.

Global interconnections: networks of trade and degrees of dependence on FDI

This flexing and fluxing global economic map reflects the major global shifts that have occurred over the past four decades or so. It is made up of complex inter-connections, most notably those constituted through networks of trade and of FDI. Figure 2.25 (p. 62) maps the network of world merchandise trade. It shows not only some of the complexity of trade flows (bear in mind that this is a rather aggregated view) but also the strong tendency for countries to trade most with their neighbours. A considerable proportion of world trade is *intra-regional*, although this does not imply that such regionalizing tendencies dominate in all cases.[20] Several features merit attention:

- Western Europe is the world's major trading region. However, two-thirds of that trade is intra-regional, that is between Western European countries themselves. Around 10 per cent of Western Europe exports go to North America and about 7 per cent to Asia (including Japan).
- Asia is the second most significant trade region, with one-third of its trade conducted internally. Just over one-fifth of Asia's external trade goes to North America and a further 17 per cent to Western Europe. Notably, 47 per cent of Japan's trade is within the region, and no less than 54 per cent of Australia and New Zealand's trade is likewise (a very clear indication of their geopolitical reorientation away from their traditional markets in Europe).
- North America conducts around 40 per cent of its trade internally, with an especially large increase in trade involving Mexico. United States' exports to Mexico increased by 134 per cent between 1993 and 2003, whilst imports from Mexico to the United States grew by 243 per cent. North America's external trade is distributed fairly evenly between Asia (20 per cent), Western Europe (18 per cent) and Latin America (15 per cent).

International trade networks, once established, tend to be fairly stable over time. However, from time to time, some major shifts do occur.

- The most important during the 1960s and 1970s involved Japan, as the country rapidly rebuilt its war-ravaged economy primarily through sustained and aggressive exporting. As a result, flows of trade between Japan and other parts of the world, notably the United States and Europe, became a marked feature of the global economic map, particularly in some industries such as automobiles (see Chapter 10) and electronics (see Chapter 11).

- The second – and most recent – major shift in the international trade network involves, inevitably, China. Table 2.3 shows how significant China has become, not only to trade within Asia itself but also for the United States and Europe.

Table 2.3   China's trade network (% of China total)

Region/country	Exports	Imports
North America	22.4	9.3
United States	21.1	8.2
Latin America	2.7	3.6
Western Europe	17.5	13.9
EU 15	16.5	12.8
C/E Europe, Baltic states, CIS	3.6	3.6
Africa	2.3	2.0
Middle East	3.0	3.5
Asia	48.4	58.0
Japan	13.6	18.0
Hong Kong	17.4	2.7
Korea	4.6	40.5
Taiwan	2.1	12.1

Source: calculated from WTO *International Trade Statistics 2004*

One measure of the extent of a country's integration in the global economy is the percentage of its goods GDP that is traded. The higher the figure, the greater is the dependence on external trade. There is huge variation between countries in such trade integration. But the reasons for this are not straightforward, not least because, other things being equal, international trade is bound to be more important for geographically small countries than for large ones (contrast, say, the United States with Singapore). Nevertheless, the trends are significant. More important is the fact that the overall degree of integration in the global economy through trade increased markedly between 1990 and 2003. Of the 91 countries for which comparable data are available, only 16 became less integrated. Significantly, 11 of the 16 were African countries. In every other case, the relative importance of trade for the national economy increased.

A second measure of global integration is the relative importance of inward and outward FDI to a country's economy, measured by its GDP. Again, we might expect there to be some correlation with country size, although other factors are significant – not least, in the case of inward FDI, differing national policies (see Chapters 6 and 7). As with the case of trade, the relative importance of FDI to national economies has increased virtually across the board, a clear indication of increased interconnectedness within the global economy. In the case of inward

FDI, only 13 of almost 200 cases experienced a (small) relative decline in the GDP share of FDI and four remained unchanged. In the case of outward FDI, only 11 of around 140 cases experienced a relative decline and a further five remained unchanged.

Table 2.4   Inward foreign direct investment as a share of gross domestic product (% share of GDP)

	1990	2004		1990	2004
Ireland	88.9	126.3	Singapore	83.1	150.2
Netherlands	23.3	74.2	Vietnam	25.5	66.3
Belgium	–	73.5	Chile	33.2	58.2
Hungary	1.7	60.7	Malaysia	23.4	39.3
Czech Republic	3.9	52.7	Argentina	6.2	35.3
Switzerland	15.0	50.6	Thailand	9.7	29.7
Sweden	5.3	47.0	Mexico	8.5	27.0
Australia	23.7	41.1	Brazil	8.0	25.2
Denmark	6.9	40.5	China	5.8	14.9
United Kingdom	20.6	36.3	Philippines	7.4	14.9
Spain	12.8	34.9	Taiwan	6.1	12.8
France	7.1	26.5	Korea	2.1	8.1
Canada	19.6	30.5	India	0.5	5.9
Poland	0.2	25.4	Indonesia	7.7	4.4
Italy	5.4	13.1			
Germany	6.6	12.9			
United States	6.9	12.6			
Japan	0.3	2.1			

Source: based on data in UNCTAD, 2001; 2005

The actual extent of FDI involvement varies markedly, as Table 2.4 shows for a sample of countries. The focus here is on the relative significance of *inward* FDI as a percentage of national GDP. The variations between countries are very striking. In some cases, such as Ireland or Singapore, there is an overwhelming dependence on inward FDI; in others, notably Japan, India, Indonesia and Korea, inward FDI is a minuscule proportion of GDP. Overall, it should be noted that just seven East Asian countries (Hong Kong, China, Singapore, Korea, Thailand, Malaysia and Taiwan) contain almost 50 per cent of all FDI located in developing countries (an increase from one-third in 1990).

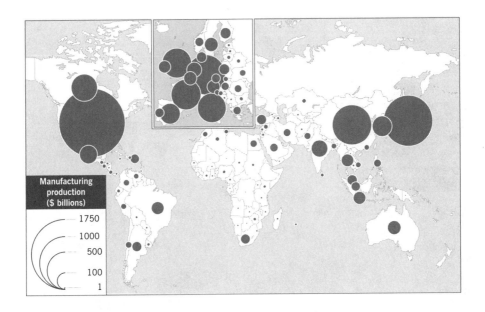

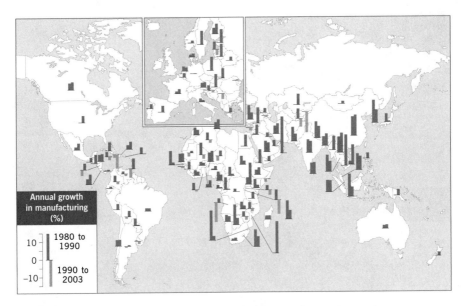

Figure 2.6   The global map of manufacturing production and growth

Source: calculated from World Bank, 2005a: Table 4.2

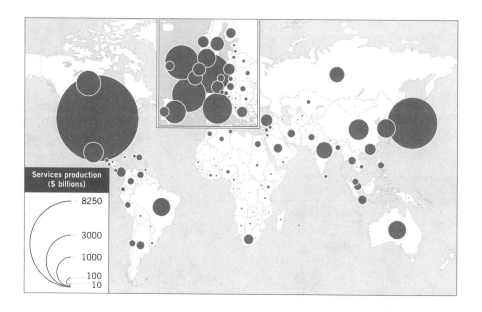

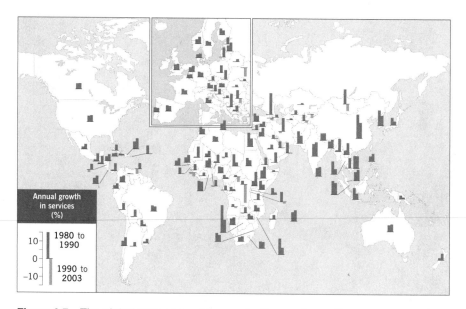

Figure 2.7   The global map of services production and growth

Source: calculated from World Bank, 2005a: Table 4.2

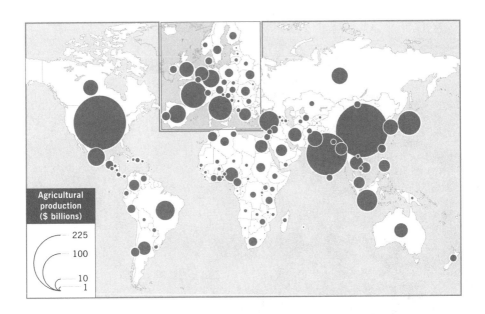

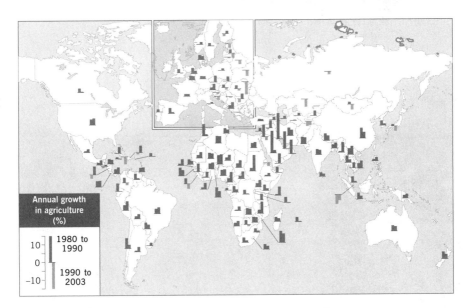

Figure 2.8   The global map of agriculture services and growth

Source: calculated from World Bank, 2005a: Table 4.2

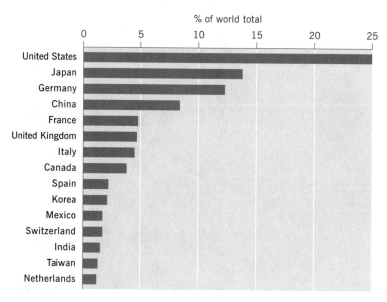

Figure 2.9   **The world's leading manufacturing producers**

Source: calculated from World Bank, 2005a: Table 4.2

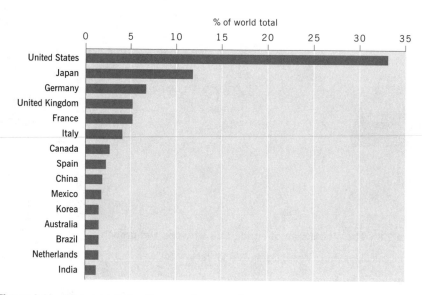

Figure 2.10   **The world's leading services producers**

Source: calculated from World Bank, 2005a: Table 4.2

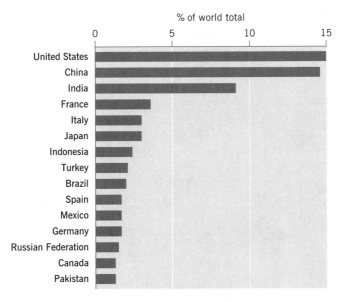

Figure 2.11   The world's leading agricultural producers

Source: calculated from World Bank, 2005a: Table 4.2

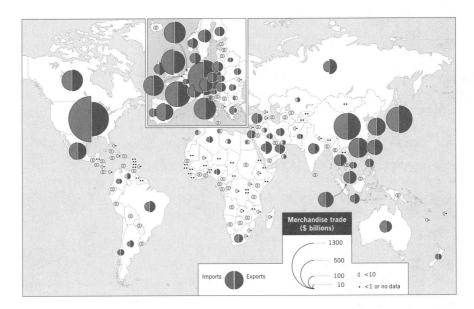

Figure 2.12   The global map of merchandise exports and imports

Source: calculated from WTO, 2004, *World Trade Report*: Tables A6, A7

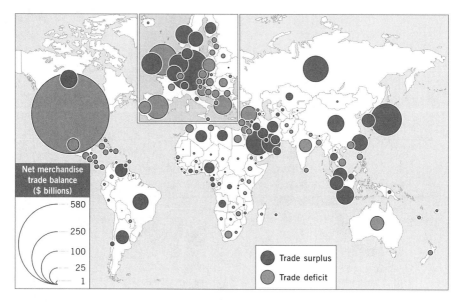

Figure 2.13   The pattern of merchandise trade surpluses and deficits

Source: calculated from WTO, 2004, *World Trade Report*: Tables A6, A7

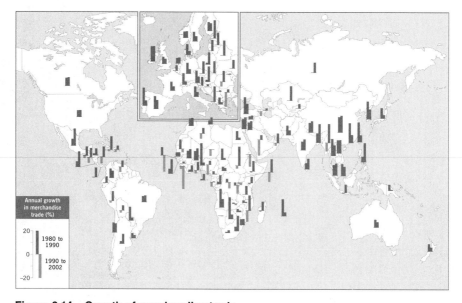

Figure 2.14   Growth of merchandise trade

Source: calculated from World Bank, 2005a: Table 4.4

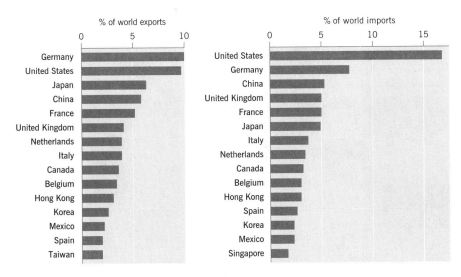

Figure 2.15   The world's leading merchandise exporters and importers

Source: calculated from WTO, 2004, *World Trade Report*: Tables A6, A7

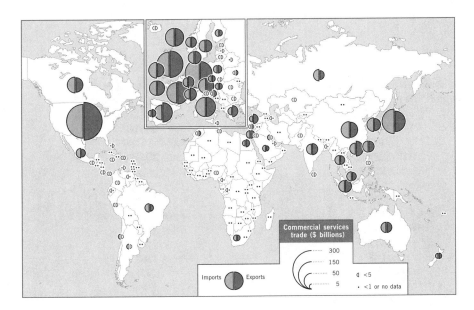

Figure 2.16   The global map of services exports and imports

Source: calculated from WTO, 2004, *World Trade Report*: Tables A8, A9

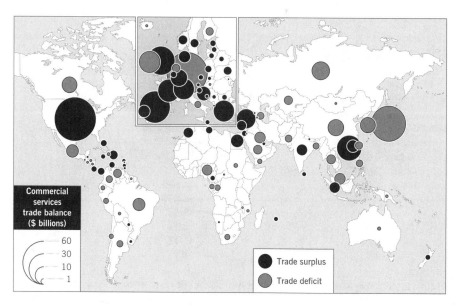

Figure 2.17   The pattern of services trade surpluses and deficits

Source: calculated from WTO, 2004, *World Trade Report*. Tables A8, A9

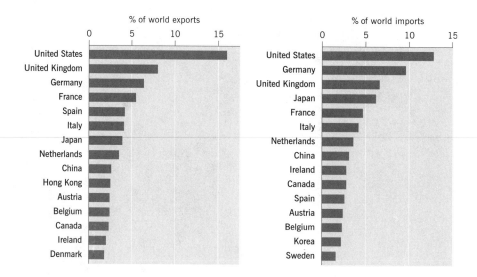

Figure 2.18   The world's leading services exporters and importers

Source: calculated from WTO, 2004, *World Trade Report*. Tables A8, A9

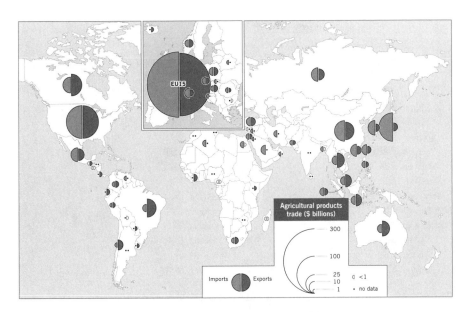

Figure 2.19   The global map of agricultural exports and imports

Source: calculated from WTO, 2004, *World Trade Report*. Tables IV.9, IV.10

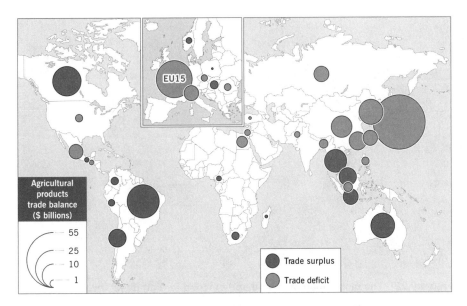

Figure 2.20   The pattern of agricultural trade surpluses and deficits

Source: calculated from WTO, 2004, *World Trade Report*. Tables IV.9, IV.10

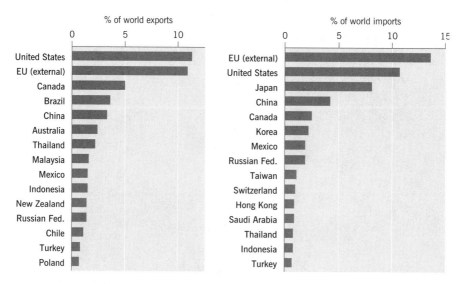

Figure 2.21   The world's leading agricultural exporters and importers

Source: calculated from WTO, 2004, *World Trade Report*: Tables IV. 9, IV.10

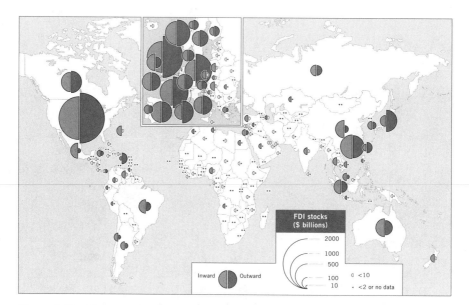

Figure 2.22   The global map of inward and outward foreign direct investment

Source: calculated from UNCTAD, 2005: Annex Table B2

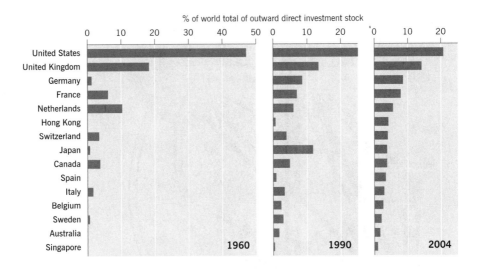

**Figure 2.23   Changing shares of leading source countries in outward for-
eign direct investment**

Source: calculated from UNCTAD *World Investment Report*, various issues

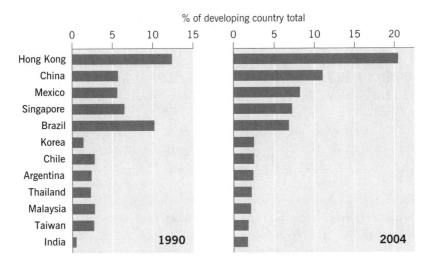

**Figure 2.24   Concentration of inward foreign direct investment in
developing countries**

Source: calculated from UNCTAD *World Investment Report*, various issues

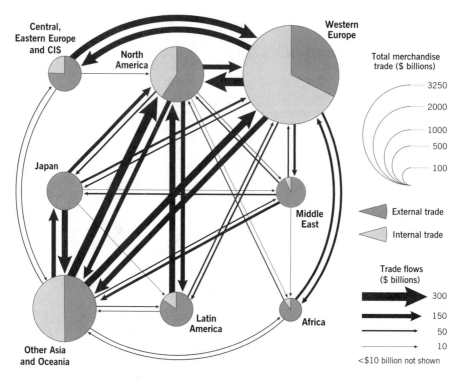

Figure 2.25   The network of world trade

Source: calculated from WTO, 2004, *World Trade Report.* Table A2

The micro-scale: cities as foci of economic activity

If we could observe the earth from a very high altitude and look at its 'economic surface' we certainly would not see the kinds of national economic boxes we have had to use as the basis of our analysis of the global economic map. Particularly if we were making the observation at night, what we would see are distinctive *clusters*, picked out by the lights of localized agglomerations of activity. Unfortunately, data disaggregated in this way, showing details of production, trade and investment, are simply not available. We have to resort to surrogate measures or individual case studies. But it is vital to stress this most fundamental fact of economic life: the place-specific and clustered nature of most economic activity. The most widely available micro-scale indicator of the localized clustering of economic activity is the map of the world's cities (Figure 2.26). Virtually all manufacturing and business service activity is located in urban places.

It is these cities, and their associated local regions, which contain a nation's economic activity, not some national statistical box. Within any individual country, there will almost certainly be considerable diversity between cities and local

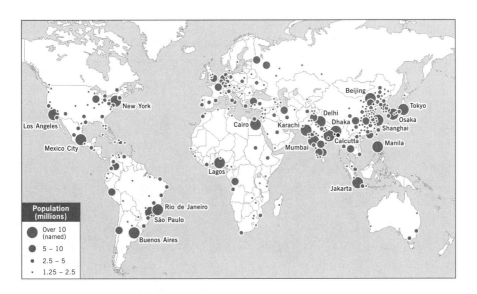

Figure 2.26   The world's major cities by size of population

Source: based on UN Centre for Human Settlements, 2003: Table A1

regions, not only in terms of their particular economic specializations but also in terms of their growth rates. In most cases, this reflects their specific historical trajectory – the 'path dependency' idea introduced in Chapter 1. In others, however, such differentials may be the outcome of very specific political decisions to develop one particular part of a country rather than another. In some countries, just one, or perhaps two, major cities dominate; in other countries there is a 'flatter' urban hierarchy and a wider spread of activity among more evenly sized cities. Increasingly, however, it is necessary to think of cities as being involved in *networks* that transcend national boundaries. In one sense, therefore, 'the city is embedded in a global economy … All cities today are world cities.'[21]

Cities differ in importance not only in terms of their population size (Figure 2.26) but also – and more importantly – in terms of the functions they perform and the influence they exert. In particular, observers of world cities emphasize the role of high-level service functions (financial and business services, in particular) and their uneven concentration in certain cities, creating a global hierarchical network. (The case of corporate control and coordination functions and their concentration in 'headquarters cities' is discussed in Chapter 5.) But it is extraordinarily difficult to produce a single map of the world city network.[22] One illustrative example is shown in Figure 2.27, based upon Taylor's analysis of business advertisements in *The Economist*. It shows the clear dominance of a relatively small number of cities. Although different criteria would yield a rather different map in detail, the dominant cities in the network recur again and again. The inset table within Figure 2.27 shows the top 10 cities ranked in terms of their global network connectivity.

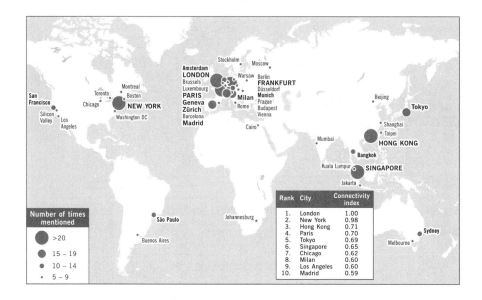

Rank	City	Connectivity index
1.	London	1.00
2.	New York	0.98
3.	Hong Kong	0.71
4.	Paris	0.70
5.	Tokyo	0.69
6.	Singapore	0.65
7.	Chicago	0.62
8.	Milan	0.60
9.	Los Angeles	0.60
10.	Madrid	0.59

Figure 2.27   Key cities in the global economy

Source: based on Taylor, 2001: Figure 2; Taylor, 2004: Table 3.5

The meso-scale: transborder clusters and corridors

Between the macro-scale of the global triad, and the highly localized concentrations of economic activity in cities, lies a meso-scale of economic-geographic organization which crosses, or sometimes aligns with, national boundaries. In some cases, this scale of organization is actually defined and created by the existence of the political boundary itself. In others it develops in spite of such boundaries and simply extends across them in a functionally organized manner. Three examples, one drawn from each of the three global triad regions, illustrate this phenomenon.

- *Europe's major economic growth axis*: within Europe, the pattern of economic activity is extremely uneven, both within and between individual countries. But as Figure 2.28 shows, we can also identify a distinctive 'growth axis' running north-west to south-east across the core area and cutting across national boundaries.

The most advanced areas of Europe and most of Europe's major international cities lie on or near an axis extending from the north-west of London through Germany to Northern Italy ... Along this axis lie two major foci: in the north-west are found the historic capitals of Europe's major colonial powers (Paris, London and Randstad-Holland) ... in the south-east are cities and regions whose faster recent economic growth has pulled the axis' centre of gravity to the south-east. A parallel axis extends from Paris to the Mediterranean – and a south-western extension stretches down to the major cities in Iberia ... another parallel axis may emerge in the east extending from Hamburg to Berlin, Leipzig, Prague and Vienna.[23]

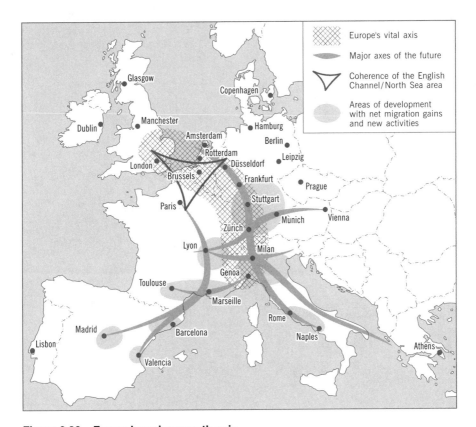

Figure 2.28   Europe's major growth axis

Source: based on Dunford and Kafkalas, 1992: Figure 1.4

- *Emerging urban corridors in Pacific Asia*: we can see a similar phenomenon, at least in embryonic form, developing in Pacific Asia. Such 'growth triangles' include the Singapore–Batam–Johor triangle and the Southern China–Hong Kong–Taiwan triangle, both of which are focused upon a distinctive major city. But some Asian urban scholars argue that much larger urban corridors are becoming evident, as Figure 2.29 suggests.

The best illustration of a mature urban corridor is … an inverted S-shaped 1,500 km urban belt from Beijing to Tokyo via Pyongyang and Seoul … [which] connects 77 cities of over 200,000 inhabitants each. More than 97 million urban dwellers live in this urban corridor, which, in fact, links four separate megalopolises in four countries in one.[24]

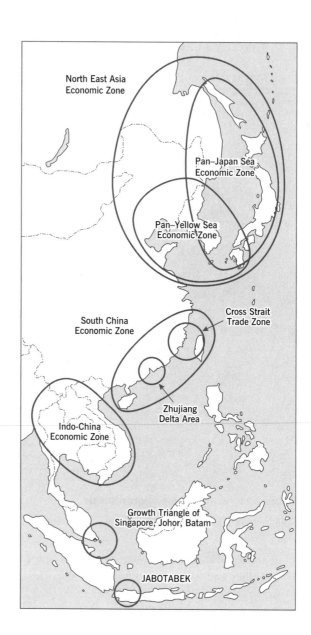

Figure 2.29   Emerging urban corridors in East Asia

Source: based on Yeung and Lo, 1996: Figure 2.8

- *The United States–Mexico border zone*: the two previous examples illustrate the development of meso-scale regions cutting across national boundaries and creating 'transnational regions'. But there are other cases where the form of economic and urban development is actually defined and created by the existence of a border between countries. Where there is a very marked differential between two adjacent countries – for example, in taxation rates or production costs – there is often a strong incentive for development to occur on one side of the border to take advantage of benefits on the other side. One of the best examples of this is the United States–Mexico border, the sharpest geographical interface between an extremely wealthy economy and a much poorer developing economy. Far from being just a line on the map, the US–Mexican border is defined in the starkest of physical terms by a whole string of towns and concentrations of manufacturing activity along its entire length (Figure 2.30).

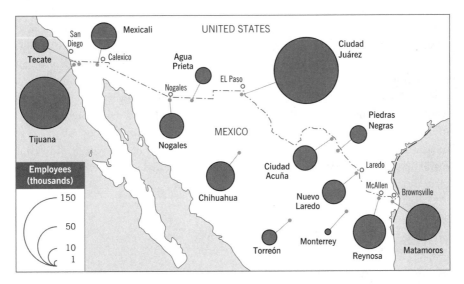

Figure 2.30   United States and Mexico: the border defines the pattern of economic activity

Source: calculated from data in Instituto Nacional de Estadística, Geografía e Informática, 2000, *Industria Maquiladora de Exportación*

Conclusion

In these first years of the new millennium, the global economic map is vastly more complicated than that of only a few decades ago. Although there are clear elements of continuity, dramatic changes have occurred. The overall trajectory of world economic growth has become increasingly volatile: short-lived surges in economic growth punctuated by periods of downturn or even recession.

Within this uneven trajectory, however, there has been a substantial reconfiguration of the global economic map. Although a handful of older core economies still dominates international trade and investment flows, the most spectacular recent growth rates have been achieved by the East Asian NIEs. Without doubt, then, the most important single global shift of recent times has been the emergence of East Asia – including the truly potential giant, China – as a dynamic growth region. Of course, in an interconnected global economy what happens in one part of the world has repercussions in other parts of the world. So, for example, the continued growth and development of the East Asian economies depends to a large extent on continued growth of their export markets – and the most important of these is the United States.

So, there have been big changes in the contours of the global economic map. But the fact remains that the actual extent of global shifts in economic activity is extremely uneven. Only a small number of developing countries have experienced substantial economic growth; a good many are in deep financial difficulty whilst others are at, or even beyond, the margins of survival. Thus, although we can indeed think in terms of a new global division of labour, its extent is far more limited than is sometimes claimed. What is clear is that a relatively simple global division of labour no longer exists. It has been replaced by a far more complex, multiscalar, structure. The global economy can perhaps best be described as 'a mosaic of unevenness in a continuous state of flux'.[25] That mosaic is, however, made up of processes which operate – and are manifested – at different but interrelated spatial scales.

Notes

1 Wallerstein (1979).

2 See Bayly (2004), O'Rourke and Williamson (1999).

3 O'Rourke and Williamson (1999: 2).

4 League of Nations (1945).

5 Frank (1998).

6 Maddison (2001: 125).

7 Webber and Rigby (1996: 6).

8 The *World Investment Report*, compiled by UNCTAD on an annual basis, is the most comprehensive source of data on foreign direct investment. Historical trends are discussed by Dunning (1993), Kozul-Wright (1995).

9 UNCTAD (2005: xix).

10 Feenstra (1998: 34).

11 Ohmae (1985).

12 Poon et al. (2000) are sceptical.

13 The data are drawn from three major sources: the World Bank's *World Development Indicators*; the WTO's *International Trade Statistics*; and UNCTAD's *World Investment Report*. Maddison (2001: Chapter 3) provides a broad perspective on development in the second half of the twentieth century.

14 The term was introduced by Friedmann (1993).

15 Beattie (2005: 17).

16 See, for example, Hofheinz and Calder (1982), Oshima (1983).

17 *The Economist* (2005b).

18 *The Economist* (20 July 2005).

19 Ramesh (2005).

20 See Poon (1997).

21 King (1983: 7, 15).

22 Taylor (2004) provides a huge variety of mappings of world city networks based on different criteria.

23 Dunford and Kafkalas (1992: 25, 27).

24 Yeung and Lo (1996: 39, 41).

25 Storper and Walker (1984: 37).

PART TWO
PROCESSES OF
GLOBAL SHIFT

Three
Technological Change: 'Gales of Creative Destruction'

Technology and economic transformation

> The fundamental impulse that sets and keeps the capitalist engine in motion comes from the new consumers' goods, the new methods of production or transportation, the new markets, the new forces of industrial organization that capitalist enterprise creates.[1]

In writing these words, Joseph Schumpeter set technological change, specifically *innovation* – the creation and diffusion of new ways of doing things – at the very heart of the processes of economic growth and development.[2] Technological change is, without any doubt, 'a fundamental force in shaping the patterns of transformation of the economy'.[3] Self-evidently, therefore, it is one of the most important processes underlying the globalization of economic activity. However, it is all too easy to regard technology as being deterministic; to be seduced by the notion that it 'causes' a specific set of changes, that it makes particular structures and arrangements 'inevitable', or that the path of technological change is linear and predictable. Two important points need to be made if we are to avoid such pitfalls.

First, technology is not independent or autonomous; it does not have a life of its own.

> Specific choices within the frontier of technological possibilities are not the product of technological change; they are, rather, the product of those who make the choices within the frontier of possibilities. *Technology does not drive choice; choice drives technology.*[4]

Technological change, therefore, is a *socially and institutionally embedded process*. The ways in which technologies are used – even their very creation – are conditioned by their social and their economic context. In the contemporary world this means primarily the values and motivations of capitalist business enterprises, operating within an intensely competitive system. Choices and uses of technologies are influenced by the drive for profit, capital accumulation and investment, increased market share and so on.

Second, technology should be seen as, essentially, an *enabling* or a *facilitating* agent – making possible new structures, new organizational and geographical arrangements of economic activities, new products and new processes, while not making particular outcomes inevitable. On the other hand, in a highly competitive environment, once a particular technology is in use by one firm, then its adoption by others may become virtually essential to ensure competitive survival.

In this chapter we focus only on those aspects of technological change that specifically influence the globalization of economic activity. The chapter is divided into four major parts.

- First, we take a broad view of technological change, identifying the key technologies and their evolution over time.
- Second, we focus on the 'time–space shrinking' technologies of transportation and communication.
- Third, we look at developments in product and process technologies.
- Fourth, we focus on the geography of innovation, on the different scales – national and local – at which innovation processes operate.

Processes of technological change: an evolutionary perspective

Technological change is a form of *learning* – by observing, by doing, by using – how to solve specific problems in a highly differentiated and volatile environment. However, it is much more than a narrowly 'technical' process. Not only does it involve the invention of new things, or new ways of doing things, but also – more importantly – it depends upon the transformation of inventions into usable *innovations,* and the subsequent adoption and diffusion or spread of such innovations. In the economic sphere, this is essentially an *entrepreneurial* process.[5]

Types of technological change

There are four broad types of technological change,[6] each of which is progressively more significant and far-reaching in its impact:

- *Incremental innovations*: small-scale, progressive modifications of existing products and processes, created through 'learning by doing' and 'learning by using'. Although individually small – and, therefore, often unnoticed – they accumulate, often over a very long period of time, to create highly significant changes.

- *Radical innovations*: discontinuous events that drastically change existing products or processes. A single radical innovation will not, however, have a widespread effect on the economic system; what is needed is a 'cluster' of such innovations.
- *Changes of technology system*: extensive changes in technology that impact upon several existing parts of the economy, as well as creating entirely new sectors. These are based on a combination of radical and incremental technological innovations, along with appropriate organizational innovations. Changes of technology system tend to be associated with the emergence of key generic technologies (for example, information technology, biotechnology, materials technology, energy technology, space technology).
- *Changes in the techno-economic paradigm*: truly large-scale revolutionary changes, embodied in new technology systems. These

> have such pervasive effects on the economy as a whole that they change the 'style' of production and management throughout the system. The introduction of electric power or steam power or the electronic computer are examples of such deep-going transformations. A change of this kind carries with it many clusters of radical and incremental innovations, and may eventually embody several new technology systems. Not only does this fourth type of technological change lead to the emergence of a new range of products, services, systems and industries in its own right – it also affects directly or indirectly almost every other branch of the economy ... the changes involved go beyond specific product or process technologies and affect the input cost structure and conditions of production and distribution throughout the system.[7]

Long waves

The notion that economic growth occurs in a series of cycles or 'waves' goes back almost 100 years. One particular type of wave – usually known as a Kondratiev wave (K-wave) – is a long wave of more or less 50 years' duration.[8] In Figure 3.1, four complete K-waves are identified; we are now in the early stages of a fifth. Each wave may be divided into four phases: prosperity, recession, depression and recovery. Each wave tends to be associated with particularly significant technological changes around which other innovations – in production, distribution and organization – swarm or cluster and ultimately spread through the economy.

Such diffusion of technology stimulates economic growth and employment, although technology alone is not a sufficient cause of economic growth. Demographic, social, economic, financial and demand conditions also have to be appropriate. At some point, however, growth slackens: demand may become saturated or firms' profits become squeezed through intensified competition. As a result, the level of new investment falls, firms strive to rationalize and restructure their operations and unemployment rises. Eventually the trough of the wave will be reached and economic activity will turn up again. A new sequence will be initiated on the basis of key technologies – some of which may be based on innovations that emerged during recession itself – and of new investment opportunities.

Although there is disagreement over the precise mechanisms and timing involved, each of the waves is generally associated with changes in the techno-economic paradigm, as one set of techno-economic practices is displaced by a new set. This is not a sudden process but one that occurs gradually and involves the ultimate 'crystallization' of a new paradigm.

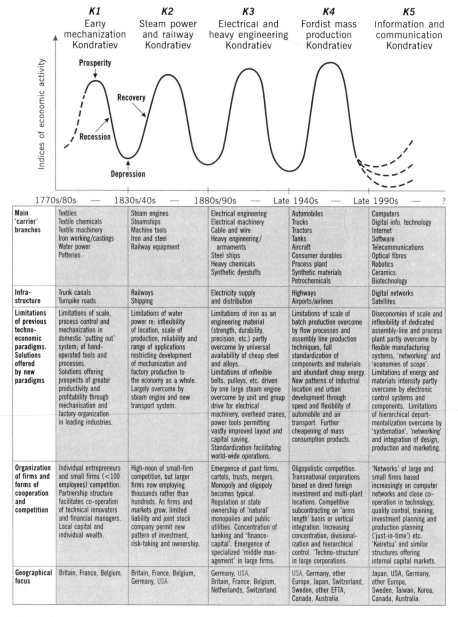

Figure 3.1   Kondratiev long-waves

Source: based, in part, on Freeman and Perez, 1988: Table 3.1

As Figure 3.1 shows, the process of change involves more than just technical change. Each phase is also associated with characteristic forms of economic organization, cooperation and competition. Organizational change has followed a path from an early focus on individual entrepreneurs in K1, through small firms (but of larger average size) in K2, to the monopolistic, oligopolistic and cartel structures of K3, the centralized, hierarchical TNCs of K4 and, it is argued, the 'network' and alliance organizational forms of K5. (These are issues we will explore in Chapter 5.) Each successive K-wave has a *specific geography* as technological leadership shifts over time, both in terms of lead nations (as the bottom row of Figure 3.1 makes clear) and also at the micro-geographical scale. In effect, 'the locus of the leading-edge innovative industries has switched from region to region, from city to city'.[9] a clear reflection of the clustering processes introduced in Chapter 1.

Information technology (IT): entering a digital world

The fifth Kondratiev cycle is associated primarily with *information technology* (IT) and, especially, with a particular kind of information technology based on *digitization*.[10]

> For the first time in history, information generation, processing and transmission have become the main commodities and sources of productivity and power and not only a means of achieving better ways of doing things in the production process. New information technologies are not simply tools to be applied but processes to be developed.[11]

Information technology, in itself, is nothing new.[12] But the current generation of information technologies has one very special characteristic, as Figure 3.2 shows. It is based upon the *convergence* of two initially distinct technologies: *communications technologies* (concerned with the transmission of information) and *computer technologies* (concerned with the processing of information). This convergence, especially through the transition from analogue to digital systems, is the key to many of the developments in the global economy. Digitization is, without doubt, the most pervasive and influential technological development of recent years. All kinds of information can now be stored in numerical (binary) form as electronic 'digits'. This means they can then be processed, manipulated and stored by computers and transmitted anywhere in the world almost instantly. In particular, the remarkable, and very recent, growth of the Internet and of mobile telephony, together with big changes in the electronic mass media, are generating major global effects at all levels, including individuals, households, local communities, nation-states and, of course, business organizations, especially transnational corporations. In the next section, we focus specifically on what can be called the 'time–space shrinking technologies'.

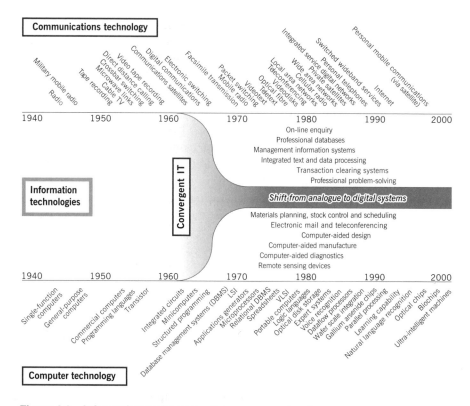

Figure 3.2   Information technology: the convergence of communications technologies and computer technologies

Source: based, in part, on Freeman, 1987: Figure 2.

The time–space shrinking technologies

As Figure 1.4 shows, *processes of circulation* are fundamental to the operation of production circuits and networks: indeed, to the operation of society as a whole. Circulation technologies – transportation and communications tech-nologies – overcome the frictions of space and time.[13] Neither can be regarded as the cause of globalization. However, without them, today's complex global economic system simply could not exist. Transportation and communications technologies perform two distinct, though closely related and complementary, roles.

- *Transportation systems* are the means by which materials, products and other tan-gible entities (including people) are transferred from place to place.
- *Communications systems* are the means by which information is transmitted from place to place in the form of ideas, instructions, images and so on.

For most of human history, transportation and communications were effectively one and the same. Prior to the invention of electric technology in the nineteenth century, information could move only at the same speed, and over the same distance, as the prevailing transportation system allowed. Electric technology broke that link, making it increasingly necessary to treat transportation and communication as distinct, though intimately related, technologies. Such developments have transformed societies in all kinds of ways. From a specifically economic and business perspective, two points are especially important.

> [First] Technologies of communications and transport shape and reshape systems of market access across *and* within geographical space ... [Second] Communications revolutions are, in effect *control* revolutions. They change the environment of profit opportunities by providing firms with different and more intensified types of control over space and time in economic activity. Such control in turn enables firms to create new routes to efficiency in the form of more innovative routines. As firms discover these routes to more efficient routines, they are compelled to assume new and different capabilities in order to capture the profit ... from doing things differently. In recasting business practices and assuming new capabilities, the innovative firm reorganizes the structure of its enterprise and in the process reshapes the territory in which it assumes control over newly crafted routines.[14]

Through their influence on how and where business organizations are able to operate, therefore, these technologies have helped progressively to transform the economic-geographical landscape, at increasing geographical scales and over shorter periods of time. Figure 3.3 summarizes this process. We will focus on how such transformations in firms' activities and their geographies are actually being worked out later in this chapter and in Chapters 4 and 5. Before that, we need to identify some of the major innovations in transportation and communications which have been so important in helping to transform the economic landscape.

Accelerating geographical mobility: innovations in transportation technologies

A shrinking world

In terms of the time it takes to get from one part of the world to another there is no doubt that the world has 'shrunk' dramatically (Figure 3.4). For most of human history, the speed and efficiency of transportation were staggeringly low and the costs of overcoming the friction of distance prohibitively high. Movement over land was especially slow and difficult before the development of the railways. Indeed, even as late as the early nineteenth century, the means of transportation were not really very different from those prevailing in biblical times.

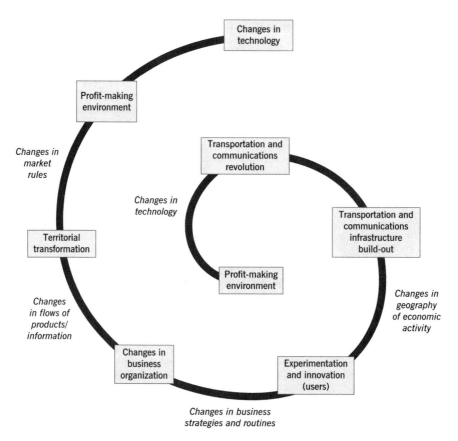

**Figure 3.3   Transportation/communications revolutions and
economic transformation**

Source: based on Fields, 2004: Figure 2.1

The major breakthrough came with two closely associated innovations: the application of steam power as a means of propulsion, and the use of iron and steel for trains, railway tracks and ocean-going vessels. These, coupled with the linking together of overland and oceanic transportation (for example, with the cutting of the canals at Suez and Panama), greatly telescoped geographical distance at a global scale. The railway and the steamship introduced a new, and much enlarged, scale of human activity. Flows of materials and products were enormously enhanced and the possibilities for geographical specialization greatly stimulated. Such innovations were a major factor in the massive expansion of the global economic system during the nineteenth century.

Moving in bulk: containerization

The past few decades have seen an acceleration of this process of global shrinkage. In economic terms, there have been two particularly important developments, both

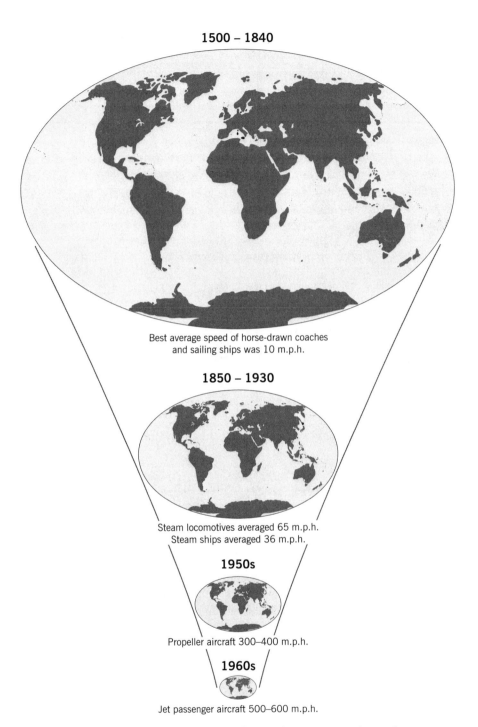

1500 – 1840

Best average speed of horse-drawn coaches
and sailing ships was 10 m.p.h.

1850 – 1930

Steam locomotives averaged 65 m.p.h.
Steam ships averaged 36 m.p.h.

1950s

Propeller aircraft 300–400 m.p.h.

1960s

Jet passenger aircraft 500–600 m.p.h.

**Figure 3.4   Global shrinkage: the effect of changing transportation tech-
nologies on 'real' distance**

Source: based on McHale, 1969: Figure 1

of them appearing for the first time during the 1950s. One development was the introduction of *commercial jet aircraft*, whose 'take-off' in the late 1950s was extraordinarily rapid. The other development was the introduction of *containerization* for the movement of ocean and land freight, an innovation that vastly simplified transhipment of freight from one mode of transportation to another, increased the security of shipments, and greatly reduced the cost and time involved in moving freight over long distances.

The first container ship, launched in 1956 to move goods from Newark, New Jersey to Houston, Texas, through the Gulf of Mexico, was merely a conventional oil tanker strengthened to take 58 boxes each 9 metres long. By the late 1990s, roughly 90 per cent of total world trade was moved in containers on purpose-built ships.[15] In fact,

> container shipping certainly is the great hidden wonder of the world, a vastly underrated business … It has shrunk the planet and brought about a revolution because the cost of shipping boxes is so cheap. People talk about the contribution made by the likes of Microsoft. But container shipping has got to be among the 10 most influential industries over the past 30 years … Before container shipping, seaborne trade was slow and unreliable. In the early sixties, unloading a ship at, say, Liverpool docks could take weeks, even months. And during that time a substantial proportion of the goods could fall prey to thieves, or the weather. Today, the goods are protected in a container, during passage and in port. With cranes specially built to lift the containers, a ship can be in and out of a port in 10 hours, saving thousands in port charges and speeding trade.[16]

For good reason, 'the box' is seen to have 'made the world smaller and the world economy bigger'.[17]

However, the very success of containerization, in a world in which trade has grown very rapidly, especially on certain routes like those from Asia to North America and to Europe, has created immense problems, especially in port bottlenecks. In 2004 a new generation of container ships, more than 300 metres long, more than 40 metres wide, and capable of carrying 8000 containers, each six metres long, entered service, with even larger ones with up to 12,000 six-metre containers expected in the near future.[18] Relatively few ports have the capacity to take such huge vessels and this will, inevitably, enhance their dominance over smaller ports. In any case, there are already massive problems of delays at most major world ports because of the sheer volume of freight traffic and the physical and human problems of handling it quickly.

The unevenness of time–space convergence

Although the world has indeed shrunk in relative terms, we need to be aware that such shrinkage has been, and continues to be, highly uneven. This is contrary to the impression given by Figure 3.4. In fact, technological developments in transportation have a very strong tendency to be geographically concentrated. The big investments needed to build transportation infrastructures tend to go where demand is greatest and financial returns are highest. Consequently, *time–space convergence* affects some places more than others. While the world's leading national economies and the world's major cities are being pulled closer together in relative time or cost terms, others – less industrialized countries or smaller towns and rural areas – are, in effect, being left behind. The time–space surface is highly plastic; some parts shrink whilst other parts become, in effect, extended. By no means everywhere benefits from technological innovations in transportation.

'Everywhere is at the same place': innovations in communications technologies

Both the time and the relative cost of transporting materials, products and people have fallen dramatically as the result of technological innovations in the transportation media. Such developments have depended on parallel developments in communications technology. However, the communications media are fundamentally significant in their own right. Indeed, communications technologies must now be regarded as *the* key technologies transforming relationships at the global scale.

> The new telecommunications technologies are the electronic highways of the informational age, equivalent to the role played by railway systems in the process of industrialization.[19]

As Figure 3.2 shows, global communications systems have been transformed radically during the past 20 or 30 years through a whole cluster of significant innovations in information technologies. In terms of communications infrastructure – the transmission channels through which information flows – two innovations have been especially significant: satellite communications and optical fibre technologies.

Transmission channels: satellites and optical fibre cables

Satellite technology began to revolutionize global communications from the mid 1960s when the Early Bird or Intelsat I satellite was launched.[20] This was the first geostationary satellite,[21] located above the Atlantic Ocean and capable of carrying 240 telephone conversations or two television channels simultaneously. Since then the carrying capacity of the communications satellites has grown exponentially. Intelsat VIII carries 22,500 two-way telephone circuits and three television channels. Its capacity can be increased to 112,500 two-way telephone circuits with the use of digital multiplication equipment.[22] The Intelsat system is a multi-nation consortium of 122 countries whose satellites are positioned to provide complete global coverage. In addition to Intelsat, there are regional systems (such as Eutelsat, which serves Europe, and others in Asia, Latin America and the Middle East) as well as private satellite systems. Today, there are around 100 geostationary satellites in orbit.

Satellite technology made possible remarkable levels of global communication of both conventional messages and the transmission of data. A message could be transmitted in one location and received in another on the other side of the world virtually simultaneously. But not quite simultaneously. A geostationary satellite system involves a short delay in the transmission and receipt of signals. One way of getting round these problems is to use a whole series of satellites that operate together in low earth orbit (LEO) systems.

In the past two or three decades, satellite communications have been increasingly challenged by *optical fibre* technology.

> Satellites were ideal for broadcasting ... as well as providing a larger number of telephone circuits than the combined capacity of all submarine cables. The life of satellites, however, is much shorter than that of cables and the number of 'parking spots' available in geosynchronous orbit is limited ... Satellites also forced a shift from analogue to digital transmission and digital signals are optimally carried by fibre-optic cables, which appeared in the 1980s, making both old telephone cables and even most satellites themselves obsolete.[23]

The first commercially viable optical fibre system was developed in the United States in the early 1970s. Since then, the speed, carrying capacity and cost of optical fibre transmission cables have changed dramatically. Optical fibre systems have a huge carrying capacity, and transmit information at very high speed and, most importantly, with a high signal strength. By the end of the 1990s, for example,

> a single pair of optical fibres, each the thickness of a human hair ... [could] carry North America's entire long-distance communications traffic. Gemini, a transatlantic undersea cable ... completed ... [in 1998] ha[d] more capacity than all existing transatlantic cables combined.[24]

Since then, technological developments in optical fibres have continued to accelerate, vastly increasing the speed and capacity of communications networks. At the same time, the geographical spread of optical fibre systems has increased.

> Only 35 countries were connected by submarine cables in 1979, with a total bandwidth of 321.4 megabits per second – less than that connecting a mid-size US city today. The most connected country, the USA, accounted for 19.6% of total world bandwidth ... By 1989, the first fibre-optic cables were being installed within Europe ... During the 1990s, 209 new submarine cables were installed, tripling the world total in 1989. The new cables provided cable links to a total of 92 countries and increased total submarine bandwidth 100-fold ... Cables installed during the 1990s focused on the North Atlantic market ... The newest cables, installed between 2000 and 2005, linked 28 countries previously unconnected by cable ... Perhaps the most important shift is the most recent one: despite a 'fibre glut' in developed countries ... new cables have continued to be installed in Asia and the Middle East.[25]

Figure 3.5 shows the enormous increase in the carrying capacity of submarine oceanic cable systems created by developments in optical fibre technologies. Figure 3.6 shows the extent to which cable systems have overtaken satellite systems in the transmission of voice traffic across the Atlantic. 'Today, 94 per cent of all international telecommunications is transmitted via cables'.[26]

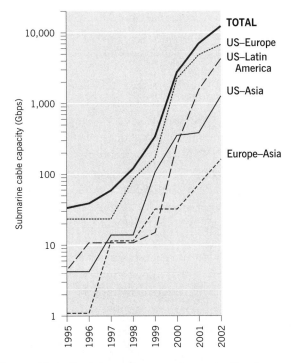

**Figure 3.5   The growth in the information carrying capacity of
submarine cable systems (Gbps: gigabits per second)**

Source: based on material in the *Financial Times*, 15 November 2000

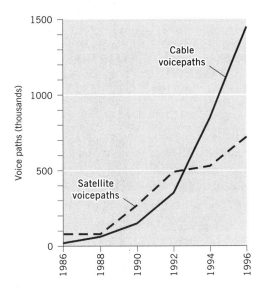

Figure 3.6   Cable overtakes satellite across the Atlantic

Source: based on Graham and Marvin, 1996: Figure 1.6

Technological developments in satellite and cable technologies have transformed the relationship between geographical distance and the cost of transmitting and receiving information. Two examples demonstrate this:

- In the 1960s the annual cost of an Intelsat telephone circuit, giving a connection from one point on the earth's surface to any other, was more than $60,000. In the late 1980s the same facility cost only $9000. Between 1973 and 1993, the price of a 3 minute international telephone call from London to New York fell by almost 90 per cent in real terms.[27]
- In 1970 the cost of transmitting the *Encyclopaedia Britannica* electronically from the east to west coasts of the United States would have been $187. In 2000, 'the entire content of the Library of Congress could be sent across America for just $40. As bandwidth expands, costs will fall further.'[28]

The rapid increase in *bandwidth* is especially important. 'Bandwidth' measures the speed (in bits per second) at which information can be transmitted. Video and data transmission require much higher bandwidth than speech.[29]

The development of faster, more sophisticated and more widely available communications infrastructures embodied in satellite and, especially, in optical fibre systems enables the transformation of all kinds of communications. Here we mention just three: the Internet, the electronic mass media and mobile telephony.

The Internet: the 'skeleton of cyberspace'

> The Internet has revolutionized the way the world communicates. In less than a decade … it has transformed from a relatively obscure computer network into a global system of hundreds of millions of networked computers (hosts) and tens of millions of formal sites for interaction and commerce (domains). Contacting someone on the other side of the world is as simple as a mouse click and billions of web pages offer a cornucopia of content, commerce, interaction, services, and products.[30]

The phenomenally rapid spread of the Internet has been one of the most remarkable developments of recent decades.[31] Its origins go back to the early 1970s and are to be found within the US Department of Defense. It spread initially through the linking of more specialized academic computer networks and, for some time, it seemed that it would remain a niche technology. Not so. As Figure 3.7 shows, growth of the Internet – measured in terms of the number of registered hosts and domains – has been exponential. Most important of all has been the rapid increase in access to broadband. Of the one billion people online in mid 2005, one-fifth were connected to broadband.[32]

Internet communication has replaced (in whole or in part) a huge swathe of conventional communications methods, notably the fax, whose life-cycle has been truncated by the use of the Internet to transmit documents, as well as some conventional telephone traffic. A large proportion of business communication

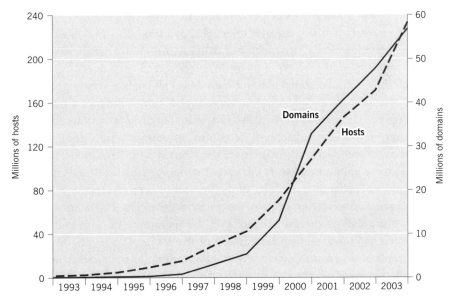

Figure 3.7 Exponential growth of the Internet

Source: based on Zook, 2005: Figure 1.1

(both inside, and between, firms) is now through the Internet. Similarly for millions of individuals, e-mail has become the preferred means of communication. At the same time, the Internet gives access to a phenomenal amount of information on virtually everything under the sun through the World Wide Web. A few years ago, hardly anybody was familiar with the term 'dotcom'; it's now part of most people's vocabulary. Without doubt, the Internet – for those with access to it – has changed the world.

The electronic mass media

The development of electronic media (radio, television) during the twentieth century provided one of the major ways in which people living in one part of the world learn about what is happening in other parts of the world. Such media are especially powerful and influential, partly because of their apparent immediacy but also because they do not require the high level of literacy of books or newspapers. The electronic media, therefore, are particularly important in making people aware of the wider world. This is, of course, very important politically but it is also very important commercially. Large business firms require large markets to sustain them; global firms aspire to global markets. The existence of such markets obviously depends on income levels, but it depends, too, on potential customers becoming aware of a firm's offerings and being persuaded to purchase them. Even where consumer incomes are low, the ground may be prepared for possible future ability to purchase by creating an aspirational image. The electronic mass media are particularly powerful means both of spreading information and of persuasion; hence their vital importance to the advertising industry and, in particular, to branded products.

Today, television is the mass medium that has the most dramatic impact on people's awareness and perception of worlds beyond their own direct experience. Although the electronic media transmit messages of all kinds, a very large proportion of these messages are *commercial* messages aimed at the consumer. Commercial advertising is a feature of most radio and television networks throughout the world. Even in state-controlled, or state-owned, systems some advertising is often included, while 'product placement' has become an art form. In a variety of ways, therefore, the communications media open the doors of national markets to the heavily advertised, branded products of the transnational producers.

These trends have been under way for several decades. The 1980s, however, saw a major 'phase shift' in the mass media with the appearance of cable and satellite broadcasting and a widespread deregulation of the media. As a result, the number of TV channels has grown dramatically, from a small number in each country to, potentially, hundreds of channels accessible through cable or satellite. This has had major effects on the ways in which TV is used. Prior to the diversification wave of the 1980s, there was a high level of standardization in the kinds of TV programme available. It was this kind of 'mutual experience' that led Marshall McLuhan to coin the metaphor of the *global village* in which certain images are shared and in which events take on the immediacy of participation. Although in one sense the world may not have shrunk for the rural peasant or the urban slum dweller with no adequate means of personal transportation, it had undoubtedly shrunk in an indirect sense. It was now possible to be aware of distant places, of lifestyles, of consumer goods through the vicarious experience of the electronic media.

But the increasing segmentation of TV messages made possible by the communications revolution of the 1980s means that the global village idea is no longer an accurate picture of reality:

> the fact that not everybody watches the same thing at the same time, and that each culture and social group has a specific relationship to the media system, does make a fundamental difference *vis-à-vis* the old system of standardized mass media ... While the media have become indeed globally interconnected, and programs and messages circulate in the global network, *we are not living in a global village, but in customized cottages globally produced and locally distributed.*[33]

Communications on the move: mobile telephony

Communications, of course, depend on a massive physical infrastructure. But within that infrastructure, one of the most significant developments of recent years has been the phenomenal growth of *mobile* communications, especially the mobile telephone. One of the first patents for a 'radio-telephone' system linked to base stations was taken out by Motorola in 1973. But development was slow. In the early 1980s, what was then still a relatively rare, and very prestigious, instrument – the mobile phone – was the size of a brick, weighed around 800 grams and cost almost $4000. By 2005, the weight was down to 90 grams and the typical cost to around $100. At the same time, the geographical range and sophistication of mobile phones and their operating systems have increased dramatically. As a consequence, there has been an explosion in ownership throughout the world. In the early 1990s, there were only a few hundred thousand subscribers to mobile systems; now there are around 1.5 billion.

We are currently on the edge of a phase shift in the capabilities of mobile phones: the so-called 3G (third-generation) system. According to the International Telecommunication Union, reporting in 2000,

> the 3G device will function as a phone, a computer, a television, a pager, a videoconferencing centre, a newspaper, a diary and even a credit card ... it will support not only voice communication but also real-time video and full-scale multimedia ... In short, the new mobile handset will become the single, indispensable 'life tool', carried everywhere by everyone, just like a wallet or purse is today.[34]

So far, the hype has exceeded the reality. Nevertheless, there is no doubt that the impact has moved beyond being the preserve of a few to the perceived necessity of the many and has the potential to revolutionize people's access to information across the world.

Digital divides: an uneven world of communications

Technological developments in communications media have transformed time–space relationships between virtually all parts of the world. We live in a *digital age*, characterized by certain key trends:[35]

- convergence between computer technologies and communications technologies
- continuing increase in the speed and capacity of these technologies and a continuing decline in their cost
- increasing ease of interfacing between different parts of the Internet
- growth of broadband communications technologies
- growth of mobile telecommunications.

However, not all places are equally connected; the 'time–space convergence' process is geographically uneven. In general, the places that benefit most from innovations in the communications media are the already 'important' places. New investments in communications technology are market related; they go to where the returns are likely to be high. The cumulative effect is both to reinforce certain communications routes at the global scale and to enhance the significance of the nodes (cities/countries) on those routes.

For example, when we look at the map of the Internet (even allowing for problems of measurement) we find a very distinctive – and very uneven – geography.[36] Figure 3.8 shows that two-thirds of all the registered domain names are in just three countries (the United States, Germany, the United Kingdom). The United

States alone has one-third of the world total, although this is less than in the past.[37] These very strong geographical concentrations are further reflected in the pattern of Internet capacity between world regions (Figure 3.9). However, in terms of the growth of broadband adoption, the United States and Europe lag some way behind Japan and Korea.[38]

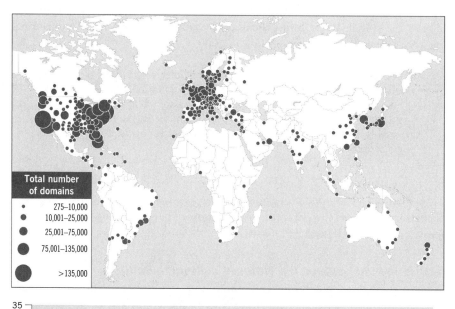

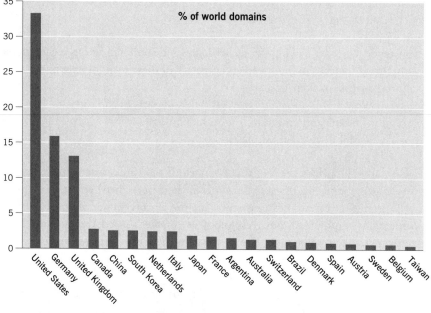

Figure 3.8   The geography of the Internet

Source: based on Zook, 2001: Figure 1; Zook 2005, Table 2.1

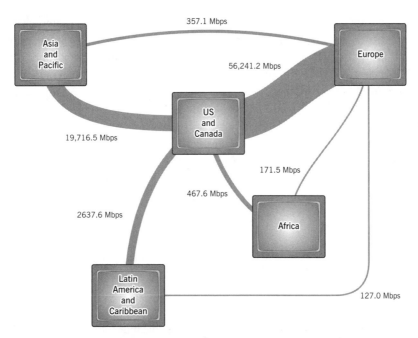

357.1 Mbps

56,241.2 Mbps

19,716.5 Mbps

171.5 Mbps

467.6 Mbps

2637.6 Mbps

127.0 Mbps

Figure 3.9   Inter-regional Internet bandwidth (megabits per second)

Source: based on material in *Telegeography*, 2002

So, the Internet is far from being the placeless/spaceless phenomenon so often envisaged. In particular, as far as provision of its basic infrastructure is concerned, it is overwhelmingly an urban – especially a big-city – phenomenon:

> This is partly an historical accident ... the Internet's fibre-optic cables often piggyback on old infrastructure where a right-of-way has already been established: they are laid alongside railways and roads or inside sewers ... Building the Internet on top of existing infrastructure in this way merely reinforces real-world geography. Just as cities are often railway and shipping hubs, they are also the logical places to put network hubs and servers, the powerful computers that store and distribute data. This has led to the rise of 'server farms', also known as data centres or web hotels – vast warehouses that provide floor-space, power and network connectivity for large numbers of computers, and which are located predominantly in urban areas.[39]

To a considerable extent, therefore, the map of the Internet mirrors the network of global cities discussed in Chapter 2 (see Figure 2.27), although the match is not perfect:

> [Although] ... the evolving infrastructure of the Internet is reinforcing old patterns of agglomeration ... At the same time, new technologies cause new 'disturbances' ... The prominence of Amsterdam and Stockholm in Europe and of Salt Lake City and Atlanta in the United States suggests that new clusters can emerge.[40]

The persistent geographical unevenness in the provision of communications infrastructure is a major problem at the global scale. There is a real and serious *digital divide* between those places and people with access to communications technologies and those without. Because such access is the key to so much information and knowledge, this poses severe developmental problems. For example, although developing countries contain around 75 per cent of the world's population they have only around 12 per cent of the world's telephone lines. Table 3.1 summarizes some of these global inequalities in access to the communications media. The range is enormous, as shown by the ratios between high- and low-income countries, although the figures need to be treated with some caution because, in most developing countries, there is a large amount of 'shared' viewing and listening in communal places.

The most promising new development for helping to bridge the digital divide is, without doubt, the mobile telephone. One of the biggest obstacles to communications growth and access in poor countries is the lack of fixed line infrastructures and the immense cost of providing them, especially in rural areas. The mobile telephone has the potential to overcome this.

Table 3.1   Digital divides: uneven access to communications media (number per 1000 population)

Region	Radios	TVs	Telephone lines	Mobile phones	Personal computers	Internet users
High income	1265	735	560	708	466.5	377
Middle income:	345	280	178	225	42.9	116
Lower middle	330	326	175	207	35.6	63
Upper middle	467	326	199	395	100.6	208
Low income	137	84	32	24	6.9	16
Low and middle income:	257	190	112	137	28.4	75
East Asia and Pacific	287	317	161	195	26.3	68
Europe and Central Asia	447		228	301	73.4	161
Latin America and Caribbean	411	289	170	246	67.4	106
Middle East and N. Africa	277	200	135	102	38.2	48
South Asia	112	84	39	23	6.8	10
Sub-Saharan Africa	198	69	11	52	11.9	20
World	419	275	183	223	100.8	150
Ratio high: low	9.2	8.8	17.5	29.5	67.6	23.6

Source: based on World Bank, 2005b: Tables 5.10, 5.11

Mobile phones do not rely on a permanent electricity supply and can be used by people who cannot read or write. Phones are widely shared and rented out by the call, for example by the 'telephone ladies' found in Bangladeshi villages. Farmers and fishermen use mobile phones to call several markets and work out where they can get the best price for their produce. Mobile phones are used to make cashless payments in Zambia and several other African countries. Even though the number of phones per 100 people in poor countries is much lower than in the developed world, they can have a dramatic impact … *The digital divide that really matters, then, is between those with access to a mobile network and those without.*[41]

Technological changes in products and processes

Product innovation

In an intensely competitive environment, the introduction of a continuous stream of new products is essential to a firm's profitability and, indeed, its very survival. 'Long-run growth requires either a steady geographical expansion of the market area or the continuous innovation of new products. In the long run, only product innovation can avoid the constraint imposed by the size of the world market for a given product.'[42] The idea that the demand for a product will decline over time is captured in the concept of the *product life cycle* (PLC).[43]

Figure 3.10 shows the major characteristics of an idealized PLC. Its essence is that the growth of sales of a product follows a systematic path from initial innovation through a series of stages: early development, growth, maturity and obsolescence. When a new product is first introduced on the market the total volume of sales tends to be low because customers' knowledge is limited and also they tend to be uncertain about the product's quality and reliability. Assuming that the new product gains a foothold in the market (and very many do not get beyond this initial stage) it then enters a phase of rapid growth as overall demand increases. Such growth is likely to have a ceiling, however; the product attains maturity in which demand levels out. Eventually, demand for the product will slacken as the product becomes obsolescent.

The kind of development path suggested by the product life-cycle concept has very important implications for the growth of firms and for their profit levels. It implies that all products have a limited life; that obsolescence is inevitable. Of course, the rate at which the cycle proceeds will vary from one product to another. In some highly ephemeral products the cycle may run its course within a single year or even less. In others the cycle may be very long. However, product cycles are becoming shorter. In order to continue to grow, and to make profits, firms need to innovate on a regular basis (or to acquire innovations from other firms). There are three major ways in which a product's sales may be maintained or increased:

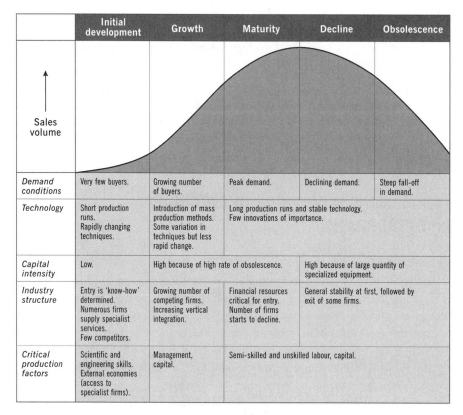

	Initial development	Growth	Maturity	Decline	Obsolescence
Demand conditions	Very few buyers.	Growing number of buyers.	Peak demand.	Declining demand.	Steep fall-off in demand.
Technology	Short production runs. Rapidly changing techniques.	Introduction of mass production methods. Some variation in techniques but less rapid change.	Long production runs and stable technology. Few innovations of importance.		
Capital intensity	Low.	High because of high rate of obsolescence.		High because of large quantity of specialized equipment.	
Industry structure	Entry is 'know-how' determined. Numerous firms supply specialist services. Few competitors.	Growing number of competing firms. Increasing vertical integration.	Financial resources critical for entry. Number of firms starts to decline.	General stability at first, followed by exit of some firms.	
Critical production factors	Scientific and engineering skills. External economies (access to specialist firms).	Management, capital.	Semi-skilled and unskilled labour, capital.		

Figure 3.10   The product life-cycle

Source: based, in part, on Hirsch, 1967: Table II (1)

- to introduce a new product as the existing one becomes obsolete so that 'overlapping' cycles occur
- to extend the cycle for the existing product, either by making minor modifications in the product itself to 'update' it or by finding new uses
- to make changes to the production technology itself to make the product more competitive.

Whichever strategy is pursued, innovation and technological change are fundamental. In so far as product cycles are shortening in many sectors, this implies increasing pressure on firms to develop new products.

Process innovation

Product innovation alone is inadequate as a basis for a firm's survival and profitability. Firms must strive to produce their products as efficiently as possible. Recent developments in technology – and, especially, in information technologies – are having profound effects upon production processes in all economic sectors. Three major, and closely interrelated, decisions are involved in all production processes, whatever the specific product may be:[44]

- *Technique of production*: the particular technology used and the way in which the various inputs or factors of production are combined. However, there are limits to such substitution of factors. Some production processes are intrinsically more capital intensive than others and vice versa.
- *Scale of production*: in general, the average cost of production tends to decline as the volume of production increases, as a result of what are known as economies of scale. These vary considerably from one industry to another. Technique and scale are, themselves, closely related.
- *Location of production*: questions of technique and scale are also intimately related to the geographical location of production. Large-scale operations require access to large markets; highly labour-intensive production processes need access to appropriate pools of labour.

The stage in the product life-cycle may well be an important factor in each of these decisions. There is often a close relationship between the progress of a product through its life-cycle and the way it is made (Figure 3.10). Each stage has particular production characteristics. One of the most important is the way in which the relative significance of the major production factors changes. In general, as the cycle proceeds, the emphasis shifts from product-related technologies to process technologies and, in particular, to ways of minimizing production costs. In this respect, the relative importance of labour costs – especially of semi-skilled and unskilled labour – increases. More generally, different types of geographical location are relevant to different stages of the product cycle.

This view of systematic changes in the production process as a product matures is appealing and has some validity. There undoubtedly are important differences in the nature of the production process between a product in its very early stages of development and the same product in its maturity. But this linear, sequential notion of change in the production process is overly simplistic and deterministic. At any stage, the production process may be 'rejuvenated' by technological innovation. There may not necessarily be a simple sequence leading from small-scale production to standardized mass production.

Changes in production systems: towards greater flexibility

This leads us to consider the major recent developments that have been occurring in the technology of production processes and, particularly, those associated with the techno-economic paradigm of information technology. Most technological developments in production processes are, as we observed earlier, gradual and incremental: the result of 'learning by doing' and 'learning by using'. But periods of radical transformation of the production process have occurred throughout history. We are now in the midst of such a radical transformation, as Figure 3.1 suggests.

Over the long time-scale of the development of industrialization, the production process has developed through a series of stages each of which represents increasing efforts to mechanize and to control more closely the nature and speed of work. Five stages are generally identified:

- *Manufacture*: the collecting together of labour into workshops and the division of the labour process into specific tasks.
- *Machinofacture*: the application of mechanical processes and power through machinery in factories together with further division of labour.
- *Scientific management* ('*Taylorism*'): the subjection of the work process to scientific study in the late nineteenth century. This enhanced the fineness of the division of labour into specific tasks together with increased control and supervision.
- '*Fordism*': the development of assembly-line processes that controlled the pace of production and permitted the mass production of large volumes of standardized products.
- *Flexible production*: the development of new production systems based upon the deep application of information technologies.

These stages in the production process map fairly closely on to the long-wave sequence shown earlier in Figure 3.1. The first Kondratiev wave was associated with the transition from manufacture to machinofacture. The application of scientific management principles to the production process emerged in the late phase of K2 and developed more fully in K3. The bases of Fordist production were established during K3 but reached their fullest development during K4. The Fordist system was epitomized by very large-scale production units, using assembly-line manufacturing techniques and producing large volumes of standardized products for mass market consumption. It was a type of production especially characteristic of particular industrial sectors, notably automobiles (see Chapter 10). Not all sectors, nor all production processes, lent themselves to such a system, but it was seen to be the main characteristic of the K4 phase.

The key to production flexibility in today's world lies in the use of *information technologies* in machines and operations. These permit more sophisticated control over the production process. With the increasing sophistication of automated processes and, especially, the new flexibility of electronically controlled technology, far-reaching changes in the process of production need not necessarily be associated with increased scale of production. Indeed, one of the major results of the new electronic and computer-aided production technology is that it permits rapid switching from one part of a process to another and allows – at least potentially – the tailoring of production to the requirements of individual customers. 'Traditional' automation is geared to high-volume standardized production; the newer 'flexible manufacturing systems' are quite different, allowing the production of small volumes without a cost penalty.

Two of the current catchphrases are *flexible specialization* and *flexible mass production*: the ability of producers to *customize* products to satisfy individual needs, without sacrificing the lower costs conventionally associated with mass production. One of the best examples of the success of mass customization is in the personal computer industry where Dell has built its entire growth on this principle. A major benefit is that it can save huge amounts of capital that would otherwise be tied up in warehouse inventories of materials, components and finished products waiting for consumers to buy them (see Chapter 14).

The potential of such flexible technologies is immense, and their implications are enormous, for the nature and organization of economic activity at all geographical scales, from the local to the global. They involve three major tendencies:[45]

- A trend towards information intensity rather than energy or materials intensity in production.
- A much enhanced flexibility of production that challenges the old best-practice concept of mass production in three central respects:
  o A high volume of output is no longer necessary for high productivity; this can be achieved through a diversified set of low-volume products.
  o Because rapid technological change becomes less costly and risky, the 'minimum change' strategy in product development is less necessary for cost effectiveness.
  o The new technologies allow a profitable focus on segmented rather than mass markets; products can be tailored to specific local conditions and needs.
- A major change in labour requirements in terms of both volume and type of labour. This involves a shift towards multitasking, rather than narrow labour specialization; a greater emphasis on team-working; and indidualized payments systems.

The current position, therefore, is that of a *diversity* of production systems (Figure 3.11), but where the relative importance of specific processes is changing. In fact, no single system is ever completely dominant. Even during the heyday of Fordist mass production there were firms and sectors in which smaller-scale, more craft-based production persisted. Today, when all the emphasis is on flexible production, there is still a good deal of mass production and, indeed, of craft production. Table 3.2 summarizes some of the contrasts between the three systems of craft production, Fordist mass production and flexible production.

Thus, we can find a trend towards:

- *increasingly fine degrees of specialization* in many production processes, enabling their fragmentation into a number of individual operations
- *increasing standardization and routinization* of these individual operations, enabling the use of semi-skilled and unskilled labour; this is especially apparent during the mature stage of a product's life-cycle
- *increasing flexibility* in the production process that is altering the relationship between the scale and the cost of production, permitting smaller production

runs, increasing product variety, and changing the way production and the labour process are organized

- *increasing modularity* of production – what Berger calls the 'Lego' model of production – involving networks of firms (as we will see in Chapter 5).[46]

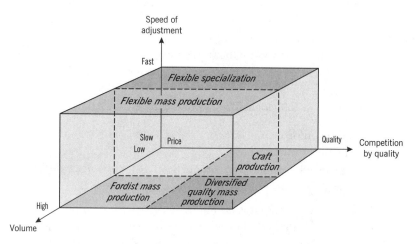

Figure 3.11   Ideal-types of production system

Source: based on Hollingsworth and Boyer, 1997: Figure 1.3

Geographies of innovation

Innovation – the heart of technological change – is fundamentally a *learning* process.[47] Such learning – by 'doing', by 'using', by observing from, and sharing with, others – depends upon the accumulation and development of relevant knowledge. Without doubt, the development of highly sophisticated communications systems facilitates the diffusion of knowledge at unparalleled speed and over unprecedented distances. As we have seen in this chapter, information can flow virtually instantaneously and globally both within and between organizations and between individuals (provided, of course, that they are 'connected' into the global communications systems). Nevertheless, 'conditions of knowledge accumulation are highly localized'.[48] Knowledge is *produced in specific places* and often used, and enhanced most intensively, in those same places. Hence,

> to understand technological change, it is crucial to identify the economic, social, political and geographical context in which innovation is generated and disseminated. This space may be local, national or global. Or, more likely, it will involve a complex and evolving integration, at different levels, of local, national and global factors.[49]

One reason for the continuing significance of 'localness' in the creation and diffusion of knowledge lies in a basic distinction in the nature of knowledge itself, which is broadly of two kinds:

Table 3.2   The major characteristics of craft production, Fordist mass production
and flexible production

Characteristic	Craft production	Mass production	Flexible/lean production
Technology	Simple but flexible tools and equipment using non-standardized components	Complex but rigid single-purpose machinery using standardized components Heavy time and cost penalties in switching to new products	Highly flexible production methods using modular component systems Relatively easy to switch to new products
Labour force	Highly skilled in most aspects of professional production	Very narrowly skilled workers design products but production itself performed by unskilled/ semi-skilled 'interchangeable' workers Each performs a relatively simple task repetitively and in predefined time sequence	Multiskilled, polyvalent workers operate in teams Responsible for several manufacturing operations plus simple maintenance and repair
Supplier relationships	Very close contact between customer and supplier Most suppliers located within single city	Distant relationships with suppliers, both functionally and geographically Large inventories held at assembly plant 'just-in-case' of supply disruption	Very close relationships with functionally tiered system of suppliers Use of 'just-in-time' delivery systems encourages geographical proximity between customers and suppliers
Production volume	Relatively low	Extremely high	Extremely high
Product variety	Extremely wide: each product customized to specific requirements	Narrow range of standardized designs with only minor product modifications	Increasingly wide range of differentiated products

Source: based in part on material in Womack et al., 1990

- *codified (or explicit) knowledge*: the kinds of knowledge that can be expressed formally in documents, blueprints, software, hardware etc.
- *tacit knowledge*: the deeply personalized knowledge possessed by individuals that is virtually impossible to make explicit and to communicate to others through formal mechanisms.

This distinction is fundamentally important to understanding the role of space and place in technological diffusion. Codified knowledge can be transmitted easily across distance. It is through such means that, throughout history, political, religious and economic organizations, for example, have been able to 'act at a distance'; to exert control over geographically dispersed activities.[50] Developments in transportation and communications technologies have enabled such 'acting' or 'controlling' to take place over greater and greater distances. Tacit knowledge, on the other hand, has a very steep 'distance-decay' curve. It requires direct experience and interaction; it depends to a considerable extent – though not completely by any means – on geographical proximity. It is much more 'sticky'. However, it is a mistake to take the 'tacit = local'/'codified = global' too far because 'both tacit and codified knowledge can be exchanged locally and globally'.[51]

Nevertheless, the specific socio-technological context within which innovative activity is embedded – what is sometimes called the *innovative milieu* – is a key factor in knowledge creation. This context consists of a mixture of both tangible and intangible elements:

- the economic, social and political institutions themselves
- the knowledge and know-how which evolve over time in a specific context (the 'something in the air' notion identified many decades ago by Alfred Marshall)
- the 'conventions, which are taken-for-granted rules and routines between the partners in different kinds of relations defined by uncertainty'.[52]

The geographical scale of such innovative milieux may range from the national through to the local.

National innovation systems

The idea underlying the notion of national innovation systems is that the specific combination of social, cultural, political, legal, educational and economic institutions and practices varies systematically between national contexts.[53] Such nationally differentiated characteristics help to influence the kind of technology system that develops there together with its subsequent development trajectory. These underlying forces help to explain the gradual shifts in national technological leadership evident in successive K-waves. Despite the claims of the hyper-globalists that national distinctiveness is declining, the evidence strongly suggests that national variations in technology systems – and, therefore in technological competence – persist.

Localized knowledge clusters

National systems of innovation, in fact, consist of aggregations of *localized* knowledge clusters, sometimes termed technology districts[54] or technopoles.[55] Table 3.3 lists some of the major technology districts or technopoles identified in empirical research. Many of them are associated with major metropolitan areas, although some have developed outside the metropolitan sphere in rather less urbanized areas. Most are the outcome of the historical process of cumulative, path-dependent growth processes although a few are the deliberate creations of national technology policy.[56] But whatever their specific origin – and this will vary from place to place because of historical and geographical contingencies – these technological agglomerations form one of the most significant features of the contemporary global economy.

Table 3.3   Some leading technology districts or technopoles

United States	Europe	Asia
Southern California	M4 Corridor, London	Tokyo
(including Silicon	Munich	Seoul–Inchon
Valley)	Stuttgart	Taipei–Hsinchu
Boston, MA	Paris-Sud	Singapore
Austin, TX	Grenoble	
Seattle, WA	Montpellier	
Boulder, CO	Nice/Sophia Antipolis	
Raleigh–Durham, NC	Milan	

The basis of localized knowledge clusters lies in several characteristics of the innovation process that are highly sensitive to geographical distance and proximity:[57]

- *Localized patterns of communication*: geographical distance greatly influences the likelihood of individuals within and between organizations sharing knowledge and information links.
- *Localized innovation search and scanning patterns*: geographical proximity influences the nature of a firm's search process for technological inputs or possible collaborators. Small firms, in particular, often have a geographically narrower 'scanning field' than larger firms.
- *Localized invention and learning patterns*: innovation often occurs in response to specific local problems. Processes of 'learning by doing' and 'learning by using' tend to be closely related to physical proximity in the production process.
- *Localized knowledge sharing*: because the acquisition and communication of tacit knowledge are strongly localized geographically, there is a tendency for localized 'knowledge pools' to develop around specific activities.
- *Localized patterns of innovation capabilities and performance*: geographical proximity, in enriching the depth of particular knowledge and its use, can reduce the risk and uncertainty of innovation.

Local innovative milieux, therefore, consist primarily of a *nexus of untraded interdependencies* set within a temporal context of *path-dependent* processes of technological change. We outlined the major elements of such processes in general terms in Chapter 1 (see Figure 1.7). The point of emphasizing the 'untraded' nature of the interdependencies within such milieux is to distinguish the social 'cement' which binds this kind of localized agglomeration from that which may be associated with the minimization of transaction costs (for example, of materials and components transfers) through geographical proximity. The 'buzz' derived from 'being there' is at the heart of these social processes.[58]

But that is not the entire story. Localized knowledge clusters cannot be sustained, and develop further, entirely through such incestuous relationships. A key additional process involves the connections between some of the actors in a given locality with outsiders (for example, firms with suppliers, customers, or sources of specific information and knowledge). In other words, as well as 'local buzz' there also have to be 'pipelines': channels of communication to other actors in other places. The processes of knowledge creation and innovation, therefore, consist of a complex set of networks and processes operating *within and across* various spatial scales, from the global, through the national and the regional, to the local.

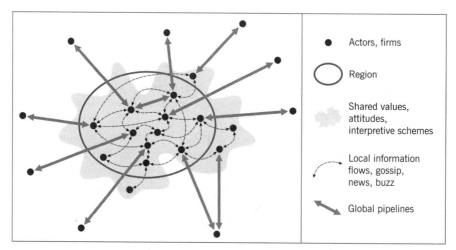

Figure 3.12 Localized knowledge clusters in a wider context: local buzz and global pipelines

Source: based on Bathelt et al., 2004: Figure 1

Figure 3.12 provides an idealized picture of this very complex process. It is based on the argument that

> the existence of local buzz of high quality and relevance leads to a more dynamic cluster ... These actors and their buzz are, however, of little relevance if firms are not 'tuned in' ... It is likely that a milieu, where many actors with related yet complementary and heterogeneous knowledge, skill and information reside, provides a great potential for dynamic interaction ...
>
> A well-developed system of pipelines connecting the local cluster to the rest of the world is beneficial for the cluster in two ways. First, each individual

firm can benefit from establishing knowledge-enhancing relations to actors outside the local cluster. Even world-class clusters cannot be permanently self-sufficient in terms of state-of-the-art knowledge creation. New and valuable knowledge will always be created in other parts of the world and firms who can build pipelines to such sites of global excellence gain competitive advantage. Second, it seems reasonable to assume that the information that one cluster firm can acquire through its pipelines will spill over to other firms in the cluster through local buzz ... That is why a firm will learn more if its neighbouring firms in the cluster are globally well connected rather than being more inward-looking and insular in their orientation.[59]

Conclusion

The aim of this chapter has been to identify some of those features of technological change that are most important in the globalization of economic activity. Technological change is the dynamic heart of economic growth and development; it is fundamental to the evolution of a global economic system. We focused on four specific aspects of this process.

First, we explored technological change as an evolutionary process in which much change is gradual and incremental, often unnoticed but, nonetheless, extremely significant. But there are periodic radical transformations of existing technologies – revolutionary developments in clusters of technologies – that not only dramatically alter products and processes in one industry but also pervade the entire socio-economic system. These are the shifts in the techno-economic paradigm that seem to be associated with the long waves of economic change. Today, information technology – especially digitization – is the dominant and all-pervasive technology.

Second, we focused on the time–space shrinking technologies of transportation and communication which lie at the heart of – but do not in themselves cause – the globalization of economic activity. Such processes have dramatically speeded up, and geographically extended, economic processes and facilitated the expansion of markets and the ability to control activities at a distance. The Internet, in particular, is especially powerful and pervasive in its effect. But the time–space shrinking technologies are highly uneven in their incidence and in their effects.

Third, we outlined some of the major transformations that have occurred in technologies of production. Here the dominant trend of recent years has been the shift away from the rigidities of Fordist mass production towards more flexible and modular forms of production. Such developments have major implications for all participants in the economy: firms, labour, consumers and the state.

Fourth, we emphasized the strongly localized nature of innovation and knowledge creation. The path-dependent nature of technological change and the social conditions within which such change occurs give major importance to the *geography* of the process. A major element in this process is the relationship between the internal characteristics and dynamics of localized knowledge clusters and the world outside.

Throughout this chapter, our concern has been with 'technical' innovations. But there are other kinds of innovation. One of these – organizational innovations – forms part of our discussion in the next two chapters, whose focus is on the development, organization and operation of transnational corporations.

NOTES

1 Schumpeter (1943: 83).
2 The term 'gales of creative destruction' is borrowed from Schumpeter (1943).
3 Freeman (1988: 2).
4 Borrus, quoted in Cohen and Zysman (1987: 183, emphasis added).
5 The approach ontlined in this section is essentially 'Schumpeterian'. See Dosi et al. (1988), Fields (2004), Freeman (1982; 1987), Metcalfe and Dilisio (1996), Perez (1985).
6 Freeman and Perez (1988).
7 Freeman (1987: 130).
8 See Freeman et al. (1982), Freeman and Perez (1988), Hall and Preston (1988), Rennstich (2002).
9 Hall and Preston (1988: 6).
10 Useful discussions of information technology can be found in Cairncross (1997), Castells (1996), Freeman (1987), Graham and Marvin (1996), Hall and Preston (1988).
11 Rennstich (2002: 174).
12 Hall and Preston (1988).
13 For broad-ranging discussions of these technologies see Brunn and Leinbach (1991), Castells (1996), Graham and Marvin (1996), Hall and Preston (1988).
14 Fields (2004: 14, 15).
15 *The Economist* (2 June 2001).
16 *The Independent* (30 August 2000).
17 Levinson (2006).
18 *Financial Times* (17 December 2004).
19 Henderson and Castells (1987: 6).
20 Malecki and Hu (2006), Warf (2006) provide recent analyses of satellite and cable systems.
21 A geostationary (or geosynchronous) satellite has to be positioned above the Equator and to orbit at an altitude of 22,300 miles.
22 These data are from Baylin (1996).
23 Malecki and Hu (2006: 7).
24 *Financial Times* (28 July 1998).
25 Malecki and Hu (2006: 14, 15).
26 Warf (2006: 10).
27 *Financial Times* (17 September 1994).
28 *The Economist* (23 September 2000).

29 Mitchell (1995).

30 Zook (2005: 1).

31 Zook (2005) provides an excellent recent analysis of the Internet from a geographical perspective. See also Batty and Barr (1994), Castells (1996), Dodge and Kitchin (2001), Malecki (2002).

32 *Financial Times* (5 August 2005).

33 Castells (1996: 341).

34 Cited in *The Economist* (4 September 2004).

35 Govindarajan and Gupta (2000: 279–80).

36 Dodge and Kitchin (2001), Malecki (2002), Zook (2001; 2005).

37 Malecki (2002: 405), Zook (2001: Table 1).

38 Fransman (2006).

39 *The Economist* (11 August 2001: 18)

40 Malecki (2002: 419).

41 *The Economist* (12 March 2005: 9).

42 Casson (1983: 24).

43 O'Shaughnessy (1995) explores this concept in a marketing context. Vernon (1966; 1979) pioneered its application to international production and international trade. See also Hirsch (1967), Wells (1972).

44 Smith (1981).

45 Perez (1985).

46 Berger (2005: 57). See also Sturgeon (2002).

47 Bunnell and Coe (2001) provide an overview of the current literature on the geographies of innovation.

48 Metcalfe and Dilisio (1996: 58).

49 Archibugi and Michie (1997: 2).

50 Fields (2004), Law (1986).

51 Bathelt et al. (2004: 32).

52 Storper (1995: 208).

53 Lundvall (1992), Nelson (1993). For recent reviews of the national innovation systems concept, see Archibugi and Michie (1997), Archibugi et al. (1999), Freeman (1997), Lundvall and Maskell (2000), Patel and Pavitt (1998).

54 Storper (1992).

55 Castells and Hall (1994).

56 This is certainly so in the case of the Hsinchu Science-Based Industry Park in Taiwan (Mathews, 1997) (see Chapter 11) and of the Xingwang Industrial Park in Beijing (Yeung et al. (2006).

57 Howells (2000: 58–9).

58 Bathelt et al. (2004), Gertler (1995), Sturgeon (2003).

59 Bathelt et al. (2004: 45–6).

Four
Transnational Corporations: The Primary 'Movers and Shapers' of the Global Economy

The significance of the transnational corporation

More than any other single institution, the transnational corporation has come to be seen as the primary shaper of the contemporary global economy and a major threat to the economic autonomy of the nation-state. As shown in detail in Chapter 2, not only has there been a massive growth of foreign direct investment (FDI) but also the sources and destinations of that investment have become increasingly diverse. But FDI is only one measure of TNC activity. Because the FDI data are based on ownership of assets they do not capture the increasingly intricate ways in which firms engage in transnational operations, through various kinds of collaborative ventures and through the different modes by which they coordinate and control transactions within geographically dispersed production networks. It is for this reason that we adopt a much broader definition of the TNC than that normally used in the conventional literature:

A transnational corporation is a firm that has the power to coordinate and control operations in more than one country, even if it does not own them.

The significance of the TNC lies in three basic characteristics:

- its ability to coordinate and control various processes and transactions within transnational production networks, both within and between different countries
- its potential ability to take advantage of geographical differences in the distribution of factors of production (for example, natural resources, capital, labour) and in state policies (for example, taxes, trade barriers, subsidies etc.)

- its potential geographical flexibility – an ability to switch and to reswitch its resources and operations between locations at an international or even a global scale.

Hence, much of the changing geography of the global economy is shaped by the TNC through its decisions to invest, or not to invest, in particular geographical locations. It is shaped, too, by the resulting flows – of materials, components, finished products, technological and organizational expertise, finance – between its geographically dispersed operations. Although the relative importance of TNCs varies considerably – from sector to sector, from country to country, and between different parts of the same country – there are now very few parts of the world in which TNC influence, whether direct or indirect, is not important. In some cases, indeed, TNC influence on an area's economic fortunes can be overwhelming.

However, TNCs are highly differentiated, not only in size and geographical extent but also in the ways in which they operate. They are, emphatically, not all of a kind: identical economic monsters stamping an identical footprint on the landscape. Indeed, far from being the 'placeless' organizations often claimed, TNCs continue to reflect many of the basic characteristics of the home country environments in which they remain strongly embedded.

The aim of this, and the next, chapter is to explore the activities of TNCs, particularly in the context of the increasingly complex networks of interrelationships within which all TNCs are embedded. The chapter is organized into three parts:

- First, we pose the question of *why* firms should attempt to transnationalize their activities beyond simply exporting their products to foreign markets.
- Second, we address the question of *how* firms transnationalize their activities, in terms of both the sequence through which they develop and the organizational architectures they have evolved.
- Third, we challenge the view that TNCs are becoming 'placeless' and converging towards a universal business model.

Why firms transnationalize

The 'natural expansion of capital': a macro-structural perspective

Most TNCs are capitalist enterprises. As such, they must behave according to the basic 'rules' of capitalism, the most fundamental of which is the drive for *profit*. Of course, business firms may well have a variety of motives other than profit, such as increasing their share of a market, becoming the industry leader, or simply making the firm bigger. But, in the long run, none of these is more important than the pursuit of profit itself. A firm's profitability is the key barometer to its business 'health'; any firm that fails to make a profit at all over a period of time is likely to go out of business (unless rescued by government or acquired by another firm). At best, therefore, firms must attempt to increase their profits; at worst, they must defend them.

Of course, a capitalist market economy is an intensely *competitive* economy. One firm's profit may be another firm's loss unless the whole system is growing sufficiently strongly to permit all firms to make a profit. Even so, some will make a larger profit than others. Two key features of today's world are: first, that competition is increasingly global in its extent and, second, that such competition is extremely volatile. 'This creates an environment of hyper-competition – an environment in which advantages are rapidly created and eroded'.[1] Firms are no longer competing largely with national rivals but with firms from across the world. Given these circumstances, therefore, one way of explaining TNCs is simply as a reflection of the 'normal' expansionary tendencies of the different *circuits of capital*.[2] In these terms, the question 'Why transnationalize?' might almost be better put as 'Why *not* transnationalize?'

Micro-level approaches: a focus on the firm

Explanations cast at the macro-level of the capitalist system, useful as they are in emphasizing the structures within which decisions are made, are in themselves insufficient. The decision to transnationalize (or not) is made by individual firms (or, more accurately, by decision-makers within firms).

Hymer's pioneering contribution

Before Hymer's pioneering study in 1960 there was no specific theory of why firms engage in transnational production: FDI was treated as just another variant of international capital theory. Hymer struck out in a new direction, drawing his inspiration from a completely different source: industrial organization theory and, especially, that part which dealt specifically with barriers to entry.[3] Hymer started from the assumption that, in serving a particular market, domestic firms would have an intrinsic advantage over foreign firms: for example, a better understanding of the local business environment, including the nature of the market, business customs and legislation, and the like. Given such domestic-firm advantage, a foreign firm wishing to produce in that market would have to possess some kind of firm-specific asset that would offset the advantages held by domestic firms. Such assets are primarily those of firm size and economies of scale, market power and marketing skills (for example, brand names, advertising strength), technological expertise (either product, or process, or both), or access to cheaper sources of finance. On these bases, then, a foreign firm would be able to out-compete domestic firms in their own home territory.

Hymer's contribution was truly seminal. It was the first time that the firm had been taken as the specific focus of explanation and that transnational production (rather than international trade) had been the explicit object of analysis. He emphasized the importance of market imperfections in stimulating the transnationalization of production. He showed how, once established, the *control* of overseas productive assets itself became a source of competitive advantage. Of course, Hymer's theory had its limitations. It was much better at explaining why and how firms might *begin* to become transnationalized as producers and less good at explaining their subsequent development from an established transnational position. But, like all theories, it needs to be evaluated in terms of the level of understanding prevailing at the time he was writing. By that criterion, his contribution was immense.

Dunning's 'eclectic' paradigm

Dunning's attempt to produce a more complete explanation of international production owes a considerable debt to Hymer.[4] According to Dunning, a firm will engage in transnational production when all of the following three conditions are present:

- A firm must possess certain *ownership-specific advantages* (O) not possessed by competing firms of other nationalities. These are assets internal to a firm over which it has proprietary right of use: for example, types of knowledge, technology, marketing (including brands), organizational and human skills. Firm size and market power are especially significant advantages.
- Such advantages must be most suitably exploited by the firm itself rather than by selling or leasing them to other firms. In other words, the firm will *internalize* (I) the use of its ownership-specific advantages. The more uncertain the environment faced by a firm (for example, in terms of the availability, price or quality of supplies, or of the price the firm can obtain for its product on the market), the more likely it is to internalize its operations. Internalization is especially likely to occur in the case of knowledge. Many firms spend huge sums of money on research and development. To ensure a satisfactory return on such investment, and to protect against predators, firms have a strong incentive to retain the technology for use within their own organizational boundaries. Rather than sell or lease the technology to another firm abroad, the firm sets up its own production facilities and exploits its technological advantage directly.
- There must be *location-specific factors* (L) that make it more profitable for the firm to exploit its internalized assets in foreign, rather than in domestic, locations. Location-specific factors are those that are 'specific in origin to particular locations and *have to be used in those locations*'.[5] Several major types of location-specific factor are especially important in the context of transnational production – for example, markets, resources, production costs, political conditions (including degrees of political risk), cultural/linguistic attributes – although their precise significance will vary according to the type of activity involved.

To some, Dunning's OLI 'paradigm' is rather overblown and rather less profound than its author claims. On the other hand, it is a useful pragmatic and comprehensive framework and also – perhaps especially – it emphasizes the critical role of geographical location in understanding the complex nature of TNC behaviour. 'Far from being the poor relative, L may well turn out to be the prince'.[6]

Motivations underlying transnationality

Although a firm's motivation for engaging in transnational operations may be highly individual we can classify them into two broad categories (although the boundary between them is by no means as sharp as this dichotomy suggests):

- market orientation
- asset orientation.

Market orientation

Most foreign direct investment, whether in production or in marketing and sales, is designed to serve a specific geographical market by locating inside that market. The good or service produced abroad may be virtually identical to that being produced in the firm's home country, although there may well be modifications to suit the specific tastes or requirements of the local market. In effect, such specifically market-oriented investment is a form of horizontal expansion across national boundaries. Three attributes of markets are especially important:

- The most obvious attraction of a specific market is its *size*, measured, for example, in terms of per capita income. Figure 4.1 shows the enormous variation in income levels per capita on a global scale. The largest geographical markets in terms of incomes, although not in terms of population, are obviously the United States and Western Europe. Such variations in per capita income provide a crude indication of how the *level* of demand will vary from place to place across the world.
- Countries with different income levels will tend to have a different *structure* of demand. As incomes rise, so does the aggregate demand for goods and services. But such increased demand does not affect all products equally. Populations in countries with low income levels tend to spend a larger proportion of their income on basic necessities while, conversely, people in countries with high income levels tend to spend a higher proportion of their income on 'higher-order' manufactured goods and services.
- Markets vary in their *accessibility*. In the past, a major barrier was the cost of transportation. Today, this is far less significant, although not totally unimportant, especially for some products. However, political constraints in the form of various kinds of trade barrier remain significant (see Chapter 6).

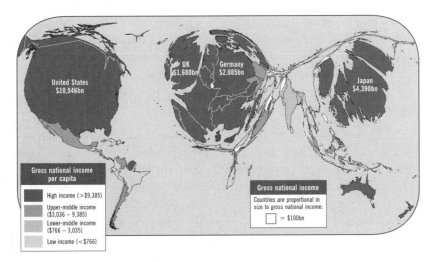

Figure 4.1   Variations in market size: gross national income per capita

Source: based on World Bank, 2005b: Table 1

Asset orientation

Most of the various assets needed by a firm to produce and sell its specific products and services are unevenly distributed geographically. This is most obviously the case in the natural resource industries, where firms must, of necessity, locate at the sources of supply. Often, such investments form the first element in an organizational sequence of vertically integrated operations whose later stages (processing) may be located quite separately from the source of supply itself. In many cases, final processing of natural resources occurs close to the final market. Natural-resource-oriented foreign investments have a very long history and remain highly significant in the global economy.

However, there are other asset-oriented foreign investments closely related to the developments in product and process technologies discussed in Chapter 3. Technological changes in production processes and in transportation have evened out the significance of location for some of the traditionally important factors of production (for example, natural resources). Many now hold the view that, at least at the global scale, the two most important location-specific factors are access to *knowledge* and access to *labour*. The strong tendency for knowledge and technological innovation processes to appear in *geographical clusters* creates a major incentive for firms to locate their relevant operations in such locations. Particularly in those activities in which technological change, whether in product or process development, is especially rapid and unpredictable, the incentive to locate 'where the knowledge and the action are' becomes very powerful. Such knowledge may be based in specific kinds of institution (such as universities, research institutes, industry associations). However, much of the attraction exerted by such knowledge clusters derives from the skills and knowledge embodied in *labour*.

The locational significance of labour as a 'production factor' is reflected in a number of ways although, of course, working people are much more than 'crude abstractions in which labour is reduced to the categories of wages, skill levels, location, gender, union membership and the like, the relative importance of which is weighed by firms in their locational decision-making'.[7] Five especially important attributes of labour show large geographical variations:

- *Knowledge and skills*: knowledge and skills depend on such conditions as the breadth and depth of education and on the particular history of an area's development. As a result, there are wide geographical variations in the availability of different types of labour. One very approximate indicator at the global scale is the variation in educational levels (for example, extent of literacy, enrolment in various stages of education, public expenditure on education etc.). As might be expected, there is a very high correlation between these measures and the distribution of per capita income shown in Figure 4.1.

- *Wage costs*: international differences in wage levels can be staggeringly wide, as Figure 4.2 shows. These figures should be treated with some caution; they are averages across the whole of manufacturing industry and are therefore affected by the specific industry mix. Some industries have much higher wage levels than others. Even so, the contrasts are striking.

- *Labour productivity*: spatial variations in wage costs are only a partial indication of the locational importance of labour as a production factor. The 'performance capacity' of labour varies enormously from place to place, a reflection of a number of influences including education, training, skill, motivation, as well as the kind of machinery and equipment in use.

- *Labour 'controllability'*: largely because of historical circumstances, there are considerable geographical differences in the degree of labour 'militancy' and in the extent to which labour is organized through labour unions. The proportion of the workers who are members of labour unions has declined markedly in some countries, notably the United States (from 35 per cent in 1955 to under 14 per cent today), the United Kingdom (from 48 per cent in 1970 to around 30 per cent) and France (from 21 per cent in 1970 to around 10 per cent). The fact that most firms are very wary of 'highly organized' labour regions is demonstrated by their tendency to relocate from such regions or to make new investments in places where labour is regarded as being more malleable.

- *Labour mobility*: labour is far less geographically mobile than capital, particularly over great distances. In general, labour is strongly *place bound*, although the strength of the tie varies a great deal between different types of labour. On average, male workers are more mobile than female workers; skilled workers are more mobile than unskilled workers; professional white-collar workers are more mobile than blue-collar workers. Clearly, there are exceptions to such generalizations, as shown by the substantial waves of labour migration at different periods of history and towards particular kinds of geographical location (see Part Four). Such flows do not, however, contradict the basic point that labour is strongly differentiated spatially and deeply embedded in local communities in distinctive ways.[8]

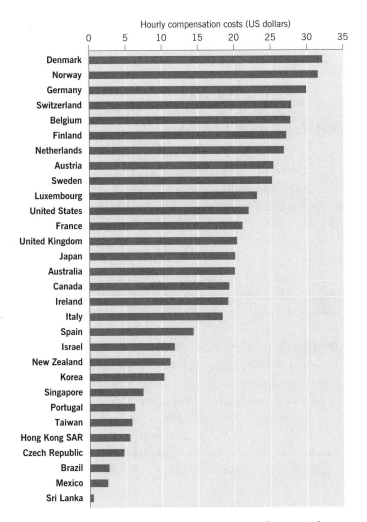

Hourly compensation costs (US dollars)

Figure 4.2 **Geographical variations in hourly compensation costs in manufacturing**

Source: based on US Department of Labor statistics

Global variations in production costs are a highly significant element in the
transnational investment-location decision. This is obviously the case for asset-
oriented investments but it is also a critical consideration for market-oriented
investments. In that case there is always a trade-off to be made between the benefits of market proximity on the one hand and locational variations in production
costs on the other. But the problem is not merely one of variations in production
costs at a single moment in time or even the obvious point that such costs change
over time. A particularly important consideration is the *uncertainty* of the level of
future production costs in different locations. One way of dealing with such
uncertainty is for the TNC to locate similar plants in a variety of different locations and then to adopt a flexible system of production allocation between plants

(see Chapter 5). However, this strategy is complicated further by the volatility of currency exchange rates between different countries. What appears to be a least-cost source with one set of exchange rates may look very different if there is a major change in these rates.

How firms transnationalize

How they get there: paths of development

Is there an identifiable *evolutionary sequence* of TNC development? Does the transition from a firm producing entirely for its domestic market to one engaged in foreign production follow a systematic development path? The answer to these questions is both 'yes' and 'no'. Yes, in the sense that certain common patterns of development are evident. No, in the sense that we should not expect all firms to follow the same sequence, or inevitably to become TNCs. It is useful to consider the broad path of TNC development, however, because it helps to give some sense of the dynamics of the processes involved.

Vernon's locational product life-cycle

The concept of the product life-cycle, already encountered in Chapter 3 (see Figure 3.10), was specifically adopted and adapted as an explanation of the evolution of international production by Raymond Vernon in 1966.[9] Vernon's major contribution was to introduce an explicitly *locational* dimension into the product cycle concept that, in its original form, had no spatial connotation at all. Figure 4.3 shows Vernon's PLC model based upon the experience of United States' TNCs, especially during the 1960s (it should be read in conjunction with Figure 3.10).

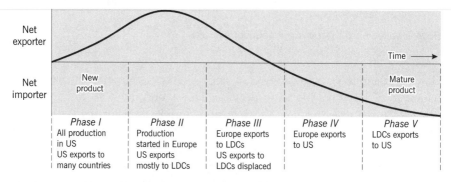

Figure 4.3   The product life-cycle as an evolutionary sequence of United States' TNC development

Source: based on Wells, 1972: Figure 15

Vernon assumed that firms are more likely to be aware of the possibility of introducing new products in their home market than producers located elsewhere. The kinds of new products introduced, therefore, would reflect the specific characteristics of the domestic market. In the United States case, high average-income levels together with high labour costs encouraged the development of new products that catered to high-income consumers and were labour saving (for both consumer and producer goods).

In this first phase of the locational product life-cycle, as Figure 4.3 shows, all production would be located in the United States and overseas demand would be served by exports. But this situation would be unlikely to last indefinitely. US firms would eventually set up production facilities in the overseas market, either because they saw an opportunity to reduce production and distribution costs or because of a threat to their market position. Such a threat might come from local competitors or from government attempts to reduce imports through tariff and other trade barriers.

It follows from the nature of the product cycle model that the first overseas production of the product would occur in other high-income markets. The newly established foreign plants would come to serve these former export markets and thus displace exports from the United States. These would be redirected to other areas where production had not yet begun (phase II in Figure 4.3). Eventually, the production cost advantages of the newer overseas plants would lead the firm to export from them to other, third-country, markets (phase III) and even back to the United States itself (phase IV). Finally, as the product became completely standardized, production would be shifted to low-cost locations in developing countries (phase V). It is interesting to note that when Vernon first suggested this possibility he regarded it as a 'bold projection'. At that time (the mid 1960s) there was still little evidence of developing country export platforms serving European and US markets.

How valid is the product life-cycle as an explanation of the locational evolution of TNCs? There is no doubt that a good deal of the *initial* overseas investment by US firms did fit the product cycle sequence quite well. But it can no longer explain the majority of transnational investment by TNCs. As these firms have become more complex globally it is unrealistic to assume a simple evolutionary sequence from the home country outwards. Even within strongly innovative TNCs, the initial source of the innovation and of its production may be from any point in the firm's global network. In addition, as we saw in Chapter 2, much of the world's FDI is reciprocal, or cross, investment between the industrialized countries. Such investment cannot easily be explained in product cycle terms.

Diverse paths of development

The most commonly identified ideal-type sequence through which a firm may move from being entirely a domestically oriented firm to a TNC is shown in the centre of Figure 4.4. It is a PLC-type view, which begins with the assumption that, initially, the firm is purely a domestic firm in terms both of production and of markets. In all national economies the majority of firms are of this type. However, the limits of the firm's domestic market may be reached and overseas markets may need to be penetrated to maintain growth and profitability. It is generally assumed that this is done initially through exports, using the services of overseas sales agents. Such agents are independent of the exporting firm. However, the benefits of internalization (discussed earlier) may eventually stimulate the firm to exert closer control over its foreign markets by setting up overseas sales outlets of its own.

This may be achieved in one of two ways: by setting up an entirely new facility or by acquiring, or merging with, a local firm (possibly the previously used sales agency itself). Merger and acquisition (M&A) is one of the most common methods of entry both to new product markets and to new geographical markets. It offers the attraction of an already functioning business compared with the more difficult, and possibly more risky, method of starting from scratch in an unfamiliar environment. Cross-border M&A activity tends to occur in waves: there was a huge wave of activity in the late 1990s, followed by a sharp decline between 2000 and 2003. It has since surged again: in 2004, cross-border M&A increased by 28 per cent.[10]

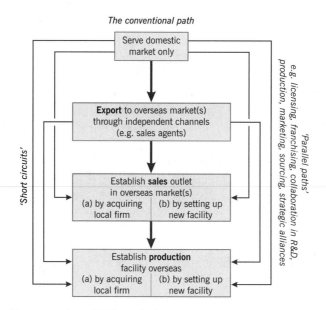

Figure 4.4   Diverse paths of TNC evolution

There is a good deal of anecdotal material to support this sequence of development among firms that actually became TNCs. Apart from Vernon's US evidence, Japanese firms investing in Europe showed a similar path. Actual manufacturing operations came rather late following a long period of development of

Japanese service investments in the form of general trading companies, banks and other financial institutions and the sales and distribution functions of manufacturing firms themselves.[11]

However, there is nothing inevitable about the progression through each of the stages. The broad definition of the TNC adopted in this book allows for other possibilities. Figure 4.4 shows a diversity of possible developmental paths. A firm may bypass the intermediate stages and set up overseas production facilities from the beginning. This is especially common among firms from small countries, such as the Netherlands, Switzerland or Sweden.

> Switzerland is a small country. Within five months of its creation, Nestlé was already manufacturing abroad. The mentality here was never to export from the home market but to produce locally.[12]

One pathway is through the acquisition of another domestic firm that already has foreign operations. Thus, a firm may become transnational almost incidentally. Increasingly, as we shall see in Chapter 5, firms are engaging in a bewildering variety of collaborative arrangements with other firms and these provide further diverse paths of TNC development. For example, a firm may tap into existing TNC networks as a supplier of specific products, functions and services. Alternatively, a firm may take on various kinds of transnational networking roles and yet remain relatively small.

There is increasing evidence of new entrepreneurial ventures actually starting out as transnational operations from the very beginning. In other words, they are what some have called 'born globals'.[13] For example,

> Momenta Corporation of Mountain View California ... [is] a start-up in the emerging pen-based computer market. Its founders were from Cuba, Iran, Tanzania, and the United States. From its beginning in 1989, the founders wanted the venture to be global in its acquisition of inputs and in its target market ... Thus, software design was conducted in the United States, hardware design in Germany, manufacturing in the Pacific Rim, and funding was received from Taiwan, Singapore, Europe, and the United States.[14]

A diversity of organizational architectures: how TNCs are coordinated

One of the basic 'laws' of growth of any organism or organization is that as growth occurs its internal structure has to change. In particular, the *functional role* of its component parts tends to become more *specialized* and the links between the parts become more *complex*. As the size, organizational complexity and geographical spread of TNCs have increased, the internal interrelationships between their geographically separated parts have become a highly significant element in the global economy. The precise manner in which TNCs organize and configure their production networks arises from a number of interrelated influences,[15] notably:

- the *firm's specific history and geography*, including

  ○ characteristics derived from its *home country embeddedness*

  ○ its *culture and administrative heritage* in the form of accepted practices built up over a period of time, producing a particular 'strategic predisposition'[16]

- the nature and complexity of the *industry environment(s)* in which the firm operates, including the nature of competition, technology, regulatory structures etc.

Coping with complexity

The traditional approach to changing organizational structures – based primarily on the hierarchical Western (that is, US) model – depicts the process as a sequential one, whereby firms transform their organizational structures from a *functional* form, in which the firm is subdivided into major functional units (production, marketing, finance etc.), into a *divisional* form (usually product based). In such a divisional structure, each product division is responsible for its own functions, particularly of production and marketing, although some functions (especially finance) tend to be performed centrally for the entire corporation. Each product division also usually acts as a separate profit centre. The main advantage of the divisional structure is a greater ability to cope with product diversity. Thus, as large US firms became increasingly diversified during the 1950s and 1960s they also tended to adopt a divisional structure.

Adoption of a divisional structure gave firms greater control over their increasingly diverse product environment. However, operating across national

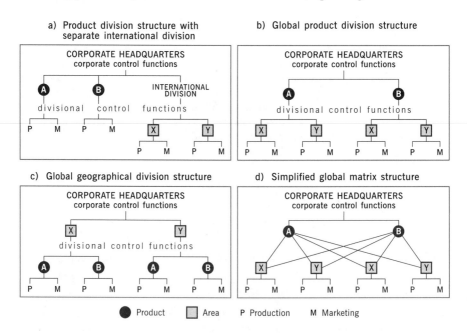

Figure 4.5   Types of transnational organizational structure

boundaries, rather than within a single country, poses additional problems of coordination and control. Largely through trial and error, TNCs have groped their way towards more appropriate organizational structures. Figure 4.5 shows four commonly used structures. Which one is actually adopted depends upon a number of factors, including the age and experience of the enterprise, the nature of its operations and its degree of product and geographical diversity.

The form most commonly adopted in the early stages of TNC development is simply to add on an *international division* to the existing divisional structure (Figure 4.5a). This has tended to be a short-lived solution to the organizational problem if the firm continues to expand its international operations because problems of coordination and tension inevitably arise between the parts of the organization operated on product lines (the firm's domestic activities) and those organized on an area basis (the international operations).

There are two obvious possible solutions. One is to organize the firm on a *global product* basis: that is, to apply the product-division form throughout the world and to remove the international division (Figure 4.5b). The other possibility is to organize the firm's activities on a *worldwide geographical* basis (Figure 4.5c). But neither of these structures resolves the basic tension between product- and area-based systems. For such reasons some of the largest TNCs have adopted sophisticated *global grid* or *global matrix* structures (Figure 4.5d), containing elements of both product and area structures and involving dual reporting links.

Although there is plenty of evidence to support such a sequence, there is also plenty of evidence to demonstrate far greater organizational diversity. In this regard, Bartlett and Ghoshal's typology is very useful.[17] Table 4.1 summarizes the major features of each type.

The *'multinational organization' model* emerged particularly during the 1930s. A combination of economic, political and social factors forced firms to decentralize their operations in response to national market differences. The result was a decentralized federation of overseas units and simple financial control systems, overlain on informal personal coordination (Figure 4.6). The company's worldwide operations are organized as a portfolio of national businesses. This was the kind of transnational organizational form used extensively by expanding European companies. Each of the firms' national units has a very considerable degree of autonomy and a predominantly 'local' orientation. It is able, therefore, to respond to local needs, but its fragmented structure lessens scale efficiencies and reduces the internal flow of knowledge.

The *'international organization' model* came to prominence in the 1950s and 1960s as large US corporations expanded overseas to capitalize on their firm-specific assets of technological leadership or marketing power. This ideal-type involves far more formal coordination and control by the corporate headquarters over the overseas subsidiaries (Figure 4.7). Whereas 'multinational' organizations are, in effect, portfolios of quasi-independent businesses, 'international' organizations see their overseas operations as appendages to the controlling domestic

Table 4.1 Some ideal-types of TNC organization: basic characteristics

Characteristics	Multinational	International	Global	Integrated network
Structural configuration	Decentralized federation. Many key assets, responsibilities, decisions decentralized	Coordinated federation. Many assets, responsibilities, resources, decisions decentralized but controlled by HQ	Centralized hub. Most strategic assets, resources, responsibilities and decisions centralized	Distributed network of specialized resources and capabilities
Administrative control	Informal HQ–subsidiary relationship; simple financial control	Formal management planning and systems control allow tighter HQ–subsidiary linkage	Tight central control of decisions, resources and information	Complex process of coordination and cooperation in an environment of shared decision-making
Management attitude towards overseas operations	Overseas operations seen as portfolio of independent businesses	Overseas operations seen as appendages to a central domestic corporation	Overseas operations treated as 'delivery pipelines' to a unified global market	Overseas operations seen as an integral part of complex network of flows of components, products, resources, people, information among interdependent units
Role of overseas operations	Sensing and exploiting local opportunities	Adapting and leveraging parent company competencies	Implementing parent company strategies	Differentiated contributions by national units to integrated worldwide operations
Development and diffusion of knowledge	Knowledge developed and retained within each unit	Knowledge developed at the centre and transferred to overseas units	Knowledge developed and retained at the centre	Knowledge developed jointly and shared worldwide

Source: based on material in Bartlett and Ghoshal, 1998

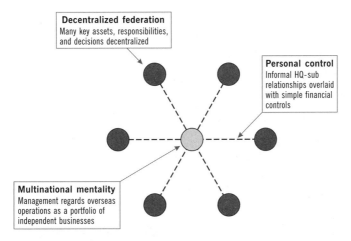

Figure 4.6 'Multinational organization' model

Source: based on Bartlett and Ghoshal, 1998: Figure 3.1

corporation. Thus, the international subsidiaries are more dependent on the centre for the transfer of knowledge and the parent company makes greater use of formal systems of control. The 'international' TNC is better equipped to leverage the knowledge and capabilities of its parent company but its particular configuration and operating systems tend to make it less responsive than the 'multinational' model. It is also rather less efficient than the next ideal-type.

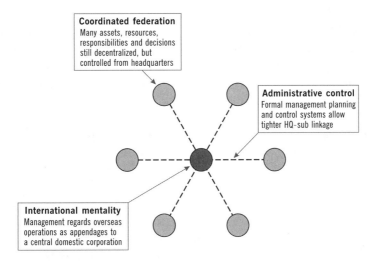

Figure 4.7 'International organization' model

Source: based on Bartlett and Ghoshal, 1998: Figure 3.2

The *'global organization' model* was one of the earliest forms of transnational business (used, for example, by Ford and by Rockefeller in the early 1900s, as well as by Japanese firms in their much later transnationalization drive of the 1970s and 1980s). It is based upon a tight centralization of assets and responsibilities in which the role of the local units is to assemble and sell products and to implement plans and policies developed at the centre (Figure 4.8). Overseas subsidiaries have far less freedom to create new products or strategies or to modify existing ones. Thus, the 'global' TNC capitalizes on economies of scale and on centralized knowledge and expertise. But this implies that local market conditions tend to be ignored and the possibility of local learning is precluded.

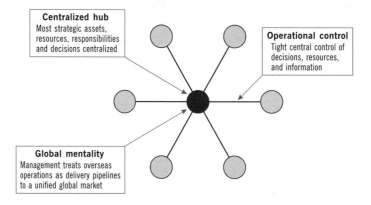

Figure 4.8   'Global organization' model

Source: based on Bartlett and Ghoshal, 1998: Figure 3.3

Although each of these three organizational types developed initially during specific historical periods, one did not simply replace the other. Because each has its strengths (as well as weaknesses), each has tended to persist, in either a pure or a hybrid form, helping to produce today's diverse TNC population.[18] There is some correlation between organizational type and nationality of parent company but it is by no means perfect; it is better to regard firms of different national origins as having a predisposition to one or other form of organization.[19]

The dilemma facing TNCs is that they need the best features of each organizational form: to be globally efficient, geographically flexible, and capable of capturing the benefits of worldwide learning, all at the same time. Hence, it is argued, we are now seeing the emergence of a fourth ideal-type TNC: the *'integrated network organization' model*, characterized by a distributed network configuration and a capacity to develop flexible coordinating processes (Figure 4.9). Such capabilities apply both inside the firm, displacing hierarchical governance relationships with what Hedlund[20] terms a *heterarchical* structure, and also outside the firm through a complex network of inter-firm relationships. We will explore this issue further in Chapter 5 in the broader context of networks of externalized relationships.

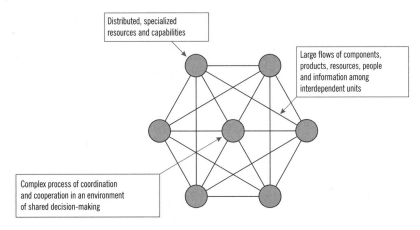

Figure 4.9 **'Integrated network organization' model**

Source: based on Bartlett and Ghoshal, 1998: Figure 5.1

The point to emphasize is the continuing diversity of organizational architectures: TNCs come in all shapes, sizes and forms of governance. Their internal architecture reflects not only the external constraints and opportunities they have to face – including the structures made possible by new communications technologies[21] – but also a strong element of path dependency. Firms organized on hierarchical principles not only still exist but also may still be in a majority. The newer, 'flatter' organizational forms tend to be confined to a limited number of firms in certain sectors.

Headquarters–subsidiary relationships

The various TNC structures discussed in the preceding section, and especially those shown in Figures 4.6 to 4.8, assume a clear distinction between a TNC's organizational centre – its headquarters – and its subsidiary operations. In a pure hierarchical model, the relationship is intrinsically top-down: the TNC subsidiaries simply perform the role allocated to them. In contrast, in a *heterarchical* organization (Figure 4.9) the position is far more complex. The roles played by a subsidiary, therefore, vary between different organizational structures and in terms of a TNC's specific strategy. Within all of this, the roles – and the powers – of subsidiary managers are in a continuous state of flux.

Three broad types of subsidiary role can be identified:[22]

- *The local implementer*: limited geographical scope and functions. Its primary purpose is to adapt the TNC's products for the local market.
- *The specialized contributor*: specific expertise tightly integrated into the activities of other subsidiaries in the TNC. Narrow range of functions and a high level of interdependence with other parts of the firm.
- The *world mandate*: worldwide (or possibly regional) responsibility for a particular product or type of business.

These different subsidiary roles have important implications for the impact of TNC activities on national and local economies (see Chapter 16). How these roles develop and possibly change – for example from local implementer to world mandate – depends upon the nature of the bargaining relationships within the TNC. Such relationships are highly contested processes which reflect internal power structures. In a similar way, the individual affiliates of a firm (its subsidiaries, branches etc.) are continuously engaged in competition to improve their relative position within the organization by, for example, winning additional investment or autonomy from the corporate centre. At the same time, the performance of each affiliate is continuously monitored against the relevant others (internal benchmarking) and this is used as an integral part of the internal bargaining processes within the firm.

In fact, the actual *geography* of a TNC influences these internal bargaining processes as well:

> Different 'places' within the firm, organizationally and geographically, develop their own identities, ways of doing things and ways of thinking over time … The firm's dominant culture, created by and expressed through the activities and understandings of top management at headquarters, necessarily contains multiple subcultures. Some of these may revolve around functions and cut across places (engineers versus sales people, for example), but some will have real geographical locations – they will have grown up in specific plants in particular places.[23]

These relationships between TNCs and 'place' are the subject of the next section of this chapter.

'Placing' firms: the myth of the 'global' corporation

> Before national identity, before local affiliation … before any of this comes the commitment to a single, unified global mission … Country of origin does not matter. Location of headquarters does not matter. The products for which you are responsible and the company you serve have become denationalized.[24]

One of the central claims of the hyper-globalists is that transnational firms are abandoning their ties to their country of origin and converging towards a universal *global* organizational form. Technological and regulatory developments in the world economy have created a 'global surface' on which a dominant organizational form will develop and wipe out less efficient competitors no longer protected by national or local barriers. Such an organization is, it is argued, 'placeless' and 'boundaryless'. In this section we challenge this view.[25]

How 'transnational' are the world's leading TNCs?

If the 'global corporation' hypothesis were valid then we would expect to find that at least the majority of the world's largest TNCs would have the overwhelming majority of their operations outside their home country. One measure of such geographical spread is the *transnationality index* (TNI) of each of the leading 100 TNCs.[26] The TNI is a weighted average of three indicators: foreign sales as a percentage of total sales; foreign assets as a percentage of total assets; and foreign employment as a percentage of total employment. The higher the value, the greater the extent of a firm's transnationality; the lower the value, the more a firm is domestically oriented.

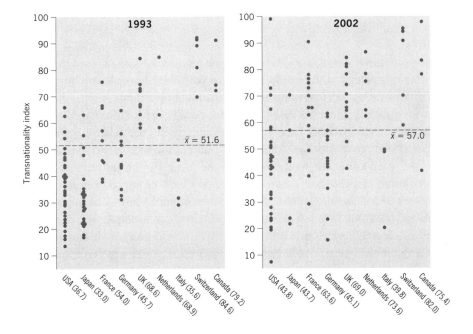

Figure 4.10   Transnationality indices by country of origin

Source: calculated from UNCTAD *World Investment Report*, various issues

Figure 4.10 compares TNIs for firms from individual countries over a 10-year period. The results are quite striking.

- The average TNI ($\bar{x}$) for all 100 TNCs in 2002 was 57.0; in 1993 it was 51.6. In other words, for the top 100 TNCs as a group (not a completely identical group because of entries and exits into the list over the period) the degree of transnationality increased only slightly. Indeed, the figure indicates that, on average, these leading TNCs have roughly half their operations at home and half abroad. This does not suggest a particularly high degree of 'globalization'. In fact, in 2002 only 61 companies had a TNI greater than 50, and only 21 had a TNI above 75 (indicating that more than three-quarters of their activities were located outside their home country). The other 16 had a TNI less than 50.

- The degree of transnationality continues to vary substantially between firms from different geographical origins. Part of this is explained by country size: the smaller the country, in general, the more transnational its companies tend to be. But that is not the entire story. Japanese firms continue to have the lowest average TNI of the major home countries, though it increased substantially between 1993 and 2002. Although the US average increased from 36.7 to 43.8, the overall extent of transnationality of US firms remains significantly lower than that of other major countries.

Thus, despite many decades of international operations, TNCs – at least in quantitative terms – remain distinctively connected with their home base. They are, in many cases, 'national corporations with international operations'.[27]

However, such quantitative analysis provides only a partial answer to the questions posed in this section. It tells us something about the relative *geographical extent* of TNC activities outside the home country and, to that degree, demonstrates the continuing emphasis on the home base. But it distinguishes only between home and foreign. A firm might have a TNI of, say, 80 (meaning that 80 per cent of its activities were outside its home country) but all of those activities might be located in just one foreign country. An example would be the large number of US firms that operate only in Canada. Neither does it help us to establish whether or not TNCs of different national origins are becoming similar in their modes of operation. It is at least possible that TNCs may retain more of their assets and employment in their home country but still be converging organizationally and behaviourally towards a universal, global form. To address this issue we need a different type of empirical evidence from that which merely measures the geographical dispersion of a firm's activities. We need evidence that explicitly compares TNCs from different countries of origin.

The geographical embeddedness of transnational corporations

> TNCs are 'locally grown'; they develop their roots in the soil in which they were planted. The deeper the roots the stronger will be the degree of local embeddedness, such that they should be expected to bear at least some traces of the economic, social and cultural characteristics of their home country ... This is not to argue a case for cultural determinism or even to argue that all firms of a given nationality are identical. Clearly they are not. But they do tend to share some common features.[28]

Although such territorial embeddedness occurs at a variety of geographical scales, the most significant is the national state, the major 'container' within which distinctive practices develop (see Chapter 6) and which helps to 'produce' particular kinds of firms.

Table 4.2 suggests some of the links that may exist between the ownership-specific advantages of firms and the location-specific characteristics of the firm's *home* country. It is this link that helps to explain the different characteristics of TNCs from different source nations. For example, the large domestic market and high level of technological sophistication of the US domestic economy have helped to produce the distinctive characteristics of US TNCs. The lack of natural resources, and the strong involvement of government and other institutions in technological and industrial affairs, help to explain the particular attributes of Japanese TNCs, at least in their earlier development.

Table 4.2 Links between selected ownership-specific advantages and country-specific characteristics

Ownership-specific advantages	Country characteristics favouring such advantages
Size of firm	Large, standardized markets
	Liberal regime towards mergers and concentration
Managerial expertise	Pool of managerial talent
	Educational and training facilities
Technology-based advantages	Good R&D facilities
	Government support of innovation
	Pool of scientific and technical labour
Labour and/or mature, small-scale intensive technologies	Large pool of labour (including technical labour)
	Appropriate consultancy services
Production differentiation	High-income national markets
Marketing economies	High income elasticity of demand
	Highly developed marketing/advertising system
	Consumer-oriented society
Access to (domestic) markets	Large national market
	No restrictions on imports
Capital availability and financial expertise	Well-developed, reliable capital markets
	Appropriate professional advice

Source: based on Dunning, 1979: Table 6

A comparison between US, German and Japanese companies

A comparison of US, German and Japanese firms provides strong arguments to counter the convergence hypothesis. In a book entitled *The Myth of the Global Corporation*,[29] Doremus and his colleagues argue that there is

> little blurring or convergence at the cores of firms based in Germany, Japan, or the United States ... Durable national institutions and distinctive ideological

traditions still seem to shape and channel crucial corporate decisions ... the domestic structures within which a firm initially develops leave a permanent imprint on its strategic behavior ... our findings underline, for example, the durability of German financial control systems, the historical drive behind Japanese technology development through tight corporate networks, and the very different time horizons that lie behind American, German, and Japanese corporate planning.[30]

Table 4.3 summarizes the basic differences between firms from these three home countries.

East Asian business organizations

With the rise of East Asia as a major dynamic growth point in the global economy, it became commonplace to contrast the 'East Asian way' of doing business from that of firms in the West.[31] There is some basis in such a comparison because we do find certain common features of East Asian business organizations that, in combination, tend to differentiate them from Western firms.[32]

- formation of intra- and inter-firm business relationships
- reliance on personal relationships
- strong relationships between business and the state.

However, it is a mistake to think in terms of *one* East Asian business model because, in fact, there is considerable *diversity* between firms from different East Asian countries. This can be illustrated by looking briefly at the cases of Japanese, Korean, Taiwanese and Overseas Chinese business organizations.

The Japanese keiretsu

Intercorporate alliances in the contemporary Japanese economy are marked by an elaborate structure of institutional arrangements that enmesh its primary decision-making units in complex networks of cooperation and competition ... There is a strong predilection for firms in Japan to cluster themselves into coherent groupings of affiliated companies extending across a broad spectrum of markets.[33]

The precise composition and structure of such business groupings is immensely varied. Here we focus on the industrial groupings generally known as *keiretsu*.[34] Five diagnostic characteristics of these groups can be identified:[35]

- Transactions are conducted through alliances of *affiliated* companies. This creates a form of organization intermediate between vertically integrated firms and arm's-length markets.
- Inter-firm relationships tend to be *long term* and stable, based upon mutual obligations.
- These inter-firm relationships are *multiplex* in form, expressed through cross-shareholdings and personal relationships as well as through financial and commercial transactions.

Table 4.3 Differences between United States, German and Japanese TNCs

	US TNCs	German TNCs	Japanese TNCs
Corporate governance and corporate financing	Constrained by volatile capital markets	Relatively high degree of operational autonomy except during crises	Bound by complex but reliable networks of domestic relationships
	Short-term perspectives	Long-term perspectives	Long-term perspectives
	Finance-centred strategies	Conservative strategies	Market-share-centred strategies
	High risk of takeover	Low risk of takeover	Very low risk of takeover, mainly confined to within network
	90% of firm shares held mainly by individuals, mutual funds pension funds	Firm shares held mainly by non-financial institutions (40%)	High degree of cross-shareholdings within group
	Less than 1% held by banks	Significant role of regional bodies	
	Banks provide mainly secondary financing, cash management, selective advisory role	Banks play a lead role	Lead bank performs a steering function
		Supervisory boards of companies are strongly bank influenced	
	Ratio of bank loan/corporate financial liabilities = 25–35%	Ratio of bank loan/corporate liabilities = 60–70%	Ratio of bank loan/corporate liabilities = 60–70%
Research and development	Corporate R&D expenditure peaked in 1985 at 2.1% of GDP; declining	Corporate R&D expenditure declined steeply in late 1980s/early 1990s	Corporate R&D grew very rapidly in 1980s
		At 1.7% of GDP, lower than US and Japan	Overtook US in 1989; peaked at 2.2% of GDP
			Real cuts made only as last resort
	Diversified pattern; innovation oriented	Narrow focus	High-tech and process orientation
	Some propensity to perform R&D abroad	Some propensity to perform R&D abroad	Very limited propensity to perform R&D abroad
Direct investment and intra-firm trade	Extensive outward investment	Selective outward investment	Extensive outward investment
	Substantial competition from inward investment	Moderate competition from inward investment	Very limited competition from inward investment
	Moderate intra-firm trade; high propensity to outsource	High level of intra-firm trade	Very high level of intra-firm and intra-group trade

Source: based primarily on material in Pauly and Reich, 1997

- Bilateral relationships between firms are embedded within a broader 'family' of *related companies*.
- Inter-corporate relationships are imbued with *symbolic significance* which helps to sustain links even where there are no formal contracts.

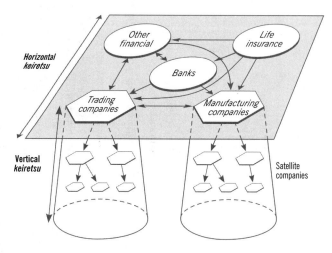

Figure 4.11   Basic elements of the Japanese *keiretsu*

Source: based, in part, on Gerlach, 1992, Figure 1.1

As Figure 4.11 shows, there are two basic types of *keiretsu*:

- *Horizontal keiretsu*: highly diversified industrial groups organized around two key institutions: a core bank and a general trading company (*sogo shosha*). Three of the horizontal *keiretsu* groups (Mitsubishi, Mitsui, Sumitomo) are the successors of the pre-war family-led *zaibatsu* groups that were abolished after 1945. The others are primarily bank centred.
- *Vertical keiretsu*: organized around a large parent company in a specific industry (for example, Toshiba, Toyota, Sony).

Although distinctive, these two types are not mutually exclusive. For example, Toshiba is itself a parent company hierarchically controlling substantial numbers of satellite companies including parts suppliers, in a vertically integrated Toshiba group. At the same time, Toshiba is also a member of the horizontally integrated Mitsui industrial group. In fact, the webs of interrelationships are extremely complex and the organizational scale of the leading *keiretsu* is immense. For example, the eight leading horizontal *keiretsu* consist of around 900 separate companies but, in effect, they control in total some 12,000 companies.

A comparison of South Korean and Taiwanese business groups

There are strong similarities in the history and developmental experiences of South Korea and Taiwan.[36] One is that their business firms share common features of Confucian-based familism. 'In all these background variables – economic, social, and cultural – Taiwan and South Korea are as nearly the same as could be imagined between any two countries in the world today. Yet the economies of these two countries are organized in radically different ways.'[37] Although business organizations in both countries are organized as networks of family-owned firms, their mode of organization differs considerably.

In Korea, the dominant type of business group is the *chaebol*, modelled on the pre-World War II Japanese *zaibatsu*, the giant family-owned firms which had been so important in the development of the Japanese economy. The *chaebol*

> are highly centralized, most being owned and controlled by the founding patriarch and his heirs through a central holding company. A single person in a single position at the top exercises authority through all the firms in the group. Different groups tend to specialize in a vertically integrated set of economic activities.[38]

As a result, the Korean economy became highly concentrated and oligopolistic, while the small- and medium-firm sector is relatively underdeveloped. Not only this, but many of these smaller firms are tightly tied into the production networks of the *chaebol,* which have developed into some of the most highly vertically integrated business networks in the world:

> the firms in the *chaebol* are the principal upstream suppliers for the big downstream *chaebol* assembly firms ... in Samsung Electronics, most of the main component parts for the consumer electronics division are manufactured and assembled in the same compound by Samsung firms.[39]

Taiwanese business networks, in contrast, have low levels of vertical integration. The more horizontal Taiwanese networks consist of two main types: 'family enterprise' networks and 'satellite assembly' networks (independently owned firms that come together to manufacture specific products primarily for export).

These contrasts between the Korean and Taiwanese business groups – despite the strong similarities between the two countries – have been explained as arising from

> differences in social structures growing out of the transmission and control of family property. In South Korea, the kinship system supports a clearly demarcated, hierarchically ranked class structure in which core segments of lineages acquire elite rankings and privileges. These are the 'great families' ... In Taiwan, however, the Confucian family was situated in a very different social order ... Unlike Korea (and in the early Chinese dynasties), where the eldest son inherited the lion's share of the estate and all the lineage's communal holdings, in late imperial China the Chinese practiced partible inheritance, in which all sons equally split the father's estate ... This set of practices preserved the household and made it the key unit of action, rather than the lineage itself ... In summary, although based on similar kinship principles, the Korean and the Chinese kinship systems operate in very different ways.[40]

Such differences in socio-cultural practices largely explain the contrasts between the ways that business firms are organized in the two neighbouring countries.

Overseas Chinese business networks

> In many Western economies, the main efficiencies in coordination derive from
> large-scale organization. In the case of the Overseas Chinese, the equivalent
> efficiencies derive from networking.[41]

A very different kind of business network is to be found within the Overseas
Chinese entrepreneurial system that underpins much of the dynamic economic
development not only of Hong Kong and Taiwan but also throughout much of
South East Asia.[42] Their essence has been described as 'weak organizations and
strong linkages'.[43] Personal relationships based on reciprocity (*guanxi*) play a central
role, in contrast to the situation in Western firms where formal contractual arrangements are the norm.[44] The basis of the Overseas Chinese business system is to be
found in the specifics of cultural and historical experience and in the set of norms
and values derived from a common historical experience and implemented in particular contexts. Three influences have been especially strong in influencing how
Overseas Chinese businesses are managed and operated: Confucian value systems of
familism and respect for authority; experience as refugees; and experience of
oppression. Family control, minimizing dependence on outsiders for key resources,
and a tight inner circle of decision-makers, are basic characteristics.[45]

By its very nature, the Overseas Chinese business is less visible than other, more
formally structured, businesses. Because of its extended network form, any
approach based solely on the legal definition of the firm will fail to capture its full
extent. It will also underestimate its significance as a form of transnational business activity. Yet the significance throughout Asia, and increasingly in other parts
of the world, of Overseas Chinese family business networks – what has been
termed 'the worldwide web'[46] or the 'bamboo network'[47] – is immense. Their
entrepreneurial drive has enabled them to occupy dominant positions in what are
otherwise non-Chinese societies, such as Indonesia and Malaysia.

The precise form of Overseas Chinese business varies between different social
and institutional contexts in different parts of Asia. In the case of Hong Kong and
Singapore, for example, the different political and institutional contexts have produced distinctive entrepreneurial characteristics even though, in both cases, their
ethnicity is comparable.

> Ethnic Chinese industrialists in Hong Kong are known for their entrepreneur-
> ship and higher propensity to engage in risky business and overseas ven-
> tures ... The peculiar neoliberal political economy in Hong Kong had several
> consequences for transnational entrepreneurship in Hong Kong. First, the
> private sector assumed a leading role in Hong Kong's economic development ...
> Second, the lack of direct state intervention in Hong Kong's industrialization
> and economic development processes has contributed to the growth of
> domestic companies in both large firm sectors and small firm networks ...
> Third, the financial system in Hong Kong ... is highly favourable for the devel-
> opment of the service sector.[48]

In contrast, the Chinese mode of business in Singapore differs significantly from that in Hong Kong. The main reason is the very different institutional structure in Singapore.

> Notably, a large proportion of local investments, particularly in the manufacturing sector, came from foreign firms and GLCs [government-linked companies] ... and their various subsidiaries. The role of indigenous private enterprises in Singapore's industrialization is rather limited ... The majority of Singaporeans have become contented with their job security and are less willing to take specific kinds of risks to launch new business ventures.[49]

Convergence or differentiation?

As we have seen, TNCs are 'produced' through an intricate process of embedding, in which the cognitive, cultural, social, political and economic characteristics of the national home base continue to play a dominant part. This is not to claim that TNCs from a particular national origin are identical. This is self-evidently not the case. Within any national situation there will be distinctive corporate cultures, arising from the firm's own specific corporate history, which predispose it to behave strategically in particular ways. But, in general, the similarities between TNCs from one country will be greater than the differences between them.

But this does not imply that nationally embedded TNCs are unchanging. On the contrary, the very interconnectedness of the contemporary global economy means that influences are rapidly transmitted across boundaries. Corporations are *learning* organizations: they strive to tap into appropriate practices wherever they occur. This will, inevitably, affect the way business organizations are configured and behave. There 'is essentially a process of *coevolution* through which different business systems may converge in certain dimensions and diverge in other attributes'.[50] The very fact that TNCs are *trans*national – that they operate in a diversity of economic, social, cultural and political environments – means that they will, inevitably, take on some of the characteristics of their host environments.

For a whole variety of reasons, non-local firms invariably have to adapt some of their domestic practices to local conditions. It is virtually impossible to transfer the whole package of firm advantages and practices to a different national environment. For example, Japanese overseas manufacturing plants tend to be 'hybrid' forms rather than the pure organizations found in Japan itself.[51] The same argument applies to US firms operating abroad. Even in the United Kingdom, where the apparent 'cultural distance' between the US and the UK is less than in many other cases, there is a very long history of American firms having to adapt some of their business practices to local conditions.

We can find plenty of evidence of change within TNCs in response to these various forces. For example, the *keiretsu* have been at the centre of Japanese economic development during the post-World War II period. But the financial crisis in Japan that has persisted since the bursting of the 'bubble economy' at the end of the 1980s has put them under considerable pressure to change at least some of their practices. In particular, the recent influx of foreign capital to acquire significant, sometimes controlling, shares in some of these companies has had a catalytic effect. The most notable example was the acquisition by the French automobile company, Renault, of almost 40 per cent of the equity of Nissan (see Chapter 10). There are strong pressures, particularly from Western (notably US) finance capital, for Japanese business groups to open up to outsiders, to reduce or eliminate the intricate cross-shareholding arrangements, and to become more like Western (that is, US) firms with their emphasis on 'shareholder value' rather than the broader, socially based 'stakeholder' interests intrinsic to Japanese companies.

While, without doubt, some changes are occurring it would be a mistake to assume that Japanese firms will suddenly be transformed into US clones. The Japanese have a very long history of adapting to external influences by building structures and practices that remain distinctively Japanese. Similarly, Korean and other East Asian firms have come under enormous pressure to change some of their business practices in the aftermath of the region's financial crisis of the late 1990s. In Korea, the *chaebol* are being drastically restructured and their relationships with the state diluted. Among Chinese businesses, the strong basis in family ownership and control is being challenged by both internal and external forces. Greater involvement in the global economy is forcing these firms to modify some of their practices.[52] Similar observations apply to firms from other home countries. In the case of Germany, for example, while some of the established characteristics of German business are under threat, the evidence suggests that many of the core elements remain in place.[53]

And yet it would be extremely surprising if the distinctive nature of nationally based TNCs were to be replaced by a standardized, homogeneous form. Continued differentiation would appear to be the most likely scenario, though undoubtedly containing elements of change and some degree of convergence. A recent comparison of Japanese and American TNCs in the Asia-Pacific, for example, showed precisely that.

> The results point to a certain degree of inertia among American and Japanese TNCs regarding organizational change in international environments. While selection and mimetic convergence may not have been completely absent … the reproduction of structures from home country would seem also to have contributed to persistent differentiation … pressures for convergence appear to have been mediated by some degree of contingency that suggests that TNCs will likely continue to exhibit national characteristics in international environments.[54]

In other words,

> The global corporation, adrift from its national political moorings and roaming an increasingly borderless world market, is a myth … The empirical evidence … suggests that distinctive national histories have left legacies that continue to affect the behavior of leading [TNCs] … The scope for corporate interdependencies across national markets has unquestionably expanded in recent decades. But history and culture continue to shape both the internal structures of [TNCs] … and the core strategies articulated through them.[55]

So, where are we heading? Are TNCs converging towards a more or less universal form or are we likely to see continued differentiation? In my view, the answer is emphatically the latter.

Conclusion

The aim of this chapter has been to explore three specific aspects of transnational corporations. First, we outlined some of the major explanations of why firms become 'transnational'. A complex mix of market and asset orientation is involved. Second, we focused on how TNCs may develop. There are distinctive – though diverse – developmental paths. Similarly, we can identify a variety of organizational architectures, designed to enable firms to implement their chosen competitive strategies. The particular type employed appears to be influenced both by the firm's specific history and geography and by the nature and complexity of the industry environment(s) in which it operates. Third, we challenged the popular conception of the 'placeless' transnational corporation. All TNCs have an identifiable home base whose characteristics continue to exert an influence on how firms behave as they develop transnational networks of operations. We showed clear differences between TNCs of different nationalities; there is little evidence of TNCs converging towards a single model. Of course, as TNCs move into new environments they have to adapt, to a greater or lesser degree, to local circumstances. On both counts, however, it is very apparent that 'geography matters' to how TNCs coordinate and configure their production networks and, by extension, to the kind of impact they have on the states and communities within which they operate.

NOTES

1 D'Aveni (1994: 2).
2 See Palloix (1975; 1977).
3 Hymer (1976). See also Pitelis and Sugden (1991).
4 Dunning introduced his 'eclectic' approach in the 1970s. He has subsequently elevated it to 'paradigm' status (see Dunning, 2000a).
5 Dunning (1980: 9, emphasis added).
6 Pitelis and Sugden (1991: 5).
7 Herod (1997: 2).
8 Peck (1996) provides an excellent discussion of the 'place' of labour within the capitalist market system. See also Castree et al. (2004), Herod (2001), Hudson (2001).
9 Vernon (1966; 1979), Wells (1972). Cantwell (1997) evaluates the applicability of the PLC concept.
10 UNCTAD (2005: 9).
11 Dicken and Miyamachi (1998), Mason (1994).
12 Peter Brabeck, CEO of Nestlé, quoted in the *Financial Times* (22 February 2005).

13 Gabrielsson and Kirpalani (2004).

14 Oviatt and McDougall (2005: 38).

15 Bartlett and Ghoshal (1998), Hedlund (1986), Heenan and Perlmutter (1979), Morgan et al. (2004), Schoenberger (1997), Whitley (2004).

16 Heenan and Perlmutter (1979).

17 Bartlett and Ghoshal (1998).

18 Harzing (2000), Malnight (1996).

19 Heenan and Perlmutter (1979).

20 Hedlund (1986: 218-30).

21 Fields (2004), Roche and Blaine (2000).

22 Birkinshaw and Morrison (1995: 732–5). See also Bartlett and Ghoshal (1998).

23 Schoenberger (1999: 210–11).

24 Ohmae (1990: 94).

25 Dicken (2000; 2003b).

26 The data are derived from the annual UNCTAD *World Investment Report*.

27 Hu (1992: 34).

28 Dicken et al. (1994: 34).

29 Doremus et al. (1998). See also Pauly and Reich (1997).

30 Pauly and Reich (1997: 1, 4, 5, 24).

31 See Mirza (2000), Orrù et al. (1997), Whitley (1992; 1999), Yeung (2000).

32 Yeung (2000: 408).

33 Gerlach (1992: xiii).

34 Aoki (1984), Fruin (1992), Gerlach (1992), Helou (1991).

35 Gerlach (1992: 4).

36 This section is based primarily on Hamilton and Feenstra (1998).

37 Hamilton and Feenstra (1998: 124).

38 Wade (2004a: 324).

39 Hamilton and Feenstra (1998: 128–9).

40 Hamilton and Feenstra (1998: 134, 135).

41 Redding (1991: 45).

42 Yeung (2004) provides a comprehensive and critical analysis of Chinese business. See also Kao (1993), Redding (1991), Whitley (1992), Yeung and Olds (2000).

43 Redding (1991: 30).

44 Yeung and Olds (2000).

45 Redding (1991: 36).

46 Kao (1993).

47 Weidenbaum and Hughes (1996).

48 Yeung (2002: 98).

49 Yeung (2002: 99).

50 Yeung (2000: 425). See also Yeung (2004).

51 Abo (1994; 1996). See also Beechler and Bird (1999).

52 Yeung (2004).

53 See, for example, Bathelt and Gertler (2005), Buck and Shahrim (2005).

54 Poon and Thompson (2004: 123).

55 Doremus et al. (1998: 3, 9).

Five
'Webs of Enterprise': The Geography of Transnational Production Networks

The 'global–local' question: an oversimplified view of the TNC's dilemma

In broad terms, the intensification of global competition in a world that retains a high degree of local differentiation creates, for all TNCs, an internal tension between globalizing forces on the one hand and localizing forces on the other. As Figure 5.1 shows, there are considerable potential advantages for a firm pursuing a globally integrated strategy. But there are also substantial disadvantages. Figure 5.2 captures this basic tension within a 'global-integration/local-responsiveness' framework. The vertical axis shows the major pressures on firms to strive for global strategic coordination of their activities; the horizontal axis shows the countervailing pressures on firms to develop locally responsive strategies.

Advantages	Costs and risks
✓ The firm's oligopoly power is increased through the exploitation of scale and experience effects beyond the size of individual national markets	✗ The TNC may be vulnerable to disruption of its entire operations (or part of them) because of labour unrest or government policy changes affecting a particular unit
✓ The TNC is placed in a better position to exploit the growing discrepancy between a relatively efficient market for goods (created by freer trade) and very inefficient markets for production factors	✗ Fluctuations in currency exchange rates may disrupt integration strategies, drastically altering the economies of intra-firm transactions of intermediate or final goods
✓ The possibility of exploiting differences in tax rates and structures between countries is increased and so, therefore, is the possibility of engaging in transfer pricing	✗ Governments may impose performance requirements or other restrictions which impede the optimal operation of the firm's integrated production chain
✓ The specialized and integrated function of individual country operations makes hostile government action less rewarding and less likely	✗ The task of managing a globally integrated operation is more complex and demanding than that of managing separate national subsidiaries

Figure 5.1   The advantages and disadvantages of a globally integrated strategy

Source: based on material in Doz, 1986b

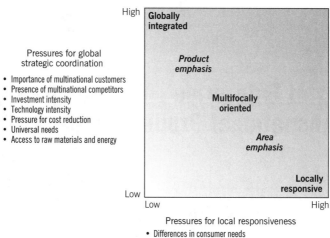

Pressures for global
strategic coordination

- Importance of multinational customers
- Presence of multinational competitors
- Investment intensity
- Technology intensity
- Pressure for cost reduction
- Universal needs
- Access to raw materials and energy

Pressures for local responsiveness

- Differences in consumer needs
- Differences in distribution channels
- Availability of substitutes and the need to adapt
- Market structure
- Host government demands

Figure 5.2   A global-integration/local-responsiveness framework

Source: based on material in Prahalad and Doz, 1987: Figure 2.2; pp. 18–21

There is plenty of evidence that TNCs are taking the 'local' more seriously, although precisely what 'local' means, in this context, is very variable. For most TNCs, it refers to individual countries or even to entire regions, such as Europe, North America or Asia, rather than to genuinely local communities. Given that, even some of the most archetypal 'global' firms have been forced to rethink their strategy. A good example is Coca-Cola, which experienced major problems in its international operations in the late 1990s.

> For a couple of years, the world was moving in one direction, and we were moving in another. We were heading in a direction that had served us very well for several decades, generally moving towards consolidation and centralized control. That direction was particularly important during the 1970s and 1980s when we were 'going global' … The world, on the other hand, began moving in the 1990s in a different direction … the very forces that were making the world more connected and homogeneous were simultaneously triggering a powerful desire for local autonomy and preservation of unique cultural identity … we had a lesson to learn. And what we learned was … that the next big evolutionary step of 'going global' now has to be 'going local'. In other words, we had to rediscover our own multi-local heritage.[1]

Other examples of self-confessed TNC local sensitivity include:

- the Swedish transnational corporation ABB, which claims to have perfected 'the art of being local worldwide'

- the American financial services TNC J.P. Morgan, which asserts that 'the key to global performance is understanding local markets'
- Anglo-Dutch consumer products firm Unilever, which describes itself as a 'multi-local multinational'
- HSBC Bank, which boasts of its 'local insight, global outlook'
- the Japanese electronics firm Sony, which was one of the first to use the term 'glocalization' to describe its international corporate strategy.

In reality, firms engaged in transnational production face far more complex and multidimensional decisions about the geographical configuration and organizational coordination of their operations than the simple global/local dichotomy suggests. Production circuits and production networks are immensely complex structures, made up of many different functions and activities. From a TNC's perspective, operating in many different economic, political, social and cultural environments, very difficult decisions have to be made about every one of them:

- Which functions are to be performed internally (in-house) and which are to be outsourced to other firms?
- Where should each of the firm's own internalized functions be located? Which functions need to be located close to each other? Which ones can be separated out and located elsewhere to enhance efficiency? What should the balance be between domestic production and producing offshore?
- Where should suppliers be located? Should they be close to the firm's home base or should they be in other countries (i.e. offshore)? Do suppliers need to be located nearby or can they be geographically dispersed to take advantage of lower costs or other locationally specific attributes?
- How is control over geographically dispersed activities – both internal and external – to be exercised?

As if these were not sufficiently complicated decisions, there is the fundamental problem that, of course, TNCs are not dealing with a 'clean surface': their geographically dispersed assemblage of functions, offices, factories and the like have evolved over time rather than having been planned. Some will have been located in particular places for reasons that may have been valid at the time the decisions were made but which, in the light of changed circumstances, may no longer be optimal for the firm's current needs. Because so much TNC growth and expansion have been through acquisition and merger, most TNCs consist of elements originally put in place by quite different firms (often of different nationalities and ways of doing things). In many cases, these have been only partially integrated into the new corporate entity, often creating a veritable 'dog's breakfast' of bits and pieces. This is one reason why so many mergers and acquisitions are far less successful than their advocates predict.

Not only is the entire system in flux because of the dynamics of competition or the uncertainties inherent in operating in uncertain, and sometimes volatile, political environments, but also each of the individual elements is continuously faced with uncertainty. One of the diagnostic characteristics of TNCs is that they continuously monitor the performance of each of their individual operations and benchmark them against some best-practice metric. Hence, transnational corporate networks are almost always in a state of rationalization and restructuring, either in whole or in part. Precisely because TNC operations are located in different countries, such adjustments – perhaps involving the closure, downsizing or functional status of individual establishments – have very sensitive political implications.

In this chapter, therefore, we try to unravel some of this complexity to provide a more grounded understanding of how TNCs operate across a spectrum of geographical scales. The primary focus is on the *networks of relationships* that exist both within and between firms as they attempt to coordinate and configure their production activities at different geographical scales. It is concerned, in particular, with how TNCs use geographical space in order to achieve their strategic objectives. Three sets of relationships are explored:

- TNCs as networks of internalized relationships
- the complex networks of externalized relationships that exist between independent and quasi-independent firms
- the extent to which such networks tend to be regional in their geographical scope.

Configuring the firm's production network: the complex internal geographies of the TNC

We begin by looking at how TNCs geographically configure their internalized operations – how they locate their own productive assets and capabilities. Different business functions have different locational needs and, because these needs can be satisfied in various types of geographical location, each part tends to develop rather distinctive spatial patterns. Some functions tend to be geographically dispersed; others are geographically concentrated and co-located with other parts of the firm. In the following sections, we look at the geographical orientations of four of the major business functions:

- control and coordination
- research and development
- marketing and sales
- production

Control and coordination

The *corporate headquarters* is the locus of overall control of the entire TNC, responsible for all the major strategic investment and disinvestment decisions that shape and direct the enterprise: which products and markets to enter or to leave, whether to expand or contract particular parts of the enterprise, whether to acquire other firms or to sell off existing parts.[2] One of its key roles is *financial*. The corporate headquarters generally holds the purse strings and decides on the allocation of the corporate budget between its component units. Headquarters offices are, above all, handlers, processors and transmitters of *information* to and from other parts of the enterprise and also between similarly high-level organizations outside. The most important of these are the major business services on which the corporation depends (financial, legal, advertising) and also, very often, major departments of government, both foreign and domestic. There is evidence that the size and complexity of corporate headquarters vary substantially between firms from different home countries.[3]

Regional headquarters offices constitute an intermediate level in the corporate organizational structure, having a geographical sphere of influence encompassing several countries. Regional headquarters perform several distinctive roles.[4] Most commonly, their primary responsibility is to *integrate* the parent company's activities within a region, that is, to coordinate and control the activities of the firm's affiliates (manufacturing units, sales offices etc.) and to act as the intermediary between the corporate headquarters and its affiliates within its particular region. Regional headquarters, therefore, are both coordinating mechanisms within the TNC and also an important part of the TNC's 'intelligence-gathering' system. A regional headquarters major role may also be *entrepreneurial*: to act as a base to initiate new regional ventures or to demonstrate to governments that the company has a commitment to the region. In either case, regional headquarters act as 'strategic windows' on regional developments and opportunities.[5] In some cases, regional headquarters are located close to the firm's major production facilities in a particular country or region. But that is not always the case. General Motors, for example, has located its European headquarters in Switzerland, which is not only not a member of the EU but also not an automobile production location.

The characteristic functions of corporate and regional headquarters define their particular *locational* requirements.

- Both require a *strategic location* on the global transportation and communications network in order to keep in close contact with other, geographically dispersed, parts of the organization.
- Both require access to *high-quality external services* and a particular range of *labour market skills*, especially people skilled in information processing.
- Since much corporate headquarters activity involves interaction with the head offices of other organizations, there are *strong agglomerative forces* involved. Face-to-face contacts with the top executives of other high-level organizations are

facilitated by close geographical proximity. Such high-powered executives invariably prefer a location that is rich in social and cultural amenities.

At the global level only a very small number of cities contain the major proportion of both corporate and regional headquarters offices of TNCs. For example, analysis of the headquarters and branch locations of the world's 500 largest TNCs[6] showed that not only are the 'Global 500 headquartered in just 125 cities' but also four cities – New York, Tokyo, London and Paris – stand head and shoulders above all the others. For such reasons, these *global cities* are sometimes described as the geographical 'control points' of the global economy.[7] Below them is a tier of other key headquarters cities in each of the three major economic regions of the world: Europe (e.g. Amsterdam, Brussels, Düsseldorf, Frankfurt); North America (e.g. Atlanta, Chicago, Houston, Los Angeles, Montreal, San Francisco, Toronto); Asia (e.g. Beijing, Hong Kong, Osaka, Seoul, Singapore, Taipei).

One of the most striking features of the geography of corporate headquarters is that very few, if any, major TNCs have moved their ultimate decision-making operations outside their home country. For example, a recent analysis of headquarters relocations of the *Fortune* Global 500, covering the period 1994–2002,[8] found only one! And this was the result of the merger between Daimler of Germany and Chrysler of the United States: the merged company's headquarters are in Germany. This is a further strong indicator of the continuing significance of the home base for corporate behaviour (see Chapter 4).

Within individual countries, on the other hand, the locational pattern of both corporate – and especially regional – headquarters is far from static, with substantial geographical decentralization of corporate headquarters out of the city centres of New York and London. In the case of London, most of these shifts are a short distance to the less congested outer reaches of the metropolitan area. In the United States, on the other hand, there appears to be a much higher degree of locational change in headquarters functions.[9] Nevertheless, corporate headquarters tend not to be spread very widely within any particular country. In the United Kingdom, for example, there are very few corporate headquarters of major firms or regional headquarters of foreign TNCs outside London and the south-east; in France few locate outside Paris. In Italy the most important centre is Milan, in the highly industrialized north, which is more important than Rome as a location for foreign TNCs.

Both the further integration and expansion of the European Union and the rapid growth of the East Asian economies have stimulated the need for regional headquarters in those areas. For example, many US TNCs that have had a presence in Europe for a very long time have now established European headquarters to coordinate their regional operations. A number of Japanese TNCs have set up regional headquarters in Europe as the scale and extent of Japanese operations within Europe have increased.[10] In East Asia, the Singapore government introduced an Operational Headquarters Scheme to attract foreign TNCs to establish their regional headquarters in Singapore.[11]

Apart from corporate and regional headquarters, there are other coordination functions that may be separated out geographically. For example, although a TNC's supply chain management is normally located at or near the corporate headquarters, or at one or more of the regional headquarters sites, there are cases where this function has been located beyond this part of the network. Recently, for example the Anglo-Dutch consumer products company Unilever announced it was to concentrate all its global supply chain coordination and management at a completely new site at Schaffhausen in Switzerland.

Creating innovation: the location of corporate research and development

The process of research and development (R&D) is a complex sequence of operations (Figure 5.3) consisting of three major phases, each of which tends to have rather different locational requirements although, in each case, the TNC has to reconcile several factors.[12] One of these is the advantage of gaining economies of scale from concentrating R&D in one or a few large establishments. Another is the possible benefit of locating R&D close to corporate headquarters or, alternatively, close to production units to enhance communications and the sharing of ideas. Yet another possible locational pull is to markets, in order to benefit from closeness to customer needs, tastes and preferences.

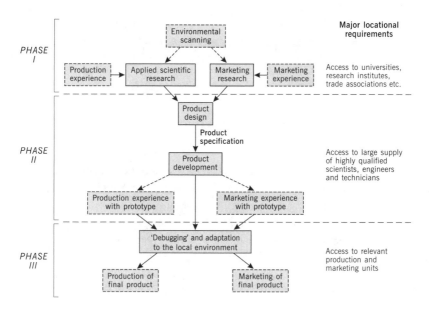

Figure 5.3   Major phases in the R&D process

Source: based, in part, on Buckley and Casson, 1976: Figure 2.7

The type of R&D undertaken by TNCs within their own transnational network can be classified into three major categories (Figure 5.4). The lowest level is the *support laboratory*, whose primary purpose is to adapt parent company technology to the local market and to provide technical back-up. It is the equivalent of phase III in Figure 5.3 and is by far the most common form of overseas R&D facility. The *locally integrated R&D laboratory* is a much more substantial unit, in which product innovation and development are carried out for the market in which it is located. It is the equivalent of phase II in Figure 5.3. The *international interdependent R&D laboratory* is of a quite different order. Its orientation is to the integrated transnational enterprise as a whole rather than to any individual national or regional market. Indeed, there may be few, if any, direct links with the firm's other affiliates in the same country. Only a small number of technologically intensive global corporations operate international interdependent laboratories.

International interdependent R&D laboratory

Function:	Basic research centre; close links with international research programme; may or may not interact with the firm's foreign manufacturing affiliates
Reason for establishment:	Operation of coordinated world R&D programmes as part of global product strategies involving the manufacture of a single product line for world markets. Units tend to be created by direct placement

Locally integrated R&D laboratory

Function:	Local product innovation and development; transfer of technology
Reason for establishment:	Improved status of subsidiaries; concept of overseas operations as fully developed business entities; identification of new business opportunities outside home country. Frequently develop out of support laboratories

Support laboratory

Function:	Technical service centre; translator of foreign manufacturing technology
Reason for establishment:	Response to market growth; differing market conditions; expectation of continuing stream of technical service projects

Figure 5.4   Three major types of corporate R&D facility

The organizational configuration of R&D varies a good deal, mainly according to the type of organizational structure in use (see Figures 4.6–4.9). But to what extent are TNCs dispersing their R&D geographically? There is considerable disagreement over this. On the one hand, TNCs continue to show a very strong preference for keeping their high-level R&D at home. Detailed empirical analyses of patent data for almost 600 firms[13] produced the following conclusions:

- Less than 8 per cent located more than half of their technological activities outside their home country.
- More than 40 per cent performed less than 1 per cent of their technological activity abroad.

- More than 70 per cent performed less than 10 per cent of this activity abroad.
- Very little of the overseas R&D activity of firms from the United States, Japan, Germany, France and Italy is located outside the 'global triad'.
- Most of the apparent increases in overseas R&D came about through merger and acquisition rather than through internal growth and geographical expansion.

Why should such home country bias persist? The answer lies in the importance of the kinds of *untraded interdependencies* discussed earlier (see Chapters 1 and 3):

> Two key features related to the launching of major innovations may help explain the advantages of geographic concentration: the involvement of inputs of knowledge and information that are essentially 'person-embodied', and a high degree of uncertainty surrounding outputs. Both of these are best handled through geographic concentration. Thus it may be most efficient for firms to concentrate the core of their technological activities in the home base with international 'listening posts' and small foreign laboratories for adaptive R&D.[14]

On the other hand, there is some evidence of increasing geographical dispersal of R&D activities within TNC networks. For example, by the late 1990s there was 'an increasing share of company-financed R&D performed abroad by US firms *as compared to domestically financed* industrial R&D … US firms' investment in overseas R&D increased three times faster than did company-financed R&D performed domestically.'[15] In this regard, Asia is playing an increasingly important role as a location for certain kinds of R&D, especially in product development.[16] There are two major reasons for Asia's increasing significance in R&D. One is personnel:

> Asia's greatest overall advantage is its huge supply of scientists and engineers, particularly in China and India, at a time when students in the west are turning away from science and engineering. Companies in the US and Europe … can exploit Asia's trained workforce by building research and development centres there or collaborating with Asian companies and universities.[17]

The other advantage of an Asian location for some kinds of R&D is cost:

> The relative costs of doing research in Asia vary enormously according to circumstances … [However] the pay of newly graduated researchers in India and China is around one-quarter of US levels. For more senior staff, it is usually at least half the US level and in exceptional circumstances may even exceed it.[18]

Support laboratories (Figure 5.4) tend to be the most widely spread geographically, in so far as they generally locate close to production units. But the larger-scale R&D activities tend to be confined to particular kinds of location. The need for a large supply of highly trained scientists, engineers and technicians, together with proximity to universities and other research institutions, confines such facilities to large urban complexes. These are often also the location of the firm's corporate headquarters. A secondary locational influence is that of 'quality of living' for the highly educated and highly paid research staff: an amenity-rich setting, including a good climate and high potential for leisure activities, as well as a stimulating intellectual environment.

Spatial patterns of corporate R&D in both the United States and the United Kingdom illustrate both of these locational influences. In the United States, corporate R&D is still predominantly a big-city activity despite recent growth in smaller urban areas. The pull of the amenity-rich environment is illustrated by the considerable concentration of R&D activities in locations such as Los Angeles, San Francisco and San Diego in California, Denver–Boulder in Colorado and the 'Research Triangle' in North Carolina. In the United Kingdom corporate R&D, like corporate headquarters and regional offices, is disproportionately concentrated in south-east England. Almost two-thirds (61 per cent) of the research undertaken by foreign affiliates in the UK is located in the south-eastern region of the country (compared with only 40 per cent of domestic firms' research).[19]

Marketing and sales

Of all the various functions within a TNC, it is the marketing, and especially the sales, units that are likely to be the most geographically dispersed. The reason is obvious. These functions need to be as close as possible to the markets served by the firm. They must be sensitive to local conditions in order to be able to feed back relevant information. They must be in a position to help to tailor the firm's products to local tastes. Not least, they must be in a position to prevent the firm from making costly, and often embarrassing, mistakes in misreading the various consumer cultures in which the firms are operating. The marketing literature is full of examples of insensitive, sometimes culturally insulting branding or packaging decisions made by foreign TNCs, which have not fully understood local conditions. It is for such reasons that firms such as the Swiss company Nestlé have developed a strategy in which

> the consumer is paramount. Every decision has to be made as close to the consumer as possible. It makes no sense for us in Vevey to decide on the taste of a soup to be sold in Chile.[20]

Apart from the obvious tendency to locate marketing and/or sales units in the firm's most important geographical markets, there is a good deal of flexibility in the precise geographical articulation of such activities. Marketing functions, in particular, are often concentrated either at corporate headquarters or, increasingly, within regional headquarters where they are responsible for all the marketing decisions in the specific region. In some cases they are located close to some of the firm's R&D activities, especially those at the development end of the spectrum, in order to create positive synergy between product development and market needs. Of course, with sophisticated internal communications systems, virtual geographical proximity may replace physical proximity. Sales units, on the other hand, tend to be smaller and very widely dispersed.

Production units

There are clearly some identifiable geographical regularities in the patterns of TNC coordination, R&D, and marketing and sales activities. This is because their locational needs are broadly similar for all firms, regardless of the particular industries in which they are involved. This is not so for production units, whose locational requirements vary considerably according to the specific organizational and technological role they perform within the enterprise and the geographical distribution of the relevant location-specific factors. It is certainly true that, compared with corporate headquarters and R&D facilities, production units of TNCs have become more and more dispersed geographically. But there is no single and simple trend or pattern of dispersal common to all activities, whether at the global scale or within individual nations. The pattern varies greatly from one industry to another. Figure 5.5 illustrates four types of geographical orientation which a TNC might adopt for its production units.

(a) Globally concentrated production

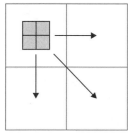

All production occurs at a single location. Products are exported to world markets

(b) Host market production

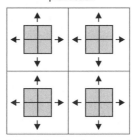

Each production unit produces a range of products and serves the national market in which it is located. No sales across national boundaries. Individual plant size limited by the size of the national market

(c) Product specialization for a global or regional market

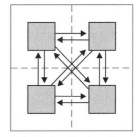

Each production unit produces only one product for sale throughout a regional market of several countries. Individual plant size very large because of scale economies offered by the large regional market

(d) Transnational vertical integration

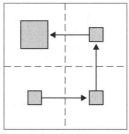

Each production unit performs a separate part of a production sequence. Units are linked across national boundaries in a 'chain-like' sequence – the output of one plant is the input of the next plant

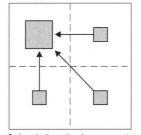

Each production unit performs a separate operation in a production process and ships its output to a final assembly plant in another country

Figure 5.5 Alternative ways of organizing the geography of transnational production

Globally concentrated production

Figure 5.5a presents the simplest case. All production is concentrated at a single geographical location (or, at least, within a single country) and exported to world markets through the TNC's marketing and sales networks. This is a procedure consistent with the classic global strategy shown in Table 4.1 and Figure 4.8.

Host market production

In Figure 5.5b, production is located in, and oriented directly towards, a specific host market. Where that market is similar to the firm's home market, the product is likely to be virtually identical to that produced at home. The specific locational criteria for the setting up of host market plants are:

- size and sophistication of the market as reflected in income levels
- structure of demand and consumer tastes
- cost-related advantages of locating directly in the market
- government-imposed barriers to market entry.

In effect, this kind of production is import substituting. The need to establish a production unit in a specific geographical market has become less necessary in purely cost terms. However, there are two reasons for the continued development of host market production:

- the need to be sensitive to variations in customer demands, tastes and preferences, or to be able to provide a rapid after-sales service
- the existence of tariff and, particularly, non-tariff barriers to trade.

Product specialization for a global or regional market

During the past four decades or so a radically different form of production organization has become increasingly prominent. Figure 5.5c shows production being organized geographically as part of a rationalized product or process strategy to serve a global, or a large regional, market (such as the European Union, North America or East Asia). The existence of a huge regional market, together with differences in locationally specific characteristics between countries within a region, facilitates the establishment of very large, specialized units of TNCs to serve the entire regional market rather than single national markets. The key locational consideration, therefore, involves the 'trade-off' between

- economies of large-scale production at one or a small number of large plants
- additional movement costs involved in assembling the necessary inputs and in shipping the final product to a geographically extensive regional market.

Transnational vertically-integrated production

A rather different kind of transnational production strategy involves geographical specialization by process or by semi-finished product, in which different parts of the firm's production system are located in different parts of the world. Materials, semi-finished products, components and finished products are transported between geographically dispersed production units in a highly complex web of flows. In such circumstances, the traditional geographical connection between production and market is broken. The output of a manufacturing plant in one country may become the input for a plant belonging to the same firm located in another country or countries. Alternatively, the finished product may be exported to a third-country market or to the home market of the parent firm. In such cases, the host country performs the role of an 'export platform'. Its role is to act as an international sourcing point for the TNC as a whole. Figure 5.5d shows two ways in which such transnational process specialization might be organized as part of a vertically integrated set of operations across national boundaries.

Such offshore sourcing and the development of vertically integrated production networks at a global scale were virtually unknown before the early 1960s. The pioneers were US electronics firms that set up offshore assembly operations in East Asia as well as in Mexico. Since then, the growth of such transnational production networks has been extremely rapid, although it is far more important in some types of activity than in others, as we shall see in the case study chapters of Part Three. These practices of transnational intra-firm sourcing have become an increasingly important mechanism of global integration of production processes.

However, the choice of location for a production unit at the global scale is by no means as simple as it is often made out to be. It is not just a matter of looking at differences in labour costs between one country and another, or at the incentives offered as part of an export-oriented government policy. Despite the enormous shrinkage of geographical distance that has occurred, the relative geographical location of parent company and overseas production unit may still be significant. The sheer organizational convenience of geographical proximity may encourage TNCs to locate offshore production in locations close to their home country even when labour costs there are higher than elsewhere. A clear example of this is Mexico, in the case of United States firms, and parts of Southern and Eastern Europe in the case of European firms.

Of course, just as geographical proximity may override differentials in labour costs, so too may other locational influences dominate in any particular case. For the largest TNCs the world is indeed their oyster. Their production units are spread globally, often as part of a strategy of *dual or multiple sourcing* of components or products. This is one way of avoiding the risk of over-reliance on a single source whose operations may be disrupted for a whole variety of reasons. In a vertically integrated production sequence, in which individual production units are tightly interconnected, an interruption in supply can seriously affect the other units, perhaps those located at the other side of the world. In an extreme case, a whole segment of the TNC's operations may be halted.

Reshaping TNCs' internal networks: continuous processes of reorganization and restructuring

Transnational corporate networks are always in a continuous state of flux. At any one time, some parts may be growing rapidly, others may be stagnating, yet others may be in steep decline. The functions performed by the component parts and the relationships between them may alter. Change itself may be the result of a planned strategy of adjustment to changing internal and external circumstances or the 'kneejerk' response to a sudden crisis.

Forces underlying reorganization and restructuring

Corporate reorganization and restructuring is driven by two, often overlapping, forces:

- *External conditions*: these may be negative pressures such as declining demand, increased competition in domestic or foreign markets, changes in the cost or availability of production inputs, militancy and resistance of labour forces in particular places, and the pressure of national governments to modify their activities or even to cede control. Conversely, changes in external conditions may be positive rather than negative, for example the growth of new geographical markets or the availability of new production opportunities. A good illustration is the formation of regional economic groupings, where the creation of a large regional market provide an unprecedented opportunity for TNCs to restructure their production activities to serve the regional market. Investments that had made sense in the context of an individual national context are no longer necessarily rational in the wider context (see Figure 5.5).
- *Internal pressures*: these may relate to the enterprise as a whole or to one or other individual part: for example, sales may be too low in relation to the firm's target, or production costs may be too high. In a TNC the performance of individual plants in widely separate locations can be continuously monitored and compared to assess their efficiency. A key influence is often the 'new broom factor': a new chief executive who undertakes a sweeping evaluation of the enterprise's activities and makes changes that stamp his/her authority on the firm.

In reality, external and internal pressures may be so closely interrelated that it is often difficult to disentangle one from the other. More than this, precisely how firms both identify and respond to changes in their circumstances is very much conditioned by the firm's culture.[21]

Complex corporate restructuring is occurring at all geographical scales, from the global to the local, as strategic decisions are made about the organizational coordination and geographical configuration of the TNC's production network. Decisions to centralize or to decentralize decision-making powers or to cluster or disperse some or all of the firm's functions in particular ways are, however, contested decisions. They are the outcome of power struggles within firms, both within their headquarters and between headquarters and affiliates. How they are resolved depends very much on the nature and the location of the dominant coalition. Such processes also have to be seen within the context of the fundamental tension facing TNCs: to globalize fully or respond to local differentiation (see Figure 5.2). Specifically, corporate restructuring may occur in a variety of ways (Figure 5.6) involving, in some cases, technological change, changes in work practices, rationalization of corporate activities, changes in the extent to which different functions are internalized, and increased transnational investment.

Major forms of corporate restructuring	
• Collaboration	• Incremental transnationalization

• Intensification
  – Contractual flexibility
  – Flexible working practices
  – Concession bargaining
	• Investment and technical change
	  – New technology
	  – Automation
	  – Flexible manufacturing systems

• Deintegration
  – Outsourcing
  – Subcontracting
  – Intrapreneuring
	• Rationalization
	  – Divestment
	  – Differential expansion (contraction)
	  – Changes in product lines
	  – Transfers of business

Figure 5.6   Forms of corporate restructuring

The geography of reorganization and restructuring

Whether corporate reorganization is the result of a consciously planned strategy for 'rational' change or simply a reaction to a crisis (internal or external), its geographical outcome may take several different forms (Figure 5.7).

*   *In situ adjustment* to the existing network of production units is by far the most common form of adjustment. The capacity of an existing plant can be increased to achieve economies of scale or reduced (partial disinvestment) to shed surplus capacity; an existing plant's capital stock can be replaced with new technology. In such ways, the importance and even the actual function of production units can be altered as the TNC reallocates tasks among its existing geographically dispersed operations. Change at an existing plant may be either a gradual process of incremental adjustment or a more sudden change to its scale or function.
*   *Locational shifts* explicitly involve abrupt change because they consist of either an increase or a decrease in the number and location of plants operated by the enterprise or even, in rare cases, the physical relocation of an entire plant.

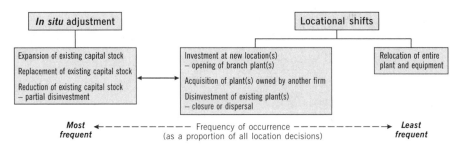

Figure 5.7   Reorganization, restructuring, and spatial change

Thus, reorganization, restructuring and the resulting geographical changes are an inevitable – and a continuous – aspect of the evolution of transnational corporations. The actual form such change takes depends upon forces both internal and external to the firm. The very large global corporations are developing into *global scanners*. They use their immense resources to evaluate potential production locations in all parts of the world. The performance of existing corporate units is monitored and evaluated against competitors, against the rest of the corporate network, and also against potential alternative locations. Those existing plants that fall short of expectations created by such *benchmarking* procedures[22] may be disposed of. As plants become obsolete in one location they are closed down; whether or not new investment occurs in the same locality depends upon its suitability for the TNC's prevailing strategy. The chances are, in many cases, that the new investment will be made at a different location – quite possibly in a different country altogether.

However, we should beware of over-exaggerating the speed and ease with which TNCs can, and do, radically restructure their operations. There are 'barriers to exit'. Production units represent huge capital investments which cannot be written off lightly. But there are other kinds of sunk costs that impose additional exit barriers. 'These are costs that cannot be recovered (for example by selling off surplus assets) or closed out in the short run (as with the case of pension liabilities) even when an operation is terminated'.[23] Political pressures may also inhibit firms from closing plants, especially in areas of economic and social stress. On the other hand, TNCs do have a highly tuned capacity to *switch* and *reswitch* operations within their existing corporate network. They also have the resources to alter the shape of their geographical network through locational shifts.

Although most such change is fairly limited, in some cases the effects are fundamental, especially where they involve the large-scale transfer of production capacity from long-established sites in the West to newly industrializing areas in Asia or Eastern Europe. For example, the Swedish electrical white goods manufacturer Electrolux recently implemented a huge programme of restructuring:

> Electrolux is closing or downsizing sites in western Europe and the US while switching production to lower-cost plants in Mexico, Poland, Romania,

Hungary, Thailand, Russia and China ... The latest phase of the programme was announced in February [2005]. At that point, Electrolux had 27 of its 44 white goods factories in high-cost countries. It said as many as half of them – 13 or 14 – could be switched to low-cost countries over the next four years ... 'The restructuring they're undertaking now is monumental. No other Swedish engineering company has undergone such a transition in the last 10 years' ... Since the beginning of last year [2004] Electrolux has announced the closure of nine factories in high-cost countries. But the process is now picking up speed ... These moves are not about cutting capacity but cutting costs. In some cases, production is being switched directly from a high-cost western plant to a lower-cost one in an emerging market. For example, last year Electrolux switched production of vacuum cleaners from Vastervik in Sweden to Hungary, where labour costs are up to eight times lower. Across the Atlantic, a refrigeration plant at Greenville in Michigan, employing 2,500 people, is being closed with production switched to a new plant at Ciudad Juárez in Mexico ... [However] not all production is being shifted ... 'Some entities in Europe and the US will still be competitive in their present locations, where the labour content is small and we are able to source components from low-cost countries.'[24]

Hence, the processes of reorganization and restructuring are complex, dynamic, and far from predictable. Overall, however, four general tendencies are especially apparent:

- redefining core activities by stripping away activities that no longer 'fit' the firm's strategy
- placing greater emphasis on downstream, service functions
- geographically reconfiguring production networks transnationally to redefine the roles and functions of individual corporate units
- redefining the boundary between internalized and externalized transactions.

It is the last point that forms the focus of the next section of this chapter.

TNCs within networks of externalized relationships

In the previous section, the focus was on how TNCs organize and geographically configure their internalized networks. But, of course, this is only a small part of the story of how the global economy is organized. No firm is completely self-sufficient. Overall, between 50 and 70 per cent of manufacturing costs are to purchase inputs and the general trend is for a greater proportion of both materials/components and services to be outsourced to supplier firms.[25] Some of these purchases will be 'off-the-shelf' sourcing from independent suppliers at the arm's-length market price. However, most are now made on the basis of longer-term relationships.

TNCs are locked into *external* networks of relationships with a myriad of other firms: transnational and domestic, large and small, public and private (see Figure 1.5). Indeed, the *boundary* between what is inside and what is outside the firm has become far more blurred.[26] TNCs, therefore, are best understood as 'a dense network at the centre of a web of relationships'[27] – in other words, as networks within networks. These are the threads from which the fabric of the global economy is woven. It is through such links that changes are transmitted between organizations and, therefore, between different parts of the global economy.

The subcontracting process

Subcontracting, in the form of one firm outsourcing some of its operations to another firm, is a kind of half-way house between complete internalization of procurement on the one hand and arm's-length transactions through the open market on the other. As a 'putting-out' process it is as old as industrialization itself. There are, broadly, two major types of subcontracting, both based on a subcontractor producing a good or service to the principal firm's specifications:

- *Commercial subcontracting*: the manufacture of a finished product. The subcontractor plays no part in marketing the product, which is generally sold under the principal's brand name and through its distribution channels. The principal firm may be either a producer firm, that is, also involved in manufacturing, or a retailing or wholesaling firm whose sole business is distribution.
- *Industrial subcontracting*:
  - *Speciality* subcontracting involves the carrying out, often on a long-term basis, of specialized functions which the principal chooses not to perform for itself but for which the subcontractor has special skills and equipment.
  - *Cost-saving* subcontracting is based upon differentials in production costs between principal and subcontractor for specific processes or products.
  - *Complementary or intermittent* subcontracting is a means adopted by principal firms to cope with occasional surges in demand without expanding their own production capacity. In effect, the subcontractor is used as extra capacity, often for a limited period or for a single operation.

Figure 5.8 summarizes some of the major features of the subcontracting relationship.

Outsourcing using subcontractors obviously has profound geographical implications. Initially, it depended on close proximity between firms and suppliers and was a major factor underlying the development and persistence of traditional 'industrial districts' – agglomerations of linked firms. However, innovations in transportation and communications have greatly increased the geographical extensiveness of

Types of relationship between principal firm and subcontractor
• Time period may be long term, short term, single batch • Principal may provide some or all materials or components • Principal may provide detailed design or specification • Principal may provide finance, e.g. loan capital • Principal may provide machinery and equipment • Principal may provide technical and/or general assistance and advice • Principal is invariably responsible for all marketing arrangements

Benefits and costs to principal firm	Benefits and costs to subcontractor
• Avoids need to invest in new production capacity • Flexible: easier to change subcontractors • Externalizes some risks, while retaining some control • Possible problems of controlling how subcontractors work (including labour issues)	• Access to markets • Continuity of orders • Access to technical knowledge, possible injection of finance • Risk of being 'expendable' if principal firm changes its priorities • Possible over-dependence on one or a small number of customers

Figure 5.8 Subcontracting relationships

subcontracting networks. Much of the increase in long-distance sourcing was driven by the desire to take advantage of the wide differentials in labour costs between different parts of the world (see Figure 4.2). As the distance between customers and suppliers increased, however, problems inevitably arose in terms of the reliability of supplies. Many firms had to establish sophisticated – and very costly – systems of stock/inventory holding to insure against interruptions in the supply of finished goods or components. This is the kind of *just-in-case* system shown on the left-hand side of Table 5.1.

However, as we saw in Chapter 3, production processes have changed dramatically. The emphasis is increasingly on rapid product turnover, speed to market, responsiveness to customer needs: on what, in the context of the automobile industry, has come to be called 'lean' production (see Chapter 10). In such a production system, holding large stocks of inventory in warehouses is anathema. Supplies must be delivered precisely when (and where) they are needed, that is, *just-in-time* (JIT). The major features of a JIT system are shown on the right-hand side of Table 5.1. When JIT first came to prominence, largely within Japanese automobile and electronics firms in the 1970s, it was widely predicted that the result would be a large-scale geographical reconcentration of suppliers near to their customers. In other words, there would be a return to the old-style industrial clusters. In fact, although there has undoubtedly been some geographical shifts towards closer links between customers and suppliers, there has not been universal reconcentration. Rather, a wide variety of geographical scales is involved, depending on firm- and industry-specific circumstances.

Table 5.1   The characteristics of 'just-in-case' and 'just-in-time' systems

'Just-in-case' system	'Just-in-time' system
Characteristics	
Components delivered in large, but infrequent, batches	Components delivered in small, very frequent, batches
Very large 'buffer' stocks held to protect against disruption in supply or discovery of faulty batches	Minimal stocks held – only sufficient to meet the immediate need
Quality control based on sample check after supplies received	Quality control 'built in' at all stages
Large warehousing spaces and staff required to hold and administer the stocks	Minimal warehousing space and staff required
Use of large number of suppliers selected primarily on the basis of price	Use of small number of preferred suppliers within a tiered supply system
Remote relationships between customer and suppliers	Very close relationships between customer and suppliers
No incentive for suppliers to locate close to customers	Strong incentive for suppliers to locate close to customers
Disadvantages	
Lack of flexibility – difficult to balance flows and usage of different components	Must be applied throughout the entire supply chain
Very high cost of holding large stocks	Reliance on small number of preferred suppliers increases risk of interruption in supply
Remote relationships with suppliers prevents sharing of developmental tasks	
Requires a deep vertical hierarchy of control to coordinate different tasks	

Source: based on material in Sayer, 1986

It is tempting to regard subcontractors as always being in a subservient position – merely small cogs in a much larger structure over which they have no influence. Often that is the case, although much depends on the position they occupy within a production network and also the kind of network involved (see the next section). However, Andersen and Christensen argue that some subcontractors act as important 'connective nodes in supply networks'.[28] They identify five such 'bridging roles':

- *Local integrator:* firms able to provide access to additional production capacity through their relationships with other firms in the same locality. 'Local integrators may take on a host of different roles in the supply network, depending on fluctuations in the demand situation ... [they] may even take over commercial contracting, functioning as end producers – e.g. act as private label producers for a trading firm or look for market opportunities in a broad range of industrial branches'.[29]

- *Export base:* such firms act as 'the gate of access to a local technological district ... These subcontractors provide ... superior knowledge regarding the competences of local suppliers ... This bridging role is most likely found in speciality subcontracting or systems supplies, through which the subcontractor adds knowledge to the products of the customer ... Export base subcontractors utilize their relations with customers to marshal relations with other subcontractors in their local area. Often, they take on the complex task of coordinating a local network of suppliers'.[30]

- *Import base*: these firms represent 'the gateway to international resources or skills for customers in the local or national area, as the international customer–supplier relations of the import base suppliers are used as generators of foreign market and product knowledge ... compared to the local integrator, this firm depends heavily on representing a specific international range of technologies to local customers ... the important task of the subcontractor is to foresee demand fluctuations and buffer them accordingly, either via extended information exchange with subcontractors or by keeping an extended inventory'.[31]

- *International spanner*: often, such firms have evolved from being an export or import base subcontractor as their supply sources and buyers have internationalized. 'The position of the international spanner is a precarious one. There are pressures at both ends of the supply chain, which seek to attract the subcontractor further into the supply source or further into the logistical basis of the buyer ... international spanners sometimes base their business potential on information asymmetries between subcontractors and buyers ... The ability to orchestrate these activities globally is the required coordinative capability of these firms'.[32]

- *Global integrator*: 'A hybrid form of subcontractor ... responsible for all bridging activities of the international supply chain ... connecting internationally dispersed buyers and subcontractors and supplying the necessary logistical infrastructure for carrying out exchange. The primary strategic asset ... is its developed infrastructure and its ability to often manage quite different streams of manufactured goods ... Compared to the roles taken by the other subcontractors in the typology, the global integrator has a strong bargaining position towards buyers as well as subcontractors'.[33]

Different ways of coordinating transnational production networks

In Chapter 1 we noted that production networks may be coordinated in a variety of different ways, involving a mix of intra-firm and inter-firm structures. It is now time to examine this more closely.[34] Figure 5.9 shows one way of categorizing types of network coordination. In this discussion I will concentrate on three of the five coordination types shown in Figure 5.9: captive, relational and modular production networks. The previous section (and Chapter 4) dealt with the hierarchical mode, whilst open market transactions seem to have become less important for the kinds of firms we are concerned with in this chapter although, of course, they continue be relevant in some cases. In all three networks, what we are interested in are the changing relationships between 'lead' firms and suppliers. We will encounter examples of these different types of network in the industry case studies of Part Three.

Coordination mechanism	Complexity of transactions	Ability to codify transactions	Capabilities of potential suppliers	Degree of explicit coordination and power asymmetry
Hierarchy Vertical integration within a firm with governance of subsidiaries and affiliates based on head-quarters' managerial control	HIGH	LOW	LOW	HIGH
Captive Small suppliers transactionally dependent on larger buyers. Suppliers face significant switching costs	HIGH	HIGH	LOW	
Relational Complex interactions between buyers and sellers often creating mutual dependence and high levels of asset specificity	HIGH	LOW	HIGH	
Modular Production to customer's specification	HIGH	HIGH	HIGH	
Market May involve repeat transactions but switching costs low for both parties	LOW	HIGH	HIGH	LOW

Figure 5.9   Different ways of coordinating transnational production networks

Source: based on material in Gereffi et al., 2005

Captive production networks

These are networks in which a lead firm is dominant and effectively controls – although it does not own – all the major components in the network.

> Lead firms seek to lock-in suppliers in order to exclude others from reaping the benefits of their efforts. Therefore, the suppliers face significant switching costs and are 'captive'. Captive suppliers are frequently confined to a narrow range of tasks – for example, mainly engaged in simple assembly – and are dependent on the lead firm for complementary activities, such as design, logistics, component purchasing, and process technology upgrading.[35]

In such networks, the instructions to suppliers are highly codifiable while power is highly asymmetrical and lies unequivocally with the lead firm.

The hierarchically organized networks of major Japanese and Korean companies, discussed in Chapter 4, are captive networks. An especially graphic (although not necessarily a typical) example is provided by the US sports footwear company Nike. Nike does not wholly own any integrated production facilities but is characterized by 'the large-scale vertical disintegration of functions and a high level of subcontracting activity'.[36] Its development displays great flexibility in adapting to changing competitive circumstances. As Figure 5.10 shows, Nike consists of a complex tiered network of subcontractors that perform specialist roles.

> Nike develops and produces all high-end products with exclusive partners, while volume producers manufacture more standardized footwear that experiences larger fluctuations in demand ... Nike acts as the production co-ordinator and three categories of primary production alliance form the first tier of subcontractors. A second tier of material and component subcontractors supports production in the first tier ... the headquarters in Beaverton, Oregon, houses Nike's research facilities.[37]

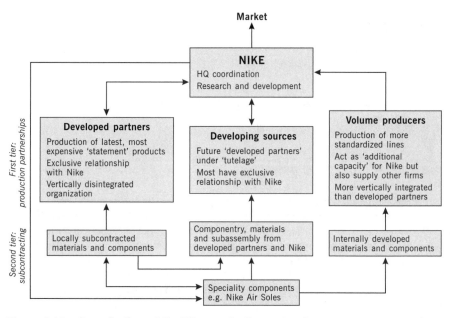

Figure 5.10 Organization of the Nike production network

Source: based on Donaghu and Barff, 1990: Figure 4; pp. 542–4

In 2005 Nike, for the first time, published a list of its suppliers worldwide, noting however that 'the active factory base is constantly in flux'. Figure 5.11 shows the geographical spread of its supplier network. In terms of its extensiveness, it is certainly global. However, there is a strong bias towards East Asia, which contains almost 60 per cent of total suppliers. Of these, the majority are located in China (31 per cent of the region's total), followed by Thailand (almost 20 per cent). Indonesia,

Vietnam and Malaysia are also significant elements in the network. Only around 8 per cent of the total suppliers are in South Asia (mostly in Sri Lanka and India); 10 per cent in Central/Latin America (mostly in Mexico and Brazil); and 12 per cent in Europe (primarily Portugal and Turkey).

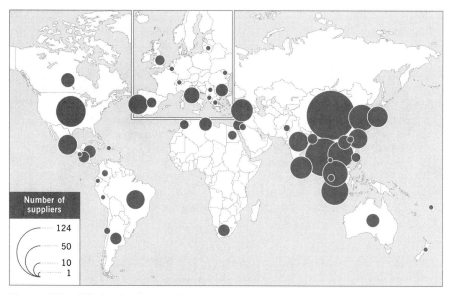

Figure 5.11   Nike's global supplier network, 2005

Source: based on data in Nike Inc., 2005

Relational production networks

> Relational production networks are governed less by the authority of lead firms, and more by social relationships between network actors, especially those based on trust and reputation. Accordingly, relational production networks tend to be embedded in larger socioeconomic systems ... relational production networks can adapt to volatile markets quite well, as the trust, personal and familial relationships of the community enable its members – individuals and small firms – to quickly respond to changing conditions.[38]

Relational production networks, therefore, have more symmetrical power relationships than captive production networks. They are the kinds of network that exist among overseas Chinese businesses (see Chapter 4) and some other ethnic/social groups. The 'technical' transnational communities that have developed in the US electronics industries around Taiwanese, Chinese and Indian immigrants have facilitated rapid and extensive growth of global production networks based on relational processes.

> Communities of technically skilled immigrants with business experience and connections in the USA are ideally positioned to accelerate the diversification

and technical upgrading of supplier networks in their home countries ...
US-educated returnees provide a direct mechanism for transferring the skill
and tacit knowledge that can dramatically accelerate industrial upgrading in
their developing countries. In addition, they frequently coordinate relationships between the network flagships and suppliers, particularly when they are
based in regions with differing languages and business cultures ...

Transnational communities likewise provide a mechanism for seeding entirely
new centers of low-cost (at least initially) supply in less developed regions.
Taiwan's leading personal computer suppliers, including Acer, Mitac and
Compaq, for example, got their initial contracts for IBM-compatible PCs in the
early 1980s from Chinese entrepreneurs in Silicon Valley. Senior engineers in
large US corporations were similarly among the first to outsource software
services to India.[39]

Within Europe, examples of relational networks have been identified in Germany
(the complex contracting relationships between small and medium firms) and in
Italy. In both cases, as well as in other parts of the world, it has been argued that it
is the close *spatial proximity* between firms and other social institutions that provides
the 'relational cement' for the networks to exist. During the 1980s, in particular, it
became extremely popular to eulogize such 'industrial districts'[40] as the way forward
from the old rigidities of Fordist mass production systems. However, important as
close spatial proximity may be in facilitating the development of relational production networks, on its own it is not sufficient, as we saw in discussing localized knowledge clusters in Chapter 3.

One current view of relational networks is that they may point the way towards
the emergence of the *virtual firm* or the *cellular network* organization[41] (Figure 5.12).
Organizationally, the entire network structure is relatively 'flat' and non-hierarchical.
Its essence is that the participants are all separate firms with no common ownership. They are cooperative, *relational* structures between independent and quasi-
independent firms that are based upon a high degree of trust, something that takes
time to develop. However, this does not mean that there are not *power* differentials
within the network. There certainly are.

Such network forms are gradually emerging in such 'knowledge businesses' as
advanced electronics, computer software design, biotechnology, design and engineering services, health care and the like. The Taiwanese computer company Acer is
moving along that path in its '21 in 21' vision: a planned federation of at least 21
self-managing, independent firms, located around the world by the twenty-first century, and held together by mutual interest rather than hierarchical control.

Modular production networks

For many industries, the changes of the past twenty years mean that
organizing production has become like playing with a set of Legos than building a model airplane or a car. In other words, it's now possible to create many
different models using the same pieces. New components can be added
on to old foundations; elements from old structures can be reused in new

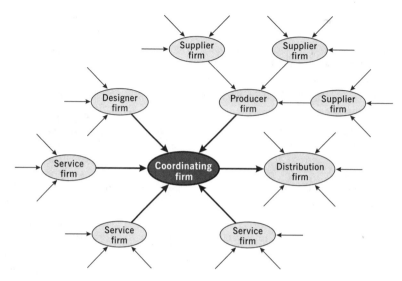

Figure 5.12   A 'pure' relational network

Source: based, in part, on Miles and Snow, 1986: Figure 1

configurations; parts can be shared by many players with different construction plans in mind ... the myriad possibilities of organizing a company have grown out of new digital technologies that create countless opportunities for using resources, organizations, and customers all over the world to build businesses that did not even exist ten years ago.[42]

The development of modular production networks depends largely on the fact that some modern production circuits have 'natural' breakpoints, where there is a transition from dependence on tacit knowledge to one where information can be codified through standard, agreed protocols.[43] This has led, in an increasing number of industries, to a situation in which lead firms concentrate primarily on product development, marketing and distribution, while what are termed *turnkey suppliers* concentrate on producing those functions outsourced by lead firms and to sell them, in effect, as services to a wide range of customers. To achieve this, turnkey suppliers develop three types of cross–cutting specialization:[44]

- '*base process*, one which is used to manufacture products sold in a wide range of end markets (e.g. pharmaceutical manufacture, semiconductor wafer fabrication, plastic injection molding, electronics assembly, apparel assembly, brewing, telecommunications backbone switching)'
- '*base component*, one that can be used in a wide variety of end products (e.g. semiconductor memory, automotive braking systems, engine controls)'
- '*base service*, one that is needed by a wide variety of end users (e.g. accounting, data processing, logistics), rather than processes services that are idiosyncratic or highly customer-specific'.

Turn-key suppliers, then, tend to develop *generic* manufacturing capacity and services that allow product variation to be very large as long as their specifications fall within the parameters of the base process. In the manufacturing sectors mentioned above, most turn-key suppliers use highly automated production systems (apparel assembly is a major exception) that can be programmed and re-programmed on short notice to produce a wide variety of products.[45]

Figure 5.13 shows how such modular production networks differ from that of a traditional vertically integrated hierarchical firm.

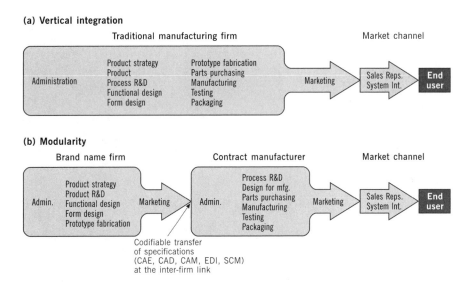

(a) Vertical integration

Figure 5.13   From vertical integration to modular production networks

Source: based on Sturgeon, 2002: Figure 1

The development of contract manufacturing in the electronics industry provides a clear example of the development of modular production networks.[46] As Sturgeon shows, the increasing scale and complexity of outsourcing by US electronics firms in the 1980s and 1990s created a demand for suppliers to develop large capabilities at a global scale in order to serve the increasingly transnationalized lead electronics firms. The result was the emergence of a small number of very large electronics contract manufacturers from North America – Solectron, Flextronics, Sanmina/ SCI, Celestica, Jabil Circuit – which now operate a global network of establishments serving the leading brand-name electronics manufacturers. Figure 5.14 maps Solectron's network.

The current largest electronics contract manufacturer, Solectron, was concentrated in a single campus in Silicon Valley until 1991, when its key customers, Sun Microsystems, Hewlett Packard and IBM, began to demand global manufacturing and process engineering support. Within ten years, the company's footprint had expanded to nearly 50 facilities worldwide ... Today, Solectron operates a set of global and regional headquarters, high and low mix manufacturing facilities, purchasing and materials management centers, new product introduction centers, after-sales repair service centers for products manufactured by Solectron and others, and technology centers to develop advanced process and component packaging technologies.[47]

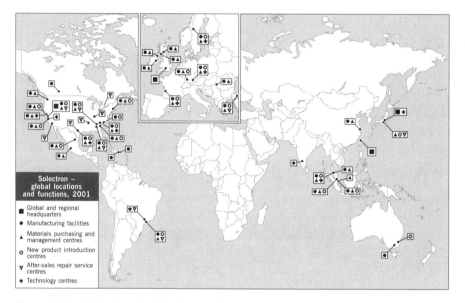

Figure 5.14   Solectron's global network

Source: based on Sturgeon, 2002: Table 2

These different types of production network – hierarchical, captive, relational and modular – coexist in varying combinations in different industries and in different parts of the world. There is some evidence to suggest that firms from particular national origins tend to adopt particular types of production network. In this context, the modular production network form has developed most clearly in the United States and reflects a relative openness of procedures and a desire to reduce the degree of mutual dependence. Key to this system is the extensive and intensive use of 'IT suppliers that provide widely applicable "base processes" and widely accepted standards that enable the codifiable transfer of specification across the inter-firm link. These preconditions lead to generic (not product-specific) capacity at suppliers that has the potential to be shared by the industry as a whole'.[48] The extent to which such a system will be adopted more widely and lead to convergence of practice is an open question and relates to the comments made about the persistence of nationally grounded variations in TNC structures and practices discussed in Chapter 4.

Benefits and costs of outsourcing

The basis of the various production networks is the increasing trend for firms to outsource some of their major functions, thus providing opportunities for supplier firms to fill the gap. For the firm doing the outsourcing it is widely seen as a way of focusing on its 'core competences' and shedding activities that do not fit. The logic is that costs will be reduced and profits enhanced through such concentration on core activities. This may well be the case. But there are also risks in outsourcing, as a number of well-publicized cases have shown. In 2005, for example, British Airways found itself mired in a very messy dispute with the firm supplying all its airline food, Gate Gourmet. Originally, BA had produced all its catering needs in-house and what became Gate Gourmet was the result of a sell-off of these activities to a US venture capital firm. Should BA have retained the catering function as one of its core activities? The company says no; others are less sure.

The potential benefits, as well as the costs, of outsourcing are even greater where it occurs across national boundaries, that is, *offshoring* outsourced activities. In fact, the trend towards outsourcing, though very strong at present, is not inevitable or irreversible. There are many example of activities that have been outsourced being brought back in later as the firm reassesses its priorities. In other words, the boundary between externalization and internalization is continuously shifting, and not always in predictable directions.

Transnational strategic alliances

Related to the development of relational and modular production networks is one of the most significant developments in the global economy in recent years: the growth and spread of transnational *strategic alliances* between firms.[49] In fact, collaborative ventures between firms across national boundaries are nothing new. What is new is their current scale, their proliferation, and the fact that they have become *central* to the global strategies of many firms rather than peripheral to them. Most strikingly, the overwhelming majority of strategic alliances are *between competitors*. In other words, they reflect a new form of business relationship, a 'new rivalry ... in the way collaboration and competition interact'.[50] Many companies are forming not just single alliances but *networks of alliances*, in which relationships are increasingly multilateral rather than bilateral, polygamous rather than monogamous. In effect, they create new *constellations* of economic power. As a result, a new component – 'collective competition' – has been added to the economic landscape. Figure 5.15 shows four alliance-based constellations in the computer industry in the 1990s.

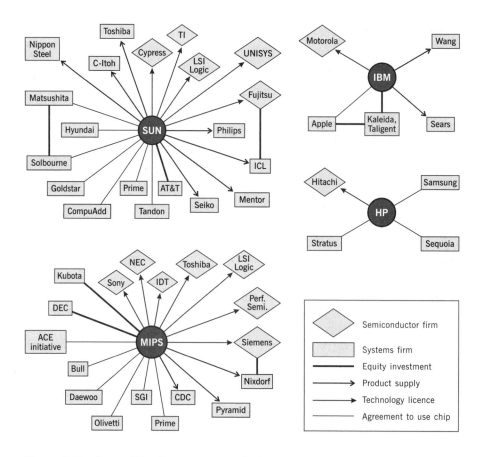

Figure 5.15   Competition between constellations of firms in the computer industry

Source: based on Gomes-Casseres. 1996: Figure 9

There has been a dramatic increase in the number of strategic alliances during the past two decades:

> The number of new strategic alliances (both domestic and international) increased more than six-fold during the period 1989–1999, from just over 1,000 in 1989 (of which around 860 are cross-border deals) to 7,000 in 1999 (cross-border deals: 4,400) … there are indications that recent alliances, particularly joint ventures, are far larger in scale and value terms than earlier partnerships … International strategic alliances accounted for 68% of all alliances (numbering 62,000) between 1990–99. On average, there are about two international strategic alliances for every domestic partnership, illustrating that globalization is a primary motivation for alliances.[51]

Strategic alliances are formal agreements between firms to pursue a *specific* strategic objective; to enable firms to achieve a specific goal that they believe cannot be achieved on their own. It involves the sharing of risks as well as rewards through joint decision-making responsibility for a specific venture. Strategic alliances are not the same as mergers, in which the identities of the merging companies are completely subsumed. In a strategic alliance only *some* of the participants' business activities are involved; in every other respect the firms remain not only separate but also usually competitors.

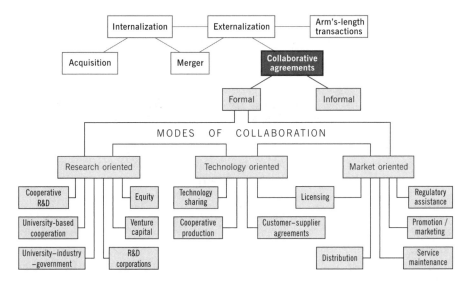

Figure 5.16   Types of inter-firm collaboration

Source: based on Anderson, 1995: Figure 1

Three major modes of collaboration are involved in strategic alliances (Figure 5.16): research oriented, technology oriented and market oriented. Alliances offer the following (potential) kinds of advantage to the participants:[52]

- overcoming problems of access to markets
- facilitating entry into new/unfamiliar markets
- sharing the increasing costs, uncertainties and risks of R&D and of new product development
- gaining access to technologies
- achieving economies of synergy, for example, by pooling resources and capabilities and by rationalizing production.

Very often, the motivations for strategic alliances are highly specific. In the case of R&D ventures, for example, cooperation is limited to research into new products and technologies while manufacturing and marketing usually remain the responsibility of the individual partners. Cross-distribution agreements offer firms ways of widening their product range by marketing another firm's products in a specific market area. Cross-licensing agreements are rather similar but they also offer the possibility of establishing a global standard for a particular technology, as happened, for example, in the case of compact disc players. Joint manufacturing agreements are used both to attain economies of scale and also to cope with excess or deficient production capacity. Joint bidding consortia are especially important in very large-scale projects in industries such as aerospace or telecommunications, where the sheer scale of the venture or, perhaps, the specific regulatory requirements of national governments put the projects out of reach of individual companies.

The majority of strategic alliances are in sectors with high entry costs, economies of scale, rapidly changing technologies and/or substantial operating risks.

> Pharmaceuticals, chemicals, electronic equipment, computers, telecommunications, and financial and business services are examples of industries characterized by a large number of strategic alliances ... Although a large number of alliances are still formed in manufacturing industries, more and more strategic alliances are taking place in the services ... As the world economy becomes more service-based, strategic alliances are playing a more important role in cross-border restructuring in service sectors.[53]

Advocates of strategic alliances claim that by cooperating, companies can combine their capabilities in ways that will benefit each partner. But not everybody shares this rosy view. Many fear that entering into such alliances will result in the loss of key technologies or expertise by one or other of the partners. More broadly, strategic alliances are clearly more difficult to manage and coordinate than single ventures; the potential for misunderstanding and disagreement, particularly between partners from different cultures, is great. Certainly many such alliances have relatively short lives. Nevertheless, the obvious attractions of transnational strategic alliances in today's volatile and competitive global economy are likely to guarantee their continued growth as a major organizational form.

Regionalizing transnational production networks

The complex organizational-geographical networks of TNCs occur at a whole spectrum of geographical scales, from the finely local at one extreme to the global at the other. However, TNCs have a very strong propensity to organize their production networks *regionally* (that is, at the multi-nation scale of groups of contiguous states).[54] For example, 'for Western core companies, regionalism has become the institutional framework of choice within which the struggle for preservation of their core positions is played out ... *the acceleration in the rate of internationalization after 1995 was an intra-regional phenomenon*'.[55]

The basis for such a regional orientation is evident in Figure 5.5c. In effect,

> a regional strategy offers many of the efficiency advantages of globalization while more effectively responding to the organizational barriers it entails ... From the perspective of a TNC, a regional strategy may represent an ideal solution to the competing pressures for organizational responsiveness and global integration.[56]

In particular:[57]

- Regional-scale manufacturing facilities may represent the limits of potential economies of scale.
- Regionalization allows for faster delivery, greater customization and smaller inventories than would be possible under globalization.
- Regionalization accommodates organizational concerns and exploits subsidiary strengths.

In some instances, such TNC regionalization is reinforced by regional political structures – as in the EU or the NAFTA – although this is not necessarily the case. Simple geographical proximity is, itself, a very powerful stimulus for integrating operations.

Transnational production networks organized at the regional scale are evident in most parts of the world but most especially in the three 'triad regions' of Europe, North America and East Asia, as we shall see in several of the case study chapters in Part Three. In North America, the establishment of the NAFTA is leading to a reconfiguration of corporate activities (especially in Mexico) to meet the opportunities and constraints of the new regional system, although it is too early yet to calculate its likely extent.[58] In Europe, the increasing integration and recent enlargement of the European Union are leading to substantial reorganization of existing corporate networks and the establishment of pan-EU systems by existing and new TNCs. 'The EU can be seen as a gigantic international production complex made up of the networks of TNCs which straddle across national boundaries and form trade networks in their own right.'[59]

There is abundant evidence of US and Japanese TNCs – as well as many European firms themselves – creating regional networks within the EU. Some Japanese companies, for example, are adopting

> a three-tier European operation, partly centralised and partly decentralised. A number of them have set up small, new, pan-European head offices, with a purely strategic role: financial control, overall direction, high-level brand management ... The real work is done, however, by the next two tiers of the business: the operational centres (production, distribution, logistics) organised on a pan-European basis and sited where convenient. For distribution, this means in the heartland of western Europe, with easy access to France and Germany; for production it will increasingly mean in eastern and central Europe, where costs are lower. Sales and tactical marketing are handled at a national level. Perhaps where two or three countries have very similar

characteristics, such as the Nordic region, they can be aggregated together. But, in general, national markets are sufficiently distinctive to require their own local sales operations.[60]

The process is complicated. On the one hand, supply-side forces are stimulating a pan-EU structure of operations to take advantage of scale efficiencies. On the other hand, demand-side forces are still articulated primarily at the country-specific level, where linguistic and cultural differences play a major role in the demands for goods and services. In effect, the strategic tensions between global integration and local responsiveness, discussed at the beginning of this chapter, are played out at the EU regional level.

Although East Asia does not have the same kind of regional political framework as the EU or NAFTA, there is very strong evidence of the existence of regional production networks organized primarily by Japanese firms, although non-Asian as well as some other Asian firms (from Korea, Hong Kong, Singapore and Taiwan, for example) also tend to organize their production networks regionally.[61] Within East Asia, a clear intra-regional division of labour has developed consisting of four tiers of countries: Japan; the so-called 'four tigers' of Hong Kong, Korea, Singapore and Taiwan; the South East Asian 'later industrializers' of Malaysia, Thailand, Indonesia, and the Philippines; and China, together with, at least potentially, countries such as Vietnam. However, the relationships between these tiers are more complex than is often suggested. The idea of a simple developmental progression starting with Japan and automatically moving on to the other Asian countries – as suggested in the so-called 'flying geese' model of development – is not tenable.[62] In East Asia, in particular, the rapid emergence of China as both a huge potential market and a production location is transforming intra-regional networks.

Such regionally focused production networks play a very important part in the current obsession with outsourcing and offshoring. Recent research has shown, for example, that

> more manufacturers in the US and western Europe are putting their investments closer to domestic operations, as part of an effort to simplify supply structures ... This trend towards so-called 'near-shoring' ... highlights difficulties facing US companies, in particular in managing their production operations in China. In the past few years, US companies have been more willing than they were at the end of the 1990s to invest in production operations in Mexico. Many see it as a low-cost region that is easier to operate in than China ... [in Europe] more companies ... are finding it more attractive to put production investments in low-wage areas in eastern Europe, as opposed to grappling with the problems of bringing in parts and finished goods over much longer distances.[63]

Of course, this should not be interpreted as heralding a complete reversal of long-distance networks. On the contrary. But what it does indicate is that what are sometimes presented as inexorable trends in one particular direction are, in fact, far less determinate. Reality consists of exceedingly complex cross-cutting tendencies both in time and over geographical space.

Conclusion

The aim of this chapter has been to explain how economic activities are organized and reorganized through dynamic networks of relationships within and between transnational corporations and other firms. We have emphasized, in particular, the immense diversity of processes and outcomes, both organizational and geographical. Production networks can be articulated through many different combinations of organizational structures and geographical configurations.

In this chapter we have focused on three specific aspects of transnational production networks. First, we focused on the geography of TNCs' internal networks. In pursuing their specific strategies, TNCs not only create organizational structures but also have to configure their component parts geographically. But, of course, these structures are dynamic, not static. Second, we explored the networks of externalized inter-firm relationships within which TNCs are embedded. Again, these take on a variety of forms. Relationships between firms and their suppliers have become increasingly complex, as has the tendency for firms to enter into strategic alliances with other – often competitive – firms. Third, we explored the tendency for TNCs to organize their transnational production networks at a regional scale.

NOTES

1 Chairman of Coca-Cola, in the *Financial Times* (27 March 2000).
2 See Baaij et al. (2004), Yeung et al. (2001), Young et al. (2000).
3 Young et al. (2000).
4 Lasserre (1996).
5 Yeung et al. (2001: 165).
6 Alderson and Beckfield (2004).
7 Friedmann (1986), Sassen (2001), Taylor (2004).
8 Baaij et al. (2004: 143).
9 Lyons and Salmon (1995: 103–4).
10 Aoki and Tachiki (1992).
11 Yeung et al. (2001: 169–70).
12 Recent studies of trends in R&D activities by TNCs include Blanc and Sierra (1999), Cantwell (1997), Hotz-Hart (2000), Patel (1995), Zanfei (2000).
13 Patel (1995)
14 Patel (1995: 152).
15 Blanc and Sierra (1999: 188).
16 *Financial Times* (23 June 2005).
17 *Financial Times* (9 June 2005).
18 *Financial Times* (9 June 2005).
19 Cantwell and Iammarino (2000: 322).
20 CEO of Nestlé, cited in the *Financial Times* (22 February 2005).
21 Schoenberger (1997: 204).

22 See Sklair (2001: Chapter 5).

23 Schoenberger (1997: 88).

24 *Financial Times* (28 July 2005).

25 Mol et al. (2005), Berggren and Bengtsson (2004).

26 Dicken and Malmberg (2001: 351).

27 Badaracco (1991: 314).

28 Andersen and Christensen (2004: 1261).

29 Andersen and Christensen (2004: 1266).

30 Andersen and Christensen (2004: 1266, 1267).

31 Andersen and Christensen (2004: 1268).

32 Andersen and Christensen (2004: 1269).

33 Andersen and Christensen (2004: 1270).

34 This section draws, in part, on Gereffi, et al. (2005), Sturgeon (2002; 2003).

35 Gereffi et al. (2005: 87).

36 Donaghu and Barff (1990: 539).

37 Donaghu and Barff (1990: 544).

38 Sturgeon (2003: 482).

39 Saxenian (2002: 186).

40 See, for example, Amin and Robins (1990), Amin and Thrift (1992), Lovering (1990), Scott (1988), Storper (1995; 1997), Sturgeon (2003).

41 Miles et al. (1999).

42 Berger (2005: 61).

43 This section draws heavily on Sturgeon (2002; 2003). See also Berger (2005).

44 Sturgeon (2002: 466–7).

45 Sturgeon (2002: 467).

46 Sturgeon (2002; 2003), Lüthje (2002).

47 Sturgeon (2002: 461–2).

48 Sturgeon (2002: 486).

49 See Anderson (1995), Gomes-Casseres (1996), Kang and Sakai (2000), Mockler (2000).

50 Gomes-Casseres (1996: 2).

51 Kang and Sakai (2000: 7).

52 See Hudson (2001: 206), Dunning (1993: 250).

53 Kang and Sakai (2000: 20).

54 Dicken (2005), Dunning (2000b), Elango (2004), Kozul-Wright and Rowthorn (1998), Morrison and Roth (1992), Muller (2004), Rugman and Brain (2003).

55 Muller (2004: ix, 219, original emphasis).

56 Morrison and Roth (1992: 45, 46).

57 Morrison and Roth (1992: 46–7).

58 Eden and Monteils (2000), Holmes (2000).

59 Amin (2000: 675).

60 *Financial Times* (8 November 2001).

61 Abo (2000), Borrus et al. (2000), Coe (2003a), Dicken and Yeung (1999), Yeung (2001), Yeung et al. (2001).

62 Bernard and Ravenhill (1995).

63 *Financial Times* (10 June 2005).

Six
'The State is Dead … Long Live the State'

'Contested territory': the state in a globalizing economy

Almost 40 years ago, the eminent economist Charles Kindleberger bluntly asserted that 'the nation state is just about through as an economic unit'.[1] Today, as we saw in Chapter 1, a central claim of the 'hyper-globalizers' is that we live in a borderless world where states no longer matter. A combination of the revolutionary technologies of transportation and communications and the increasing power of TNCs has, it is argued, shifted economic power out of the control of nation-states. This is a highly misleading view. While recognizing that the position of the state is being redefined, I emphatically reject the view that it is no longer a major player.

While some of the state's capabilities are being reduced, and while there may well be a process of 'hollowing out'[2] of the state, the process is not a simple one of uniform decline on all fronts.[3]

> While smaller or less influential states might have experienced a decline in power … other states (such as the USA) experience no diminution, and perhaps even an enhancement of geopolitical and geoeconomic power … *Much of the 'end of the state' or 'reasserting the state' literature focuses on western notions of statehood and experiences* … Implicit is a common experience of the emergence of the state in the nineteenth century and its zenith in the postwar Fordist regime of accumulation … *In many parts of the world, however, experiences of statehood have followed a quite different trajectory* and are, in a postcolonial context, *still being actively constructed, strengthened and extended rather than weakened.*[4]

The state, therefore, remains a most significant force in shaping the world economy. It has played, and continues to play, a fundamental role in the economic development of *all* countries. Every government, whatever its political complexion, intervenes to varying degrees in the operation of the market. I agree with Porter that 'while globalization of competition might appear to make the nation less important, instead it seems to make it more so'[5] and with Wade that 'reports of the death of the national economy are greatly exaggerated'.[6]

In fact, the more powerful states can actually *use* globalization as a means of increasing their power:

> States actively construct globalization and use it as soft geo-politics and to acquire *greater* power over, and autonomy from, their national economies and societies respectively … [for example] The US and the G-7's other dominant members design and establish the international trade agreements, organizations, and legislation that support and govern the trans-border investments, production networks, and market-penetration constitutive of contemporary economic globalization. Advanced capitalist states, particularly, use these political instruments to shape international economic decision-making and policy in their interests.[7]

We need to be clear about what we mean by the terms 'state', 'nation' and 'nation-state':[8]

- A *state* is a portion of geographical space within which the resident population is organized (i.e. governed) by an authority structure. States have externally recognized sovereignty over their territory.
- A *nation* is a 'reasonably large group of people with a common culture, sharing one or more cultural traits, such as religion, language, political institutions, values, and historical experience. They tend to identify with one another, feel closer to one another than to outsiders, and to believe that they belong together. They are clearly distinguishable from others who do not share their culture.' A nation is an *imagined community*. Note that whereas a state has a recognized and defined territory, a nation may not.
- A *nation-state* is the condition where 'state' and 'nation' are coterminous. 'A nation-state is a nation with a state wrapped around it. That is, it is a nation with its own state, a state in which there is no significant group that is not part of the nation.'

Although it is often regarded as a natural institution (for all of us it has always been there), the nation-state is actually a relatively recent phenomenon. It emerged from the particular configuration of power relationships in Europe following the Treaty of Westphalia in 1648. Since then, the map of nation-states has been redrawn continuously, sometimes peacefully and incrementally, often violently through revolution. During the second half of the twentieth century, two particular events had a profound effect on the map of nation-states. First, the waves of decolonization that swept through Africa and Asia in the 1960s created a whole new set of nation-states. Second, the collapse of the former Soviet Union, after 1989, resulted in the creation not only of a new Russian Federation but also of a number of newly independent states throughout Eastern Europe, including the fragmentation of the former Yugoslavia. As a result, the number of nation-states has grown dramatically to around 190 at the present time (Figure 6.1).

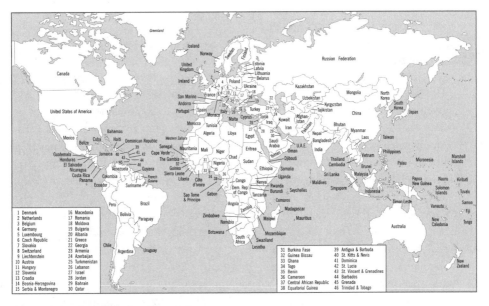

Figure 6.1   A world of nation-states

But that isn't all. An important feature of the contemporary world is the tension that exists between the triad of nation, state and nationalism. Increasingly, it seems, there are more and more 'nations without states', manifested in separatist movements engaged in conflict with the state in which they are (wrongly in their view) embedded (obvious examples include the Basques in Spain and France, the indigenous groups in Chiapas, Mexico, the East Timoreans in Indonesia, the Palestinians in Israel).

In this chapter and in Chapter 7, we explore the roles of the nation-state in the contemporary global economy. In this chapter the focus is set at a deliberately general level on four key roles of nation-states:

- as *containers* of distinctive institutions and practices
- as *regulators* of economic activities and transactions
- as *competitors* with other states
- as *collaborators* with other states.

States as *containers* of distinctive cultures, practices and institutions

The cultural context

All economic activity is *embedded* in broader cultural structures and practices.[9] However, 'culture' is an extremely slippery concept to define. Here it is taken to be a learned, shared, compelling, interrelated set of symbols whose meanings provide a set of orientations for members of a society. These orientations, taken together, provide solutions to problems that all societies must solve if they are to remain viable.[10]

The nation-state is one of the primary *containers* of such cultural structures and practices – of distinctive 'ways of doing things'.[11] The term 'container' should not be taken too literally. It is used here as a fairly loose metaphor to capture the idea that nation-states are *one* of the major ways in which distinctive institutions and practices are 'bundled together'. Of course, such containers are not (except in very rare cases) hermetically sealed off from the outside world. The container is permeable or leaky to varying degrees. Most obviously, a major impact of modern communications systems, especially the Internet, is to make national containers even more permeable. But that does not mean that the container no longer exists at all. Indeed, there is a good deal of compelling evidence to show the persistence of national distinctiveness – although not necessarily uniqueness – in structures and practices which help to shape local, national and global patterns of economic activity.

From an economic perspective, there are relatively few comprehensive and robust analyses of how cultures vary between countries. A classic study, still relevant today, is Hofstede's massive survey of more than 100,000 workers employed by the American company IBM in 50 different countries.[12] The great strength of Hofstede's study is that, by focusing on a controlled population within a common organizational environment, he was able to isolate *nationality* as a variable. Hofstede identified four distinct cultural dimensions:

- *Individualism versus collectivism*: societies vary between those in which people, in general, are motivated to look after their own individual interests – where ties between individuals are very loose – and those in which ties are very close and the collectivity (family, community etc.) is the important consideration.
- *Large or small power distance*: societies vary in how they deal with inequalities (for example, in power and wealth) between people. This is reflected in the extent to which authority is centralized and in the degree of autocratic leadership within society.
- *Strong or weak uncertainty avoidance*: In some societies, the inherent uncertainty of the future is accepted; each day is taken as it comes, that is, the level of uncertainty avoidance is weak. In other societies, there is a strong drive to try to 'beat the future'. Efforts (and institutions) are made to try to create security and to avoid risk. These are strong uncertainty avoidance societies.
- *Masculinity versus femininity*: societies can be classified according to how sharply the social division between male and females is drawn. Societies with a strong emphasis on traditional masculinity allocate the more assertive and dominant roles to men. They differ substantially from societies where the social sex role division is small and where such values are less evident.

Hofstede went on to show how different countries could be characterized in terms of their positions on varying combinations of these four dimensions. Figure 6.2 summarizes the results in terms of eight country clusters. Although it is always rather dangerous to classify phenomena into statistical boxes, most of the categories identified by Hofstede seem intuitively reasonable. Most of us would be able to recognize our own national contexts, whilst also realizing the danger of using simple stereotypes without due care and without being sensitive to changing attitudes and circumstances.

Group	1 Anglo	2 Germanic	3 Nordic	4 More developed Asian
Characteristics	Low power distance Low to medium uncertainty avoidance High individualism High masculinity	Low power distance High uncertainty avoidance Medium individualism High masculinity	Low power distance Low to medium uncertainty avoidance Medium individualism Low masculinity	Medium power distance High uncertainty avoidance Medium individualism High masculinity
Countries	Australia Ireland Britain New Zealand Canada USA	Austria Italy Germany South Africa Israel Switzerland	Denmark Norway Finland Sweden Netherlands	Japan
Group	**5 Less developed Asian**	**6 Near Eastern**	**7 More developed Latin**	**8 Less developed Latin**
Characteristics	High power distance Low uncertainty avoidance Low individualism Medium masculinity	High power distance High uncertainty avoidance Low individualism Medium masculinity	High power distance High uncertainty avoidance High individualism Medium masculinity	High power distance High uncertainty avoidance Low individualism Whole range on masculinity
Countries	India Singapore Pakistan Taiwan Philippines Thailand	Greece Iran Turkey	Argentina France Belgium Spain Brazil	Chile Peru Colombia Portugal Mexico Venezuela

Figure 6.2 National variations in cultural characteristics

Source: based on Hofstede, 1980: 336

Varieties of capitalism

Over time, and under specific historical circumstances, societies have developed distinctive ways of organizing their economies, even within the apparently universal ideology of capitalism. In fact, capitalism comes in many different varieties,[13] of which three are especially significant:

- *neo-liberal market capitalism*: exemplified by the United States and, to a lesser extent, the United Kingdom
- *social market capitalism*: exemplified by Germany, Scandinavia and many other European countries
- *developmental capitalism*: exemplified by Japan, South Korea, Taiwan, Singapore and most other East Asian countries.

Figure 6.3 summarizes the major characteristics of each of these three major varieties of capitalism.

Characteristics	United States	Germany	Japan
	Neo-liberal market capitalism	**Social-market capitalism**	**Development capitalism**
Dominant ideology	Free-enterprise liberalism	Social partnership	Technonationalism
Political institutions	Liberal democracy Divided government Interest-group liberalism	Social democracy Weak bureaucracy Corporatist legacy	Developmental democracy Strong bureaucracy Reciprocity between state and firms
Economic institutions	Decentralized, open markets Unconcentrated, fluid capital markets Antitrust tradition	Organized markets Tiers of firms Dedicated, bank-centred capital markets Certain cartelized markets	Guided, closed, bifurcated markets Bank-centred capital markets Tight business networks Cartels in sunset industries

Figure 6.3   Three major varieties of capitalism: the US, the German and the Japanese models

Source: based on Doremus et al., 1998: Table 2.1

The essence of these governance models is their differing conception of the 'proper' role of government in regulating the economy.

- In *neo-liberal market capitalism*, market mechanisms are used to regulate all, or most, aspects of the economy; individualism is a dominant characteristic; short-term business goals tend to predominate; and the state does not overtly attempt to plan the economy strategically. The dominant philosophy is 'shareholder value' – facilitating maximum returns to the owners of capital.
- In *social market capitalism*, in contrast, a higher premium is placed upon collaboration between different actors in the economy, with a broader identification of 'stakeholders' beyond that of owners of capital.
- In *developmental capitalism*, the state plays a much more central role (although not usually in terms of public ownership of productive assets). The state sets substantive social and economic goals within an explicit industrial strategy.

To these three varieties of capitalism we should now add a fourth:

- The *communist-capitalist* system of China. Here, uniquely, a highly centralized political system is combined with an increasingly open capitalist market system.

Although change undoubtedly occurs in all social systems, unless they are completely insulated from the outside world, such distinctive forms of capitalism tend to persist over time:

> There are inherent obstacles to convergence among social systems of production of different societies, for where a system is at any one point in time is influenced by its initial state ... Existing institutional arrangements block certain institutional innovations and facilitate others ... There are critical turning points in the history of highly industrialized societies, but the choices are

limited by the existing institutional terrain. Being path dependent, a social system of production continues along a particular logic until or unless a fundamental societal crisis intervenes.[14]

States as *regulators* of trade, foreign investment and industry

Recognizing that countries continue to differ as 'containers' of distinctive structures and practices is important in emphasizing that we do not live in a homogenized world. In this section, we focus specifically on some of the ways states *regulate* how their economies operate as they attempt to control what happens within, and across, their boundaries. Of course, states do not merely 'intervene' in markets: 'The institutions of the state apparatus are not simply involved in regulating economy and society, for *state activity is necessarily involved in constituting economy and society* and the ways in which they are structured and territorially organized.'[15] Although a high level of contingency may well be involved (no two states behave in exactly the same way), certain regularities in basic policy stance can be identified. These will reflect the kinds of cultural, social and political structures, institutions and practices in which the state is embedded. The precise policy mix adopted by a state will be influenced by:

- its political and cultural complexion and the strength of institutions and interest groups
- the size of the national economy, especially that of the domestic market
- the nation's resource endowment, both physical and human
- the nation's relative position in the world economy, including its level of economic development and degree of industrialization.

At one extreme, the *macroeconomic* policies pursued by governments to control domestic demand or to manage the money supply have extremely important implications for the distribution and redistribution of economic activity. Two basic types of macroeconomic policy tend to be used by the state to manage its national economy:

- *Fiscal policies* to raise or lower taxes on companies and/or individual citizens and to determine appropriate levels and recipients of government expenditure. Raising taxes dampens down domestic demand; lowering taxes stimulates demand. (Although, as the Japanese experience during the 1990s showed, such automatic responses to changes in fiscal policy do not always occur.) Similarly, raising or lowering public expenditure – or targeting specific types of expenditure – can influence the level of economic activity in the economy.
- *Monetary policies* aimed at influencing the size of the money supply within the country and at either speeding up, or slowing down, its rate of circulation (its

velocity). The main mechanism employed is manipulation of the interest rate on borrowing. Lowering interest rates should stimulate economic activity through increased investment or private expenditure while, conversely, raising interest rates should dampen down activity. Again, however, rapid and automatic adjustment does not always occur. In the international context, exchange rates also have to be taken into account because their level and volatility affect the costs of exports and imports.

At a more tangible and material level, governments generally provide – or at least secure the provision of – those 'conditions of production that are not and cannot be obtained through the laws of the market'.[16] One example is the *physical infrastructure* of national economies – roads, railways, airports, seaports, telecommunications systems – without which private sector enterprises, whether domestic or transnational, could not operate. They are the providers, too, of the *human infrastructure*: in particular of an educated labour force as well as of sets of laws and regulations within which enterprises must operate. Between these two modes of government involvement in the workings of the economy – macroecnomic policies on the one hand, provisions of physical and human infrastructures on the other – lie those policies whose explicit purpose is to influence the level, composition and distribution of production and trade.

National governments possess an extensive kit of regulatory tools with which to control and to stimulate economic activity and investment within their own boundaries and to shape the composition and flow of trade and investment at the international scale. They may be employed as part of a deliberate, cohesive, all-embracing national economic strategy as in the *developmental capitalist* state or, alternatively, individual policy measures may be implemented in an *ad hoc* fashion with little attempt at coordination as in the *market-oriented capitalist* state model.

Trade policies

Of all the measures used by nation-states to regulate their international economic position, policies towards trade have the longest history. The shape of the emerging world economy of the seventeenth and eighteenth centuries was greatly influenced by the mercantilist policies of the leading European nations. Trade policy is unique in that, since the late 1940s, it has been set within an *international* institutional framework. We will discuss this international regulatory framework in some detail in Chapter 19. Here, we just need to note that national policies have to operate within an international system.

Figure 6.4 summarizes the major types of trade policy pursued by national governments. In general, policies towards imports are restrictive whereas policies towards exports, with one or two exceptions, are stimulatory.

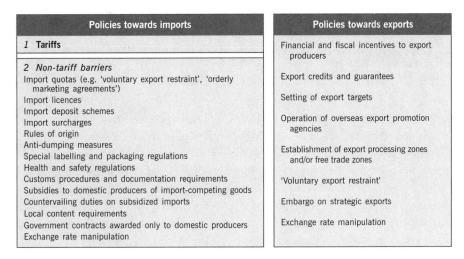

Policies towards imports	Policies towards exports
1 **Tariffs**	Financial and fiscal incentives to export producers
2 *Non-tariff barriers* Import quotas (e.g. 'voluntary export restraint', 'orderly marketing agreements') Import licences Import deposit schemes Import surcharges Rules of origin Anti-dumping measures Special labelling and packaging regulations Health and safety regulations Customs procedures and documentation requirements Subsidies to domestic producers of import-competing goods Countervailing duties on subsidized imports Local content requirements Government contracts awarded only to domestic producers Exchange rate manipulation	Export credits and guarantees Setting of export targets Operation of overseas export promotion agencies Establishment of export processing zones and/or free trade zones 'Voluntary export restraint' Embargo on strategic exports Exchange rate manipulation

Figure 6.4   Major types of trade policy

Policies on *imports* fall into two distinct categories:

- *Tariffs* are taxes levied on the value of imports that increase the price to the domestic consumer and make imported goods less competitive (in price terms) than otherwise they would be. In general, the tariff level tends to rise with the stage of processing, being lowest on basic raw materials and highest on finished goods. The purpose of such 'tariff escalation' is to protect domestic manufacturing industry whilst allowing for the import of industrial raw materials. Thus, although tariffs may be regarded simply as one means of raising revenue, their major use has been to *protect* domestic industries: either 'infant' industries in their early delicate stages of development or 'geriatric' industries struggling to survive in the face of external competition.
- *Non-tariff barriers (NTBs)*. While tariffs are based on the value of imported products, non-tariff barriers are more varied: some are quantitative, some are technical. Although, in general, tariffs have continued to decline, the period since the mid 1970s witnessed a marked increase in the use of non-tariff barriers. Indeed, today NTBs are probably more important than tariffs in influencing the level and composition of trade between nation-states. It has been estimated that NTBs affect more than a quarter of all industrialized country imports and are even more extensively used by developing countries. Certainly much of what has been termed the 'new protectionism' consists of the increased use of NTBs.

Foreign direct investment policies

In a world of transnational corporations and of complex flows of investment at the international scale, national governments have a clear vested interest in the effects of FDI, whether positive or negative. From a national viewpoint, such investment is of two types: *outward* investment by domestic enterprises and *inward* investment by foreign enterprises. Few national governments operate a totally closed policy towards FDI, although the degree of openness varies considerably.

Figure 6.5 summarizes the major types of national policy towards foreign direct investment.[17] Most national policies are concerned with inward investment although governments may well place restrictions on the export of capital for investment (for example, through the operation of exchange control regulations) or insist that proposed overseas investments be approved before they can take place. Historically, there have been very large differences in the policy positions adopted by countries towards inward direct investment. At the broadest level, developed countries tended to adopt a more liberal attitude towards inward investment than developing countries, although there were exceptions within each broad group. For example, among developed countries France had a much more restrictive stance than most other European countries. Among developing countries, Singapore had a particularly open policy, far more so than most other Asian countries. In the past two decades, however, national FDI policies have tended to converge in the direction of liberalization.

Although national differences still exist, therefore, they are now rather less stark than in the past. Figure 6.6 summarizes the major regulatory changes towards FDI between 1991 and 2004. The proportion of regulatory changes that are more favourable to FDI continues to far outweigh those that are unfavourable.

Policies relating to inward investment by foreign firms
Entry Government screening of investment proposals Exclusion of foreign firms from certain sectors or restriction on the extent of foreign involvement permitted Restriction on the degree of foreign ownership of domestic enterprises Compliance with national codes of business conduct (including information disclosure)
Operations Insistence on involvement of local personnel in managerial positions Insistence on a certain level of local content in the firm's activities Insistence on a minimum level of exports Requirements relating to the transfer of technology Locational restrictions on foreign investment
Finance Restrictions on the remittance of profits and/or capital abroad Level and methods of taxing profits of foreign firms
Incentives Direct encouragement of foreign investment: competitive bidding via overseas promotional agencies and investment incentives

Policies relating to outward investment by domestic firms
Restrictions on the export of capital (e.g. exchange control regulations) Necessity for government approval of overseas investment projects

Figure 6.5   Major types of FDI policy

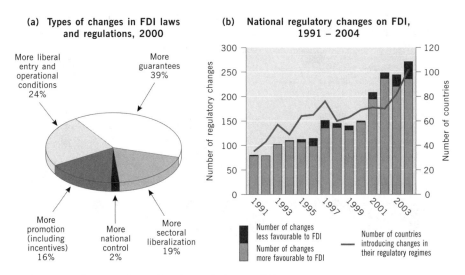

Figure 6.6   Changes in national regulation of FDI

Source: based on UNCTAD *World Investment Report*, various issues

Industry policies

National policies towards trade or foreign direct investment are explicitly concerned with international or cross-border issues and, therefore, are most obviously relevant to our interest in the globalization of economic activity. But there is a third policy area – industry policy – that, although essentially concerned with internal issues, also has broader international implications. Indeed, it is becoming increasingly apparent that the boundaries between trade, foreign direct investment and industry policies are extremely blurred. Figure 6.7 lists the major types of regulatory industry policies that may be used by national governments.

As Figure 6.7 suggests, the various stimulatory and regulatory policies may be applied *generally* across the whole of a nation's industries or they may be applied *selectively*. Such selectivity may take a number of forms: particular sectors of industry, particular types of firms (including, for example, the efforts to attract foreign firms), particular geographical locations. For example, in most countries there has been great interest in trying to encourage the development of *growth clusters*: an attempt to capture the virtuous circle of growth that has come to be associated with the kinds of clusters described in Chapter 3. In addition, as we will see in Chapter 17, states have become increasingly involved in labour market policies, especially in terms of attempting to make labour markets more flexible.

The range of potential industry policies	
Investment incentives: 　- Capital related 　- Tax related	Merger and competition policies
	Company legislation
Labour market policies: 　- Subsidies 　- Training	Taxation policies
	Labour market regulation: 　- Labour union legislation 　- Immigration policies
State procurement policies	
Technology policies	National technical and product standards
Small-firm policies	State ownership of production assets
Policies to encourage industrial restructuring	Environmental regulations
Policies to promote investment	Health and safety regulations

Some or all of these policies may be applied either generally or, more commonly, selectively. Selectivity may be based on several criteria:

1　*Particular sectors of industry, e.g.*
　(a) to bolster declining industries
　(b) to stimulate new industries
　(c) to preserve key strategic industries

2　*Particular types of firm, e.g.*
　(a) to encourage entrepreneurship and new firm formation
　(b) to attract foreign firms
　(c) to help domestic firms against foreign competition
　(d) to encourage firms in import-substituting or export activities

3　*Particular geographical areas, e.g.*
　(a) economically depressed areas
　(b) areas of 'growth potential' ('cluster' policies)

Figure 6.7　Major types of industry policy

States as *competitors*

Do states compete? Is it correct to think of nations as being in competition with each other, just as firms compete with other firms? Paul Krugman argues that the very idea of 'competitive' states is a 'dangerous obsession'.[18] However, the generally accepted view among both policy makers and academics is very different:

> The transformation of the nation-state into a 'competition state' lies at the heart of political globalization.[19]

Engaging in economic competition is another manifestation of the exercise in 'soft power' by states.[20] Books, government reports, newspaper articles, television programmes in virtually all countries resound with the language and imagery of the competitive struggle between states for a bigger slice of the global economic pie.

Prior to the East Asian crisis of the late 1990s, much of the concern focused on the perceived loss of economic standing by the United States and European countries *vis-à-vis* Japan and the East and South East Asian NIEs. Today, although the focus may be a little different – the big threat is now seen to be China – the rhetoric remains the same. Indeed, the Swiss business school IMD publishes an annual *World Competitiveness Yearbook* with a 'competitiveness scoreboard' (or 'league table') of 49 countries based on no fewer than 286 individual criteria!

States compete to enhance their international trading position in order to capture as large a share as possible of the gains from trade. They compete to attract productive investment to build up their national production base which, in turn, enhances their international competitive position. Indeed, one of the most graphic expressions of competition between states is their intense involvement in what have been called 'locational tournaments': the attempts to entice investment projects into their own national territories. There has been an enormous escalation in the extent of *competitive bidding* between states (and between local communities within the same state) to attract the relatively limited amount of geographically mobile investment (see Chapter 8).

Michael Porter argues that national competitive advantages are created through highly localized processes internal to the country. Porter conceives of this set of processes as a 'diamond': an interconnected system of four major determinants (Figure 6.8). Connecting these four components by double-headed arrows emphasizes the fact that the 'diamond' is a mutually reinforcing system.

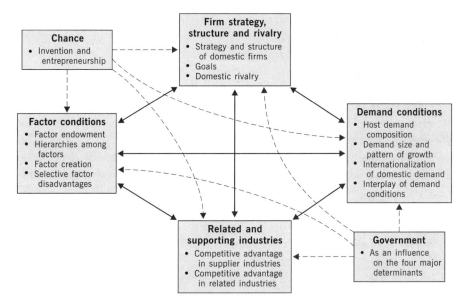

Figure 6.8 National competitive advantage: the Porter 'diamond'

Source: based on material in Porter, 1990: chapter 3

Each of the four determinants can be broken down into several subcomponents:

- *Factor conditions* in a particular country are a combination of 'given' and 'created' factors. 'The factors most important to competitive advantage in most industries, especially the industries most vital to productivity growth in advanced economies, are not inherited but are *created within a nation, through processes that differ widely across nations and among industries … Thus, nations will be competitive where they possess unusually high quality institutional mechanisms for specialized factor creation'.*[21] The most important include the levels of skills and knowledge of the country's population and the provision of sophisticated physical infrastructure, including transport and communications.

- *Demand conditions*, especially those in the home market, are particularly important. 'The home market usually has a disproportionate impact on a firm's ability to perceive and interpret buyer needs … proximity to the right type of buyers … [is] of decisive importance in national competitive advantage'.[22] However, the extent to which a nation's firms are connected into international markets also increases national competitiveness.

- *Related and supporting industries* that are internationally competitive constitute a third major determinant of national competitive advantage. 'Perhaps the most important benefit of home-based suppliers … is in the *process of innovation and upgrading*. Competitive advantage emerges from close working relationships between world-class suppliers and the industry'. [23]

- *Firm strategy, structure and rivalry*: 'The way in which firms are managed and choose to compete is affected by national circumstances … Some of the most important are attitudes toward authority, norms of interpersonal interaction, attitudes of workers toward management and vice versa, social norms of individualistic or group behaviour, and professional standards. These, in turn, grow out of the education system, social and religious history, family structures, and many other often intangible, but unique, national conditions'.[24] Porter lays great emphasis on the importance of intense rivalry between domestic firms, arguing that this creates strong pressures on firms to innovate in both products and processes, to become more efficient and to become high-quality suppliers of goods and services. 'Vigorous local competition not only sharpens advantages at home but pressures domestic firms to sell abroad in order to grow … Toughened by domestic rivalry, the stronger domestic firms are equipped to succeed abroad. It is rare that a company can meet tough foreign rivals when it has faced no significant competition at home'.[25]

In addition to the four primary competitive determinants that, in combination, form his 'diamond', Porter attributes secondary importance to two other components (see Figure 6.8):

- *The role of chance*: for example, the occasionally random occurrence of innovations or the 'historical accidents' that may create new entrepreneurs are seen to be important.

- *The role of government*: Porter explicitly refuses to regard government as a competitive determinant of the same order as the four primary determinants of his 'diamond'. While describing the various policies that governments might implement, he sees government as merely an 'influence' on his four determinants, a contingent rather than a central factor – a contributor to the environment in which, as in the biological realm, the selective survival of species occurs.

Porter's analysis of the competitive advantage of nations has attracted much attention. On the one hand its 'recipes' have been adopted by many governments, both national and local, in their attempts to improve their competitive position. On the other hand, it has attracted considerable criticism:

- It is highly reductionist in compressing immense complexity into a simple four-pointed 'diamond'. Hence, as a policy prescription, it needs to 'carry a public policy health warning'.[26]
- Its underplaying of the role of the state in pursuit of national competitiveness is a significant omission. *All* states perform a key role in the ways in which their economies operate, although they differ substantially in the specific measures they employ and in the precise ways in which such measures are combined.
- It neglects the influence of the transnationalization of business activity on national 'diamonds': 'there is ample evidence to suggest that the technological and organizational assets of TNCs may be influenced by the configuration of the diamonds of the foreign countries in which they produce and that this, in turn, may impinge upon the competitiveness of the resources and capabilities in their home countries'.[27]

States as *collaborators*: the proliferation of regional integration agreements

While there is controversy over whether states do, or should, see themselves as competitive states, there is no doubt that states *collaborate* with other states to achieve specific economic and welfare goals.[28] Such collaborations can take many forms. Here we focus on one dimension: the tendency for states to develop political-economic relationships at the *regional* scale through regional integration agreements (RIAs). Indeed, regionalism has become one of the dominant features of the contemporary global economy.

The acceleration of regional integration agreements
The basis of regional integration agreements is the *preferential trading arrangement* (PTA). PTAs simply involve states agreeing to provide preferential access to their markets to other members of the regional group – primarily through tariff reductions, at least initially. Thus, preferential trading arrangements have a two-sided

quality: they liberalize trade between members whilst, at the same time, discriminating against third parties.[29] There has been an especially marked acceleration in PTA formation since the early 1990s, as Figure 6.9 shows. Around 176 of the 300 PTAs notified to the WTO up to the end of 2004 occurred after 1995. Of these, 150 were in force and a further 70 were believed to be operational but not yet notified.[30] At least one-third of total world trade occurs within PTAs.

Such acceleration in trade-based RIAs reflects a number of motivations by national governments. Most have a strongly defensive character; they represent an attempt to gain advantages of size in trade by creating large markets for their producers and protecting them, at least in part, from outside competition. There is also an undoubted 'bandwagon' effect: a 'fear of being left out while the rest of the world swept into regionalism, either because this would be actually harmful to excluded countries or just because "if everyone else is doing it, shouldn't we"'.[31]

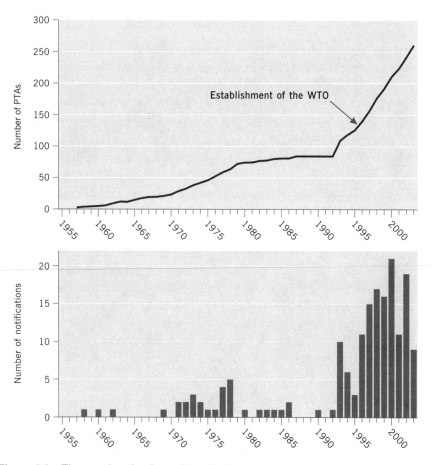

Figure 6.9   The acceleration in preferential trading arrangements

Source: WTO data

The classic analysis of the trade effects of regional integration agreements identifies two opposing outcomes:

- *Trade diversion* occurs where, as the result of regional bloc formation, trade with a former trading partner (now outside the bloc) is replaced by trade with a partner inside the bloc.
- *Trade creation* occurs where, as the result of regional bloc formation, trade replaces home production or there is increased trade associated with economic growth in the bloc.

In addition, regional trading blocs have a major influence on *flows of investment* by transnational corporations. The effects of regional integration on direct investment, like those on trade, can also be conceptualized in terms of 'creation' and 'diversion'. In the latter case, the removal of internal trade (and other) barriers may lead firms to realign their organizational structures and value-adding activities to reflect a regional rather than a strictly national market (see Figure 5.5). This, by definition, 'diverts' investment from some locations in favour of others.

Despite a widespread view that regional blocs are a relatively new phenomenon they have, in fact, been an important feature of the global economic landscape since the middle of the nineteenth century. But their basis and their nature have changed over time. Four 'waves of regionalism' can be identified:[32]

- During the second half of the nineteenth century there were a number of trade agreements in place, especially in Europe: for example, the German *Zollverein*, the customs unions between the Austrian states, and those between several of the Nordic countries. 'As of the first decade of the twentieth century, Great Britain had concluded bilateral arrangements with forty-six states, Germany had done so with thirty countries, and France had done so with more than twenty states'.[33]
- After the disruptive effects of World War I (1914–18) a new wave of regional arrangements occurred, but this time in a more discriminatory form. 'Some were created to consolidate the empires of major powers, including the customs union France formed with members of its empire in 1928 and the Commonwealth system of preferences established by Great Britain in 1932. Most, however, were formed among sovereign states ... The Rome Agreement of 1934 led to the establishment of a PTA involving Italy, Austria and Hungary. Belgium, Denmark, Finland, Luxembourg, the Netherlands, Norway and Sweden concluded a series of economic agreements throughout the 1930s ... Outside of Europe, the United States forged almost two dozen bilateral commercial agreements during the mid-1930s, many of which involved Latin American countries'.[34]
- Since the end of World War II (1939–45) there have been two distinct waves of regionalism. 'The first took place from the late 1950s through the 1970s and was marked by the establishment of the EEC, EFTA, the CMEA and a plethora of regional trade blocs formed by developing countries. These

arrangements were initiated against the backdrop of the Cold War, the rash of decolonization following World War II, and a multilateral commercial framework, all of which colored their economic and political effects'.[35]

- The fourth wave of economic regionalism – from the late 1980s onwards – occurred in the drastically changed geopolitical circumstances of the collapse of the Soviet-led system and the increased uncertainties of a more fragmented political and economic situation. 'Furthermore, the leading actor in the international system (the United States) is actively promoting and participating in the process. PTAs also have been used with increasing regularity to help prompt and consolidate economic and political reforms in prospective members, a rarity during prior eras. And unlike the interwar period, the most recent wave of regionalism has been accompanied by high levels of economic interdependence, a willingness by the major economic actors to mediate trade disputes, and a multilateral (that is, the GATT/WTO) framework'.[36]

Types of regional economic integration

All of the regional collaborative arrangements that have been established over the years have been based on the principle of preferential trading arrangements. However, there are, in fact, several different types of politically negotiated regional integration agreements, involving different degrees of economic and political integration. The following four types of regional arrangement are especially important, and they are listed in order of increasing economic and political integration:

- A *free trade area*, in which trade restrictions between member states are removed by agreement but where member states retain their individual trade policies towards non-members.
- A *customs union*, in which member states operate a free trade arrangement with each other and also establish a common external trade policy (tariffs and non-tariff barriers) towards non-members.
- A *common market*, in which not only are trade barriers between member states removed and a common external trade policy adopted but also the free movement of factors of production (capital, labour etc.) between member states is permitted.
- An *economic union*, which involves the highest form of regional economic integration short of full-scale political union. In an economic union, not only are internal trade barriers removed, a common external tariff operated and free factor movements permitted, but also broader economic policies are harmonized and subject to supranational control.

As Figure 6.10 shows, the progression is cumulative: each successive stage of integration incorporates elements of the previous stage, together with the additional element that defines each particular stage.

Levels of economic integration	Free trade area	Customs union	Common market	Economic union
Removal of trade restrictions between member states	✓	✓	✓	✓
Common external trade policy towards non-members		✓	✓	✓
Free movement of factors of production between member states			✓	✓
Harmonization of economic policies under supranational control				✓

Figure 6.10   Types of regional economic integration

The vast majority of the regional integration agreements fall into the first two categories shown in Figure 6.10: the free trade area and the customs union. Indeed, around 90 per cent of regional integration agreements are free trade areas. There is a small number of common market arrangements, but only one group – the European Union – comes close to being a true economic union. In fact, not only is there enormous variation in the scale, nature and effectiveness of these regional integration agreements but also there is, in some cases, a considerable overlap of membership of different groups. Table 6.1 shows the major regional integration agreements currently in force, while Figure 6.11 maps the intricate webs of preferential trading relationships that involve, but also go beyond, the major regional blocs.

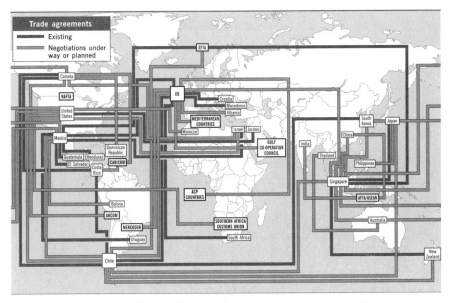

Figure 6.11   The tangled web of major preferential trade agreements

Source: based on WTO data; *Financial Times*, 19 November 2003

Table 6.1 Major regional integration agreements

Regional group	Membership	Date(s)	Type
EU (European Union)	Austria, Belgium, Cyprus, Czech Republic, Denmark, Estonia, France, Finland, Germany, Greece, Hungary, Ireland, Italy, Latvia, Lithuania, Luxembourg, Malta, Netherlands, Poland, Portugal, Slovakia, Slovenia, Spain, Sweden, United Kingdom	1957 (European Common Market) 1992 (European Union)	Economic union
NAFTA (North American Free Trade Agreement)	Canada, Mexico, United States	1994	Free trade area
EFTA (European Free Trade Association)	Iceland, Norway, Liechtenstein, Switzerland	1960	Free trade area
Mercosur (Southern Cone Common Market)	Argentina, Brazil, Paraguay, Uruguay, Venezuela (2006)	1991	Common market
ANCOM (Andean Common Market)	Bolivia, Colombia, Ecuador, Peru, Venezuela	1969 (revived 1990)	Customs union
CARICOM (Caribbean Community)	Antigua & Barbuda, Bahamas, Barbados, Belize, Dominica, Grenada, Guyana, Haiti, Jamaica, Montserrat, St Kitts & Nevis, St Lucia, St Vincent & the Grenadines, Suriname, Trinidad & Tobago	1973	Common market
AFTA (ASEAN Free Trade Agreement)	Brunei Darussalam, Cambodia, Indonesia, Laos, Malaysia, Myanmar, Philippines, Singapore, Thailand, Vietnam	1967 (ASEAN) 1992 (AFTA)	Free trade area

Regional integration within Europe, the Americas, East Asia and the Pacific

In Chapter 2 we observed the strong tendency for a disproportionate share of global production, trade and investment to be concentrated in three 'mega-regions' – the so-called global triad of North America, Europe and East Asia. Such concentrations reflect, first and foremost, the basic economic-geographical processes of preference for proximity to markets and suppliers and a general tendency to 'followership' in location decision-making. But there are also rather different kinds of regional integration agreement in each of the three major regions. In this section we briefly explore these different manifestations of regional integration.

The European Union

The EU is by far the most highly developed and structurally complex of all the world's regional economic blocs. Although initially established as a six-member European *Economic* Community (EEC) in 1957, it was always more than simply an economic institution. Indeed, the initial stimulus was the desire to bring together France and Germany in such a way that traditional enmities could no longer find their outlet in another round of European wars, and also to strengthen Western Europe in the face of the perceived Soviet threat.

We can identify six major stages of development of the EU:

- *1958–1968.* Elimination of customs duties between the six founder member states (Belgium, France, West Germany, Italy, Luxembourg, Netherlands) and, in 1968, the introduction of a common external tariff.
- *1973–1986.* (a) Initial enlargement of the Community, with the accession of Denmark, Ireland and the United Kingdom in the 1970s and of Greece, Portugal, Spain in the 1980s. (b) Establishment of preferential trading agreements with EFTA countries; with countries around the Mediterranean rim; with countries in Africa, the Caribbean and the Pacific (the so-called ACP nations – all former European colonies).
- *1986–1992.* (a) A renewed attempt to complete the Single European Market. (b) A drive by the 'core' members of the EC to move towards full economic and monetary union. (c) The signing of the Treaty on European Union at Maastricht in December 1991 created the European Union.
- *1992–1999.* (a) Further enlargement of the EU to 15 with the accession of Austria, Finland and Sweden. (b) Attempts to implement fully both the Single European Market and the Maastricht Treaty in the context of both internal divisions and the pressures to enlarge the EU.
- *1999–2004.* (a) Introduction of the European single currency in 11 (later 12) of the 15 member states. European Central Bank established with responsibility for setting EU-wide interest rates. (b) Negotiations with Eastern European countries wishing to become members of the EU. (c) Inter-governmental conference at Nice in 2001 to reform the EU's voting system.

- *2004–present.* (a) Enlargement to incorporate 10 new member states (Cyprus, Czech Republic, Estonia, Hungary, Latvia, Lithuania, Malta, Poland, Slovakia, Slovenia). (b) Negotiations with other possible members (Bulgaria, Romania, Croatia, Turkey). (c) Failure to agree a new EU Constitution.

Figure 6.12 shows how the EU has grown geographically from its original six member states in 1957 to 25 in 2004, together with possible further enlargement in the next 10 years or so.

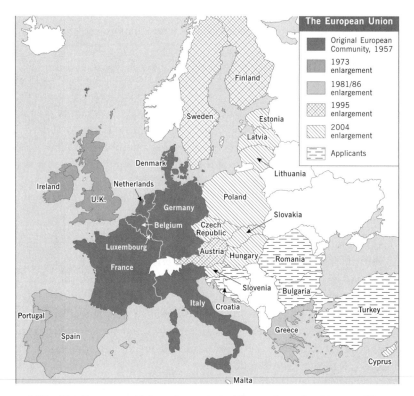

Figure 6.12   The European Union: from six to 25 members (and beyond?)

Since the early 1990s, three developments have been especially important for the EU. The first was the drive to complete the *Single European Market* in 1992. Almost 40 years after the Treaty of Rome, individual countries were still resorting to tactics which prevented, or delayed, the import of certain products from other member nations through the use of various kinds of non-tariff barrier. Each individual member state was guilty of some such practices, although some were more guilty than others. The Commission argued that the costs of 'non-Europe' amounted to a significant loss of potential GDP and of jobs, as well as a lessened ability to compete with the United States and Japan.

The Single European Act proposed the removal of three major sets of barriers – physical, technical and fiscal – together with the liberalization of financial services, the opening of public procurement and other measures. Such internal liberalization and deregulation, it was argued, would create a virtuous circle of growth for the European Community as a whole, its member states and those business firms successfully taking advantage of the changes. It is virtually impossible to measure precisely the actual effects of the Single Market process. However, a review of a range of evaluative studies concludes that

> There is no convincing empirical evidence of European integration having led to either short-term or sustained economic growth effects. The regulatory changes of the Single Market project were part of a global process of economic restructuring and mainly served to enhance the competitiveness of world-market oriented European countries.[37]

The second major event in the EU since the early 1990s was the *Treaty on European Union* (TEU), signed at Maastricht in 1991. This was significant because it marked a much more ambitious political agenda, aimed at creating a fully fledged *economic union*. In particular it

- strengthened social provisions by (a) incorporating the Social Charter; (b) the enlargement of the EC Structural Funds; (c) the creation of a new Cohesion Fund to assist poorer areas of the Union
- set out the mechanisms for the creation of a single European currency and monetary union (EMU).

European Monetary Union came into effect in 1999, when 11 of the 15 member states not only agreed to join the system but also met the technical criteria for doing so. The countries opting out of the EMU at that time were Denmark, Greece, Sweden and the United Kingdom (Greece subsequently joined in 2001). The issue of monetary union and the adoption of a single European currency (the *euro*) crystallize some of the most difficult political problems within the EU, notably the sensitive issue of national sovereignty. Within the EMU, national control over monetary policy – notably the setting of interest rates – has been passed upwards to the European Central Bank based in Frankfurt. The ECB, therefore, has an immense influence over the economies of individual member states. Each member state in the EMU has to comply with the Stability and Growth Pact, which sets limits on permissible budget deficits and debts.

The pros and cons of a single European currency are finely balanced. The major benefits are the reduced costs and uncertainties associated with having to deal with many separate currencies within the Single Market and the overall stability this is intended to produce. Set against this is the fact that an individual state's ability to use monetary mechanisms to deal with periodic economic crises is greatly reduced. It is still too early to judge the effect and the effectiveness of the EMU, although the Stability and Growth Pact restrictions were contravened by both Germany and France in 2003. In its first two years of existence, the euro was very weak, especially against the US dollar (and against the UK pound). The fact that three EU member states – Denmark, Sweden, the United Kingdom – remain outside the EMU is a major source of uncertainty. Denmark has voted in a referendum to remain outside whilst opinion in both Sweden and the United Kingdom is strongly divided. Continued existence of 'insiders' and 'outsiders' will inevitably pose problems for the EU.

The third major development has been the dramatic *enlargement of membership* to 25 states, with the likelihood of further enlargement. Of course, enlargement of the EU is nothing new (Figure 6.12). But now the circumstances are very different. The majority of the new members were, until recently, embedded within the Soviet-dominated system. They are the 'transitional market economies' of Eastern Europe, with very different recent histories and socio-political structures from the existing EU members. Others are smaller countries like Cyprus and Malta. Significantly, the income gap between existing and new members is much wider than in previous rounds of enlargement. The average GDP per head of the 10 new members in 2004 was only 46.5 per cent of the existing EU average. This compares with the average of 95.5 per cent for Denmark, Ireland and the UK when they joined in 1973, and the 103.6 per cent for Austria, Finland and Sweden on their accession in 1995.

Such huge income differences pose massive problems for the already problematical EU budget. However, perhaps the most contentious issue concerns the potential entry of Turkey. Agreement was finally reached in 2005 to begin membership talks, which will last for 10 years. However, the outcome is by no means certain: the terms are much more stringent than those adopted for other applicants because opinion within the EU is polarized between those states which wish to see the involvement of a Muslim country in the EU and those which see this as a threat to an essentially Christian entity.

Finally, the increasing extent of enlargement makes the process of decision-making even more difficult. Such increased complexity was the reason for the recent attempt to agree a new EU Constitution, a process which suffered a major setback in 2005 when referendums in both France and the Netherlands rejected it and other states (such as the UK) put the process on hold. Not surprisingly, therefore, many continue to hold the view that a 'variable geometry' or 'variable speed' European Union is likely to emerge, with a core of states, fully integrated economically and financially, surrounded by various groups of countries with different degrees of integration.

The Americas

Whereas the history of political-economic integration in Europe has been one of progressive deepening and widening – albeit with many interruptions and uncertainties now about the future structure of the EU – the history of attempts to create regional integration agreements in the Americas have been far more fragmented and shallow. To a great extent, this reflects the overwhelming dominance of the United States in the region; the fact that, until very recently, the US had chosen not to enter into bilateral or regional trading arrangements; and the limited success among Latin American countries in creating robust and lasting regional agreements. The picture in the Americas, therefore, is of a mosaic of regional agreements of different type and scope (Figure 6.13), together with the external agreements shown in Figure 6.11.

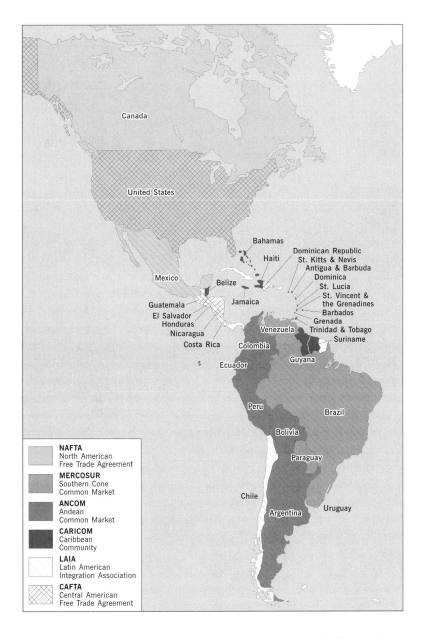

Figure 6.13   The mosaic of regional integration agreements in the Americas

By far the most important regional integration agreement in the Americas – at least in terms of its wider impact – is the *North American Free Trade Agreement* (NAFTA), between the United States, Canada and Mexico. By integrating two highly developed countries and one large developing country into a single free trade area it radically changed the economic map of North America. The income gap between the United States and Mexico is very much greater than that between the richest and poorest states within the EU. The NAFTA came into force in 1994, but its origins can be traced back into the 1980s.

One important building block, although this was not its intent, was the Canada–United States Free Trade Agreement (CUSFTA), signed in 1988 and implemented in 1989. As the CUSFTA was being signed, two other developments were also occurring. President George Bush (Senior) had made freer trade with Mexico a campaign issue in 1988. At the same time, President Carlos Salinas of Mexico made clear his determination to negotiate a free trade area with the United States. Within a short time of bilateral talks starting, Canada had joined in an obvious defensive response.

The arguments in favour of creating the NAFTA varied between the three parties. For the United States, it formed part of its long-term objective of ensuring stable economic and political development in the western hemisphere and also gave access to Mexican raw materials (especially oil), markets and low-cost labour. It also promised further US leverage in a world of increasing regional integration. The Canadian government was anxious to consolidate the recent CUSFTA. The motives of the Mexican government were primarily to help to lock in the economic reforms of the previous few years, to create a magnet for inward investment, not only from the United States but also from Europe and Asia, and to secure access to the United States and Canadian markets.

The main provisions of the NAFTA are summarized in Figure 6.14. Note that, in addition to the various trade provisions, two 'side agreements' (on the environment and on labour standards) were incorporated to meet US and Canadian concerns. However, in contrast to the EU, there are no social provisions. The aims of the NAFTA were gradually to eliminate most trade and investment restrictions between the three countries over a 10- to 15-year period. The possibility of other countries joining the NAFTA was left open to negotiation but it is important to stress that NAFTA is not a customs union. It does not incorporate a common external trade policy. Each of the three NAFTA members is free to make free trade agreements with other states outside the NAFTA (as Mexico has recently done, for example, with the EU).

The NAFTA was – and remains – a highly controversial issue in all three member countries. Against the claimed benefits of an enlarged economic space (from both a production and a marketing point of view) is set a number of concerns. In the United States, there are particular worries about environmental and labour impacts. In the latter case, a former presidential candidate Ross Perot offered the spectre of a 'giant sucking sound' as jobs left the United States for Mexico.[38]

Major provisions of the North American Free Trade Agreement

General provisions

- Tariffs reduced over a 10- to 15-year period, depending on the sector
- Investment restrictions lifted (except for oil in Mexico; cultural industries in Canada; airline and radio communications in the United States)
- Immigration is not covered, with the exception that movement of some white-collar workers to be eased
- Any member state can leave the Agreement with 6 months' notice
- The Agreement allows for the inclusion of any additional country
- Government procurement to be opened up over 10 years
- Dispute resolution panels of independent arbitrators to resolve disagreements
- Some 'snap-back' tariffs allowed if surge in imports hurts a domestic industry

Sector-specific provisions

- *Agriculture:* most tariffs between US and Mexico removed immediately. Tariffs on 6% of products – corn, sugar, some fruits and vegetables – fully eliminated after 15 years. For Canada, existing agreement with US applies
- *Automobiles:* tariffs removed over 10 years. Mexico's quotas on imports lifted over same period. Cars eventually to meet 62.5% local content rule in order to be free of tariffs
- *Energy:* Mexican ban on private sector exploration continues, but procurement by state oil company opened up to US and Canada
- *Financial services:* Mexico gradually to open up financial sector to US and Canadian investment. Barriers to be eliminated by 2007
- *Textiles:* agreement eliminates Mexican, US and Canadian tariffs over 10 years. Clothes eligible for tariff breaks to be sewn with fabric woven in North America
- *Trucking:* North American trucks can be driven anywhere in the three countries by year 2000

Side agreements

- *Environment:* the three countries liable to fines, and Mexico and the US sanctions, if a panel finds repeated pattern of not enforcing environmental laws
- *Labour:* countries liable to penalties for non-enforcement of child, minimum wage and health and safety laws

Other arrangements

- The US and Mexico to set up a North American Development Bank to help finance the clean-up of the US–Mexico border
- The US to spend roughly $90m in the first 18 months retraining workers losing their jobs because of the Agreement

Figure 6.14   Major provisions of the North American Free Trade Agreement (NAFTA)

A similar fear was expressed in Canada. One politician saw the NAFTA as a 'nightmare of US continentalists come true: Canada's resources, Mexico's labour, and US capital'.[39] In Mexico, the fear was expressed that the country would become even more dominated by the United States.

> A mere ten years' experience has settled few of these quarrels. Today, most trade economists read the evidence as saying that NAFTA has worked: intra-area trade and foreign investment have expanded greatly. Trade sceptics and anti-globalists look at the same history and feel no less vindicated ... Politically, the sceptics ... can fairly claim victory. NAFTA is unpopular in all three countries. In Mexico ... the agreement is widely regarded as having been useless or worse ... In all three countries, the perceived results of NAFTA seem to have eroded support for further trade liberalization.[40]

Not surprisingly, then, attempts to create a *Central American Free Trade Agreement* (CAFTA) between the United States and five Central American countries (Costa

Rica, El Salvador, Honduras, Nicaragua plus the Dominican Republic) were far from smooth (the legislation was passed in the US in July 2005). The major opposition has come from US labour organizations and sugar farmers fearing job relocations to the cheap labour economies (and poorer working conditions) of Central America. On the other hand, CAFTA is seen as being a way for Central American producers of sugar and of garments to gain better access to their biggest markets. In fact, unlike the NAFTA, which removed most US barriers to imports from Mexico and Canada, 'CAFTA largely makes permanent the access Central America already has to the US market … under the Caribbean Basin Initiative … in exchange for significantly greater access to the Central American market'.[41]

Whereas US interest in regional integration agreements did not emerge until the late 1980s, Latin America has a long history of attempts to create free trade areas and customs unions, dating back to 1960 with the establishment of the Latin American Free Trade Area (LAFTA).[42] As Figure 6.13 shows, there has been a complex overlapping of bilateral and multilateral agreements between Latin American countries. Some of these agreements have failed to develop, notably the LAFTA , despite its reinvention as LAIA (Latin American Integration Association) in 1980.

Two Latin American regional integration agreements have had rather more staying power: the Andean Community and Mercosur. Of the two, *Mercosur* is the more widely significant.[43] Its four members (Argentina, Brazil, Paraguay, Uruguay) have a combined population of 224 million (compared with 119 million in the Andean Community) and a total GPD of $693 (compared with $258). Mercosur was established in 1991 with the intention of liberalizing trade between the member states, establishing a common external tariff, coordinating macroeconomic policy, and adopting sectoral agreements. Economically, Mercosur has certainly increased the degree of internal trade. Between 1989 and 1997, intra-Mercosur trade grew from 8.3 to 23.7 per cent of the group's total trade. Venezuela joined Mercosur in mid 2006.

In some respects Mercosur has features in common with the EU. Like the EU, one of its primary motivations was to deal with security relationships between Argentina and Brazil (a parallel with the Franco-German relationship in Europe). It certainly goes some way beyond a simple free trade area (such as the NAFTA). On the other hand, Mercosur does not have any of the supranational institutions that are at the heart of the EU.

> Conflict continues to plague the organization because of a lack of coordinated economic policies and supranational institutions. Deepening of the integration process has slowed because member states have not established common mechanisms for coordinated macro-economic policy nor have they truly committed themselves, despite the rhetoric, to establishing a regional institutional framework … rather, loose regulations and shallow institutionalism have been maintained at a relatively low political cost … Put simply, the member states of Mercosur want the maximum economic and political benefits from integration while forgoing as little sovereignty as possible.[44]

Looming over all attempts to create a more vigorous regional economy in Latin America is the United States, which now aspires to create a pan-hemispheric Free Trade Area of the Americas (FTAA), encompassing North, Central and South America. So far, progress in the negotiations involving 34 countries have stalled. Partly to resist an FTAA, there are counter-moves to create a *South America Community of Nations*, whose core would be a merger, over 15 years, between Mercosur and the Andean Community.

East Asia and the Pacific

Both the nature and the scale of regional economic collaboration in the Asia–Pacific region are very different from the situation in Europe and the Americas. In general, regional arrangements in the Asia–Pacific are much looser, less formalized and more open.[45] There are, in effect, two main regional economic collaborations. One – AFTA – is confined to South East Asia. The other – APEC – is a much wider and looser arrangement. In addition, as Figure 6.11 shows, there is an increasing number of bilateral agreements (both inside and outside the region) involving Asian countries.

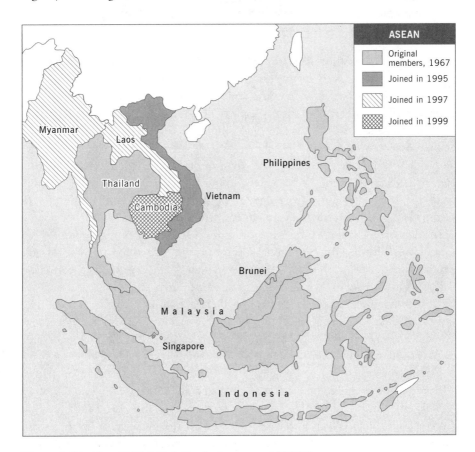

Figure 6.15   The ASEAN Free Trade Agreement (AFTA)

Figure 6.15 shows the current membership of the *ASEAN Free Trade Agreement* (AFTA). AFTA was initiated in 1992 (ASEAN itself had been established in 1967 as a group of four, then six South East Asian countries: Singapore, Malaysia, Thailand, Indonesia, the Philippines, Brunei). ASEAN's membership grew substantially in the second half of the 1990s (Figure 6.15). Ten countries are now involved although they are extremely varied in their political and economic structures and levels of economic development.

> ASEAN as an intergovernmental institution established to promote regional cooperation, offers a striking contrast to the Western institutions such as EU and NAFTA ... it is ... based on a different concept of institutionalisation –
>
> Paying full respect for the sovereignty and independence of each member state is one of the fundamental principles of the Association – most of the decisions have been made by consensus through the 'consultation based on the ASEAN tradition', which means to negotiate and consult thoroughly till achieving an agreement –
>
> [The] mechanism for dispute settlement also reflects ASEAN's preference for an informal approach. This is a striking contrast with the Western approach to dispute settlement in which preference is clearly on the side of judicial settlement based on clear rules and binding decisions.[46]

Such a system has both its strengths and its weaknesses. One of its strengths is that it has helped what is a very diverse group of countries to maintain positive relationships. One of its weaknesses is that firm and rapid response to problems is often difficult, especially in the light of the principle of non-interference in domestic matters of member states. ASEAN has had only limited success in stimulating economic activity. As a consequence, in 1992 the original six member states agreed to initiate an ASEAN Free Trade Agreement with the aim of removing all internal trade barriers by 2008. Subsequently, it was agreed to move the deadline for completion forward to 2003. However, incorporating the four very different economies of Cambodia, Laos, Myanmar and Vietnam has been a difficult task.

In addition, increasing competitive pressures on the ASEAN region from other East Asian countries (notably China) has forced the Association to look towards making agreements with other countries in East Asia. In particular, Korea and ASEAN have agreed to establish a free trade area by 2006, and China and ASEAN by 2010, while Japan and ASEAN have begun negotiations to liberalize trade. The notion of 'ASEAN plus 3' has clearly become a serious prospect. An agreement has also been negotiated with India to establish an Indo-ASEAN free trade area by 2012. At the same time, some ASEAN members, notably Singapore, have negotiated bilateral trade agreements with China, South Korea and Japan, and with the EU, the United States, Canada, Mexico and Chile (see Figure 6.11).

The other major regional economic organization is the *Asia–Pacific Economic Cooperation* forum (APEC), established in 1989 on the initiative of the Australian government. Figure 6.16 shows the extremely diverse composition of APEC. It includes not only the obvious East and South East Asian states themselves (including

China and Taiwan) but also Australia and New Zealand on the one hand and the United States, Canada, Mexico, and Chile on the other. However, APEC is, as yet, little more than a broadly based 'forum'. Indeed, cynics translated the APEC acronym as '*A Perfect Excuse to Chat*'. Certainly, little real progress has been made in fulfilling APEC's stated goal of 'open regionalism'. Particularly following the Asian financial crisis of 1997, APEC has become increasingly criticized by Asian participants.

> APEC's failure to provide any meaningful response to the biggest economic crisis in the Asia–Pacific region since 1945 made it, if not irrelevant, then less important for many Asian members … Increasingly, Asian observers evaluated APEC as a tool of American foreign economic policy. And the resistance of Asian policy makers to a strengthened APEC was caused by their fear of US dominance … APEC has not been successful in creating a joint identity as the basis for further pan-Pacific cooperation and the lack of tangible benefits has been progressively criticized … APEC has failed to provide much-needed political legitimacy for the wider regional liberal economic project.[47]

This failure of APEC has led to various initiatives within East Asia to create a more robust regional economic (and financial) framework. It is suggested that

> We are seeing the emergence of a new regionalism in Asia that exhibits three overlapping and complex trends:
>
> 1.   An interest in monetary regionalism arising from the desire … to combat financial volatility.

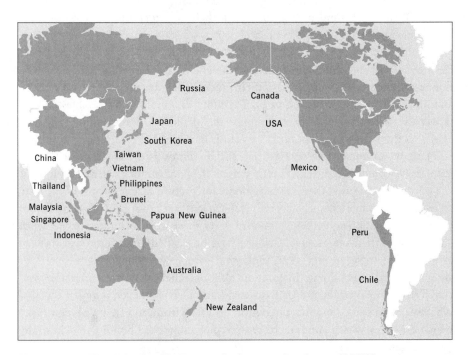

Figure 6.16   The Asia–Pacific Economic Cooperation forum (APEC)

2. An interest in bilateral trade initiatives within the context of the wider multilateral system, largely at the expense of the APECH style open regionalism of the 1990s.
3. The emergence of a *voice* of region beyond that of the sub-regions – Southeast and Northeast Asia – but more restricted than that of the Pacific as a mega-region ... the voice of region in the global political economy that is emerging is a new one, an 'East Asian' one.[48]

Conclusion

The aim of this chapter has been to assert the continuing significance of the state as a major influence in the global economy and to describe some of the general ways in which it operates. This does not mean, of course, that the roles and the functions of the state are unchanged. On the contrary, the position of the state is certainly being redefined in the context of a polycentric political-economic system in which national boundaries are more permeable than in the past. Nevertheless, the nation–state continues to contribute significantly towards the shaping and reshaping of the global economic map. We focused on four major aspects of the role of the state.

First, we explored the role of the state as *container* of distinctive 'ways of doing things'. The persistence of national differences in cultural, political, social and economic processes and institutions, that have evolved in a path dependent manner over time, helps to perpetuate 'varieties of capitalism' rather than convergence to a single form. Such fundamental differences help to explain the other aspects of state behaviour discussed in this chapter.

Second, we focused on the state as *regulator* of economic activity. All states operate a broad variety of policies aimed at influencing the level and nature of economic activities within, and across, their borders. But the precise mix of policies and how they are implemented varies according to a number of variables. In particular, the policy stance varies according to the ideological complexion of the state as well as the nature of the country's position in the global economic system.

Third, we examined the state as *competitor*: a controversial issue. Whether or not states *should* compete, the fact is that they think they do and they behave accordingly.

Fourth, we addressed the issue of regional integration as an example of the state as *collaborator*. While there has certainly been an acceleration in the number of regional integration agreements in recent years, most are very limited in the depth and extent of their integration. Very few have proceeded beyond the first stage of a simple free trade arrangement. Only the European Union has gone very far along the road of economic integration, although there is much 'regionalization' activity in the Americas and in East Asia. Ultimately, however, regional blocs of whatever degree of economic integration originate from, and are given legitimacy by, nation-states, which continue to be extremely important building blocks in the global economy.

NOTES

1 Kindleberger (1969: 207).
2 Jessop (1994).
3 See Garrett (1998), Gilpin (2001), Gritsch (2005), Hirst and Thompson (1999), Hudson (2001), Jessop (1994; 2002), Wade (1996), Weiss (1998; 2003).
4 Kelly (1999: 389–90, emphasis added).
5 Porter (1990: 19).
6 Wade (1996: 60).
7 Gritsch (2005: 2–3). Nye (2002) also addresses the question of 'soft power' in the context of the United States' current geopolitical position.
8 Glassner (1993: 35–40).
9 See Granovetter and Swedberg (1992), Smelser and Swedberg (2005).
10 Terpstra and David (1991: 6).
11 Agnew and Corbridge (1995), Taylor (1994) discuss the general notion of states as 'containers' and the nature and significance of territoriality and space in geopolitics.
12 Hofstede (1980; 1983). A recent study of 700 managers across a large number of countries confirmed the persistence of significant cultural differences (*Financial Times,* 15 October 2004).
13 See Berger and Dore (1996), Hall and Soskice (2001), Hollingsworth and Boyer (1997), Peck (2005), Redding (2005), Turner (2001), Whitley (1999; 2004).
14 Hollingsworth (1997: 266, 267–8).
15 Hudson (2001: 48–9, emphasis added).
16 Hudson (2001: 76).
17 Mortimore and Vergara (2004) and Mytelka and Barclay (2004) discuss FDI policies with particular reference to developing countries.
18 Krugman (1994).
19 Cerny (1997: 251).
20 Gritsch (2005: 2).
21 Porter (1990: 74, 80, emphasis added).
22 Porter (1990: 86, 87).
23 Porter (1990: 103).
24 Porter (1990: 109).
25 Porter (1990: 119).
26 Martin and Sunley (2003: 5).
27 Dunning (1992: 142).
28 This section draws on Cable and Henderson (1994), Gamble and Payne (1996), Gibb and Michalak (1994), Hoekman and Kostecki (1995: Chapter 9), Lawrence (1996), Mansfield and Milner (1999), Schiff and Winters (2003).
29 Mansfield and Milner (1999: 592).
30 WTO (2005).
31 Schiff and Winters (2003: 9).
32 Mansfield and Milner (1999: 595–602).
33 Mansfield and Milner (1999: 596).
34 Mansfield and Milner (1999: 597).
35 Mansfield and Milner (1999: 600).

36 Mansfield and Milner (1999: 601).
37 Ziltener (2004: 953).
38 Lawrence (1996: 72–3).
39 Quoted in McConnell and Macpherson (1994: 179).
40 *The Economist* (3 January 2004).
41 *Financial Times* (23 February 2005).
42 See Grugel (1996), Gwynne (1994), Kaltenthaler and Mora (2002).
43 This discussion of Mercosur is based upon Kaltenthaler and Mora (2002).
44 Kaltenthaler and Mora (2002: 92, 93).
45 Bowles (2002), Dieter and Higgott (2003), Haggard (1995), Hamilton-Hart (2003), Higgott (1999).
46 Liao (1997: 150–1).
47 Dieter and Higgott (2003: 433).
48 Dieter and Higgott (2003: 446).

Seven
'Doing It Their Way': Variations in State Economic Policies

From the general to the specific

The basic theme of Chapter 6 was that states not only continue to be very important actors in the global economy but also continue to differ in their characteristics. This is because states have histories that embed elements of path dependency in their contemporary behaviour. The result, from an economic perspective, is the existence – and persistence – of different varieties of capitalism. In this chapter, we look at some concrete examples of how states of different kinds attempt to influence both their domestic economies and also their position within the global political-economic system.

The chapter is organized into four parts:

- First, we identify some common patterns in state behaviour.
- Second, we focus on the older industrialized countries: the United States and Europe.
- Third, we examine the case of Japan, arguably the archetypal 'developmental state'.
- Fourth, we explore the diversity of policy positions adopted by some of the newly industrializing economies in East Asia and Latin America.

A degree of convergence

For several decades after the end of World War II in 1945, the role of the state expanded considerably, notably through the provision of welfare benefits for particular segments of the population and the development of a considerable (though varied) degree of public ownership of productive assets. The majority of economies, outside the command economies of the state-socialist world, became *mixed economies*. Certain economic sectors, such as telecommunications, railways, energy, steel and the like, became state owned or controlled in many countries. As a result, government spending as a percentage of GDP rose very substantially. In the OECD countries, such spending increased from less than 20 per cent of GDP in the early 1960s to 35 per cent in the early 1990s. In the developing countries, the average growth was from around 15 per cent to 27 per cent. Of course, the pattern varied a lot between countries.

Since the mid 1980s, many states have reduced their direct involvement in their economies. In fact, this has not reduced government expenditure as much as might have been expected; the rhetoric has often been stronger than the reality, as Figure 7.1 shows. The average GDP share of government spending in the 22 countries shown was 45 per cent in 2004. But there has certainly been a broadening and deepening *marketization* of the state's activities, extending the principles of market transactions into more and more aspects of public life. This is apparent not only in the older industrialized countries but also in many developing countries and, most dramatically of course, in the former state-socialist countries of Eastern Europe, in the former Soviet Union, and in China. Such *market liberalization* consists, primarily, of two processes: deregulation and privatization.

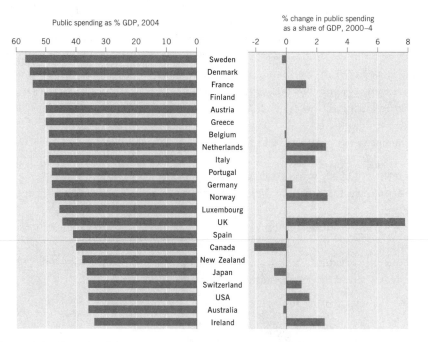

Figure 7.1   Growth of the state: central government spending as a share of GDP

Source: OECD data

To varying degrees, virtually all industrialized countries have jumped aboard the *deregulation* bandwagon. However, the issues are far less simple than the 'deregulationists' claim. Because no activity can exist without some form of regulation (otherwise anarchy ensues), 'deregulation cannot take place without the creation of new regulations to replace the old'.[1] In effect, what is often termed *de*regulation is really *re*regulation. As we saw in Chapter 6, both trade and foreign direct investment regimes have been strongly liberalized over the past two decades or so. Processes of deregulation have also spread to most economic sectors, notably in financial services and telecommunications. But they are, of course, much contested by non-business interest groups who (rightly) see major dangers in the increasing pervasiveness of a neo-liberal agenda. The *labour market* has also become a particularly significant focus of deregulation. These are issues we will return to in Part Four.

Parallel to the processes of deregulation are the processes of *privatization*. The state has been pulling out of a whole range of activities in which it was formerly centrally involved and transferring them to the private sector. The selling of state-owned assets, and the greater participation of the private sector in the provision of both 'private' goods (the denationalization of state-owned economic activities) and 'public' goods (such as health care, education, water, energy supplies), have typified a pervasive, though uneven, movement. As Figure 7.2 suggests, around the long-term secular trend shown by the middle curve in the diagram we can see cyclical variations in state involvement. However, there is a lower limit to state involvement below which the state's ability to govern by consent would be threatened. The state cannot completely wither away.

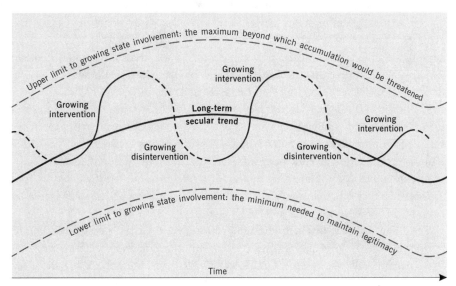

Figure 7.2   Limits to, and variations in, the extent of state involvement in the economy

Source: based on Hudson, 2001: Figure 3.3

The older industrialized economies: the United States and Europe

As we saw in Chapter 6, the continental European countries on the one hand, and the United States on the other, represent distinctively different types of capitalism. Historically, a major difference has been the centrality of some kind of industrial policy, together with a greater degree of social accountability of business in the former, and the absence of such policy and accountability in the latter.[2] The United Kingdom occupies a somewhat intermediate position between the virtually pure market capitalism of the United States and the kinds of social market capitalism practised in continental Europe, but with a tendency in some areas (notably labour market policy) to be closer to the United States. One policy thread, common to all the European states, including the United Kingdom, has been that of *regional policy*: of designating specific geographical areas for special assistance. All the EU states operate a regional economic policy to varying degrees. In contrast, direct federal involvement in regional or area economic development in the United States is relatively limited.

The United States

The policy stance of the United States reflects both the sheer scale and wealth of its domestic economy and also a basic philosophy of non-intervention by the federal government in the private economic sector. As far as industry is concerned the role of the federal government has generally been *regulatory*: to ensure the continuation of competition. Action at the federal level has been based primarily on macro-economic policies of a fiscal and monetary kind. The aim has been to create an appropriate investment climate in which private sector institutions could flourish. This has not, however, prevented the federal government from rescuing specific firms – especially very large ones – from disaster. At the other end of the size spectrum, the Small Business Administration has provided aid to stimulate new and small firms. Federal procurement policies are generally non-discriminatory but the sheer size of federal government purchases, particularly in the defence and aerospace industries, has exerted an enormous influence on US industry. Entire economic sectors, regions and communities are heavily dependent on the work created by federal defence and other procurement contracts and on subsidies in such sectors as agriculture (see Chapter 12).

US policy towards international trade during most of the post-World War II period has been one of urging liberalization and the reduction of tariffs in *multilateral* negotiations through the GATT/WTO. However, since the 1980s there has been an increasing willingness to develop bilateral trading arrangements with other countries (see Figure 6.11).

US trade policy is complicated by the structure and composition of the US Congress and the ways that new trade policies have to be negotiated with domestic interest groups. As the strongest economy the US, like Britain in the nineteenth century, has been the leading advocate of free trade. Even so, the federal government has intervened with the use of tariff and non-tariff barriers to protect particular interests. It has, for example, negotiated orderly marketing arrangements with various countries in sectors such as clothing (see Chapter 9), automobiles (Chapter 10) and electronics (Chapter 11).

The use of non-tariff barriers on a bilateral basis by the United States was explicitly specified in the Trade Act of 1974. This Act in effect heralded the emergence of a 'new protectionism' in the United States and the beginnings of a movement towards a *strategic trade policy* (STP).[3] The demand in the United States has been, increasingly, for a shift away from 'free' trade towards 'fair' trade – 'fairness' being defined by the United States itself. According to this view, because other countries allegedly engage in unfair trade practices it is only reasonable for the United States to do the same.

A major new step in US trade policy came with the introduction of the Omnibus Trade and Competitiveness Act of 1988 (OTCA). The OTCA incorporated a strongly *unilateralist* approach to trade negotiations, rather than the multilateralist approach enshrined in the GATT, and hitherto strongly supported by the United States. The key clause was the so-called 'Super 301' clause, whose aim was to achieve reciprocal access to what the United States defines as unfairly restricted markets. The difference between the 1974 and 1988 Acts, in this respect, was that the Super 301 clause was directed to *entire countries,* not just individual industries.

In the eyes of the world, the United States is increasingly being seen as having a strong *unilateralist* tendency, very much at odds with its traditional multilateral trading stance. The United States has become increasingly embroiled in a whole series of trade disputes – with Japan, with the East Asian NIEs (increasingly this means China) and with the European Union. There is also concern that the United States has a tendency to introduce *extra-territorial* trade legislation to achieve its broader political objectives. One example was the Helms–Burton law that penalized foreign companies doing business with Cuba.

The European Union

As European political-economic integration has deepened (see Chapter 6), some – though not all – of the economic policies of individual member states have been 'relocated' to the supranational EU level. For example, there is just one EU trade commissioner representing the EU in the WTO and in all other international trade negotiations. There are EU-wide policies on competition, on subsidies (both industrial and agricultural) and on investment incentives, amongst others. On the other hand, there are significant areas where policy is set at the national level: for example, in labour markets and taxation.

However, even in such areas of 'common' EU policy, the ideological positions of individual member states within the EU clearly have an effect, not least on the difficulties of reaching consensus. Trade negotiations, for example (including issues relating to the Common Agricultural Policy), have become increasingly contested within the EU, with a sharp divide opening up between states with a more protectionist stance (notably France, but also Poland) and those espousing more open trade policies (notably the UK and some of the Northern European states). In the sphere of competition policy, as well, there is much heated argument over the acquisition of domestic firms even by firms from other EU member states.

The recent debate over the proposed new EU Constitution brought into the open some of the big economic policy differences between member states. At the broadest level is the ideological conflict between the so-called 'market' and 'social' models of how the economy should be organized: between the UK's more 'neo-liberal' position and that of France and Germany, where the principle of the 'social market' remains strongly entrenched. However, the lines are not always quite as clear as is often claimed.

Although the *United Kingdom's* policy position does contrast in a number of ways with that of the continental European countries, and is closer to that of the United States, the UK remains more interventionist than the United States. Nevertheless, the UK has strongly adopted economic policies of privatization and deregulation. Its labour market policies, in particular, are far closer to the US model than to the EU model and, indeed, the UK has opted out of some of the social provisions of the EU. In 2001, the UK government published a strategy document on 'competitiveness and the knowledge economy' designed to position the UK more strongly within the global economy. The policy focused on five specific areas:

- encouraging investment in key technologies and IT infrastructure
- reforming the regulatory and financial environment to facilitate companies' access to capital
- building 'dynamic clusters' of companies in the regions (the influence of Michael Porter was considerable in this respect)
- improving labour force skills, especially technical and IT skills
- ensuring a reduction in EU bureaucracy.

Despite a more general consensus, there are considerable differences in policy emphasis between continental European member states. Of the leading EU states, *France* maintains the most 'nationalistic' economic position, having long had the most explicit state industrial policy, a reflection of a tradition of strong state involvement dating back to the seventeenth century. A major component of French industrial policy has been the promotion of 'national champions' in key industrial sectors, often through state ownership of large-scale enterprises. France's current policy position retains many of these traditional qualities (and an especially strong antagonism to the Anglo-American neo-liberal economic model). Indeed, in some respects (as in the new labour laws reducing working hours introduced by the Jospin government in 2001) this stance continues to be reinforced. The stated intention of the French prime minister M. de Villepin, in 2005, was to 'defend France's model of social protection from rampant market forces'.[4] Despite considerable privatization the French state retains a very considerable direct involvement in the economy. Key clusters, associated with specific regions in France, have been identified as France's most globally competitive industries to benefit from government funding: 'neuro-sciences and complex systems in the Ile de France; aeronautics and space in Toulouse and Bordeaux; health in Lyons; nanotechnology in Grenoble; and secure communication systems in Provence–Alpes–Côte d'Azur'.[5]

The major exception among the continental European nations to a centralized approach to industrial policy has been *Germany*. In part, at least, this reflects the fact that Germany is a federal political unit with power divided between the federal government and the provinces (*Länder*). But although often described as 'light', the federal government's role has been far from insubstantial. It has pursued policies of active intervention in industrial matters, including a substantial programme of financial subsidy. Such involvement has to be seen within the German model of a social market economy.[6] The German economy is characterized both by a considerable degree of competition between domestic firms and also by a high level of consensus between various interest groups, including labour unions, the major banks and industry. The major challenge facing Germany since 1990, of course, has been to cope with the fundamental transformation of the economy brought about by reunification. Putting together the strongest economy in Europe (the former West Germany) with one that, for half a century, had existed in a completely different ideological system, has been an immense undertaking. It has put enormous strains on the federal budget because of the huge problems of rebuilding infrastructure and dealing with problems of unemployment brought about by restructuring.[7]

In 2000, at the Lisbon summit, EU leaders proclaimed their intention to transform the EU 'into the most competitive and dynamic knowledge-based economy in the world' by 2010. No fewer than 15 goals for modernizing welfare systems, strengthening investment in education and combating social exclusion were announced. In the light of the EU's weak overall economic performance since then, the strategy is being revised. Its 'social market' critics fear that such revision may move the EU further towards a neo-liberal agenda. National tensions remain at the heart of EU decision-making. A proposal in 2005 to double the EU's R&D budget over six years to meet competition from the shift of R&D investment to East Asia was defeated.

The expansion to 25 members has made such decisions even more difficult and contentious. The new members from *Eastern Europe* were, of course, formerly command economies, subject to the rigours of Soviet-inspired centralized planning. Regionally, the economies were integrated within the framework of the Council for Mutual Economic Assistance (CMEA) established in 1949. By the 1980s, the rigid centralized planning system had become far less rigid, although to a varying degree from one country to another. Nevertheless, the system remained, in essence, a centrally planned system in which the state set the overall goals and objectives and the bounds to what was allowed in terms of enterprise autonomy and foreign involvement.

In 1989, following Mikhail Gorbachev's 'revolution from the top' in the Soviet Union,[8] the entire state-socialist system in Eastern Europe began to unravel.

> With the zeal of religious converts, most of the countries that had languished behind the Iron Curtain set about transforming themselves into market economies.[9]

But this was far from easy. Among the many problems were those relating to property rights (the ownership of productive assets), how – and how quickly – former state-owned enterprises were to be transformed into privately owned enterprises (the creation of a previously non-existent entrepreneurial class), the extent and nature of the involvement of foreign capital, the position of labour, the nature of the legal system, and the reconciliation of market forces with social goals.

Although the problems facing the so-called transition economies of Eastern Europe were essentially the same in each case, the precise way they are being approached varies, depending on their individual histories before and during the era of centralized control.[10] However, their incorporation into the EU from 2005 means that, like the other EU members, many of their policies will be determined at the EU level. Their continued transformation into market economies now occurs within a different framework

Japan

> Historically, the State assumed an active economic role in Western economies in order to correct what were considered to be the private sector's economic and social failures. Japanese historical tradition, on the other hand, grants to government a legitimate role in shaping and helping to carry out industrial policy.[11]

Japan can be regarded as the archetypal *developmental* state in which the government's economic role has been very different from that in most Western countries.[12] There has long been a high level of consensus between the major interest groups in Japan on the need to create a dynamic national economy. To some extent, this consensus has been seen as a cultural characteristic of Japanese society, with its deep roots in familism. But it also reflected the poor physical endowment of Japan and the limited number of options facing the country when, in the 1860s, it suddenly emerged from its feudal isolation. In other words, consensus was also a pragmatic stance built up over more than a hundred years. Given virtually no natural resources and a poor agricultural base, Japan's only hope of economic growth lay in building a strong manufacturing base, both domestically and internationally through trade. In this process, the state played a central role not through direct state ownership but rather by *guiding* the operation of a highly competitive domestic market economy. Indeed, there has been relatively little state-owned enterprise in Japan and a generally much smaller public sector than in most Western economies.

For more than 50 years after the end of World War II, the key government institution concerned with both industry policy and trade policy – the two are seen to be inextricably related in Japan – was the *Ministry of International Trade and Industry* (MITI), renamed the *Ministry for Economy, Trade and Industry* (METI) in 2001. After its establishment in 1949, MITI became the real 'guiding hand' in Japan's economic resurgence, although its role has often been misunderstood and exaggerated in the West. Figure 7.3 identifies the major roles played by MITI in the Japanese economy. Until the 1960s Japan operated a strongly protected economy and it was not until 1980 that full internationalization of the Japanese economy was reached. During the 1950s and early 1960s MITI, together with the Ministry of Finance, exerted very stringent controls on all foreign exchange, foreign investment and the import of technology.

1	**Constructed medium-term econometric forecasts of the development of, and needed changes in, the Japanese industrial structure**

- Established indicative plans or 'visions' of the desirable goals for the private sector
- Made specific comparisons of cost structures of Japanese and foreign competitors

2	**Arranged for preferential allocation of capital to selected strategic industries**

- Involved governmental and semi-governmental banks
- Ministry of Finance guided commercial banks to coordinate their lending policies with MITI's industrial strategy
- Financial support of an industry implied guidance (though not control) by MITI

3	**Targeted key industries for the future and put together a package of policy measures to promote their development**

- Before early 1980s the major measure wasprotection against foreign competition in the Japanese domestic market
- Protectionism abandoned in early 1980s. Emphasis then on financial assistance, tax breaks, incentives given through administrative guidance, anti-trust relief (to facilitate 'research cartels')

4	**Formulated industrial policies for 'structurally recessed industries'**

- MITI designated a specific industry as 'structurally recessed'
- The ministry responsible for that sector formulated a stabilization plan specifying how the capacity to be scrapped should be shared between enterprises. The plans were drawn up in consultation with the Industrial Structure Council of MITI
- Costs of scrapping production facilities were shared between the government and the private sector

Figure 7.3   The role of MITI in Japan's industrial development

Source: based on Johnson, 1985: pp. 66–7

Initially, MITI focused its energies on the basic industries of steel, electric power, shipbuilding and chemical fertilizers but then progressively encouraged the development of petrochemicals, synthetic textiles, plastics, automobiles and electronics. Japan was transformed from a low-value, low-skill economy to a high-value, capital-intensive economy. The foundation of this transformation was the clearly targeted, selective nature of Japanese industry policy together with a strongly protected domestic economy.

A key element in Japanese economic policy was the specific treatment of foreign direct investment which, for much of the post-war period, was extremely tightly regulated. The technological rebuilding of the Japanese economy was based on the purchase and licensing of foreign technology and *not* on the entry of foreign branches or subsidiaries. However, the inward investment laws have been liberalized and foreign firms do indeed operate within Japan in increasing numbers and, in some cases, in highly significant ways (for example Renault's effective acquisition of Nissan: see Chapter 10). As far as outward investment by Japanese firms is concerned, the situation changed dramatically during the 1970s and 1980s when Japanese overseas investment grew at a spectacular rate (Chapter 2). This was consistent with MITI's policy of internationalizing the economy and reflected the economically strategic role that Japanese overseas investment came to play.

Two major developments in the external environment were especially important. The first was the upsurge in protectionist measures in North America and Europe in such industries as automobiles and electronics from the mid 1970s (see Chapters 10 and 11). The 'laser-like' targeting of Japanese industrial policy resulted in a major backlash from various Western economies. As a direct result of the erection of non-tariff barriers Japanese firms began to invest heavily in production facilities overseas. The second external stimulus to increased overseas direct investment was *endaka*: the major rise in the value of the Japanese yen which resulted from the 1985 Plaza Agreement among the Group of Five international finance ministers. This political decision stimulated an upsurge in overseas investment by Japanese firms to take advantage of lower production costs, particularly in East Asia.

Since the early 1990s, Japanese policy has been especially exercised by the problem of a high-value currency, by contentious trading relationships with the United States and Europe, and especially by the deep domestic recession which accompanied the collapse of the so-called 'bubble economy' at the end of the 1980s. Currently, there is much debate about whether Japan can recover its economic dynamism and what kind of role the government should play. Certainly attempts throughout the 1990s by successive Japanese governments to stimulate the domestic economy through fiscal mechanisms were not successful. Questions over the kind of capitalism that might develop in Japan abound. Changes are certainly occurring but, as we argued in Chapter 4, the core characteristics of the Japanese economy are unlikely to disappear.

Newly industrializing economies

Although they are frequently grouped together, the world's newly industrializing economies are a highly heterogeneous collection of countries. They vary enormously in size (both geographically and in terms of population); in their natural resource endowments; and in their cultural, social and political complexions. But they all tend to have one feature in common: the central role of the national state in their economic development.

Despite many popular misconceptions, none of today's NIEs is a freewheeling market economy in which market forces have been allowed to run their unfettered course. They are, virtually without exception, *developmental* states: market economies in which the state performs a highly interventionist role.[13] Having said that, the precise role of the state – the degree and nature of its involvement, as well as the extent of its success – varies greatly from one NIE to another. In some cases, state ownership of production is very substantial; in others it is insignificant. In some cases, the major policy emphasis is upon attracting foreign direct investment; in others such investment is tightly regulated and the policy emphasis is upon nurturing domestic firms. Thus, although the recurring central theme running through the development of all NIEs is the role of the state, each individual NIE performs a specific variation on that general theme – a reflection of its specific historical, cultural, social, political and economic complexion.[14]

Types of industrialization strategy

Broadly speaking, a developing country may pursue one or more of three basic types of strategy:

- exports of indigenous commodities
- import-substituting industrialization (ISI) – the manufacture of products that would otherwise be imported, based upon protection against such imports
- export-oriented industrialization (EOI).

Which of these strategies is, or can be, pursued depends upon a number of factors: the economy's resource endowment (both physical and human); its size (particularly of its domestic market); its international context (especially the rate of growth of world trade and the policies of TNCs); and the attitude of the national government.

The general pattern of industrialization, beyond the commodity export phase, has generally been one of an initial emphasis on import substitution followed eventually by a shift to export-oriented policies. The aim of import substitution is to protect a nation's infant industries so that the overall industrial structure can be developed and diversified, and dependence on foreign technology and capital can be reduced. To this end, many of the policies listed in Figures 6.4, 6.5 and 6.7 have been employed.

The import-substituting strategy, in theory, is a long-term *sequential* process involving the progressive domestic development of industrial sectors through a combination of protection and incentives. The realization that an import-substituting strategy cannot, on its own, lead to the desired level of industrialization began to dawn in a growing number of countries; some during the 1950s, rather more during the 1960s. Generally it was the smaller industrializing countries that first began to shift towards a greater emphasis on *export orientation* because of the constraints imposed upon such a policy by their small domestic market. Increasingly, an export-oriented, outward-looking industrialization strategy became the conventional wisdom among such international agencies as the Asian Development Bank and the World Bank.

A shift towards export-based industrialization was made possible by:

- the rapid liberalization and growth of world trade during the 1960s
- the 'shrinkage' of geographical distance through the enabling technologies of transportation and communications
- the global spread of the transnational corporation and its increasing interest in seeking out low-cost production locations for its export platform activities.

Such export orientation was based upon a high level of government involvement in the economy. The usual starting point was a major devaluation of the country's currency to make its exports more competitive in world markets. The whole battery of export trade policy measures shown in Figure 6.4 was invariably employed by the newly industrializing economies. In effect, these amounted to a subsidy on exports that greatly increased their price competitiveness. Of course, the major domestic resource on which this export-oriented industrialization rests has been that of the labour supply – not only its relative cheapness but also its adaptability and, very often, its relative docility. Indeed, in many cases, the activities of labour unions have been very closely regulated.

In fact, the 'paths of industrialization' followed by individual NIEs have been rather more complex than is often suggested. Figure 7.4 sets out a five-phase sequence of industrialization based upon the experiences of the Latin American and East Asian NIEs. A number of important points can be made:[15]

- The distinction commonly drawn between inward-oriented Latin American industrialization strategies and outward-oriented East Asian industrialization strategies is misleading.
- The initial stages of industrialization were common to NIEs in both regions; 'the subsequent divergence in the regional sequences stems from the ways in which each country responded to the basic problems associated with the continuation of primary ISI'.[16]
- 'The duration and timing of these development patterns varied by region. Primary ISI began earlier, lasted longer, and was more populist in Latin America than in East Asia ... The East Asian NIEs began their accelerated export of manufactured products during a period of extraordinary dynamism in the world economy ... [after 1973] the developing countries began to encounter stiffer protectionist measures in the industrialized markets. These new trends were among the factors that led the East Asian NIEs to modify their EOI approach in the 1970s'.[17]
- Some degree of convergence in the strategies of the Latin American and East Asian NIEs began to occur in the 1970s and 1980s. Each 'coupled their

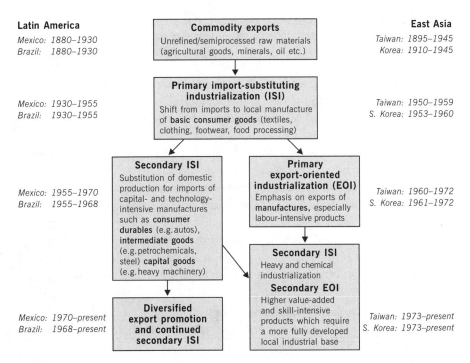

Latin America

Mexico: 1880–1930
Brazil: 1880–1930

Mexico: 1930–1955
Brazil: 1930–1955

Mexico: 1955–1970
Brazil: 1955–1968

Mexico: 1970–present
Brazil: 1968–present

East Asia

Taiwan: 1895–1945
Korea: 1910–1945

Taiwan: 1950–1959
S. Korea: 1953–1960

Taiwan: 1960–1972
S. Korea: 1961–1972

Taiwan: 1973–present
S. Korea: 1973–present

Commodity exports
Unrefined/semiprocessed raw materials
(agricultural goods, minerals, oil etc.)

Primary import-substituting industrialization (ISI)
Shift from imports to local manufacture of **basic consumer goods** (textiles, clothing, footwear, food processing)

Secondary ISI
Substitution of domestic production for imports of capital- and technology-intensive manufactures such as **consumer durables** (e.g. autos), intermediate goods (e.g. petrochemicals, steel) **capital goods** (e.g. heavy machinery)

Primary export-oriented industrialization (EOI)
Emphasis on exports of **manufactures,** especially labour-intensive products

Secondary ISI
Heavy and chemical industrialization
Secondary EOI
Higher value-added and skill-intensive products which require a more fully developed local industrial base

Diversified export promotion and continued secondary ISI

Figure 7.4   Paths of industrialization in Latin America and East Asia: common and divergent features

Source: based on material in Gereffi, 1990: Figure 1.1, p. 17

previous strategies from the 1960s (secondary ISI and primary EOI respectively) with elements of the alternate strategy in order to enhance the synergistic benefits of simultaneously pursuing inward- and outward-oriented approaches'.[18]

The attraction of *foreign direct investment* has been an integral part of both import-substituting and export-oriented industrialization in many developing countries, although the extent to which particular countries have pursued this strategy varies considerably. In general, the Latin American NIEs have been more restrictive in their attitudes to foreign direct investment than the Asian NIEs, although there have been recent shifts towards more liberal investment policies in Latin America. In general, ownership requirements in Latin America have tended to be stricter than in most East Asian countries and the number of sectors in which foreign involvement is prohibited rather greater.

Export processing zones

Amongst all the measures used by developing countries to stimulate their export industries and to attract foreign investment, one device in particular – the export processing zone (EPZ) – has received particular attention.[19] The ILO defines EPZs as

Industrial zones with special incentives set up to attract foreign investors, in
which imported materials undergo some degree of processing before being
(re-)exported again.[20]

Table 7.1 shows the rapid growth in EPZs, especially during the past 10 years. Some
90 per cent of all EPZs in the developing countries are located in Latin America,
the Caribbean, Mexico and Asia. However, in terms of employment, Asia is by far
the most important region for EPZs, with 87 per cent of the total.

EPZs come in a number of different forms: 'free trade zones, special economic
zones, bonded warehouses, free ports, and *maquiladoras*'.[21] Within developing countries, EPZs have been located in a variety of environments. Some have been incorporated into airports, seaports or commercial free zones or located next to large
cities. Others have been set up in relatively undeveloped areas as part of a regional
development strategy. EPZs themselves vary enormously in size, ranging from geographically extensive developments to a few small factories; from employment of
more than 30,000 to little more than 100 workers.

Table 7.1 Growth in export processing zones

	1975	1986	1995	1997	2003
No. of countries with EPZs	25	47	73	93	116
No. of EPZs	79	176	500	845	5000
Employment (millions):	–	–	–	22.5	43
Asia					36.8
Central America and Mexico					2.2
Middle East					0.7
North Africa					0.4
Sub-Saharan Africa					0.4
North America					0.3
South America					0.3
Transition economies					0.3
Caribbean					0.2
Indian Ocean					0.1
Europe					0.05
Pacific					0.01

Source: ILO, 2003: Table 1; *Financial Times*, 12 May 2005

EPZs in developing countries share many features in common. The overall pattern of incentives to investors is broadly similar, as is the type of industry most commonly found within the zones. Historically, the production of textiles and clothing
and the assembly of electronics – both employing predominantly young female
labour – has dominated. However, the position is not static:

Zones have evolved from initial assembly and simple processing activities to
include high tech and science zones, logistics centres and even tourist
resorts. Their physical form now includes not only enclave-type zones but also

single-industry zones (such as the jewellery zone in Thailand or the leather zone in Turkey); single-commodity zones (like coffee in Zimbabwe); and single-factory (such as the export-oriented units in India) or single-company zones (such as in the Dominican Republic).[22]

Having outlined some of the major features of NIE policies in general terms, we need to acknowledge the substantial diversity that exists between individual countries. In order to do this, the following sections present a sketch of four NIEs: Korea, Singapore, China and Mexico.

Korea

South Korea (officially the Republic of Korea) came into being in 1948, following the partition of Korea into two parts. From 1910 to 1945, Korea had been a Japanese colony, very tightly integrated into the imperial economy. Between 1948 and 1988, when political liberalization occurred, South Korea was governed by a succession of authoritarian, military-backed and strongly nationalistic governments. These governments, particularly that led by Park Chung Hee (1961–79), operated a strong state-directed economic policy articulated through a series of five-year plans.[23] As Figure 7.4 shows, the emphasis changed over time from primary ISI, through primary EOI, secondary ISI and secondary EOI. Two important developments during the 1950s helped to provide the basis for these strategies: the land reform of 1948–50, which removed the old landlord class and created a more equitable class structure; and the redistribution of Japanese-owned and state properties to well-connected individuals which helped to create a new Korean capitalist class.[24]

A powerful economic bureaucracy was created, with a key role played by a new Economic Planning Board (EPB). At the same time, the financial system was placed firmly in the hands of the state; the banks were nationalized, and the Bank of Korea was brought under the control of the Ministry of Finance. This highly centralized 'state-corporatist' bureaucracy, in effect, 'aggressively orchestrated the activities of "private" firms'.[25] In particular, the state made possible – and actively encouraged – the development of a small number of extremely large and highly diversified firms – the *chaebol* – that continue to dominate the Korean economy.

By controlling the financial system, particularly the availability of credit, the Korean government was able to operate a strongly interventionist economic policy. The *chaebol* were consistently favoured through their access to finance (including the preferential allocation of subsidized loans) and very strong, long-term relationships were developed between them and the state. From the 1960s Korean policy had a strong sectoral emphasis as the state decided which particular industries should be supported through a battery of measures, including financial subsidy and protection against external competition. The precise sectoral focus changed over time, as Table 7.2 indicates.

Like Japan at a similar stage in its development, Korea, for the most part, eschewed the channel of inward foreign investment to acquire technology. Indeed,

Korea adopted the most restrictive policy towards inward foreign investment of all the four leading Asian NIEs. Until 1983 it operated strict rules on foreign direct investment that restricted the permitted level of foreign ownership, and specified a minimum export performance and local content level. Korean government policy has been to build a very strong domestic sector. As a consequence, the share of FDI in the Korean economy has been extremely low (see Table 2.4).

Table 7.2   Changing sectoral focus in Korea's developmental strategy

Developmental phase	Major industries
Primary import-substituting industrialization	Food; beverages; tobacco; textiles; clothing; footwear; cement; light manufacturing (e.g. wood, leather, rubber, paper products)
Primary export-oriented industrialization	Textiles and apparel; electronics; plywood; wigs Intermediate goods (chemicals, petroleum, paper, steel products)
Secondary import-substituting industrialization and secondary export-oriented industrialization	Automobiles Shipbuilding Steel and metal products Petrochemicals Textiles and apparel Electronics Videocassette recorders Machinery

Source: based on material in Gereffi, 1990: Table 1.6

Starting in the early 1980s, however, the emphasis of Korean economic policy shifted towards a greater degree of (restricted) liberalization. State control of the financial system was eased in 1983. The domestic market was opened up to a greater degree of imports. Inward foreign direct investment began to be encouraged. Some relaxation of the country's extremely stringent labour laws occurred. Korea has had the most restrictive – often repressive – labour laws and practices of all the East Asian NIEs.[26] In 1988, the military regime was replaced by a democratically elected government. Some attempts were made to persuade the *chaebol* to change some of their practices, but with only limited success. Most significantly, in the mid 1990s Korea applied to join the OECD, membership of which exerted increased pressure on the state to make major changes to its financial and economic system, and especially to make its markets more open and to adopt a US-style neo-liberal system.

During the 1990s, in effect, much of Korea's traditional industry policy was diluted.[27] Major changes were made in policies of financial regulation, exchange rate management and investment coordination. The formerly tightly controlled financial sector was significantly liberalized and the policy of exchange rate management virtually abandoned. The central pillar of South Korean industrial policy for 40 years – the coordination of investment – began to be dismantled.

When the East Asian financial crisis of 1997 hit Korea, the country's problems – as for the other affected East Asian economies – were attributed by the IMF, and by the Western financial community in general, to the existence of an over-regulated, state-dominated economy with excessively close (even corrupt) relationships between government and business. Yet, in the case of Korea, that was no longer entirely the case. It could be argued, in fact, that the Korean government had already gone too far in abandoning the principles on which its spectacular economic growth had been based.

Clearly, certain reforms were needed as both the Korean economy itself and the broader global environment were changing. Not least was the need to reform the *chaebol*, which had come to distort the economy and which were, themselves, in great financial difficulty. That battle is still being fought. The *chaebol* argue that the proposed reforms will leave them vulnerable to foreign takeover; the government argues that reform of cross-shareholdings will make them more competitive. The issue of 'foreign takeover' is, however, a very sensitive issue in Korea as a whole.

Singapore

Singapore is by far the smallest of all the East Asian NIEs.[28] Like both Korea and Taiwan, Singapore had a very long history as a colony (within the British system). Unlike the two larger Asian countries, however, it was less tightly integrated into its imperial system, although it played a highly significant role as a commercial *entrepôt*, a reflection of its strategic geographical position. Singapore became fully independent in 1965 when it separated from Malaysia. Since then, although Singapore is a parliamentary democracy, it has been governed by one political party (the People's Action Party). For the first 30 years of its existence, it was led by one powerful individual, Lee Kuan Yew.[29]

From the very beginning, the Singapore government pursued a very aggressive policy of export-oriented, labour-intensive manufacturing development. Concentration on manufacturing – especially labour-intensive manufacturing – was adopted because of the need to reduce a very high unemployment rate in a society that, at the time, had one of the fastest population growth rates in the world. The twin pillars of the policy were those of complementary economic and social planning, the latter being much more overt than in other East Asian NIEs.

The particular ways in which Singapore has operated its export-oriented policy have been substantially different from those of Korea and Taiwan. Most significantly, the central pillar was a strategy of attracting foreign direct investment. As a result, the Singaporean economy is overwhelmingly dominated by foreign firms (Table 2.4). The most explicit industrialization measures, therefore, were those of incentives to inward investors, using a sectorally selective process, with particular attention being devoted initially to electronics, petroleum and shipbuilding. The government agency responsible was the Economic Development Board (EDB), which still plays an extremely influential role in the Singapore economy. With a few exceptions, Singapore operated a free port policy with little use of trade protectionist measures. The second set of direct measures used to promote industrial development was the establishment of a high-quality physical infrastructure.

At the same time, a series of social policy measures was introduced aimed at creating an amenable environment for foreign investment. Major housing programmes were undertaken, partly funded through the state's compulsory savings scheme (the Central Provident Fund). More specifically, the government effectively incorporated the labour unions into the governance system by establishing a National Trades Union Council. 'Strikes and other industrial action were declared illegal unless approved through secret ballot by a majority of a union's members. In essential services, strikes were banned altogether ... These labour market regulations resulted in the creation of a highly disciplined and depoliticised labour force in Singapore'.[30] Thus, through a whole battery of interlocking policies, the Singapore government has created a very high-growth, increasingly affluent, industrialized society in which foreign firms have played the dominant economic role in production but within a highly regulated political and social system.

In response to the unexpected (but, in fact, short-lived) economic shock of 1985, the government initiated a 'New Direction' for the Singaporean economy, based upon reduced dependence on the manufacturing sector and a shift towards the business services sectors. Incentive packages were introduced to attract foreign firms prepared to set up service operations (especially headquarters and R&D activities) in Singapore.

Singapore sets out to market itself as a global business centre on the basis of the very high quality of its physical and human infrastructure and its strategic geographical location. Government policies are geared towards this goal, which also includes an explicit strategy to 'regionalize' the Singaporean economy by encouraging domestic firms to set up operations in Asia while Singapore develops as the 'control centre' of a regional division of labour. The government is driving a series of initiatives using government-linked corporations to develop major infrastructural projects in Asia and, more broadly, to develop international networks.[31] At the same time, the emphasis on research and development and technological upgrading continues, with a specific emphasis on biotechnology to enhance its already significant role as a pharmaceuticals centre, and on IT.[32]

As Singapore entered the third millennium, its government faced a number of major challenges. One set of challenges arises from the country's position in South East Asia. What has so far been a major source of strength looks less so in the light of the potential shift in the regional centre of gravity towards the north-east. Most problematical is the emergence of China as East Asia's dominant economy (other than Japan). The relative shift in patterns of foreign direct investment towards China poses a major threat to Singapore, whose economy has been based on inward FDI above all else.

Two policy initiatives have been developed by the Singaporean government to meet these challenges (in addition to those already in place). First, Singapore's financial system has been significantly reformed and liberalized. Second, Singapore has taken the lead in pushing for greater 'Asian regionalism'. The other set of challenges are internal to Singapore itself and concern the extent to which this highly paternalistic state is able to loosen its grip on the country's political and social life without damaging its economic influence.

China

In discussing the development of Korea and Singapore (as well as the other emerging NIEs in Asia) it is always necessary to emphasize the key role played by the state. In the case of China, of course, such an assertion is unnecessary. As a centrally controlled command economy there is no doubt about the state's centrality. The point about China is that, after several decades of self-imposed separation from the world economy, it has become an immensely significant global player. What makes China so significant, in the long run, is its sheer size – some 1.2 billion people – and its massive economic potential. Whether that potential will be realized is a matter for speculation. What is important, here, is to outline the dramatic changes in Chinese policy towards the rest of the world since 1979.[33]

The People's Republic of China (PRC) came into being in 1949 with the replacement of the nationalist government by a communist government led by Mao Zedong. For the next 30 years, China followed a policy of economic self-reliance. This policy was pursued through a series of major – often extreme – initiatives. Initially, the new government followed the example of the Soviet Union in establishing a Five-Year Plan (1953–57). This relatively successful policy was jettisoned in 1958 when Mao announced the 'Great Leap Forward': a total transformation of economic planning, with the emphasis on small-scale and rural development. Although this initiated the notion of rural industrialization, the GLF was disastrous in its consequences, with mass famine one of the results. In 1966, policy changed again with the introduction of the 'Cultural Revolution', a phase that lasted for some 10 years with, again, disastrous human and social implications.

The period after Mao's death in 1976 was one of political hiatus that was eventually resolved by the emergence of Deng Xiaoping as leader. It was under Deng's leadership that China began to jettison the self-reliance policy of the previous 30 years and to make links with the world market economies. This has been done, however, without substantial political change. In the words of the new Party Constitution of 1997, it is 'Socialism with Chinese characteristics'.

The pivotal year was 1979, when China began its 'open policy' based upon a carefully controlled trade and inward investment strategy. This was set within the so-called 'Four Modernizations' (concerned with agriculture, industry, education, and science and defence). A central element of the new policy has been the opening up of the Chinese economy to foreign direct investors. As we saw in Chapter 2, FDI has grown very rapidly indeed in China since the early 1980s and now accounts for 15 per cent of GDP (Table 2.4), although a large proportion of this FDI emanates from Hong Kong (since 1997 part of China under the 'one country, two systems' arrangement) and Taiwan. The organizational form of these investments varies from wholly owned foreign subsidiaries to equity joint ventures with Chinese partners and other partnership arrangements.

The most distinctive feature of the open policy, however, is the explicit use of geography in its implementation. Partly in order to control the spread of capitalist market ideas and methods within Chinese society, and partly to make the policy more effective through external visibility and agglomeration economies, FDI has been steered to specific locations. Initially, the foci were the four Special Economic Zones (SEZs)[34] established in 1979 at Shenzhen, Zhuhai, Shantou and Xiamen (Figure 7.5). Significantly, each of these was located to maximize their attraction to investors from Overseas Chinese, notably in Hong Kong, Macau and Taiwan.[35] The Chinese SEZs offered a package of incentives, including tax concessions, duty-free import arrangements and serviced infrastructure. The original SEZs were located well away from the major urban and industrial areas in order to control the extent of their influence. However, since the mid 1980s, there has been considerable development of Economic and Technological Development Zones (ETDZs), as Figure 7.5 shows.

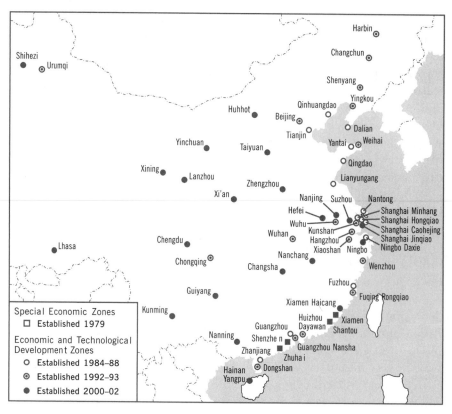

Figure 7.5   The geography of China's 'open policy'

Despite massive inflows of foreign capital and technologies, China remains a centrally controlled command economy in which state-owned enterprises (SOEs) predominate, despite halving in numbers. The SOE share of China's industrial output fell from 75 per cent in the late 1970s to around 50 per cent in 2002. But SOEs still account for about 35 per cent of China's urban employment and a substantial share of government revenues.[36] Reform of the SOEs is an immense task and one surrounded by massive controversy. A major problem for a country trying economically to 'modernize' is the sheer inefficiency (by Western standards) and high levels of corruption of the SOEs. SOEs are embedded within the Communist Party system and this fact pervades their operations.[37] A survey carried out by the Shanghai Stock Exchange in 2000 found that 'in listed companies 99 per cent of the main business and staffing decisions, including board appointments and salaries, are made with the approval of internal party committees'.[38]

The problems posed by the SOEs have been intensified with China's accession to the WTO. Although this, arguably, greatly enhances China's economic potential it also imposes severe stresses on the domestic economy and institutions. Not only are tariff levels falling from their previously high levels, thus exposing Chinese enterprises to intense competition, but also non-tariff barriers, matters relating to intellectual property rights, safety regulations, financial and telecommunications regulations are all affected. It is notable that the Chinese government now actively encourages Chinese businesses to invest overseas and there have been a number of significant Chinese acquisitions of foreign businesses, notably the IBM PC business by Lenovo in 2004. Financially, China continues to be under pressure to revalue the *renminbi* (a modest revaluation occurred in 2005).

Finally, we must not forget the problems posed for the Chinese state by the sheer geographical immensity of the country. It is no coincidence that China's economic policies have had a strong element of geographical concentration (Figure 7.5). The bulk of investment spending has gone, so far, to a relatively small part of the country, primarily on or near the coast. Despite two decades of economic reform in China, immense regional inequalities in economic well-being remain. As a consequence, in 2000 the Chinese government announced that it was to divert most of its investment spending to the deprived western provinces in a new 'Go West Strategy'.[39] A new programme to assist poor farmers was announced in 2006, recognition of the fact that some 800 million people live in the countryside. A recent report by the Chinese Academy of Sciences[40] envisages the need to relocate 500 million rural dwellers to cities and to facilitate the movement of 600 million city dwellers into the suburbs. A parallel report by the State Council announced the intention to invest heavily in clean energy sources, recognition of the immense environmental problems faced by China as it industrializes and urbanizes at such rapid rates.

Overall, it is clear that the institutional – and geographical – structure of the Chinese economy is in a state of flux, with a much increased variety of forms. As in the past, however, the key lies in the internal political power struggles between the 'modernizers', who wish to sustain and develop the open policies of the recent past, and those who wish to retain a degree of isolation. Following the death of the initiator of the open policy, Deng Xiaoping, in 1997, his successors continued the broad open policy direction. So far, at least in the eyes of the OECD, China's reform policies are proving successful.[41] But the key test of the survival of such a policy is its continued success in delivering economic growth and raising incomes for the majority of Chinese, and not just those in the more developed parts of the country. For a country so large and so populous, this is a very tall order indeed.

Mexico

Our final example of national industrialization policies is drawn from a different part of world: the Americas. Following the financial collapse of 1982 – when its economic turmoil precipitated the international debt crisis – Mexico has undergone dramatic political, social and economic change as the state has attempted to integrate the national economy more strongly into the global economy.[42] But the path of transition from a strongly inward-oriented industrialized policy position to an export-oriented position has been far from smooth. Two basic characteristics are important to an understanding of the Mexican case. The first is its location next door to the world's dominant political and economic power, the United States. The second is that from 1929 to 2000 Mexico was governed by a single party (the PRI). However, reformist pressures both inside and outside the governing party intensified, particularly from the 1980s.

For almost 60 years from 1929, Mexico pursued a predominantly import-substituting industrialization policy. This policy was, for the most part, quite successful. The large domestic market and a strong physical and human resource base, together with proximity to the US market, allowed Mexico to grow at high rates, especially during the period from the 1950s to the 1970s. The high returns from the country's oil reserves, in particular, had a major effect on the economy. It was the collapse of oil prices in 1981 that precipitated the country's financial crisis. One component of Mexico's export policy during the import-substituting phase was the *maquiladora* programme that created a very specific form of industrial growth along the Mexico–United States border (see Figure 2.30).

Between 1982 and 1985, government policy emphasized stabilization rather than structural adjustment, but external pressures from the IMF resulted in a pronounced shift in policy emphasis.

> This round of crisis and reconciliation led to important shifts within the Mexican administration. Economic difficulties had already strengthened the hand of the technocrats within the administration ... the stabilization efforts of 1985–86 were accompanied by the initiation of trade reform and negotiations that led to accession to GATT. In the Solidarity Pact of 1988 ... the government used the much-needed stabilization programme to counter strong

resistance from the private sector to further trade policy reform. The government cut the maximum tariff level from 100 per cent to 20 per cent; the average tariff level dropped to just over 10 per cent.[43]

During the second half of the 1980s, the Mexican government pursued a wide-ranging programme of liberalization involving both the deregulation of some areas of the economy and the privatization of state-owned enterprises. By 1993, some 90 per cent of Mexico's state-owned enterprises had been privatized, either wholly or partially, although the government retained control of some key companies, including PEMEX, the state oil company. The single most important act of the Salinas administration, which came to power in 1988, was to take Mexico into the NAFTA. This symbolized the government's headlong attempt to tie the domestic economy firmly into the global system and, in particular, to gain greater benefits from a formal association with the United States. One result of the major domestic reforms and the opening up of the economy was a very rapid build-up of speculative foreign capital ('hot money') that drove up the value of the peso to unsustainable levels. In December 1994, the new Zedillo administration was forced into a massive currency devaluation that heralded a new economic crisis for the country.

The most significant recent change for Mexico's economy has been political: the election in July 2000 of Vicente Fox of the PAN as President of Mexico, thus ending 71 years of control by the PRI. Fox, a former executive of Coca-Cola, set out an ambitious agenda of internal reform (to begin to remove the decades of corrupt practices) and of further opening Mexico to the global economy. Fox put forward proposals to improve Mexico's position within the NAFTA and to address the perennial problem of illegal migration to the United States (both of which will require considerable changes of attitude in the United States itself). He also entered into a free trade agreement with the European Union. For the first time in many years, Fox's accession to the presidency was not accompanied by the usual financial crisis. But the presidential term of office in Mexico is six years with no second term. It remains to be seen whether the extremely high expectations created by the defeat of the PRI can be delivered, or even whether the country will revert to the political leadership that dominated for most of the twentieth century.

One of the major problems that undoubtedly needs to be addressed by the government is Mexico's relatively weak technological and skills base.

> In general there was a tradition of excessive technological dependence in Mexican industry in comparison to countries like the Republic of Korea or Taiwan Province of China. There was a high level of reliance on imported capital goods for a country of its industrial size and sophistication. This was accompanied by a similar and widespread reliance on inflows of foreign know-how, licences and expertise through much of industry. Despite the nationalistic stance of the government, there was a relatively large presence of foreign multinational enterprises in the advanced sectors of Mexican industry ... It

restricted the ability of Mexican industry to move into technologically more dynamic or sophisticated industries. While the Republic of Korea used import substitution to foster industrialization, it also pursued a strategy of independent industrial technological development. Mexico was never as export-oriented, nationalistic, State-led and technologically ambitious as the Republic of Korea.[44]

Conclusion

The aim of this chapter has been to outline the variety of ways that states occupying different positions within the global economy, and having different political-ideological stances, have attempted to govern their national economies. Despite the undoubted shift towards a greater degree of 'marketization' of virtually all national systems since the 1980s (through processes of deregulation and privatization), very significant variety of state involvement in the economy persists. Convergence is limited by the interaction between the specific histories of individual states (their 'path dependency') and economic-political processes operating at the global scale. The result is a mosaic of specific national systems displaying elements of both convergence and diversity.

Thus, there is a clear and continuing difference between the United States on the one hand and the countries of continental Europe on the other. The United Kingdom occupies a rather uneasy position between the two, being closer to the United States in some respects but closer to continental Europe in others. Even though the existence of the EU provides a common regulatory framework for its member states, significant variety still exists in the precise manner of national state involvement in individual EU economies. The same point of continuing diversity, rather than homogeneity, can be made of the East Asian NIEs. These have too often been grouped together under the umbrella of an 'East Asian development model' whereas, as we have seen, the mode of state involvement has differed widely between them.

NOTES

1 Cerny (1991: 174).
2 See Gilpin (2001: Chapter 7).
3 See Krugman (1986; 1990), Richardson (1990), Yoffie (1993).
4 Quoted in the *Financial Times* (9 June 2005).
5 *Financial Times* (13 July 2005).
6 Contributors to Vitols (2004) explore the extent to which the 'German model' is sustainable.
7 Gretschmann (1994: 471).

8 Offe (1996: 30).

9 Tudor (2000: 39).

10 Whitley (1999: 209).

11 Magaziner and Hout (1980: 29).

12 Accounts of Japanese economic policy are provided by Dore (1986), Johnson (1985), Porter et al. (2000).

13 Wade (2004a). The chapters in Gereffi and Wyman (1990) analyse the diverse 'paths of industrialization' in Latin America and East Asia. See also Brohman (1996), Stallings (1995: Part II).

14 Douglass (1994: 543).

15 Gereffi (1990: 18).

16 Gereffi (1990: 21).

17 Gereffi (1990: 21).

18 Gereffi (1990: 22).

19 ILO (1998; 2003).

20 ILO (2003: 1).

21 ILO (2003: 1).

22 ILO (2003: 2).

23 Especially useful accounts of Korean economic policy are by Amsden (1989), Koo and Kim (1992), Wade (1990; 2004a).

24 Koo and Kim (1992).

25 Wade (2004a: 320). See also Amsden (1989).

26 Deyo (1992).

27 See the detailed analysis provided by Chang (1998a).

28 Singapore's developmental policies are discussed by Lall (1994), Ramesh (1995), Rodan (1991), Yeung (1998).

29 Lee Kuan Yew's autobiography (Lee 2000) provides a unique (though certainly not unbiased) account of Singapore's rapid economic development.

30 Yeung (1998: 392).

31 The 'regionalization' strategy of Singapore is discussed in detail by Yeung (1998; 1999).

32 Coe (2003b).

33 Useful accounts of Chinese economic development policy are provided in Benewick and Wingrove (1995), Crane (1990), Nolan (2001).

34 The Special Economic Zones are analysed by Crane (1990), Phillips and Yeh (1990), Thoburn and Howell (1995), Wong and Chu (1995).

35 Thoburn and Howell (1995: 173).

36 *The Economist* (20 March 2004).

37 Nolan (2001).

38 *Financial Times* (2 July 2001).

39 *Financial Times* (8 May 2000).

40 Reported in *The Guardian* (10 February 2006).

41 OECD (2005a).

42 Mexican industrialization policies are discussed in Haggard (1995), Harvey (1993), Lall (1994).

43 Haggard (1995: 81).

44 Lall (1994: 81).

Eight
Dynamics of Conflict and Collaboration: The Uneasy Relationship between TNCs and States

The ties that bind

In the preceding four chapters I have outlined the major characteristics of TNCs and of states as separate actors in shaping the global economy. However, so far, the *relationships* between TNCs and states have been merely hinted at (see Figure 1.6). In this brief chapter I focus explicitly on these relationships because they are, in many ways, at the very centre of the processes of global shift and of global economic transformation. In the case studies of Part Three we will see how these relationships are played out in different industries, whilst in Chapter 16 we examine the potential impact of TNCs on home and host economies. Here, the aim is simply to outline the major general features of the relationships between TNCs and states.

There is a widespread view that the major cause of the state's (alleged) demise is the counterpoised growth of the TNC. Indeed, one of the most common arguments is to compare the *size* of TNCs and national states. The typical argument is as follows:[1]

- 'Of the 100 largest economies in the world, 51 are corporations; only 49 are countries (based on a comparison of corporate sales and country GDPs) ... To put this in perspective, General Motors is now bigger than Denmark; DaimlerChrysler is bigger than Poland; Royal Dutch/Shell is bigger than Pakistan.'
- 'The 1999 sales of each of the top five corporations (General Motors, Wal-Mart, Exxon Mobil, Ford Motor, and DaimlerChrysler) are bigger than the GDPs of 182 countries.'

These are, of course, very striking comparisons. But are they really meaningful? The answer is that, beyond their value as polemic, they are not. They are superficial and misleading, although they certainly make eye-catching headlines. However, the statistics do not measure the same thing quantitatively, and they certainly do not capture the qualitative differences between TNCs and states.

> GDP is a measure of value added, not sales. If one were to compute total sales in a country one would end up with a number far bigger than GDP. One would also be double-, triple-, or quadruple-counting ... if corporations, too, are measured by value added as national economies are ... they tend to shrink by between 70 and 80 per cent ... Properly measured, Denmark's economy is more than three times bigger than GM. Even impoverished Bangladesh has a bigger economy than GM.[2]

Relationships between TNCs and states are, in fact, far more complex and ambiguous than the popular view would have us believe:

> It is perhaps most useful ... to view the relationship between [trans]nationals and governments as both cooperative and competing, both supportive and conflictual. They operate in a fully dialectical relationship, locked into unified but contradictory roles and positions, neither the one nor the other partner clearly or completely able to dominate.[3]

This quotation captures the essence of the intricate relationships between TNCs and states: they contain elements of both rivalry and collusion.[4] On the one hand, there is no doubt that the fundamental goals of states and TNCs differ in important respects. In ideal-type terms, whereas the basic goal of business organizations is to maximize profits and 'shareholder value', the basic economic goal of the state is to maximize the material welfare of its society. Figure 8.1 indicates some of the dimensions of these conflicting objectives of TNCs and states.

	TNC objectives	State objectives
Performance	Maximize profits and shareholder value Minimize cost base consistent with customer need	Maximize growth of GDP Maximize quantity and quality of employment opportunities
Technology	Undertake R&D at locations optimal to the needs of the firm as a whole Gain access to all necessary technology	Stimulate the development of locally rooted technology
High-order functions	Locate headquarters and other high-order functions to fit optimal pattern of the firm's overall operations	Maintain indigenous headquarters Attract and retain key operations of TNCs
Responsiveness	Retain flexibility to move profits in optimal manner Retain flexibility to modify the geographical configuration of the firm's production network to meet changing conditions Retain flexibility to use the labour force as required	Retain power to gain a fair return on local operations of TNCs through taxation policies Maximize the extent and benefits of local supplier linkages Prevent the closure or scaling down of local TNC operations Develop a flexible, high-skill, high-earning labour force

Figure 8.1   Some conflicting objectives of TNCs and states

Source: based, in part, on Hood and Young, 2000: Table 1.1

On the other hand, although their relationships may be conflictual in certain circumstances, *states and firms need each other*. Clearly, *states need firms* to generate material wealth and provide jobs for their citizens. They might prefer such firms to be domestically bounded in their allegiance but that is not an option in a capitalist market economy. Indeed, many regard TNCs as important extensions of their state foreign policy. For example, in addition to ensuring control of key natural resources,

> American political leaders have believed that the national interest has also been served by the foreign expansion of US corporations in manufacturing and services. Foreign direct investment has been considered a major instrument through which the United States could maintain its relative position in world markets, and the overseas expansion of multinational corporations has been regarded as a means to maintain America's dominant world economic position.[5]

Conversely, *TNCs need states* to provide the infrastructural basis for their continued existence: both physical infrastructure in the form of the built environment and also social infrastructure in the form of legal protection of private property, institutional mechanisms to provide a continuing supply of educated workers, and the like. TNCs, in particular, look to their home country government to provide them with diplomatic protection in hostile foreign environments.

> In the last resort … [a TNC's] directors will always heed the wishes and commands of the government which has issued their passports and those of their families.[6]

As we saw in Chapter 6, states are both *containers* of distinctive business practices and cultures – within which firms are embedded – and also *regulators* of business activity. National boundaries, therefore, create significant differentials on the global political-economic surface; they constitute one of the most important ways in which location-specific factors are 'packaged'. They create discontinuities in the flow of economic activities that are extremely important to the ways in which TNCs can operate. In particular, states have the potential to determine two factors of fundamental importance to TNCs:[7]

- the terms on which TNCs may have *access* to markets and/or resources
- the *rules of operation* with which TNCs must comply when operating within a specific national territory.

At the same time the fact that TNCs not only span national boundaries but also, in effect, incorporate *parts* of national economies within their own organizational boundaries (Figure 8.2) creates major potential problems for states. The nature and the magnitude of the problem vary considerably according to the kinds of strategies pursued by TNCs. Most important is the extent to which TNCs pursue globally integrated strategies within which the roles and functions of individual units are related to that overall global strategy.[8] As we saw in Chapter 5, geographical segmentation and fragmentation of transnational production networks have become increasingly common. States tend to be fearful about the autonomy and stability of those TNC units located within their national territory as well as concerned about the leakage of tax revenues. At the extreme, of course, TNCs have the potential capability to move their operations out of specific countries.

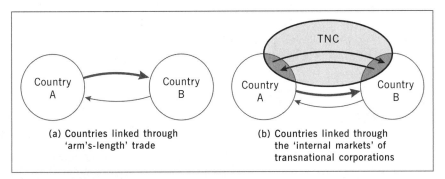

(a) Countries linked through 'arm's-length' trade

(b) Countries linked through the 'internal markets' of transnational corporations

Figure 8.2   Territorial inter-penetration: the 'incorporation' of parts of a state's territory into a TNC

At first sight it seems obvious that TNCs would seek the removal of all regulatory barriers that act as constraints and impede their ability to locate wherever, and to behave however, they wish. Such barriers include those relating to entry into a national market, whether through imports or a direct presence; freedom to export capital and profits from local operations; freedom to import materials, components and corporate services; freedom to operate unhindered in local labour markets. Certainly, given the existence of differential regulatory structures in the global economy, TNCs will seek to overcome, circumvent or subvert them. Regulatory mechanisms are, indeed, constraints on a TNC's strategic and operational behaviour.

Yet it is not quite as simple as this. TNCs may perceive the very existence of regulatory structures as an *opportunity*, enabling them to take advantage of regulatory differences between states by shifting activities between locations according to differentials in the regulatory surface – that is, to engage in *regulatory arbitrage*.[9] One aspect of this is the ability of TNCs to stimulate competitive bidding for their mobile investments by playing off one state against another as states themselves strive to outbid their rivals to capture or retain a particular TNC activity (see the next section). More generally, TNCs seem to have a rather ambivalent attitude to state regulatory policies:

> TNCs have favoured minimal international coordination while strongly supporting the national state, since they can take advantage of regulatory differences and loopholes ... While TNCs have pressed for an adequate coordination of national regulation, they have generally resisted any strengthening

of international state structures ... Having secured the minimalist principles of national treatment for foreign-owned capital, TNCs have been the staunchest defenders of the *national state*. It is their ability to exploit national differences, both politically and economically, that gives them their competitive advantage.[10]

More specifically, it has been argued that TNCs will tend increasingly to support a strategic trade policy in their home country, with the expectation that this will open up market access in foreign countries and enable them to benefit from large economies of scale and learning curve effects.[11]

Bargaining processes between TNCs and states

It is clear that the relationships between TNCs and states are exceedingly complex. In the final analysis, such relationships revolve around their relative bargaining power: the extent to which each can implement their own preferred strategies. The situation is especially complex when TNCs pursue a strategy of transnational integration – but geographical fragmentation – of their activities, in which individual units in a specific host country form only a part of the firm's overall operations. In such circumstances, governments have a number of legitimate concerns:[12]

- that integrated TNCs might relocate their operations to other countries because of relative differences in factor costs
- that integrated TNCs use their operations to engage in transfer pricing to reduce the taxes they pay
- that integrated TNCs retain their key competencies outside host countries (typically in the firm's home country) and locate only lower-skill, lower-technology operations in host countries
- that national decision centres no longer operate within an integrated TNC, making negotiations difficult between host country governments and the local affiliate.

Although the degrees of freedom of TNCs to move into and out of territories at will are often exaggerated, the *potential* for such locational mobility is obviously there. This helps to make the TNC–state bargaining process immensely complex and highly variable from one case to another. Despite such contingency, however, we need to try to understand some of the *general* features of the bargaining process. Here, for the sake of simplicity, we assume that a host country can be regarded as a single entity in the bargaining relationship. In fact, of course, many competing interest groups are involved, including domestic business interests, labour organizations and other civil society organizations, each of which will have

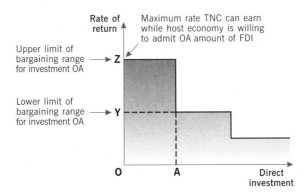

Figure 8.3   A simplified model of the bargaining relationship between a TNC and a host country

Source: based on Nixson, 1988: Figure 1

different attitudes towards TNCs. In other words, it is, in reality, a *multiparty* bargaining situation.[13]

Figure 8.3 is a highly simplified, hypothetical example. The vertical axis of the graph shows the rate of return a TNC may seek for a given level of investment OA on the horizontal axis. The bargaining range for this level of TNC investment is shown to vary between

- a lower limit (OY), which is the minimum rate of return that the TNC is prepared to accept for the amount of investment OA
- an upper limit (OZ), which is determined by the cost to the host economy of either developing its own operation, or finding an alternative investor, or managing without the particular advantages provided by the TNC.

OZ is the maximum return the TNC can make for the amount of direct investment OA permitted by the host economy. It is in the interests of the TNC to try to raise the upper limit OZ; conversely, it is in the interests of the host economy to try to lower that upper limit: 'The higher is the cost to the host economy of losing the proposed [investment], the greater are the possibilities for the TNC of setting the bargain near the maximum point'.[14]

On the other hand, the more possibilities the host economy has of finding alternatives, the greater are its chances of lowering that upper limit. The greater the competition between TNCs for the particular investment opportunity, the greater are the opportunities for the host country to reduce both the upper and lower limits: 'In addition, the host economy has an interest in lowering the lower limit, through the creation of an advantageous "investment climate" (political stability, constitutional guarantees against appropriation etc.) which might persuade the TNC to accept a lower rate of return'.[15]

Seducing investors: 'locational tournaments' and competitive bidding

The greater the competition between potential host countries for a specific investment, the weaker will be any one country's bargaining position, because countries will tend to bid against one another to capture the investment. Indeed, one of the most striking developments of the last few decades has been the development of so-called *locational tournaments*. There has been an enormous intensification in *competitive bidding* between states (and between communities within the same state) for the relatively limited amount of internationally mobile investment. Such cut-throat bidding undoubtedly allows TNCs to play off one state against another to gain the highest return for their investment. It has also become increasingly common for TNCs to try to lever various kinds of state subsidies in order to persuade them to keep a plant in a particular location. Otherwise, it is threatened, the plant will be closed or much reduced in scale.

In fact, much of the actual investment capital may be provided by the host government itself in the form of various kinds of financial and fiscal (tax) deals as well as in the form of physical and social infrastructure. The use of tax incentives continues to be especially common.

> About 20 economies reduced their corporate income tax rates during 2004 … Studies show that location of FDI is becoming more sensitive to taxation, and that corporate income tax rates can influence a TNC's decision to undertake FDI, especially if competing jurisdictions have similar enabling conditions. For instance, EU investors were found to increase their FDI positions in other EU member states by approximately 4% if the latter reduced their corporate income tax rates by one percentage point relative to the European mean.[16]

It has become quite common, therefore, for TNCs to threaten to leave a particular country because of perceived high tax rates, although often the threat is more apparent than real. The problem is, states do not necessarily know that.

The problem of transfer pricing

One of the most problematical – yet most opaque – issues in the relationships between TNCs and states is that of how a TNC's internal transactions, and its profits, are actually taxed by the states in which a TNC has a presence. By definition, a TNC moves both tangible materials and products (finished and semi-finished), and also various kinds of corporate services, across international borders to the various parts of its operations. In external markets, prices are charged on an 'arm's-length' basis between independent sellers and buyers. In the internal 'market' operated by TNCs, however, transactions are between *related* parties – units of the *same* organization. The rules of the external market do not apply. The TNC itself sets the *transfer prices* of its goods and services within its own organizational boundaries and, therefore, has very considerable flexibility in setting those transfer prices to help achieve its overall goals.

The ability to set its own internal prices – within the limits imposed by the vigilance of the tax authorities – enables the TNC to adjust transfer prices either upwards or downwards and, therefore, to influence the amount of tax or duties payable to national governments. For example, as Figure 8.4 suggests, it would be in a TNC's interest to charge more for the goods and services supplied to its subsidiaries located in countries with high tax levels and vice versa. A similar incentive exists where governments restrict the amount of a subsidiary's profits that can be remitted out of the country.

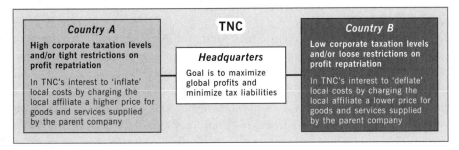

Figure 8.4   The incentives for TNCs to engage in transfer pricing

In general, the greater the differences in levels of corporate taxes, tariffs, duties and exchange rates, the greater will be the incentive for the TNC to manipulate its internal transfer prices. The very large, highly centralized, global TNC has the greatest potential for doing so. But it has proved extremely difficult for governments (and researchers) to gather hard evidence on its actual extent:

> Picture a General Motors plant in Windsor, Ontario, producing hundreds of items for assembly in [autos] … that will be sold in the Canadian market as well as for assembly by its sister plants in Michigan. No independent public market exists for many of the items, since no other firm produces these products. Nor is it obvious what the production cost may be of the items that cross the US–Canadian border – that kind of estimate will depend heavily on how the fixed costs of the Windsor plant are allocated among the many items produced, an allocation that cannot fail to be arbitrary. Without an obvious selling price or an indisputable cost price, all the ingredients exist for a pitched battle over the transfer price.
>
> When the item crossing the border is intangible, such as a right bestowed by the parent on a foreign subsidiary to use the trademark of the parent or to draw on its pool of technological know-how, the indeterminateness of a reasonable price becomes even more apparent. How much is the use of the IBM trade name worth to its subsidiary in France? How valuable is the access granted to a team of engineers in an Australian subsidiary to the databank of a parent in Los Angeles?[17]

A US House of Representatives study claimed that more than half of almost 40 foreign companies surveyed had paid virtually no taxes over a 10 year period. The Internal Revenue Service estimated that some $53 billion was lost through the transfer pricing mechanism in 2001 alone.[18] In the United Kingdom, a study of 210 TNCs showed that 83 per cent had been involved in a transfer pricing dispute.[19] But even in developed economies, like the United States or the United Kingdom, it is extremely difficult for government to assess the actual extent of transfer pricing. It is infinitely more difficult for developing countries to do so.

Relative bargaining powers of TNCs and states

In general, TNCs wish to maximize their locational flexibility to take advantage of geographical differences in the availability, quality and cost of production inputs in serving their existing and new markets. Their ideal would be to pursue such goals without any hindrance from the regulatory practices of states. States, on the other hand, strive to capture as much as possible of the value created from production within their territories. In this latter sense, a primary aim of a host state is to try to *embed* a TNC's activities as strongly as possible in the local/national economy.

One way of thinking about this specific process is to conceive of two ideal-types of embeddedness: *active* embeddedness and *obligated* embeddedness.[20] Active embeddedness is where a TNC seeks out localized assets and incorporates them, *as a matter of choice*, within its operations. Where such localized assets are widely available in different geographical locations then the power of such choice rests primarily with the TNC. However, the less widely available the assets (or where access to them is controlled by the state), the more likely the state is to have a greater degree of bargaining power over the terms on which the TNC can utilize them. In such circumstances, obligated embeddedness is likely to occur; that is, the TNC is forced to comply with state criteria in order to gain access to, and use of, the desired asset. Obligated embeddedness, therefore, is likely to occur where two conditions are satisfied. First, there must be a localized asset that is highly important to a TNC (this may include a natural resource, a human resource and/or a significant market) and to which it needs access in order to achieve its business goals. Second, access to that resource must be controlled by the state within whose territory the asset is located and the state must have the power to exert that control.

The extent to which a state feels the need to offer large incentives to attract a foreign investment or to retain an existing investment, or feels able to impose access or performance requirements, will depend on its relative bargaining strength in any specific case. Conversely, the extent to which a TNC is able to obtain such incentives, or to operate as it wishes, will depend on its relative bargaining strength. The outcome will depend on a number of factors.[21] On the one hand, the price that a *host country* will ultimately pay is a function of:

- the number of foreign firms independently competing for the investment opportunity
- the recognized measure of uniqueness of the foreign contribution (as against its possible provision by local entrepreneurship, public or private)
- the perceived degree of domestic need for the contribution.

On the other hand, the terms the *TNC* will accept are a function of:

- the firm's general need for an investment outlet
- the attractiveness of the specific investment opportunity offered by the host country, compared to similar or other opportunities in other countries
- the extent of prior commitment to the country concerned (e.g. an established market position).

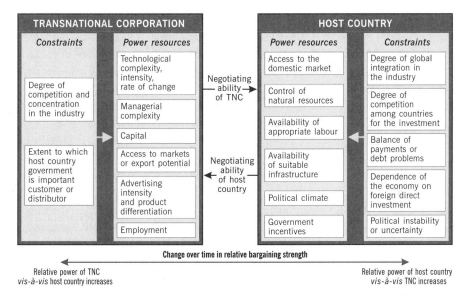

Figure 8.5 Components of the bargaining relationship between TNCs and host countries

Source: based on material in Kobrin, 1987

Figure 8.5 sets out the major components of the bargaining relationship between TNCs and host countries. Both possess a range of 'power resources' that are their major bargaining strengths. Both operate within certain constraints that will restrict the extent to which these power resources can be exercised. The relative bargaining power of TNCs and host countries, therefore, is a function of three related elements:

- the *relative demand* by each of the two participants for resources which the other controls
- the *constraints* on each which affect the translation of potential bargaining power into control over outcomes
- the *negotiating status* of the participants.

Figure 8.5 suggests that, in general, host countries are subject to a greater variety of constraints than are TNCs, a reflection of the latter's greater potential flexibility to switch their operations between alternative locations. Nevertheless, the extent to which a TNC can implement a globally integrated strategy is constrained by nation-state behaviour. Where a company particularly needs access to a given location and where the host country does have leverage, then the bargain that is eventually struck may involve the TNC in making concessions. In general, the scarcer the resource being sought (whether by a TNC or a host country), the greater the relative bargaining power of the controller of access to that resource and vice versa.

For example, states that control access to large, affluent domestic markets have greater relative bargaining power over TNCs pursuing a market-oriented strategy than states whose domestic markets are small (China is the obvious case). It is in this kind of situation that the host country's ability to impose performance requirements – such as local content levels – on foreign firms is greatest. It may also give the host country sufficient leverage to persuade the inward investor to establish higher-level functions such as R&D facilities. On the other hand, the nature of the domestic market may not be a consideration for a TNC pursuing an integrated production strategy. Where, for example, the TNC's need is for access to low-cost labour that is very widely available then an individual country's bargaining power will be limited. On the whole, cheap labour is not a scarce resource at a global scale.

On the other hand, 'host countries wield the power to limit the extent of, or even to dismantle, the [T]NC integrated manufacturing and trade networks with more regulations and restrictions on foreign investments and market access'.[22] At the extreme, of course, both institutions – TNCs and governments – possess sanctions which one may exercise over the other. The TNC's ultimate sanction is not to invest in a particular location or to pull out of an existing investment:

> The meta-power that global business interests have in relation to nation-states is based on the *exit option* … It is the experience … of actual or threatened *exclusion of states* from the world market that demonstrates and maximizes the power of global business in contrast to isolated individual states.[23]

A nation-state's ultimate sanction against the TNC is to exclude a particular foreign investment or to appropriate an existing investment.

The problem, of course, is that the whole process is *dynamic*. The bargaining relationship changes over time, as the bottom section of Figure 8.5 suggests. In most studies of TNC–state bargaining the conventional wisdom is that of the so-called 'obsolescing bargain' in which once invested, fixed capital becomes 'sunk', a hostage and a source of bargaining strength. The high risk associated with exploration and development diminishes when production begins. Technology, once arcane and proprietary, matures over time and becomes available on the open market. Through development and transfers from FDI the host country gains technical and managerial skills that reduce the value of those possessed by the foreigner.[24]

In this view, after the initial investment has been made, the balance of bargaining power shifts from the TNC to the host country, in other words, it moves to the right in Figure 8.5. But although this may well be the case in natural-resource-based industries, it is far less certain that it applies in those sectors in which technological change is frequent and/or where global integration of operations is common. In such circumstances, 'the bargain will obsolesce slowly, if at all, and the relative power of [T]NCs may even increase over time'.[25]

Not surprisingly, there are few detailed studies of TNC–state bargaining processes. The participants regard them as being far too sensitive (and possibly embarrassing). A rare example is Seidler's study of Ford's strategy to enter the Spanish market in the early 1970s.[26] This case demonstrates just how powerful a large TNC can be in persuading a host country government to change its existing regulations. The Spanish automobile market in the early 1970s was heavily protected. Not only were tariffs on imports very high (81 per cent on cars, 30 per cent on components) but also cars built in Spain had to have 95 per cent local content. In addition, no foreign company could own more than 50 per cent of a company operating in Spain. Such restrictions were very much in conflict with Ford's own preferences for a Spanish operation. Ford's aim was not only to penetrate the local market but also to create an export platform from which to serve the entire European market. As such, it wanted the lowest possible import tariffs on components and a minimal level of local content so that it could source components from other parts of its transnational network. Ford's preferred policy was also to have complete ownership of its foreign affiliates. On the other hand, the Spanish government was very anxious to build up its automobile industry and, especially, to increase exports.

Two years of negotiations at the highest political level ensued, a reflection of the 'new' diplomacy of the global economy in which heads of TNCs talk directly to heads of government. The eventual agreement showed just how powerful Ford's position was. The fact that virtually all other European governments were trying to entice Ford to locate in their countries gave the company substantial negotiating leverage. On the other hand, Ford regarded a Spanish location as vital to its future European operations, although the Spanish government could not be sure of this. Under the agreement finally signed, virtually all of Ford's demands were met. In particular, for Ford the tariff on imported components was reduced from 30 to 5 per cent; the local content requirement was reduced from 95 to 50 per cent, provided that two-thirds of production was exported (precisely what Ford wanted to do anyway); and Ford was allowed 100 per cent ownership of its Spanish subsidiary. But, of course, the benefits were not all one way. In return, Spain gained a massive boost to its automobile industry, which was subsequently enhanced by the entry of other TNCs, including General Motors. As a result, as we shall see in Chapter 10, Spain became one of the world's leading automobile producers.

However, states are not always as weak as is often assumed. Or at least in certain circumstances this is the case. A prime example is China and its policy towards automobile firms.[27] The prospect of access to the world's largest and fastest-growing market led many automobile firms to try to enter China. But the Chinese government has complete control over such entry and has adopted a policy of limited access for foreign firms. Here, then, we have the obverse of the usual situation. Whereas in most cases TNCs play off one country against another to achieve the best deal, in the Chinese case it is the state whose unique bargaining position enables it to play off one TNC against another. Of course, China is something of a special case. But although 'some developing countries have few attractive productive assets or locational advantages for which TNCs will compete with each other, and as a result may not be able to play off one TNC against another ... equally there are many others who can play this game, as they have at least some "bargaining chips" '.[28]

It is important, therefore, not to fall into the usual trap of assuming that the bargaining advantage always lies with the TNC and that the state is always in a weak position. Neither should we assume that a state's bargaining power remains unchanged. The transitional economies of Eastern Europe illustrate this very clearly.[29] As highly centralized, state-controlled economies (though to differing degrees) before 1989, they were in a position to determine the terms on which TNCs could enter, and operate within, their economies. With political liberalization after 1989 came a headlong rush into neo-liberal, market-driven economic policies. This considerably reduced their relative bargaining power as individual states in relation to TNCs.

> Western multinationals enjoyed more favourable terms of entry in Eastern Europe than in other capital-importing regions during earlier phases of FDI. The small size, economic weakness, and geopolitical vulnerability of the East European states prompted local officials to offer foreign investors unusually generous tax holidays and profits repatriation allowances. The international economic conditions prevailing at the time of Eastern Europe's opening further bolstered MNCs' bargaining position. The global ascent of economic liberalism simultaneously lowered national barriers to FDI and intensified bidding for foreign investment among capital-importing countries, allowing Western companies to obtain local-content waivers and other concessions from post-communist governments.[30]

However, the increasing *political* integration of the Eastern European states into the European Union, with its particular regulations on the concessions and incentives that can be granted to TNCs, has enabled those states to retrieve some of their bargaining power. Ford's experience in Hungary is a case in point: 'The dramatic developments of 1989 thwarted the company's plan to use Ford Hungária as a trade-balancing instrument, while Hungary's subsequent convergence towards

Western trade norms hindered Ford's attempts to extract concessions from post-communist governments'.[31] But this was only possible because, in effect, the EU acted as a 'strong state'. Left alone, the post-communist Eastern European countries would have been relatively powerless. As it is, their degrees of bargaining freedom should not be over-exaggerated. As experience throughout Europe shows, the intensity of competition between states for mobile investment is extremely high. There are far more substitutable locations within Europe for potential investors to retain considerable bargaining strength.

Conclusion

The aim of this chapter has been to explore the interrelationships between TNCs and states, relationships that form such an important nexus within the global economy. TNC–state relationships may be both conflictual and cooperative; TNCs and states may be rivals but, at the same time, they may collude with one another. In a real sense, states need firms to help in the process of material wealth creation while firms need states to provide the necessary supportive infrastructures, both physical and institutional, on the basis of which they can pursue their business goals.

TNCs and states are continuously engaged in intricately choreographed negotiating and bargaining processes. On the one hand, TNCs attempt to take advantage of national differences in regulatory regimes (such as taxation or performance requirements, like local content). On the other hand, states strive to minimize such 'regulatory arbitrage' and to entice mobile investment through competitive bidding against other states. The situation is especially complex because while states are essentially fixed and bounded geographically, a TNC's 'territory' is more fluid and flexible.[32] Transnational production networks slice through national boundaries (although not necessarily as smoothly as some would claim). In the process, parts of different national spaces become incorporated into transnational production networks (and vice versa).

There is, in other words, a *territorial asymmetry* between the continuous territories of states and the discontinuous territories of TNCs and this translates into complex bargaining processes in which, contrary to much conventional wisdom, there is no unambiguous and totally predictable outcome. TNCs do not always possess the power to get their own way, as some writers continue to assert. In the complex relationships between TNCs and states – as well as with other institutions – the outcome of a specific bargaining process is highly contingent. States still have significant power *vis-à-vis* TNCs, for example to control access to their territories and to define rules of operation. In collaboration with other states, that power is increased (the EU is an example of this). So, the claim that states are universally powerless in the face of the supposedly unstoppable juggernaut of the 'global corporation' is nonsense; the question is an empirical one.

NOTES

1   Anderson and Cavanagh (2000: 3).
2   Wolf (2002: 9).
3   Gordon (1988: 61).
4   Pitelis (1991).
5   Gilpin (1987: 242). See also Lynn (2005).
6   Stopford and Strange (1991: 233).
7   See Reich (1989).
8   Doz (1986a) provides an extensive discussion of these issues.
9   Leyshon (1992).
10   Picciotto (1991: 43, 46).
11   Yoffie and Milner (1989).
12   Doz (1986a: 231–4).
13   See Levy and Prakash (2003).
14   Nixson (1988: 379).
15   Nixson (1988: 380).
16   UNCTAD (2005: 22–3). See also *Financial Times* (20 June 2006: 3)
17   Vernon (1998: 40).
18   *Financial Times* (3 February 2005).
19   The study was conducted by Ernst & Young and reported in the *Financial Times* (23 November 1995).
20   Liu and Dicken (2006).
21   Gabriel (1966).
22   Doz (1986b: 39).
23   Beck (2005: 53).
24   Kobrin (1987: 611–12).
25   Kobrin (1987: 636).
26   Seidler (1976).
27   Chang (1998b), Liu and Dicken (2006).
28   Chang (1998b: 234).
29   Bartlett and Seleny (1998).
30   Bartlett and Seleny (1998: 320).
31   Bartlett and Seleny (1998: 328).
32   Dicken and Malmberg (2001).

PART THREE
THE PICTURE IN DIFFERENT SECTORS

Nine
'Fabric-ating Fashion': The Clothing Industries

On 1 January 2005, the clothing industries entered a new era. The special international framework which had regulated virtually all trade in the industries for four decades – the Multi-Fibre Arrangement (MFA) – ceased to exist. Trade in clothing (as well as in textiles) was no longer to be subject to import quotas. Cries of anguish emanated from developed country producers, fearing annihilation through competition from developing country producers, especially in Asia and, most of all, from China. On the other hand, developed country retailers were more sanguine, viewing with enthusiasm the prospect of being able to buy their garments more cheaply. But it was not only developed country producers that feared the repercussions of the MFA abolition. Many developing countries had been able to survive in these industries because they had some degree of quota protection. Without that, they too feared the Chinese dragon.

In fact, the clothing industries exemplify many of the intractable issues facing today's global economy, particularly the trade tensions between developed and developing economies. It is not difficult to see why. Millions of workers are officially employed worldwide in the clothing industries, in addition to countless unregistered workers, employed both in factories and in their own homes. These industries were the first to take on a global dimension. They continue to be important sources of employment in the developed economies, employing many of the more 'sensitive' segments of the labour force: females and ethnic minorities, often in tightly localized communities. In developing countries the industries employ predominantly young female workers in conditions that sometimes recall those of the sweatshops of nineteeth century cities in Europe and North America.

The clothing production circuit

The clothing industries form part of a larger production circuit involving textile production, in which each stage has its own specific technological and organizational characteristics and particular geographical configuration (Figure 9.1). The clothing industries are far more fragmented organizationally than textiles and far less sophis-ticated technologically. They are also industries in which subcontracting is especially prominent. Very often the design and even the cutting processes are performed quite separately from the sewing process, the latter being particularly amenable to out-sourcing. The clothing industries produce an enormous variety of often rapidly changing products for a very diverse, and often unpredictable, consumer market. Increasingly, it is the distributors of clothing – particularly the retailers – which have come to play the dominant role in shaping the organization and the geography of the clothing industries. These are strongly *buyer-driven* industries.[1]

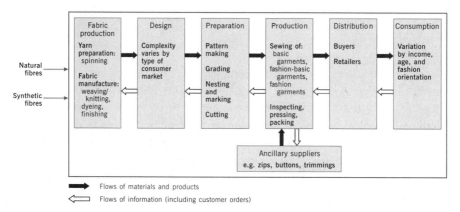

Figure 9.1   The clothing production circuit

Figure 9.2 suggests a generalized sequence of development through which indi-vidual producing countries tend to have passed. Six stages are shown, beginning with the embryonic stage typical of the least developed countries through to the maturity and decline of the older industrialized countries, together with the type of production likely to be characteristic of each stage. The sequence is a useful sum-mary of what has happened so far but, like all such sequential models, it should not be regarded necessarily as being predictive of what will happen in the future. Although many countries have passed through some or all of these stages, the pre-cise path of development depends upon a number of factors that, together, produce specific geographical patterns of these industries.

Global shifts in the clothing industries

The low barriers to entry into clothing manufacture make it one of those activities accessible to virtually any country, even at low levels of economic development. Because of its continuing labour intensity, the map of employment (Figure 9.3) pro-vides a useful indicator of global clothing production. Asia dominates the world map.

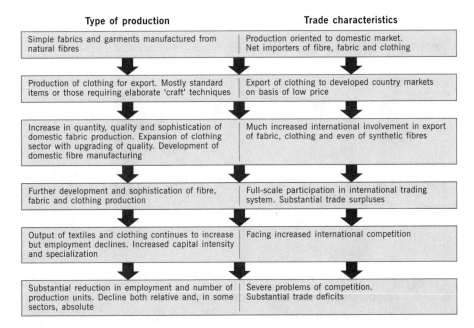

Type of production	Trade characteristics
Simple fabrics and garments manufactured from natural fibres	Production oriented to domestic market. Net importers of fibre, fabric and clothing
Production of clothing for export. Mostly standard items or those requiring elaborate 'craft' techniques	Export of clothing to developed country markets on basis of low price
Increase in quantity, quality and sophistication of domestic fabric production. Expansion of clothing sector with upgrading of quality. Development of domestic fibre manufacturing	Much increased international involvement in export of fabric, clothing and even of synthetic fibres
Further development and sophistication of fibre, fabric and clothing production	Full-scale participation in international trading system. Substantial trade surpluses
Output of textiles and clothing continues to increase but employment declines. Increased capital intensity and specialization	Facing increased international competition
Substantial reduction in employment and number of production units. Decline both relative and, in some sectors, absolute	Severe problems of competition. Substantial trade deficits

Figure 9.2   An idealized sequence of development in the textiles and clothing industries

China is now the world's biggest producer, employing some 2.7 million workers, followed a long way behind by Indonesia, India, Japan, Vietnam and Thailand. Outside Asia, clothing production is highest in Mexico, the United States and Brazil in the Americas, in Italy in Western Europe, in Romania and Poland in Eastern Europe, and in the Russian Federation. There has, indeed, been a major global shift in the clothing industries, away from the old-established producers towards newer ones in the developing world (especially Asia), Mexico and Eastern Europe.

Such geographical shifts are even more apparent when we look at patterns of trade. Figure 9.4 shows the broad regional picture. The overwhelming dominance of Asia as a source of clothing exports, and of Western Europe and North America as destinations for clothing imports, are abundantly clear. The world's leading exporting and importing countries are shown in Table 9.1. China is, by far, the leading individual clothing exporter, a position it has achieved through very high rates of growth over a very short period of time. In 1980, China generated a mere 4 per cent of world clothing exports; today its share is around one-quarter (more if Hong Kong's exports are included in the Chinese total). Very high export growth rates have also occurred recently in Turkey, Mexico (to 2000), Romania and, especially, Vietnam.

In terms of clothing imports, the most striking feature is the spectacular increase in the United States' share. In 1980, 16 per cent of the world total went to the US; in 2003, its share was 30 per cent. The net result is that both the United States and the EU have huge trade deficits in clothing. The US deficit amounts to $66 billion, that of the EU to $41 billion. Between 2004 and 2005, US imports of clothing from

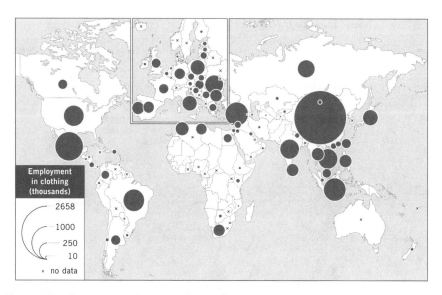

Figure 9.3   Employment in the global clothing industries

Source: calculated from UNIDO *International Yearbook of Industrial Statistics*, 2005; ILO 2005

China accelerated by a remarkable 63 per cent, from India by 38 per cent, and from Bangladesh by 25 per cent. Conversely, imports from Korea and Taiwan declined by 29 and 20 per cent respectively.[2] US clothing imports from Mexico fell by over 5 per cent between 2004 and 2005. Such spectacular shifts appear to be related to China's accession to the WTO and to the abolition of the MFA in 2005 (see later section). Chinese clothing imports to the EU have also increased significantly: from 14 per cent in 1995 to 20 per cent in 2002.

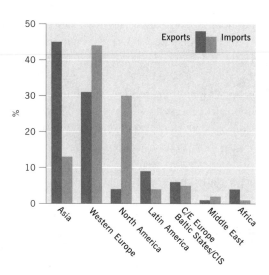

Figure 9.4   Regional shares of world trade in clothing

Source: calculated from WTO, 2004, *World Trade Report*: Chart IV.12

Table 9.1 The world's leading exporters and importers of clothing

(a) Exporters

Exporter	% share world exports		Annual % change			
	1980	2003	1995–2000	2001	2002	2003
China	4.0	23.0	8	2	13	26
EU15 external	10.4	8.4	0	7	5	15
Turkey	0.3	4.4	1	2	21	23
Hong Kong	11.5	3.6	1	−7	−10	−1
Mexico	0.0	3.2	26	−7	−3	−5
India	1.7	2.9	8	−11	10	7
United States	3.1	2.5	5	−19	−14	−8
Bangladesh	0.0	1.9	16	2	−6	8
Indonesia	0.2	1.8	7	−4	−13	4
Romania	–	1.8	11	19	17	25
Thailand	0.7	1.6	−6	−5	−6	7
Korea	7.3	1.6	0	−14	−9	−8
Vietnam	–	1.6	–	3	41	35
Morocco	0.3	1.3	–	−2	4	16
Pakistan	0.3	1.2	6	0	4	22

(b) Importers

Importer	% share world imports		Annual % change			
	1980	2003	1995–2000	2001	2002	2003
United States	16.4	30.2	10	−1	1	7
EU external	23.0	25.6	3	2	6	18
Japan	3.6	8.3	1	−3	−8	11
Canada	1.7	1.9	7	6	2	12
Switzerland	3.4	1.7	−3	0	7	14
Russian Federation	–	1.6	–	13	27	−4
Mexico	0.3	1.3	14	−3	−5	−9
Korea	0.0	1.1	4	25	38	11
Australia	0.8	0.9	8	−12	11	20
United Arab Emirates	0.6	0.8	1	9	15	–
Norway	1.7	0.6	−2	−4	10	12
China	0.1	0.6	4	7	6	5
Hong Kong	0.9	0.4	14	11	−16	−38
Saudi Arabia	1.6	0.4	−2	6	6	13
Singapore	0.2	0.2	−6	−18	18	−2

Source: WTO, 2005: Table IV.69

Changing patterns of consumption

At the most basic level, clothing satisfies one of the most fundamental human needs. But beyond that basic level, demand for clothing becomes more discretionary and subject to a whole variety of complex social and cultural forces, including people's desire to express themselves through their choice of clothing. Clothing can be a highly symbolic good, suggestive of certain self-perceptions and external self-projections. Such variables as income, age, social status, gender and ethnicity, play very important roles. It is a market full of uncertainty and volatility. Much of the business of producing and selling clothing, therefore, depends upon firms' abilities to predict, or to influence, what consumers wish to buy.

Clothing can be divided into three major types: basic garments, fashion-basic garments and fashion garments. The fastest growth is occurring in the fashion-basic segment.[3] The major general determinant of both the *level* of demand and the *composition* of demand (in terms of these three basic categories) is the level and distribution of personal income. Since personal incomes are so very unevenly distributed geographically at the global scale, it is the affluent parts of the world that largely determine the level and the nature of the demand for garments. It is in these markets that demand for fashion-basic clothing is growing most rapidly. Figure 9.5 shows that 45 per cent of the dollar volume of US clothing sales is in the basic product category, while the remaining expenditure is split more or less evenly between fashion and fashion-basic products. On the other hand, the generally low incomes in developing countries clearly restrict the size of their domestic garments markets and produce a consumer focus on the basic segment, although with aspirations among many consumers to move into the fashion-basic segment.

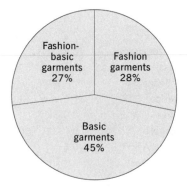

Figure 9.5   Composition of demand for different clothing categories in the United States

Source: based on Abernathy et al., 1999: Figure 1.1

The conventional economic wisdom is that, beyond the level of basic necessities, demand for clothing increases less rapidly than the growth of incomes. This poses a major problem for clothing manufacturers and retailers: they need to stimulate demand through *fashion change*, that is, they need to shift consumer demand away from low-margin basic garments to higher-margin garments. Enormous expenditure goes into promoting fashion products and creating 'designer' labels. Designer labelling is basically a device to *differentiate* what are often relatively similar products and to cater to – and to encourage – the segmentation of market demand for garments. Such a practice covers a very broad spectrum of consumer income levels from the exceptionally expensive to the relatively cheap.

Consumer behaviour in the clothing industries is not just about fashion choice. It is also about concerns that some producers (and retailers) are utilizing dubious labour practices to reduce costs. Some segments of the clothing industry, and some high-profile retailers, have become the target of large-scale anti-sweatshop campaigns. Consumer resistance has come to be a major feature of these industries.[4] As a result of pressure from groups such as Oxfam, from labour unions, and from other anti-sweatshop groups, the major clothing companies have given undertakings to monitor the operations of their suppliers and subcontractors to remove illegal practices, especially employment of child labour. The major UK retailers have promised to end contracts with firms that contravene their guidelines. Similarly in the United States, leading clothing firms (including Nike, Liz Claiborne, Nicole Miller, L.L. Bean and Reebok) subscribe to a voluntary code of conduct to eliminate domestic and overseas sweatshop conditions and to back the Fair Labor Association.

But the process of monitoring and detection is difficult. It is even more difficult to monitor the practice of home-working which, again, tends to be highly exploitative of the most disadvantaged groups who work at home for minimal rates of pay and no benefits. But in an industry as fragmented and organizationally complex as clothing this is an immense task: 'Codes of conduct are awfully slippery. Unlike laws, they are not enforceable.'[5] Despite such confusion, and continuing evasion of such codes by some companies, there is no doubt that some progress has been made in improving conditions in these industries, although problems certainly remain. A recent initiative known as 'Better Factories Cambodia', for example, linked to the ILO and supported financially by such companies as Nike, Reebok, Levi Strauss, Wal-Mart and H&M, is being heralded as a model initiative in the industry. These issues of company codes of conduct will be discussed again in Chapter 19.

Production costs and technology

In clothing manufacture capital intensity is generally low, labour intensity is generally high, the average plant size is small, and the technology is relatively unsophisticated. These characteristics contrast markedly with other parts of the textiles–clothing production circuit, as Figure 9.6 shows.

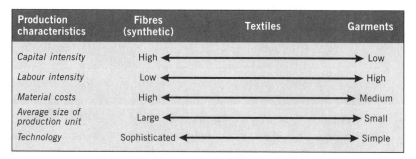

Production characteristics	Fibres (synthetic)	Textiles	Garments
Capital intensity	High ←	→	Low
Labour intensity	Low ←	→	High
Material costs	High ←	→	Medium
Average size of production unit	Large ←	→	Small
Technology	Sophisticated ←	→	Simple

Figure 9.6   Variations in production characteristics between major components of the textiles–clothing production circuit

Labour

Variations in labour costs

Labour costs are the most significant production factor in the clothing industries: Figure 9.7 shows just how wide the labour cost gap can be between different countries. The highly uneven geography of labour costs, and the increased ability of manufacturers to take advantage of such differences because of improvements in the speed and relative costs of transportation and communications, drive most of the locational shifts in the clothing industries. The major advantage of low-labour-cost producers lies in the production of basic items, which sell largely on the basis of price, rather than in fashion garments in which style is more important. The difference between the two is one of *rate of product turnover*. Fashion garments have a rapid rate of turnover, reflecting the idiosyncrasies of particular markets. Geographical proximity to such markets is vital and this helps to explain the survival of many developed country clothing manufacturers. It also partly explains the relative advantage of low-cost countries located close to the major consumer markets of the United States (e.g. Mexico, the Caribbean), Europe (e.g. Central and Eastern Europe, the Mediterranean rim) and Japan (the Asian countries). We will return to this regional situation towards the end of the chapter.

Characteristics of the labour force and conditions of work

Some 80 per cent of the workers in the clothing industries are female. A substantial proportion of the labour force is also relatively unskilled or semi-skilled. The specific socio-cultural roles of women, in particular their family and domestic responsibilities, also make them relatively immobile geographically. A further characteristic of the clothing workforce in the older industrialized countries is that a large number tend to be immigrants or members of ethnic minority groups. This is a continuation of a very long tradition. The early clothing industries of New York, London, Manchester and Leeds in the late nineteenth and early twentieth centuries were major foci for poor Jewish immigrants. Subsequent migrants from other

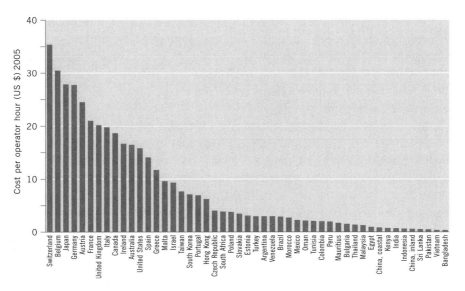

Figure 9.7   Hourly labour costs in the textile and clothing industries 2005

Source: Werner International

origins have also seen the industry as a key point of entry into the labour market. The participation of Italians and Eastern Europeans in both the United States and the United Kingdom has been followed more recently by the large-scale employment of blacks, Hispanics and Asians in the United States and by non-white Commonwealth immigrants in the United Kingdom.

The history of these industries is one of appalling working conditions in sweatshop premises. At least in the clothing industries of the developed economies such conditions are now relatively rare; factory and employment legislation have seen to this. But the sweatshop has certainly not disappeared from the clothing industries of the big cities of North America and Europe. The highly fragmented and often transitory nature of much of the industry makes the regulation of such establishments extremely difficult. The result has been a major resurgence of clothing sweatshops in some big Western cities.[6]

A survey of clothing manufacturing establishments in San Francisco and Oakland, California, in the mid 1990s found 'more than half of them in violation of minimum wage standards. Sewing jobs for Esprit, Liz Claiborne, Izumi and other glittering names were being done by underpaid workers'.[7] Similar problems were uncovered in the United Kingdom in a series of investigations of garments workshops in the big cities: 'Workers earn less than £2 an hour for a 50-hour week. Yet some of the UK's best-known high-street retail chains buy from these manufacturers, even though they appear to break their own guidelines'.[8]

In the rapidly growing clothing industries of the developing countries the labour force is similarly distinctive. Employment tends to be geographically concentrated in the large, burgeoning cities and in the export processing zones. The labour force is overwhelmingly female and predominantly young. Many workers are first-generation factory workers employed on extremely low wages and for very long hours: a seven-day week and a 12- to 14-hour day are not uncommon. Employment in the clothing industry tends to fluctuate very markedly in response to variations in demand.

Hence, a very large number of outworkers is used: women working as machinists or hand-sewers at home on low, piecework, rates of pay. Such workers are easily hired and fired and have no protection over their working conditions. Many are employed in contravention of government employment regulations. Yet there is no shortage of candidates for jobs in these fast-growing industries in some developing countries. Factory employment is often regarded as preferable to un- or underemployment in a poverty-stricken rural environment. A factory job provides otherwise unattainable income and some degree of individual freedom. Often the wages earned are a crucial part of the family's income and there is much family pressure on young daughters to seek work in the city clothing factories or in the EPZs.

Technological change

Both the cost of production and the speed of response to changes in demand are greatly influenced by the technologies used. Technological innovation can reduce the time involved in the manufacturing process and make possible an increased level of output with the same size – or even smaller – labour force. As global competition has intensified in the clothing industries the search for new, labour-saving technologies has increased, especially among developed country producers. Two kinds of technological change are especially important:

- those that increase the speed with which a particular process can be carried out
- those that replace manual with mechanized and automated operation.

The nature of the clothing production process means that the potential for such innovation varies very considerably between the different stages shown in Figure 9.1.[9] In fact, there was relatively little change in clothing technology between the industry's initial emergence in the late nineteenth century and the early 1970s. The manufacture of clothing remains a complex sequence of related *manual* operations, especially in those items in which production runs are short.

> The basic reason is the nature of the production process itself, where two-dimensional materials, i.e. cloth that is rather soft and limp in nature, are subjected to a series of individual labour-intensive handling/assembly steps, culminating in a product which then fits/drapes a three-dimensional human body.[10]

Hence, most of the recent technological developments in the industry, including those based on microelectronic technology, have been in the non-sewing operations:

grading, laying out and cutting material in the pre-assembly stage, and warehouse management and distribution in the post-assembly stage. The application of computer-controlled technology to these operations can achieve enormous savings on materials wastage and greatly increase the speed of the process. For example, the grading process may be reduced from four days to one hour; computer-controlled cutting can reduce the time taken to cut out a suit from one hour to four minutes. But these developments do not reach the core of the problem. The sewing and assembly of garments account for 80 per cent of all labour costs in clothing manufacture. So far only very limited success has been achieved in mechanizing and automating the sewing process.

Current technological developments in the manufacture of clothing are focused on three areas:

- Increasing the *flexibility* of machines, to enable them to recognize oddly shaped pieces of material, pick the pieces up in a systematic manner and align the pieces on the machine correctly, whilst also being able to sense the need to make adjustments during the sewing process.
- Addressing the problem of *sequential operations,* particularly the difficulty of transferring semi-finished garments from one workstation to the next while retaining the shape of the limp material.
- Developing the *unit production system* which will deliver individual pieces of work to the operator on a conveyor belt system. This greatly reduces the amount of (wasted) production time spent by the operator on unbundling and rebundling work pieces. The handling process has been estimated to take up to 60 per cent of the operator's total time.

The drive to introduce such new technologies has been stimulated by very low-cost competition from developing countries. But cost reduction is not the only benefit derived from the new technologies. At least as important, if not more so, are the *time savings* that result from automated manufacture. This has two major benefits:

- Speeding up the production cycle reduces the cost of working capital by increasing the velocity of its use.
- It becomes possible for the manufacturer to respond more quickly to customer demand.

In addition, electronic point-of-sale (EPOS) technologies permit a direct, real-time link between sales, reordering and production. As the production circuit has become increasingly buyer-driven, these IT-based innovations have become extremely important. They not only permit very rapid response to sales and demand at the point of sale but also enable the buyer firm to pass on the costs of producing and holding inventory to the manufacturer.

The role of the state and the Multi-Fibre Arrangement

In developing economies, textiles and clothing manufacture have occupied a key position in national industrialization strategies. Hence, the kinds of import-substituting and export-oriented measures outlined in Chapters 6 and 7 have been applied extensively. But it is in the older-established producing countries of Europe and North America and, more recently, Japan, faced with increasingly severe competition from low-cost producers, that government intervention has been especially marked. The political sensitivity of these industries has forced governments to intervene in three major ways:

- to encourage *restructuring and rationalization* through the use of subsidies and adjustment programmes
- to *stimulate* through offshore assembly (for example, by granting tariff concessions on imports of products assembled abroad using domestic materials) and through preferential trading agreements
- to *protect* from competition from low-cost producers in developing countries.

This third strategy is intimately bound up with the Multi-Fibre Arrangement.

An international regulatory framework: the Multi-Fibre Arrangement

For more than 40 years, the textiles and clothing industries were subject to an industry-specific *international regulatory framework*. Initially formulated, in 1962, as the Long-Term Arrangement to cover cotton textiles, the framework was broadened in 1973 as the *Multi-Fibre Arrangement* (MFA).[11] From that time until January 2005, the MFA regulated most of the world trade in textiles and clothing. Its provisions and their implementation – and avoidance – were major factors in the changing global pattern of production and trade.

The MFA was initially negotiated for a limited period of four years from January 1974. Its principal aim was to create 'orderly' development of trade in textiles and clothing that would benefit *both* developed and developing countries. Access to developed country markets was to increase at an annual average rate of 6 per cent, although this was far below the 15 per cent sought by the developing countries. At the same time, the developed countries were to have safeguards to protect the 'disruption' of their domestic markets. Within the MFA, individual quotas were negotiated setting precise limits on the quantity of textiles and clothing that could be exported from one country to another. For every single product, a quota was specified beyond which no further imports were allowed.

In practice it was the disruptive, rather than the liberalizing, aspect which was at the forefront of trading relationships in these industries. The MFA was renegotiated, or extended, four times (in 1977, 1982, 1986, 1991). Progressively, the MFA became more, rather than less, restrictive. Both the EU and the United States negotiated much tighter import quotas on a bilateral basis with most of the leading developing country exporters and, in several cases, also invoked anti-dumping procedures.

The effects of the MFA on world trade in textiles and clothing have been immense. Without doubt, it greatly restricted the rate of growth of exports from developing countries. A major initial beneficiary of this dampening of the relative growth of developing country exports was the United States, which greatly increased its penetration of European textiles and clothing markets during the 1970s. During the early 1980s, however, it was the European producers who greatly increased their penetration of the US market.

An inevitable consequence of the increased restrictiveness of developing country exports of textiles and garments was a parallel increase in efforts to circumvent the restrictions. Such evasive action has taken a variety of forms. Examples include:

- A producing country which had reached its quota ceiling in one product would switch to another item.
- False labelling was used to change the apparent country of origin (an illegal act).
- Firms relocated some of their production to countries which were not signatories to the MFA or whose quota was not fully used by domestic producers.

As a result, the entire clothing industry of some developing countries was, in effect, created by MFA quotas (Nepal is a case in point).

In 1995, the regulation of trade in textiles and clothing was incorporated into the WTO (see Chapter 19), with the MFA being phased out over a 10-year period (1995–2004), but in three stages. However, the process was 'heavily back-loaded, putting most of the difficult liberalization off to the future'.[12] In fact, both the US and the EU 'integrated' first those products which already entered their markets freely – hardly a major concession. The United States' 10-year liberalization schedule, in effect, left the integration of 70 per cent of imports by value to the very end of the transition period.[13] Not surprisingly, developing countries were extremely unhappy with what they regarded as a deliberate dragging of feet by the world's two largest textiles and garments markets. In response, European and US producers argued that developing countries needed to be more positive in increasing access to imports into their own domestic markets. Finally, on 1 January 2005, the MFA was abolished. But, of course, this was not the end of the story. Both the United States and the EU set up monitoring procedures and negotiated new import quotas with China (in the EU case after a chaotic response to surging imports) to last until 2008.

Inevitably, most of the concern, voiced by both developed and developing country clothing producers, has focused on China, which appears to be the most likely beneficiary of MFA abolition. In a preliminary assessment in mid 2005, the ILO found that 'Bangladesh, Sri Lanka, Cambodia and Indonesia have been able to increase their market shares, while Pakistan and Thailand have been able to maintain theirs'.[14] However, it is still far too early to assess the long-term impact of the abolition of the MFA on the global clothing industries.

Corporate strategies in the clothing industries

The clothing industries are relatively rare instances of globally significant industries that are important in many developing countries, rather than in just a few. But although vast numbers of, mostly small, developing country firms are involved in clothing production, the industry's globalization has been driven, primarily, by developed country firms. Indeed, it is paradoxical that a significant proportion of the clothing imports that are the focus of such concern in developed countries are, in fact, organized by the international activities of those very countries' own firms. But the processes and strategies involved are both complex and dynamic. One fundamental point needs to be made: the globalization of the clothing industries cannot be explained simply as a relocation of production in search of low labour costs. Other factors are involved including, in particular, orientation to specific markets.

The manufacture of clothing is heavily fragmented and far less dominated by large firms than textiles. Even in this archetypal small-firm industry, however, large firms are becoming increasingly important. Only they can afford to invest in the new technologies and to build a worldwide brand image based on mass advertising. Thus, although the clothing industry of most countries is made up of a myriad of very small firms, many of which operate as subcontractors, there is an undoubted trend towards increased concentration.

Three broad categories of clothing company can be identified:[15]

- Producers of basic goods for large markets utilizing economies of scale to lower costs and to be price competitive.
- Operators of small workshops – often in the form of 'sweatshops' in large cities – using immigrant and sometimes undocumented labour. Such firms generally work as short-term subcontractors producing lower-quality garments.
- 'Factoryless' firms that organize entire systems of clothing production. Such major international *retail chains* and *buying groups* exert enormous purchasing power and leverage over clothing manufacturers.

Although the production and retailing of clothing may be fragmented in individual markets, international buying operations are highly concentrated. For example, the Chinese trading and logistics company Li & Fung controls and coordinates all stages of the clothing supply chain, from design and production planning, through finding suppliers of materials and manufacturers of products, to the final stages of quality control, testing and the logistics of distribution.

> It works like this. Say a European clothes retailer wants to order a few thousand garments. The optimal division of labour might be for South Korea to make the yarn, Taiwan to weave and dye it, and a Japanese-owned factory in Guangdong Province to make the zippers. Since China's textiles quota has

already been used up under some country's import rules, Thailand may be the best place to do the sewing. However, no single factory can handle such bulk, so five different suppliers must share the order. The shipping and letters of credit must be seamless and the quality assured ... [organization of the supply chain] requires knowledge. Village women with sewing-machines in Bangladesh are not on the Internet. Finding the best suppliers at any given time, therefore, takes enormous research ... [companies] outsource the knowledge-gathering to Li & Fung, which has an army of 3,600 staff roaming 37 countries.[16]

The company also 'produces' private-label brands for retailers who lack the resources to do this, especially smaller companies. By organizing and managing supply chains over the Internet for such smaller retail chains, Li & Fung can combine many small orders and so achieve economies of scale in production and distribution.[17]

Within the clothing industries, the production circuit has become increasingly dominated by the purchasing policies of the major multiple retailing chains.[18] In the United States, the five largest retailers (including Wal-Mart, Sears, J.C. Penney, Dayton Hudson) account for a huge proportion of clothing sales, as do Daiei, Mitsukoshi, Daimaru and Ito Yokado in Japan. In Germany, the leading garments retailers include Karstadt, Kaufhof, Schickendanz; in the Netherlands, C&A; in France, Carrefour; in Britain, Marks and Spencer and Tesco. However, the extent and the nature of multiple retailer dominance over the clothing market vary a good deal from one country to another.

There has certainly been a 'retailing revolution' in clothing since the 1960s. The result has been the rapid emergence in the United States of specialist clothing retailers – like The Gap, The Limited, Liz Claiborne – serving niche markets. A similar process has occurred in Europe. In the United Kingdom, specialist chains such as Next and New Look have emerged, together with local branches of some of the US firms (notably The Gap). During the 1990s, this trend intensified further with the spread of such firms as Jigsaw, DKNY, Zara, H&M and, more recently, the Japanese company Uniqlo catering to affluent young consumers.

These kinds of development have profound implications for clothing manufacturers. The highly concentrated purchasing power of the large retail chains gives them enormous leverage over clothing manufacturers. When the market was dominated largely by the mass market retailers, demand was for long production runs of standardized garments at low cost. As the market has become more differentiated, and more frequent fashion changes have become the rule, manufacturers have been forced to respond far more rapidly to retailer demands. Under such circumstances, the *time* involved in meeting orders becomes as important as the cost.

In a world where manufacturers must supply an increasing number of products with fashion elements, speed and flexibility are crucial capabilities for firms wrestling with product proliferation, whether they are retailers trying to offer a wide range of choices to consumers or manufacturers responding to retail demands for shipments.[19]

(a) **Traditional apparel retailer–supplier relations**

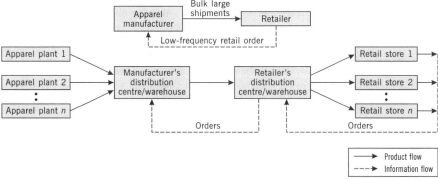

(b) **'Lean' apparel retailer–supplier relations**

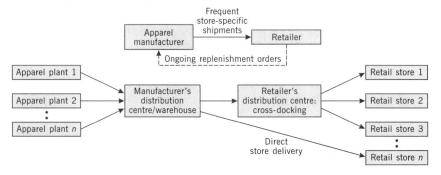

Figure 9.8   Changing relationships between garments manufacturers and retailers

Source: based on Abernathy et al., 1999: Figures 3.1, 4.1

This basic shift in the structure of marketing is having repercussions throughout the clothing industries, and influencing both the adoption of new technologies and also corporate strategies (as we shall see in later sections of this chapter). Although relatively few retail chains are themselves manufacturers of garments, they are very heavily involved in subcontracting arrangements. The production circuit in these industries is becoming transformed into a *buyer-driven* circuit. Figure 9.8 shows how the modern 'lean' manufacturer–retailer system differs from the traditional system. In the new system, deliveries of new/replenishment items are very frequent, based on real-time sales information. Rather than large consignments of garments, much smaller quantities are delivered as needed; little, if any, back-up stock is held by the retailer. Shipments are, for the most part, direct from manufacturer to retailer. Most of the big retailers also operate their own global buying operations.

> An example is Mast Industries (Andover, MA), the captive sourcing arm of The Limited, one of the world's largest multi-brand specialty retailers, which owns brands such as Express, Abercrombie and Fitch, and Victoria's Secret. In 2000, Mast was responsible for the production of $1.5bn worth of apparel

products ($2.5–3bn at retail) and had built more than 400 factory relationships around the world, some through joint ventures with local producers.[20]

The boundary between production and retailing, therefore, is becoming increasingly blurred as the power within the production circuit shifts further towards the buyers (including the department stores, mass merchandisers, discount chains, fashion-oriented firms, as well as the specialized buyers). Precisely how these organize the sourcing of their garments varies according to the position they occupy within the market.

International subcontracting, licensing and other forms of non-equity international investment are extremely pervasive and influential in these industries. In some cases, clothing production is part of a firm's vertically integrated activities. As in the case of textiles, Japanese firms initiated the extensive use of international subcontracting in clothing. During the 1960s and 1970s Japanese companies established subcontracting arrangements in Hong Kong, Taiwan, South Korea and Singapore. Their production was mostly exported to the United States and not to Japan's own domestic market. The Japanese general trading companies (*sogo shosha*) were at the leading edge of these international subcontracting developments, often using minority investments in local firms. Probably 90 per cent of the remaining Japanese overseas garments operations are still located in East and South East Asia, most having been set up in the 1970s.

The strategies developed by US clothing firms to cope with the intensified competition of the 1970s and 1980s were either to focus on the leading edge of the fashion market or to cut costs and raise productivity. The larger US firms increased their level of offshore processing using suppliers in developing countries. Several strategic mixes have been used by those clothing firms that compete in mass markets, but on the basis of brand names supported by extensive advertising. A good example is the jeans manufacturer Levi Strauss. Even by the late 1970s Levi Strauss was spending $50 million a year on worldwide advertising. But increasingly it faced the problem of how to adapt to the demographic change which was altering the size of its traditional market segment of 15- to 24-year-olds without moving too far from its core product, the denim jean.

One strand of Levi Strauss's global strategy was to develop its own branch factories in Western and Eastern Europe, Latin America and Asia. At its peak, Levi Strauss employed directly some 40,000 workers worldwide, of whom 28,000 were in North America, 7000 in Europe and 2000 in Asia. But in the late 1980s, the company began to make massive cuts in its operations in the United States and Europe and to shift more of its operations to lower-cost locations. In 1998, the company closed 13 of its plants in the United States and four in Europe, shedding 7400 jobs. The following year (1999) Levi Strauss closed half of its remaining 22 US factories and eliminated 30 per cent of its US labour force (almost 6000 jobs). At the same time, the company reduced the proportion of its production manufactured in-house to 30 per cent (in 1980, Levi Strauss had manufactured 90 per cent of its own production).[21] In 2002, a further six manufacturing plants were closed in the United States, with the loss of more than 3000 jobs. Finally, in 2003, the company announced the closure of its last four remaining North American manufacturing and finishing plants, with a loss of a further 2000 jobs.[22]

Levi Strauss, therefore, has become an entirely offshore producer. It has, at the same time, entered into licensing agreements with other firms, including Li & Fung, which will sell its own Levi-branded tops directly to US retailers.[23] Levi has also, very reluctantly, had to agree to supply Wal-Mart, something it had resisted for a long time, fearing it would devalue the brand. Retailer power won, once again.

The adoption of offshore production strategies by European clothing firms has been most pronounced among German and British companies. Already by the 1970s around 70 per cent of all the (then West) German clothing firms, including some quite small ones, were involved in some kind of offshore production. Roughly 45 per cent of the arrangements involved international subcontracting; a further 40 per cent involved varying degrees of equity involvement by West German firms in local partners.

The case of the German fashion company, Hugo Boss, provides a good example. Faced with high domestic production costs, Hugo Boss has long used offshore sub-contractors. In 1989, Boss acquired an American garments producer, Joseph & Feiss of Cleveland, Ohio. In 1991, Hugo Boss itself was acquired by Marzotto, the Italian textiles and clothing group. In addition to sourcing an increasing proportion of its garments overseas, the company also moved strongly into retailing through franchising its brand name in around 200 stores worldwide. According to its (German) chairman, 'we are no longer a production-oriented company. Today, we are a company with a strong emphasis on creativity and design, marketing and logistics'.[24] Similarly, the leading British clothing companies have developed a strong focus on offshore production and subcontracting.

Until very recently, Italian firms have been the major exception to this strong shift of production to low-cost foreign locations by European producers. Italy is the only major European country whose clothing industry has continued to perform relatively well in the teeth of intensive global competition. In general, the Italian producers have pursued a strategy of product specialization and fashion orientation with the aim of avoiding dependence upon those types of garment most strongly affected by low-cost competition. This has involved mainly small firms in a decentralized production system, capitalizing on the traditional reputation of specific towns or regions, such as Como, Prato and the like. More recently, however, some Italian firms have established international licensing or production agreements for high-fashion and designer-label products. Armani, for example, is now using some Chinese firms, although the company claims that most of its production remains in Italy.

The best-known Italian company to have developed an especially distinctive strategy, of course, is Benetton.[25] Benetton sees itself not as a manufacturer or retailer of garments but as a 'garments services company'. But it very much sells itself as an 'Italian' company. Whereas most European firms shifted much of their production to Asia, most of Benetton's garments were, until recently, still manufactured in Europe, mostly in Italy, but not by Benetton itself. The company used around 500 subcontractors for its actual production, many of which are located in the Veneto region of north-east Italy.[26] This system gave it considerable flexibility in responding to changing demand for its garments; Benetton itself performed only those functions – mainly design, cutting, dyeing and packing – considered crucial to maintain quality and cost efficiency.

For a long time, Benetton was the only major European clothing firm in the fashion-basic sector to have retained its manufacturing operations in a higher-cost European location rather than relocating to low-cost Asian locations. It did so by producing a relatively limited range of garments, but differentiating them primarily on the basis of colour. Most of the others, including the Swedish company H&M, have followed the Asian route. However, a much newer clothing company, Zara, has made spectacular progress by producing a very wide range of fast-changing fashion-basic garments from its domestic production base. Zara, owned by the Spanish company Inditex, is located in La Coruña in north-west Spain, the traditional focus of the Iberian textile industry.

> At Inditex's heart is a vertical integration of design, just-in-time production, delivery and sales. Some 300 designers work at the firm's head office ... Fabric is cut in-house and then sent to a cluster of several hundred local co-operatives for sewing ... Production is deliberately carried out in small batches to avoid oversupply. While there is some replenishment of stock, most lines are replaced quickly with yet more new designs rather than more of the same. This helps to create a scarcity value ... The result is that Zara's production cycles are much faster than those of its nearest rival, Sweden's Hennes & Mauritz (H&M). An entirely new Zara garment takes about five weeks from design to delivery; a new version of an existing model can be in the shops within two weeks. In a typical year, Zara launches some 11,000 new items, compared with the 2,000 to 4,000 from companies like H&M or America's giant casual-fashion chain, GAP.[27]

In other words, not only has Zara persisted with a manufacturing model long ago jettisoned by the major US and European garments companies (that is, producing most of its garments in-house) but also it operates within a highly volatile part of the fashion-basic sector of the industry. Zara has achieved dramatic results by combining highly efficient production and distribution logistics with a continuous monitoring of the fashion scene. In 2006, Zara's parent company Inditex overtook H&M to become Europe's largest clothing retailer.[28]

Regionalizing production networks in the clothing industries

Much of the explanation for the recent global shifts in the clothing industries can be explained in terms of the trade-off between labour costs on the one hand and the need for market proximity on the other, set within the regulatory constraints of the MFA. The result is increasing *regionalization* of clothing production networks. Figure 9.9 shows that a significant, though varying, proportion of clothing trade is *intra-regional*. In the light of these developments, let us look briefly at the three major global regions: East Asia, the Americas and Europe.

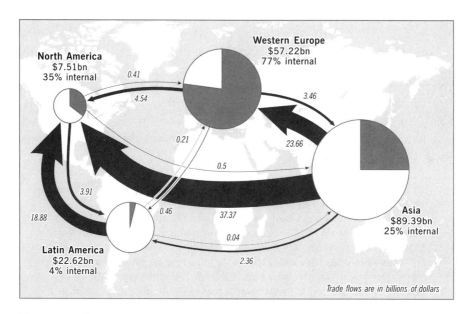

Figure 9.9   Global trade network in clothing

Source: calculated from WTO, 2004, *World Trade Report*. Table A10

Regional restructuring of the East Asian clothing production network

The key to the internal transformation of the industry in Asia lies in the changing strategies of the northern tier East Asian NIEs.[29] In the face of increasing competitive pressures from the newer wave of Asian producers (notably China, Malaysia, Thailand, Indonesia and, more recently, Vietnam and Cambodia), as well as restrictions on their trade with North America and Europe through the MFA, firms from Hong Kong, South Korea and Taiwan have progressively shifted their production offshore. In fact Hong Kong firms began to shift clothing production to other Asian countries as early as the mid 1960s. With the tightening grip of the MFA in the 1970s, Hong Kong firms set up plants in the Philippines, Thailand, Malaysia and Mauritius and, subsequently, in Indonesia and Sri Lanka, to get round quota restrictions. In the past decade, a huge number of investments have been made in China. East Asian firms have also begun to establish plants in Europe and North America (including the Caribbean) to serve developed country markets directly.

This was more than simple geographical relocation; specifically it involved upgrading to focus on design and marketing functions through the building of networks of buyers and sellers.[30] The upgrading process has consisted of three broadly sequential phases:

- simple assembly of basic garments for export
- subcontract manufacturing to design specified by the buyer with the product sold under the buyer's brand name (original equipment manufacturing, OEM)
- development of own-brand manufacturing (OBM) capability.

Of course, not all firms have followed this sequence. The most successful appear to have been Hong Kong firms. For example,

> The women's clothing chain, Episode, controlled by Hong Kong's Fang Brothers Group, one of the foremost OEM suppliers for Liz Claiborne in the 1970s and 1980s, has stores in 26 countries, only a third of which are in Asia. Giordano, Hong Kong's most famous clothing brand, has added to its initial base of garments factories 200 stores in Hong Kong and China, and another 300 retail outlets scattered across Southeast Asia and Korea. Hang Ten, a less-expensive line, has 200 stores in Taiwan, making it the largest foreign-clothing franchise on the island.[31]

The organization of the East Asian garments production complex has a particular 'geometry', that of *triangle manufacturing*:

> The essence of triangle manufacturing ... is that US (or other overseas) buyers place their orders with the NIE manufacturers they have sourced from in the past, who in turn shift some or all of the requested production to affiliated offshore factories in low-wage countries (e.g. China, Indonesia, or Guatemala). These offshore factories can be wholly owned subsidiaries of the NIE manufacturers, joint-venture partners, or simply independent overseas contractors. The triangle is completed when the finished goods are shipped directly to the overseas buyer ... Triangle manufacturing thus changes the status of NIE manufacturers from established suppliers for US retailers and designers to 'middlemen' in buyer-driven commodity chains that can include as many as 50 to 60 exporting countries.
>
> Triangle manufacturing is socially embedded. Each of the East Asian NIEs has a different set of preferred countries where they set up their new factories ... These production networks are explained in part by social and cultural factors (e.g. ethnic or familial ties, common language), as well as by unique features of a country's historical legacy.[32]

Geographical proximity is also an influential factor. For example, Hong Kong and Taiwanese firms are the major investors in China while Singaporean firms are strongly represented in Malaysia and Indonesia.

As Figure 9.9 shows, Asia remains the most globally connected region of clothing production. Around one-quarter of its clothing trade is intra-regional compared with much higher levels in North America and, especially, in Europe. However, Asia has undoubtedly become a major market in its own right: in 1980 less than 5 per cent of Asian garments trade was intra-regional.

US-focused regional production networks in the Americas

Traditionally, the US clothing market was served primarily by domestic production. But, in the past few decades, the market has become increasingly dominated by imports from low-cost producing countries, especially in Asia.[33] Most of these imports are organized through a buyer–retailer–supplier complex that has become increasingly important. Table 9.2 shows a close association between the type of retailer (and the market served) and the locational sources of garments. In general, fashion-oriented companies source from Europe and first-generation NIEs while the discount stores source from lower-cost countries.

As we have seen, China has emerged as the leading supplier of garments for the US market. But China – or any other East Asian producer – is unlikely to dominate completely for two reasons:

- The trade-off between minimizing production costs and maximizing speed of access to consumer markets has become more critical. Proximity to markets has become a key factor in determining the geography of clothing production as the dominant buyers/retailers insist on fast product turnover.
- The development of regional economic initiatives by the United States, in particular the signing of the NAFTA and the preferential arrangements with the Caribbean countries (the Caribbean Basin Initiative), has reinforced the benefits of geographical proximity driven by changing buyer–supplier relationships.

Initially, apart from the rapid growth in Chinese imports, the other major growth area for imports into the United States was the Caribbean Basin, whose share of the total grew from less than 3 per cent in 1981 to 13 per cent in 1995. During the same period, Mexico's share of US clothing imports grew from 3 per cent to 7 per cent. By 2000, Mexico had overtaken China to become the leading source of clothing imports into the United States – a dramatic turnaround indeed, or so it seemed. Today, around one-third of the North American clothing trade is internal, compared with one-quarter only five years ago.

Part of the explanation for this increased intra-regional trade lies in Mexico's emergence as the United States' single most important source of clothing imports. This, in turn, derives from the signing of the NAFTA in the early 1990s. Tariffs and quotas on clothing imported from Mexico to the United States have been eliminated. In addition, under the rules of origin for garments in the NAFTA, clothing must be cut and sewn from fabric made from fibre originating in North America in order to qualify for duty-free access. This provided a stimulus for the development of a more integrated industry within North America. Thus, 'through NAFTA,

Table 9.2 Global sourcing patterns of US clothing retailers

Type of retailer	Types of order	Main sourcing countries
Fashion-oriented companies (e.g. Armani, Donna Karan, Boss, Gucci, Polo/Ralph Lauren)	Expensive 'designer' products High level of skill Orders in small quantities	Italy, France, UK, Japan South Korea, Taiwan, Hong Kong, Singapore
Department stores, specialty stores, brand-name companies (e.g. Bloomingdales, Saks Fifth Avenue, Neiman-Marcus, Macy's, The Gap, The Limited, Liz Claiborne, Calvin Klein)	Top-quality, high-priced garments sold under variety of national brands and private labels (store brands) Medium to large orders, often coordinated by store buying groups	South Korea, Taiwan, Hong Kong, Singapore Malaysia, Indonesia, Philippines, southern China, India, Turkey, Egypt, Brazil, Mexico, Thailand Dominican Republic, Jamaica, Haiti, Guatemala, Honduras, Costa Rica, Colombia, Chile, Poland, Hungary, Czech Republic, Bulgaria, Kenya, Zimbabwe, Mauritius, Macao, Pakistan, Sri Lanka, Bangladesh, interior China, Tunisia, Morocco, UAE, Oman
Mass merchandisers (e.g. Sears, Mongomery Ward, J.C. Penney, Woolworth)	Good-quality, medium-priced goods Mostly sold under private labels Large orders	South Korea, Taiwan, Hong Kong, Singapore Malaysia, Indonesia, Philippines, southern China, India, Turkey, Egypt, Brazil, Mexico, Thailand Dominican Republic, Jamaica, Haiti, Guatemala, Honduras, Costa Rica, Colombia, Chile, Poland, Hungary, Czech Republic, Bulgaria, Kenya, Zimbabwe, Mauritius, Macao, Pakistan, Sri Lanka, Bangladesh, interior China, Tunisia, Morocco, UAE, Oman

(Continued)

Table 9.2 (Continued)

Type of retailer	Types of order	Main sourcing countries
Discount chains (e.g. Wal-Mart, Kmart, Target)	Low-priced goods Store-brand names Very large orders	Malaysia, Indonesia, Philippines, southern China, India, Turkey, Egypt, Brazil, Mexico, Thailand Dominican Republic, Jamaica, Haiti, Guatemala, Honduras, Costa Rica, Colombia, Chile, Poland, Hungary, Czech Republic, Bulgaria, Kenya, Zimbabwe, Mauritius, Macao, Pakistan, Sri Lanka, Bangladesh, interior China, Tunisia, Morocco, UAE, Oman Qatar, Peru, Bolivia, El Salvador, Nicaragua, Vietnam, Russia, Lesotho, Madagascar, North Korea, Myanmar, Cambodia, Laos, Yap, Maldives, Fiji, Cyprus, Bahrain
Small importers	Pilot purchase and special items Sourcing done for retailers by small importers acting as 'industry scouts' in seeking out new sources of supply Relatively small orders initially but could grow rapidly	Dominican Republic, Jamaica, Haiti, Guatemala, Honduras, Costa Rica, Colombia, Chile, Poland, Hungary, Czech Republic, Bulgaria, Kenya, Zimbabwe, Mauritius, Macao, Pakistan, Sri Lanka, Bangladesh, interior China, Tunisia, Morocco, UAE, Oman Qatar, Peru, Bolivia, El Salvador, Nicaragua, Vietnam, Russia, Lesotho, Madagascar, North Korea, Myanmar, Cambodia, Laos, Yap, Maldives, Fiji, Cyprus, Bahrain

Source: based on Gereffi, 1994: Figure 5.2 and Table 5.3

apparel and textile manufacturers are acquiring the freedom and flexibility to create – duty and quota free – transborder production networks that best suit their individual needs'.[34]

Because Mexico's comparative advantage lies in clothing production while the United States' comparative advantage lies in textile manufacture, synthetic fibre production and retailing, a clear division of labour is emerging between the two countries. Certainly, the combination of the NAFTA and the benefits of geographical proximity together with low production costs have stimulated many clothing firms in the United States to source more of their garments from Mexico. A study of clothing manufacturers in southern California, for example, found that 60 per cent of those involved in offshore sourcing in 1998 used Mexico, compared with only 10 per cent in 1992.[35]

The precise form of this geographical division of labour is still evolving. Traditionally, Mexico's clothing industry was dominated by *maquiladora* production: simple sewing of garments made from imported fabrics and using extremely cheap labour (see Figure 2.30). In other words, it was dominated by the very basic operations in a vertically integrated system coordinated and regulated by US manufacturers and retailers. Although this is still the dominant mode of operation, there are signs of rather more sophisticated arrangements in which Mexican firms perform some of the higher-level functions in the production chain. There is some evidence of the development of *full-package production* in which selected local manufacturers are responsible for the entire process of clothing production. Figure 9.10 shows an example of this development in the Torreón district of Coahuila in northern Mexico.

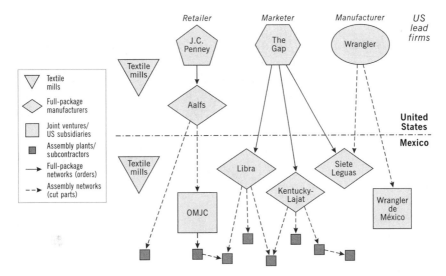

Figure 9.10   Development of 'full-package' garments production in Torreón, Mexico

Source: based on Bair and Gereffi, 2001: Figure 2

At the same time, the geographical pattern of clothing production within Mexico is changing.

> There appears to be a significant shift in southern California sourcing from the western border region to central Mexico and, to a lesser extent, the northern states (excluding the border cities) and the Yucatán peninsula. The central states of Guanajuato, Puebla, Tlaxcala and the greater Mexico City area figure prominently as new production sites. Also included is the west coast state of Jalisco, located due west of Mexico City.[36]

Major reasons for the relative shift away from the border are that labour costs in the interior of Mexico tend to be lower than along the border while the quality of production tends to be higher. There is also a much lower rate of labour turnover away from the border zone.

The spectacular and rapid growth of Mexico as a source of clothing imports into the United States has significant implications for the Caribbean Basin producers, to whom the level of preferential access to the North American market under NAFTA is denied. In particular, the Caribbean countries must still pay import duty on the value added in the clothing assembly process, a contentious issue for these producers. Currently, the Caribbean producers are striving to redress this unequal treatment. However, with the abolition of the MFA, the huge shadow of China hangs over the North American clothing industries. China's share of United States' clothing imports had already once again overtaken those of Mexico by 2003, two years before the end of the MFA. In this case, the battle between low production costs and market proximity is still being waged.

Reshaping the clothing map of 'Greater Europe'

As Figure 9.9 shows, more than three-quarters of Europe's clothing trade is intra-regional, more that twice the level of North America and three times that of Asia.[37] Historically, most European clothing production was located in the leading European economies themselves, notably France, Germany, Italy and the United Kingdom. During the past three decades, however, with the partial – though declining – exception of Italy, the clothing industries of these countries have experienced massive decline and restructuring. Some of this has been caused, of course, by the rise of the low-cost Asian producers, but much of it is the result of the geographical reconfiguration of clothing production within what might be called 'Greater Europe': the EU15, together with countries in Eastern and Central Europe (ECE) and the Mediterranean rim. In Europe, of course, the high level of politico-economic integration embodied in the EU is a critical influence.

Through the 1980s and 1990s, therefore, European clothing production networks became geographically more extensive, but within an expanded regional context. This is a situation created by the intersection of the changing sourcing strategies of clothing firms and the changing political agreements with ECE and

Mediterranean countries, some of which became – or will become in the future – members of the EU. It is a pattern with some clear geographical consistencies but also with considerable dynamism, as some supplier countries lose their dominance and others emerge.

As in the case of Mexico and the Caribbean countries, the advantages of geographical proximity in the clothing industries of Europe (in terms of their effect on speed of delivery) can offset lower production costs at more distant locations: 'It takes 22 days by water to reach the UK from China, while products from Turkey can take as little as five days to arrive'.[38] Of course, both sets of forces operate. The continued attraction of low-cost sourcing of garments is shown most graphically by China's increased share of the EU clothing market, from 14 per cent in 1995 to 20 per cent in 2002, and rising with the abolition of the MFA, as we have seen. Thus, although a substantial proportion of the EU's sourcing of clothing still takes place in Asia, the countries on the immediate geographical periphery of the EU have become tightly integrated into the production networks of European clothing manufacturers and the purchasing networks of European retailers.

This process of regionalization of European clothing production networks became increasingly common from the early 1980s, when Outward Processing Trade (OPT) provisions were introduced by the EU. These established import quotas between the EU and individual countries in Eastern and Southern Europe. Such provisions facilitated significant patterns of production relationship between EU clothing firms (both producers and retailers) and clothing manufacturers in lower-cost countries nearby. German firms were especially heavily involved in such activities. Overall, more than 80 per cent of all West German garments imports manufactured under subcontracting arrangements came from the former East Germany, Poland, Hungary, Romania, Bulgaria and the former Yugoslavia.[39] Conversely, Italian producers, until very recently, continued to keep production at home. However, this has changed, though less drastically. Benetton plans to produce 80 per cent of its clothing items outside Italy by 2007, mostly in Hungary, Croatia and Tunisia. This compares with 60 per cent currently and around half that five years ago.[40] Similar trends have occurred among the clothing manufacturers and retailers in the other EU countries, often with distinctive geographical sourcing patterns with preferred supplier countries.

The regional reconfiguration of the clothing industries in Europe between 1989 and 2000 can be summarized as follows: [41]

- The phasing out of quotas in the final stages of the MFA (from 1994) favoured the Eastern and Central European (ECE) and Mediterranean countries far more than the Asian suppliers. 'Between 1991 and 1995 OPT quotas for the ECE grew at the rate of 36.2%, whereas those for Asia were growing at only 6.9%'.[42]
- A series of preferential trade agreements was signed between the EU and applicant countries in the early 1990s, which facilitated greater integration of the region's clothing industries.

- 'Among the top ten suppliers ... ECE and the Mediterranean countries increased their share of EU apparel imports from 26.8% in 1989 to 30.8% in 2000, with Romania, Tunisia, Morocco, and Poland being the largest suppliers from the region'.[43]
- 'ECE producers have become – with the exception of the former Yugoslavia – much more important sources for the EU apparel market during the 1990s. Romania, Hungary, and Poland played the leading role in the early part of the 1990s; Bulgaria and, to a lesser extent, Slovakia, have also become increasingly important suppliers to EU apparel markets, albeit from smaller bases'.[44]
- Turkey has emerged as the second most important individual clothing supplier to the EU after China, with more than 10 per cent of EU clothing imports.

Conclusion

The strategies adopted by clothing producers are extremely varied and complex. The combinations of technological innovation, different types of internationalization strategy, the relationship with retailers, and the constraints of the Multi-Fibre Arrangement have combined to produce a more complex global map of production and trade than a simple explanation based on labour cost differences would suggest.

The manufacture of clothing is an ideal candidate for international subcontracting. It is highly labour intensive; uses low-skill or easily trained labour; and the process can be fragmented and geographically separated, with design and often cutting being performed in one location (usually a developed country) and sewing and garments assembly in another location (usually a developing country). Although international subcontracting in clothing manufacture knows no geographical bounds – with designs and fabrics flowing from the United States and Europe to the far corners of Asia, and finished garments flowing in the opposite direction – there are strong and intensifying regional biases in the relationships. These industries are becoming *globally regionalized*. At all geographical scales, however, it is the big buyers and dominant retailers who call the shots in clothing production.

NOTES

1 See Gereffi (1994) and subsequent writings.
2 ILO (2005: Table 3.2).
3 Abernathy et al. (1999: 9).
4 See, for example, Klein (2000), Ravoli (2005).
5 Klein (2000: 430).
6 Ross (2002).

7 *The Economist* (12 February 1994).

8 *Financial Times* (2 October 1996).

9 Detailed accounts of technology in the clothing industries are provided by Abernathy et al. (1999), OECD (2004: Chapter 4).

10 OECD (2004: 139).

11 See Hoekman and Kostecki (1995: Chapter 8), ILO (2005), Nordås (2004), OECD (2004), Ravoli (2005).

12 Hoekman and Kostecki (1995: 209).

13 *Financial Times* (10 January 1996).

14 ILO (2005: 13).

15 Glasmeier et al. (1993: 23–4). See also Bair and Gereffi (2002).

16 *The Economist* (2 June 2001).

17 Schary and Skjøtt-Larsen (2001: 383–4).

18 The increasingly significant role of the retailers in the clothing industries is discussed by Abernathy et al. (1999; 2004), Gereffi (1999), Taplin (1994).

19 Abernathy et al. (1999: 9).

20 Sturgeon and Lester (2002: 47–8).

21 *Financial Times* (23 February 1999).

22 *Financial Times* (26 September 2003).

23 *Financial Times* (20 August 2003).

24 *Financial Times* (9 January 1996).

25 Jarillo (1993), Schary and Skjøtt-Larsen (2001) provide accounts of Benetton's operations.

26 Schary and Skjøtt-Larsen (2001: 103).

27 *The Economist* (18 June 2005).

28 *Financial Times* (30 March 2006).

29 Gereffi (1996; 1999) and Khanna (1993) provide detailed analyses of the changing industry in Asia. Dicken and Hassler (2000) analyse Indonesian clothing networks within East Asia.

30 Gereffi (1999: 51).

31 Gereffi (1999: 56).

32 Gereffi (1996: 97–8).

33 Abernathy et al. (1999; 2004), Bair (2002), Bair and Gereffi (2003), Gereffi and Memedovic (2004), Gereffi et al. (2002), Kessler (1999).

34 Kessler (1999: 569).

35 Kessler (1999: 577–9).

36 Kessler (1999: 584).

37 This section is based primarily on Begg et al. (2003). See also Smith et al. (2005), Palpacuer et al. (2005).

38 *Financial Times* (30 August 2005).

39 Fröbel et al. (1980).

40 *Financial Times* (23 July 2004).

41 Begg et al. (2003).

42 Begg et al. (2003: 2202).

43 Begg et al. (2003: 2194).

44 Begg et al. (2003: 2194–5).

Ten
'Wheels of Change':
The Automobile Industry

Few industries have attracted as much attention as automobiles. For most of the twentieth century it was *the* key manufacturing industry; what Peter Drucker called 'the industry of industries'.[1] The internal combustion engine was, quite literally, the major engine of growth until the middle 1970s and is still seen as a key contributor to industrial development. The industry's significance lies both in its scale and in its linkages to many other manufacturing industries and services. Around 4 million people are directly employed around the world in making automobiles and a further 9–10 million are employed in supplier industries. If we add those involved in selling and servicing vehicles, we reach a total of around 20 million workers. Not only this:

> Its products are responsible for almost half the world's oil consumption, and their manufacture uses up nearly half the world's output of rubber, 25% of its glass and 15% of its steel.[2]

In contrast to the clothing industries, the world automobile industry is predominantly an industry of very large corporations, which have increasingly organized their activities on transnationally integrated lines. In so doing, they engage very closely – sometimes collaboratively, sometimes conflictually – with national governments, themselves anxious to establish, nurture or enhance automobile production within their territories. It is not surprising, therefore, to find that competitive bidding for investment is especially prevalent in this industry.

The automobile production circuit

The automobile industry is essentially an *assembly* industry. It brings together an immense number and variety of components. At the centre of the automobile production circuit (Figure 10.1) is the complex set of relationships between the assemblers of vehicles and the suppliers of components, which account for between 50 and 70 per cent of the cost price of the average car.[3] As Figure 10.1

shows, there are three major processes prior to final assembly: the manufacture of bodies, of components, and of engines and transmissions, which may be performed by the assemblers as part of a vertically integrated sequence. However, there is a strong trend towards the deverticalization of automobile production as assemblers pass more responsibility to the suppliers. Figure 10.1 shows just three tiers of suppliers, although there can be more.[4] The first-tier suppliers supply major components systems direct to the assemblers and have significant research and development and design expertise. Second-tier suppliers generally produce to designs provided by the assemblers or by the first-tier suppliers, while third-tier suppliers provide the more basic components.

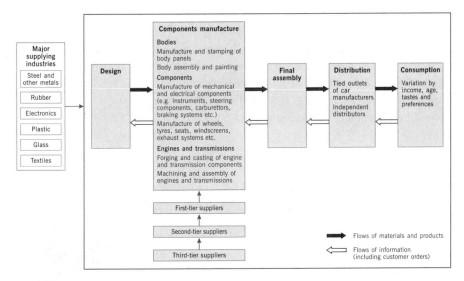

Figure 10.1   The automobile production circuit

Historically, the development of a country's automobile industry has followed a generalized sequence, although it is by no means inevitable that all countries will actually pass through each stage.

- *Stage 1*, the import of complete vehicles, tends to be limited in extent because of high transportation costs (cars are essentially large empty boxes) and by government import restrictions.
- *Stage 2*, local assembly of vehicles from a full 'kit' of component parts, permits transportation cost savings and provides the opportunity to make minor product modifications for the local market.
- *Stage 3*, assembly involving a mix of imported and locally sourced components, both depends upon and encourages the development of a local components industry. Not surprisingly, it is a stage strongly favoured by national governments as a potential entry to stage 4.

- *Stage 4*, the full-scale manufacture of automobiles. This stage is restricted to a far smaller number of countries than stages 2 and 3. It is not inevitable that countries that have reached stage 3 will then move to full-scale local manufacture. At the same time, some that have may even regress from the status of full-scale local manufacturer back to that of mere assembler.

Global shifts in the automobile industry

Since 1960, the global production of cars has increased more than threefold. At the same time, its geography has changed substantially. Figure 10.2 maps current production, which is very strongly concentrated in Europe (43 per cent), East Asia (32 per cent) and North America (18 per cent). These three regions contain 93 per cent of the global total. In fact, the level of geographical concentration is even greater than these regional figures suggest. More than two-thirds of global production is concentrated in just seven countries. Of these, Japan is, by a large margin, the world's leading automobile producer, followed by Germany, the United States, France, Korea, Spain and China.

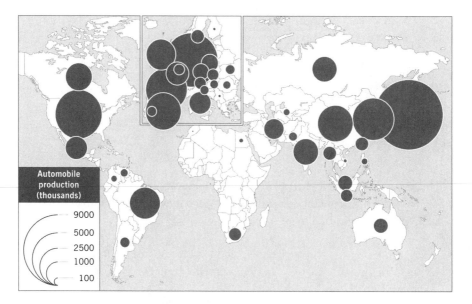

Figure 10.2   Global production of passenger cars

Source: calculated from SMMT, 2005, *World Automotive Statistics*

Today's global production map is the outcome of some profound developments over the past four decades as new centres of production have emerged and as older centres have declined in importance (Figure 10.3). By far the most dramatic development was the spectacular growth of the Japanese automobile industry. Lower, but nevertheless very impressive, rates of growth also occurred in Korea, Spain and China. There has also been significant growth of automobile production in the emerging market economies of Eastern Europe, notably in the Czech Republic, Poland, Hungary and Slovakia. Conversely, the former dominant producer, the United States, saw its share of world automobile production fall dramatically.

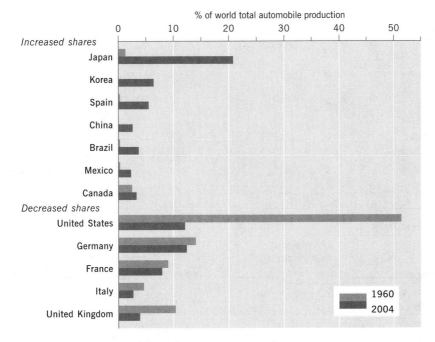

Figure 10.3   Major changes in the relative importance of automobile producing countries

A high level of geographical concentration is also evident in the pattern of automobile *trade*. Table 10.1 shows the positions of the leading exporting and importing countries in 1980 and 2003. Almost two-thirds of automobile exports originate from 15 countries (for these purposes, the EU is treated as a single entity). The United States has an enormous trade deficit in automobiles of $112 billion, while Japan has an almost comparable trade surplus of $92 billion. Such imbalances are the cause of considerable political friction.

In summary, an industry dominated in 1960 by the United States and, to a much lesser extent, Europe was transformed during the 1970s and 1980s by the spectacular growth of Japan as a leading automobile producer. This was reflected in terms of growth of production in Japan itself, of Japanese exports to the rest of the world, and of the increasing proportion of Japanese automobile production located abroad. Japan has the most geographically extensive export network of all

Table 10.1   The world's leading exporters and importers of automotive products

(a) Exporters

Exporter	% share world exports		Annual % change			
	1980	2003	1995–2000	2001	2002	2003
EU15 external	19.5	17.3	4	5	17	22
Japan	19.8	14.2	2	−9	15	11
United States	11.9	9.6	5	−6	6	3
Canada	6.9	7.9	7	−9	2	1
Mexico	0.3	4.2	17	0	1	−3
Korea	0.1	3.1	11	2	10	31
Czech Republic	−	1.1	−	19	16	23
Hungary	0.6	1.0	−	12	12	20
Poland	0.6	0.7	−	6	23	37
Brazil	1.1	0.9	10	3	2	33
Slovakia	−	0.8	−	−5	23	111
Turkey	0.0	0.7	19	54	35	55
Thailand	0.0	0.5	38	11	8	38
China	0.0	0.5	21	20	41	33
Taiwan	−	0.4	6	−2	14	23

(b) Importers

Importer	% share world imports		Annual % change			
	1980	2003	1995–2000	2001	2002	2003
United States	20.3	24.7	10	−3	7	3
EU15 external	5.3	9.1	10	2	12	30
Canada	8.7	6.7	7	−9	11	5
Mexico	1.8	2.7	35	−2	9	−5
China	0.6	1.7	8	29	42	84
Japan	0.5	1.5	−4	−7	7	13
Australia	1.3	1.5	7	−15	18	30
Switzerland	1.8	1.0	0	3	−1	11
Poland	0.9	0.9	−	7	8	35
Turkey	−	0.8	28	−64	31	122
Saudi Arabia	2.7	0.8	12	36	5	11
Russian Federation	−	0.8	−	57	19	27
Czech Republic	−	0.7	−	23	19	26
Hungary	0.4	0.5	−	2	22	27
Slovakia	−	0.5	−	25	25	66

Source: WTO, 2005: Table IV.53

the major producing countries. Its tentacles spread throughout the world to a greater extent than those of the other major producers. The recent emergence of China as a major producer of automobiles may soon be enhanced by its increasing importance as an exporter of cars. In 2005, for the first time, China exported more cars than it imported; its sights are set on breaking into the North American market in 2007.[5]

More broadly, the industry has become increasingly concentrated in the three major global regions. In the case of North America, this explains the recent fast growth of Mexico as a major automobile producer and exporter and of Canada's continued significance. In Europe, it explains the emergence of Spain as a major producer and the growth of automobile production in some of the emerging market economies of Eastern Europe, notably the Czech Republic, Poland, Hungary and Slovakia. In Asia, Korea emerged virtually from nowhere to become an important player, joined, more recently, by the very rapid growth of an automobile industry in China. Much of this reconfiguration of global automobile production is related to developments at the 'macro-regional level', as we shall see in the final section of this chapter.

Changing patterns of consumption

The intoxicating allure of private transportation makes the automobile one of the most significant of all *aspirational goods*. Car ownership offers (at least in theory) immense personal freedom to travel to places otherwise inaccessible (including travel to work, to shop, to engage in all kinds of recreation). It also – and just as importantly – embodies a complex range of symbolic attributes through which people can project their self-images and their social position, and indulge in the fantasies of driving. Certain kinds of vehicle become emblematic of particular lifestyles. The obvious example today is the sports utility vehicle (SUV), which is rarely used for those purposes for which it was originally designed (how many SUV drivers ever use them off-road, or even know how to do so?). At one stage, it was estimated that SUVs (known as 'axles of evil' by critics of their environmental damage) accounted for three out of every four vehicles sold in the United States. In one respect, therefore, car ownership is a *discretionary* good. However, in some countries – notably the United States and the United Kingdom – the failure to continue to invest in high-quality, accessible public transportation systems means that car ownership has actually become an absolute *necessity* for people to be able to go about their daily lives.

Changes in personal income levels and in consumer tastes and preferences over time, as well as the extent to which car ownership has already spread through the population, are key variables in influencing the demand for automobiles. Such demand has always been volatile. However, it has become significantly more volatile – and more complex – in recent years. Three interrelated characteristics of the market for new automobiles are especially important:

- It is highly cyclical.
- There are long-term (secular) changes in demand.
- There are signs of increasing market segmentation and fragmentation.

Demand for automobiles is geographically very uneven, growing most slowly in Western Europe and North America. Even the demand for SUVs in the United States has suddenly plummeted in response to increasing gasoline prices.[6] In both of these mature markets there is huge excess capacity (25–30 per cent, equivalent to several million automobiles). The major automobile manufacturers, therefore, are pinning their hopes on continuing high levels of growth in demand elsewhere, especially in Asia, Eastern Europe and Latin America.

Of these, it is in Asia that the major growth is expected to occur: 'As disposable incomes across the region rise, the 50cc motorcycles that swarm the streets of most of the region's cities will eventually be upgraded to cars'.[7] This can be seen already, especially in the cities of China where automobile sales have grown at such a phenomenal rate that it is now the third largest automobile market in the world. However, the growth potential of Asia, at least in the short-to-medium term, may be exaggerated.

The slow growth in demand for automobiles in the mature markets is more than merely cyclical. There are deeper *secular* or structural characteristics in these markets that limit future growth in car sales. Rapid growth in demand is associated with new demand. But as a market 'matures' and automobile ownership levels approach 'saturation', more and more car purchases become replacement purchases. In the mature automobile markets today some 85 per cent of total demand for automobiles is replacement demand. Such demand is generally slower growing, and also more variable, because it can be postponed.

The third characteristic of today's automobile market is its increasing *segmentation and fragmentation*, as affluent consumers demand different types of vehicle for different purposes or want more sophisticated specifications, as the demographic profile of consumer markets changes, and as emerging customers demand basic (low-cost) cars.

> The need to respond to an increasingly diverse set of customers generated a large proliferation of segments and models ... the number of different vehicle models offered for sale in the US market alone doubled from 1980 to 1999, reaching 1,050 different models in 2000. In addition to the different models, there is also a myriad of features that can be added to each of the models.[8]

Taken together, the volatile and geographically uneven nature of demand for, and consumption of, automobiles creates huge problems for the manufacturers. In particular, the continuing problems of excess capacity facing many of the companies result in changes in both *how* vehicles are manufactured and also *where* they are manufactured.

From mass production to lean production: technological change in the automobile industry

The basic method of manufacturing automobiles changed very little between 1913, when Henry Ford introduced the moving assembly line, and the early 1970s. It was the mass production industry *par excellence*. This certainly brought the automobile within the reach of millions of customers. To do so, however, it had to produce a limited range of standardized products at huge production volumes – around 2 million vehicles per year – to obtain economies of scale, together with a very high level of worker specialization. It was the antithesis of craft production: automobile workers were, literally, cogs in the continuously running assembly-line machine.

This situation changed dramatically in the early 1970s. Highly efficient, and cost-competitive, Japanese automobile firms, led by Toyota, emerged as world players. This new competition totally transformed the automobile industry. What had appeared to be a stable, technologically mature industry, based on well-established technologies and organization of production, entered a phase of change (not unlike the first transformation in the early twentieth century when the mass production system displaced craft-based production). The basis of this second transformation was the displacement of mass production techniques by a system of *lean production* (see Table 3.2).

Within the broad framework of lean production systems, two of the most significant technological developments are related to the *architecture* of the vehicle. The first is the increasing use of *shared platforms* between different vehicle models. Hitherto, each model produced by an individual manufacturer aimed at different market segments was distinctive not only on, but also below, the surface. By using a smaller number of common platforms, it is possible to share many components across what are, on the surface, very different vehicles (often in different price segments of the market). So, one of the paradoxes of the modern automobile industry is that, although the number of models has increased, such diversity is based on a much smaller number of platforms. Beauty, it seems, really may only be skin deep as far as automobiles are concerned.

Platform sharing has become endemic among most automobile manufacturers. For example, VW reduced the number of platforms used in its Audi, Seat, Skoda and VW models from 16 to four, although recently VW has appeared to be moving away from this practice.[9] GM reduced its number of platforms from 25 to eight, and used the same platform for seven vehicles across its Buick, Chevrolet, Oldsmobile and Pontiac ranges.[10] Among Japanese producers, Nissan reduced its number of platforms from 24 to five, and Toyota from 20 to seven.

The second significant technological development linked to vehicle architecture is the *modularization* of certain components and the development of component *systems* (see Chapter 5). In the case of automobiles, a *module* is a group of components arranged close to each other within a vehicle, which constitute a coherent unit. A component *system* is a group of components 'located throughout a vehicle that operates together to provide a specific vehicle function. Braking systems, electrical systems and steering systems are examples'.[11] A modular and system-based architecture has become the norm, although it is not as easy to implement in the automobile industry as in some other industries such as electronics.[12]

Many of the most significant developments in the technology of automobile manufacture, and of the automobiles themselves, are based upon the increasing use of *electronics*.

The modern car has become completely dependent on electronics for engine management, satellite navigation, suspension controls and a raft of other enhancements from memory seats to rain-activated windscreen wipers. The next big step in the integration of electronics in the vehicle is the connection of all computers on a 'vehicle intranet' which will provide a simple and flexible installation with a minimum of wiring ... it is believed that electronics will continue to grow in all cars, accounting for more than 30% of a vehicle's value in the executive class to around 20% in 3-door hatchbacks.[13]

The increased use of electronics in vehicles also relates to current attempts to build more fuel-efficient models, such as the Toyota Prius, a hybrid petrol (gasoline)/ electric car. Big increases in oil prices in 2005, like those in the 1970s, intensified the competitive need to produce vehicles that use less fuel (and produce less environmentally harmful emissions) whilst, at the same time, being attractive to the consumer. However, the very rapid introduction of complex electronic systems into vehicles poses problems for an industry whose expertise is in different areas. Not only does this make the automobile manufacturer more dependent on electronics and software suppliers but also 'the electronics in the car bring six or seven times more faults than normal mechanical parts'.[14] Problems of reliability, and their impact on brand image, have become important again, as in earlier stages of technological change.

The role of the state

Throughout the history of the automobile industry the state has played a major role, notably in two major respects:[15]

- determining the *degree of access* to its domestic market that the state allows, including the terms under which foreign firms are permitted to establish production plants
- establishing the kind of *support provided by the state* to its domestic firms and the extent to which the state *discriminates* against foreign firms.

Use of tariff and non-tariff barriers against automobile imports has been pervasive in virtually all countries at various times, although the level of tariffs has fallen enormously, though unevenly. In general, 'automotive industry trade regimes were significantly more open by the end of the 1990s'.[16] Today, few of the developed market economies operate particularly high tariffs against automobiles, although there are significant differences between the EU's common external tariff of 11 per cent, the United States' tariff of 3 per cent, and Japan's 0 per cent. Tariffs are substantially higher, though unevenly so, in the developing markets. Far more prevalent has been the continued use of various non-tariff barriers, including import quotas.

The specific geographical configuration of the automobile industry has been, and still is, influenced not just by the *level* of tariffs or quotas but also by frequently used *differential* tariffs and quotas between assembled vehicles and components. States may, for example, levy high tariffs on imported vehicles but, at the same time, charge lower tariffs on imported components in order to stimulate local production, especially where there is an insufficiently well-developed local components sector. In particular, *local content* regulations have become pervasive. Such policies were an integral part of the import-substituting industrialization policies pursued in most Latin American and some Asian countries from the 1950s onwards. In the face of growing Japanese competition in the 1970s, most European countries, as well as the United States, also adopted them with some vigour.

Such local content requirements have been especially influential in affecting automobile firms' policies towards their suppliers and in influencing the geographical configuration of the automobile components industries. Within the emerging market economies – for many of which the development of an automobile industry has become a key policy objective – there has been considerable change in the policy environment since the 1990s. Three types of automobile regime have become apparent in emerging markets, each of which has evolved distinctive policy emphases:[17]

- *Protected autonomous markets* (PAMs): 'countries which continue to provide strong protection to the national market and the domestic industry'.[18] Current examples include China and Malaysia. South Korea also developed its automobile industry in this way until very recently,[19] as did Japan at an earlier period. Indeed, this was the model for most major European countries (especially France) for many years.
- *Integrated peripheral markets* (IPMs): countries which have chosen to develop their automobile industry through integration with a geographically contiguous core automobile market. The relevant example here is Eastern Europe (notably Poland, the Czech Republic, Hungary and Slovakia), where national automobile industry policies are directed towards integration with the EU.
- *Emerging regional markets* (ERMs): groups of emerging market countries which have 'sought to increase the efficiency of their motor industries by reducing protection and increasing competitive pressures and by using access to the domestic market as a lever to promote investments by transnational companies'.[20] Examples include Brazil within Mercosur and Thailand within ASEAN.

The key point, then, is that the state continues to be a major player in the automobile industries through a varying combination of regulatory and stimulatory policies. In the latter case, states have been especially active in offering huge financial and other incentives to automobile firms to invest in their territories. Table 10.2 gives some examples. The sheer scale of the incentives is staggering.

Table 10.2   Seducing automobile investment: examples of incentive packages

Location	Company	Company investment ($ millions)	State investment ($ millions)	State's financial investment per employee ($)
Smyrna, TN, USA (1983)	Nissan (Japan)	745–848	22 road access 7.3 training 33 total	25,384
Flat Rock, MI, USA (1984)	Mazda (Japan)	745–750	19 training 5 road improvement 3 on-site works 21 economic development grant/loan 5 water system 53 total	13,857
Georgetown, KY, USA (1985)	Toyota (Japan)	823.9	12.5 land purchase 20 site preparation 47 road improvement 65 training 5.2 Toyota families' education 149.7 total	49,900
Tuscaloosa, AL, USA (1993)	Mercedes-Benz (Germany)	300	68 site development 77 infrastructure 15 private sector/ goodwill 90 training 250 total	166,667
Spartanburg, SC, USA (1994)	BMW (Germany)	450	130 total	108,333
Setubal, Portugal (1991)	Auto Europa (Ford/VW) (USA/Germany)	2603	483.5	254,451
West Midlands, UK (1995)	Ford/Jaguar	767	128.72	128,720
North-east England, UK (1994–5)	Samsung (Korea)	690.3	89	29,675
Lorraine, France (1995)	Mercedes-Benz (Germany)	370	111	56,923

Source: based on UNCTAD, 1995: Table VI.3

The state is also heavily involved through environmental and vehicle safety policies, each of which has profound implications for the design, technology and materials used in cars and, therefore, for their cost. Complying with changes in legislation can be especially problematical where it involves fundamental design changes. Legislation to control noxious emissions from automobile engines has become increasingly stringent. A more recent development within the EU is policy towards 'end-of-life' vehicles. Here, the EU has issued a Directive, to come into force in 2007, under which automobile manufacturers will have to cover the cost of recycling the vehicles they have manufactured. It is estimated that the annual cost of this operation in Europe will be around 2.1 billion euros. Manufacturers will have to ensure that recyclable components account for 85 per cent of each vehicle's weight. Also in the EU, significant changes are being implemented to the permitted relationships between manufacturers and distributors in an attempt to increase competition.

Corporate strategies in the automobile industry

Consolidation and concentration

At first sight, the history of the automobile industry appears to be an inexorable progress towards increased concentration: the dominance of production by a smaller and smaller number of firms. That was certainly the trend between the 1920s and the 1960s. In 1920, there were more than 80 automobile manufacturers operating in the United States, more than 150 in France, 40 in the United Kingdom, and more than 30 in Italy. By the 1960s, following successive waves of consolidation through both merger and acquisition and also the closure of inefficient firms, around 50 per cent of world automobile production was concentrated in just three firms: GM, Ford and Chrysler. Since then, however, such big three dominance has declined, largely because of the emergence, since the 1970s, of Japanese, German and, to a lesser extent, French automobile firms. In fact, as Figure 10.4 shows, it has been the spectacular rise of Japanese firms – notably Toyota, Nissan and Honda – that has created the biggest changes, although an unexpected recent development has been the very rapid emergence of the Korean firm Hyundai. Table 10.3a ranks the leading automobile assemblers by the number of cars sold. Although GM remains in first place, its dominance is now seriously threatened by Toyota, now the world's second biggest automobile producer, and likely to become number one in the near future. What is more, Toyota actually makes profits on its automobile production while GM is mired in debt.

In fact, the degree of automobile industry concentration may well be increasing again. In the last few years, a new wave of cross-border mergers and acquisitions has occurred. Both GM and Ford acquired firms in the luxury market segments. The biggest acquisition, by far, was of the American firm Chrysler by the German-owned Daimler-Benz in 1998. In 1999, Ford acquired Volvo of

Table 10.3   **The world league tables of automobile assemblers and component manufacturers**

(a) Assemblers

Rank	Company	Headquarters country	No. of cars produced
1	General Motors	United States	5,436,750
2	Toyota	Japan	5,068,063
3	Volkswagen	Germany	4,946,771
4	Ford	United States	3,514,589
5	Hyundai	Korea	2,806,577
6	Peugeot-Citroën	France	2,550,384
7	Honda	Japan	2,537,117
8	Nissan	Japan	2,326,574
9	DaimlerChrysler	Germany	2,118,609
10	Renault	France	2,075,246

(b) Component manufacturers

Rank	Company	Headquarters country	Sales ($bn)
1	Delphi	United States	27.3
2	Visteon	United States	18.5
3	Bosch	Germany	15.6
4	Denso	Japan	12.6
5	Lear	United States	12.4
6	Johnson	United States	11.1
7	TRW	United States	11.0
8	Dana	United States	10.1
9	Magna	Canada	9.0
10	Valeo	France	7.7
11	Arvin	United States	7.6
12	Aisin	Japan	7.5
13	Yakaki	Japan	6.4

Source: calculated from SMMT, 2005, *World Automotive Statistics*; ABN-AMRO, 2000: 12–13

Sweden, while the French company Renault acquired 44 per cent of equity in the Japanese firm Nissan. In 2000, DaimlerChrysler acquired 34 per cent of Mitsubishi Motors (subsequently sold in 2005); in 2002, GM acquired the Korean assets of the bankrupt Korean firm Daewoo. In mid-2006, there was even talk of a potential tie-up between GM, Nissan and Renault. At the same time, some acquisitions have unravelled, for example, BMW's short-lived ownership of the British company Rover, and GM's disposal of its stake in Isuzu.

Figure 10.5 shows some of the significant groupings that have recently emerged, including VW's acquisitions of the Czech firm Skoda and the Spanish firm SEAT, as well as the fragile equity relationship between GM and Fiat

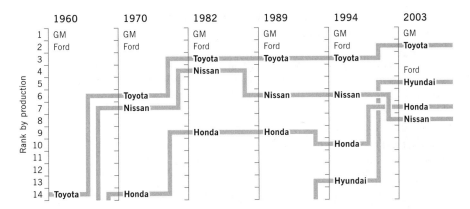

Figure 10.4   The rise of Japanese and Korean automobile manufacturers

(terminated in 2005). However, not all growth has been through acquisition and merger. The world's most successful automobile company, Toyota, has grown entirely organically. 'Apart from scooping up Daihatsu to get small-car engineering and engines years ago, Toyota has concentrated solely on improving its own offering, with a relentless focus on efficiency, cost-cutting and a flood of new variations of successful models brought to market at an increasingly rapid rate'.[21]

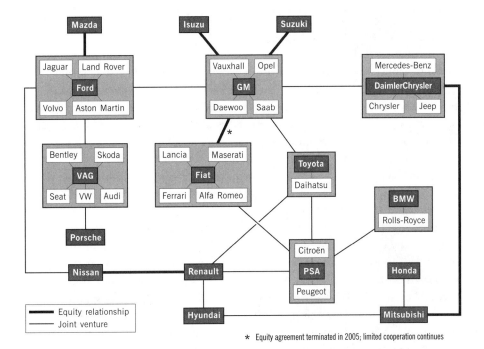

Figure 10.5   Webs of relationships in the automobile industry

In addition, all the world's automobile manufacturers are deeply embedded in *collaborative agreements* with other manufacturers:

> In recent years, there have been about 100 new alliances in the automobile industry per year. The majority of these are manufacturing joint ventures … indicating the high degree of globalization in this sector.[22]

Technology joint ventures continue to be very important for all automobile manufacturers. A recent example is the agreement between BMW, GM and DaimlerChrysler to develop and manufacture hybrid (petrol/electric) engines and transmissions.[23]

Similar consolidation and concentration trends have occurred in the components industries as leading companies have tried to develop positions as global suppliers. The list of leading companies is dominated by United States' firms, led by Delphi and Visteon (Table 10.3b). Both were formerly in-house component divisions of GM and Ford respectively before being hived off. Both have experienced serious problems of surviving as independent companies (Delphi filed for Chapter 11 bankruptcy protection in 2005). As with the assemblers, much of the continuing consolidation among suppliers is being driven by mergers and acquisitions.

Changing relationships between automobile assemblers and component manufacturers

The fundamental driving force behind consolidation amongst automobile component manufacturers is the increasing pressure being exerted by assemblers to deliver quickly (just-in-time), to deliver at lower cost on a continuous basis, and to raise the quality of components. Such pressures are manifested in two important ways. One is the pressure on suppliers to take on more of the design, research and risk of developing component modules and systems. The second is the pressure on suppliers to locate geographically close to assembly plants.

Taken together, such pressures have resulted in a massive decline in the number of supplier companies. In 1990 there were some 30,000 suppliers in North America, and 10,000 by the year 2000, and there is a predicted further decline to between 3000 and 4000 by the year 2010. The 'Ford 2000' strategy envisaged reducing the company's total number of component suppliers in North America by more than 50 per cent over 10 years – from more than 2000 to less than 1000. Of that 1000, a mere 180 companies would be awarded around two-thirds of the orders. Amongst European firms, Peugeot-Citroën has reduced its suppliers from 900 to less than 500, BMW from 1400 to 600. In turn, the major component suppliers themselves are reducing the number of their suppliers. Visteon, for example, recently announced that 'it would, in future, do business with only two or three companies in "each segment" of business for the next five years'.[24]

As part of the process of supplier consolidation, the precise roles played by suppliers are changing. The supply system is becoming more *functionally segmented*. In place of the myriad of specialist raw materials and component suppliers, four major strategies seem to be evolving, as Figure 10.6 shows. The raw material and component specialist strategies are, of course, not new. What is new is the emergence of other categories of supplier, notably the standardizers and the integrators, both of which have significantly greater design and manufacturing responsibilities and have a different kind of relationship both with assemblers and also with their own suppliers. This latter characteristic is especially significant in the case of the integrators.

	Raw material supplier	Component specialist	Standardizer	Integrator
Focus	A company that supplies raw materials to the OEM or their suppliers	A company that designs and manufactures a component tailored to a platform or vehicle	A company that sets the standard on a global basis for a specific component or system	A company that designs and assembles a whole module or system for a car
Market presence	• Local • Regional • Global	• Global for tier 1 • Regional or local for tiers 2 and 3	• Global	• Global
Critical capabilities	• Material science • Process engineering	• Research, design and process engineering • Manufacturing capabilities in varied technologies • Brand image	• Research, design and engineering • Assembly and supply chain management capabilities	• Product design and engineering • Assembly and supply chain management capabilities
Types of components or systems	• Steel blanks • Aluminium ingots • Polymer pellets	• Stampings • Injection moulding • Engine components	• Tyres • ABS • ECU	• Interiors • Doors • Chassis

Figure 10.6 Supplier strategies in the automobile industry

Source: based on Veloso and Kumar, 2002: Table 1

As a consequence of these changing roles and responsibilities of suppliers, and their relationships with the assemblers, the overall supply chain of the automobile industry is being transformed. Figure 10.7 shows one possible trajectory. The relatively simple tiered hierarchy that has developed in recent years is metamorphosing into a structure in which the connection between tier 1 suppliers and the assemblers is being mediated by a new layer of module and system integrators – what some analysts term a 'tier 0.5' to signify its closer relationship with the assemblers. The precise configuration of the future is still far from clear and may well contain more variety than this picture suggests. But there is no doubt that a significant reconfiguration of the automobile production circuit is taking place.

In effect, organizationally 'distant' relationships have been replaced by much 'closer' relationships. Much greater degrees of organizational interdependence between automobile manufacturers and components suppliers have developed. Relationships with key suppliers have become longer term. At the same time, the

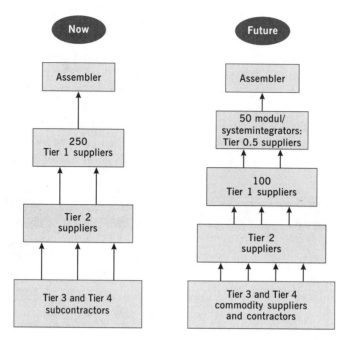

Figure 10.7   The changing structure of the automobile supply chain

Source: based on ABN-AMRO, 2000: 10

need for suppliers – especially 'first-tier' suppliers of complex modules and systems – to locate geographically close to their customers has intensified. The intensified drive to shorten the time involved in vehicle production means that 'just-in-time' supply has become the norm in the automobile industry. So, too, has the inexorable pressure throughout the production circuit continuously to reduce costs. Even so, enormous diversity persists in supplier–assembler relationships: a mixture of long- and short-distance arrangements, reflecting the path dependency of present patterns on those that evolved in earlier periods. This is especially the case for second- and third-tier component suppliers. Geographical change tends to be incremental rather than radical.

However, in the case of first-tier suppliers, especially modular suppliers, there is no doubt that profound changes are occurring in the geography of their relationships with their customers. The most developed situation involves their *co-location*, particularly where new plants are involved in modular production. What have been called *industrial condominiums* are in their early stages and, so far, are more common in certain parts of the world than in others:

> The emergence and diffusion of modular production has followed different paths in Brazil, Europe and North America. It continues to be relatively limited in Asia, with Japanese car-makers having resisted the outsourcing of modules.[25]

Three such projects, each in Brazil, have been set up by VW, GM and Ford, respectively:

- VW's assembly plant at Resende: component makers fit their products directly on the assembly line, rather than simply delivering them from external factories. Suppliers assume responsibility for putting together and installing four major modules: the chassis, axles and suspension, engines, and transmissions; and driver's controls. Management responsibilities are shared between VW and its main partner-suppliers.[26]
- GM's 'Blue Macaw' project at Gravataí, near Pôrto Alegre: consists of 17 plants, 16 of which are occupied by major suppliers. Both the plant and the car were designed collaboratively. The major suppliers' production facilities are located inside the complex, at the place where its products are needed on the assembly line: '85 per cent of the car produced at the plant is based on components assembled on-site compared with the usual position where around 60 per cent of a car's components are sourced from outside'.[27]
- Ford's Amazon project at Bahia, on the north-east coast: has 19 suppliers all within the factory. Each is responsible for complete modules. A further 12 component suppliers are located adjacent to the assembly plant. Sixty per cent of the car's components are produced locally.[28]

Whether or not these 'assembly lines of the future' become accepted practice elsewhere is a matter for the future. Certainly it will be much harder to introduce such revolutionary practices in the old-established manufacturing heartlands of the automobile industry. It is significant that each of these three examples is located well away from traditional automobile manufacturing areas. A diluted version of the industrial condominium, more common in Europe, is the *supplier park*: 'complexes that bring suppliers together in close proximity to one another, contiguous to the assembly site – a contiguity that is sometimes materialized through the overhead tunnels that connect the various plants'.[29]

Clearly, profound changes are occurring in the nature of the assembler–supplier relationship, driven primarily by the time, price and technology/design pressures exerted by the assemblers on suppliers. Suppliers have been driven to consolidate and to take on enhanced roles. That, in turn, changes the balance of power between assemblers and the mega-suppliers upon which the assemblers now depend for a much larger part of their business. So, although in general the assemblers have more bargaining power than the suppliers, there are clear exceptions to this: those mega-suppliers which have developed particularly valuable capabilities and which are able to provide a global supply service to their geographically dispersed customers.

Contrasting transnationalization strategies of the major automobile producers

In view of the kinds of technological and competitive pressures facing all automobile producers, it is not surprising to find that their strategies have some similarities. However, they are far from being identical. The big producers remain creatures of their specific histories: as noted in Chapter 4, firms develop a 'strategic predisposition', built up over time, which leads them to favour some kinds of approaches rather than others. For example, some firms have preferred to grow organically, some have grown mainly by merger and acquisition, whilst others have grown through a varying mixture of the two. They also remain creatures of their particular geographies. Where they come from, where they are still headquartered matters a lot in terms of the precise ways in which they pursue their objectives (see Chapter 4). In this section, therefore, we look very briefly at the contrasting ways in which the major automobile producers have transnationalized and, especially, at current developments in their strategies. In the final section, we will focus on the *regionalized* patterns of automobile production. Table 10.4 summarizes the transnational profiles of the leading United States, Japanese and European producers.

Table 10.4   Transnational profiles of leading automobile producers (foreign as % of total)

Company	TNI[a]	Employment	Assets	Sales
US companies				
Ford	45.5	42.3	57.1	37.0
GM	32.5	35.4	34.4	27.8
Japanese companies				
Honda	72.0	70.6	68.3	77.0
Nissan	48.5	41.1	38.9	65.6
Toyota	47.3	33.8	49.7	58.6
European companies				
BMW	54.0	25.0	62.5	74.5
Volkswagen	52.9	47.9	38.5	72.4
Renault	40.9	26.7	31.3	64.5

[a]TNI is the average of three ratios: foreign assets to total assets; foreign sales to total sales; foreign employment to total employment.

Source: calculated from UNCTAD, 2005: Table A.1.9

The US big two

GM and Ford dominated the world automobile industry for several decades. This was mainly, of course, due to the massive size of the North American market for cars. In addition, however, they were the first automobile firms to transnationalize their production, initially (and logically) in Canada and then in Europe. Ford built its first European manufacturing plant in Manchester in 1911 (subsequently replaced in 1931 by the massive integrated plant at Dagenham, near London), and spread into France and Germany. GM expanded transnationally through acquiring existing companies in both Canada and Europe (Opel in Germany, Vauxhall in the UK).

These early transnational ventures were triggered by the existence of protective barriers around major national markets as well as by the high cost of transporting assembled automobiles from the United States. Subsequently, both Ford's and

GM's transnational strategies have been concerned initially with expanding and integrating, and then with rationalizing their operations on a global scale. Figure 10.8 shows the broadly similar geography of Ford's and GM's transnational activities, although they reached their current positions by rather different routes. But, as Table 10.4 shows, Ford is much more transnationalized than GM. Indeed, although GM is the world's largest automobile producer it is actually the least transnationalized of all the leading producers.

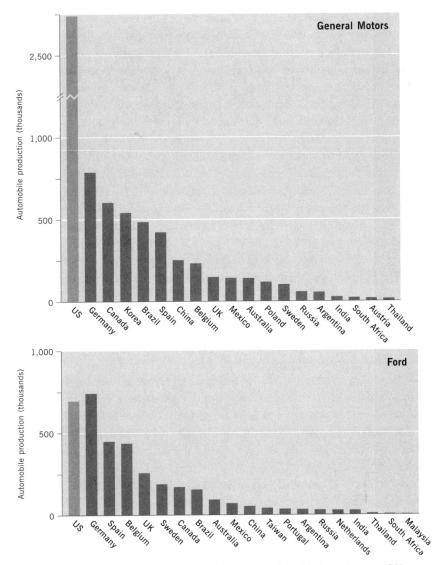

Figure 10.8 The transnational production geography of US producers: GM and Ford

Source: calculated from SMMT, 2005, *World Automotive Statistics*

From being the dominant players in the world automobile industry, both GM and Ford are currently in crisis. Their shares of the United States market have plummeted sharply and they face intensified competition in world markets. Both face huge 'legacy' costs of health care and pension provision for their unionized domestic labour forces. Both, therefore, have undertaken a whole series of rationalization and restructuring measures in the past few years – so far with little real success.

Ford's response in the mid 1990s to intensified competition in all its major markets was to embark on a hugely ambitious process of global reorganization ('Ford 2000'). The North American and European operations were merged into a single organization, with the intention of ultimately integrating its Asian, Latin American and African operations as well. A single worldwide product development organization was created with five automobile centres, each having responsibility for the global development of specific products. Four of the five automobile centres were located in the United States and one – responsible for small and medium front-wheel-drive cars – was in Europe (jointly in the UK and Germany).

Four years later (in 1999), Ford announced yet another major reorganization of its global management structure, replacing the strongly global emphasis of 'Ford 2000' with a system of strategic business units organized around brands and regions: Europe, Asia–Pacific, South America and North America. In 2001, further major restructuring was implemented, involving closure of five plants and the loss of 22,000 jobs in North America. In early 2006, Ford announced plans to close seven plants by 2012 and reduce its workforce by 25,000 to 30,000 (one-quarter of the total).[30]

During the 1990s, GM adopted a more cautious approach to global reorganization. While Ford was implementing its ambitious 'Ford 2000' reorganization, GM merely established a global strategy board consisting of the heads of GM's North American and international operations and its component division. In 1998, however, GM moved to merge its North American and its international operations into a single worldwide group, aiming to realize its potentially enormous economies of scale. But, as in Ford, the process of global reorganization has been problematical, not least because of cultural differences between the centre and the geographically dispersed operations derived from what one observer called 'the ingrained Midwestern focus of GM's top executives'.[31]

In late 2005, GM announced the closure of another five North American plants (in Oklahoma, Michigan, Tennessee, Georgia and Ontario) – in addition to the three big factories closed in the previous five years and the 21 factories closed in the early 1990s.[32] More than half a million jobs have been lost in GM over 20 years. A further 30,000 jobs will go by 2008.[33] In the cases of both GM and Ford, the mighty have indeed fallen.

Japanese producers

In contrast to the tribulations of the US big two, the leading Japanese automobile producers continue to thrive. Whereas both Ford and GM have had international operations for many decades, the spectacular rise of the leading Japanese companies up the world league table (Figure 10.4) was achieved almost entirely without any actual overseas production, apart from small-scale, local assembly operations, using imported kits. Beyond such operations, Toyota had no overseas production facilities for cars before the early 1980s, while less than 3 per cent of Nissan's total production was located outside Japan. The biggest Japanese producer, Toyota, was the slowest to transnationalize. Toyota did not build its first European plant until 1992, six years after Nissan. Paradoxically, it was one of the smaller Japanese automobile manufacturers, Honda, which was the first to build production facilities outside Asia (in Ohio in 1982). However, the transnationality of the Japanese producers changed dramatically during the 1980s. By 1989, 28 per cent of Honda's output was produced outside Japan; today the proportion is 53 per cent. Toyota more than quadrupled its overseas production share (from 8 to 36 per cent); Nissan now produces 48 per cent of its cars outside Japan compared with 14 per cent in 1989. As Table 10.4 shows, the three leading Japanese producers are now far more transnationalized than either Ford or GM.

Before the early 1980s, Japanese automobile manufacturers were perfectly capable of serving the North American and European markets by exports from Japan. Their production cost efficiencies more than offset any transport cost penalties arising from Japan's geographical distance from major markets. From the early 1980s, however, Japanese companies began to change their global strategy by locating production plants in their major markets. The primary stimulus was the increasingly powerful political opposition, in both North America and Western Europe, towards the growth of Japanese imports. But trade protectionist measures were not the only stimulus. Given the increasing possibilities of tailoring automobiles to variations in customer demand, it was becoming more important for Japanese firms to locate close to their major markets. The primary objective of the Japanese automobile producers, therefore, has been to develop a major direct presence in each of the world's major automobile markets. Figure 10.9 shows the current geographical pattern of Toyota's, Nissan's and Honda's production.

After a long period of seemingly unstoppable success, the major Japanese producers began to face increasingly difficult problems during the 1990s, especially in Japan itself. The deep and prolonged recession in Japan led to steep declines in sales so that, for the first time, Japanese firms began seriously to restructure their domestic operations. The form of such restructuring has centred primarily on reducing costs: for example, by squeezing suppliers, by rationalizing procurement and, in some cases, by actually closing plants in Japan (hitherto an unthinkable option).

Among the top three Japanese automobile producers, Nissan adopted the most radical response by selling almost 40 per cent of its equity to a foreign manufacturer – Renault. Although some of the smaller Japanese automobile firms already had tie-ups with foreign firms (see Figure 10.5), the Nissan case was in a different league. Five factories were closed, purchasing costs were drastically

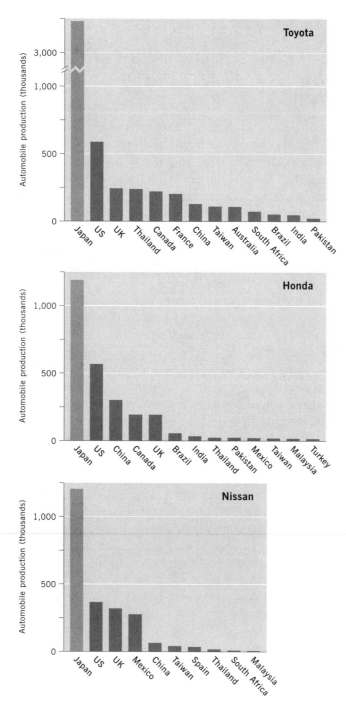

Figure 10.9   The transnational production geography of Japanese producers: Toyota, Nissan and Honda

Source: calculated from SMMT, 2005, *World Automotive Statistics*

reduced, and the dealer network was consolidated. Nissan's recovery has been dramatic. But by far the most impressive performance is that of Toyota, which not only has become the world's second biggest automobile producer but also has developed (albeit more slowly than others) a high degree of transnationality. As a recent report observed,

> There is the world car industry, and then there is Toyota. Since 2000, the out-put of the global industry has risen by about 3m vehicles … of that increase, half came from Toyota alone. Toyota has undergone a dramatic growth spurt all round the world … [and] will soon be making more cars abroad than at home.[34]

European producers

The major European automobile producers are strongly Eurocentric in their pro-duction geographies (Figure 10.10). Indeed, while the Japanese were busy build-ing large manufacturing plants in the United States during the 1980s, the Europeans did not do this. In fact, both VW and Renault pulled out of their earlier involvement in North America. Of all the European producers, VW has pursued by far the most extensive and systematic transnational strategy. Outside Europe, VW is a major producer in Brazil, Mexico and, especially, China.

VW embarked on a major restructuring programme, which involved reducing the number of automobile platforms used by its four component companies. In 2001, the company announced a major reorganization of its entire operations.[35] More recently, VW has been struggling to reduce the size of its workforce in Germany, aiming to reduce by one-fifth over three years.[36] One ploy was to agree to build a new model at Wolfsburg rather than, as planned, in Portugal. Mercedes (DaimlerChrysler) is also attempting to reduce its German labour force (in addi-tion to tackling its massive structural problems in the United States). BMW, Germany's third largest automobile producer, remains a privately owned company focused on high-value markets. For a time in the early 1990s, BMW aspired to extend its range into the mass market segment by acquiring Rover, a deeply troubled British company. BMW finally withdrew from its ownership of Rover, although it retained, and has very successfully developed, a small-car assembly plant and an engine plant in the UK. BMW has extended its operations in North America and has a growing presence in Asia. However, BMW chose to locate its large new assembly plant in Germany (at Leipzig) rather than in the Czech Republic.

The two French automobile companies, Peugeot-Citroën (PSA) and Renault, have both traditionally been strongly home country oriented in their produc-tion. Peugeot-Citroën was formed by a state-induced merger in 1975. It has no production plants outside Europe and yet has managed to be very successful in serving the world market. Renault has been, for more than 40 years, the French government's national champion, supported by massive state aid, which served to

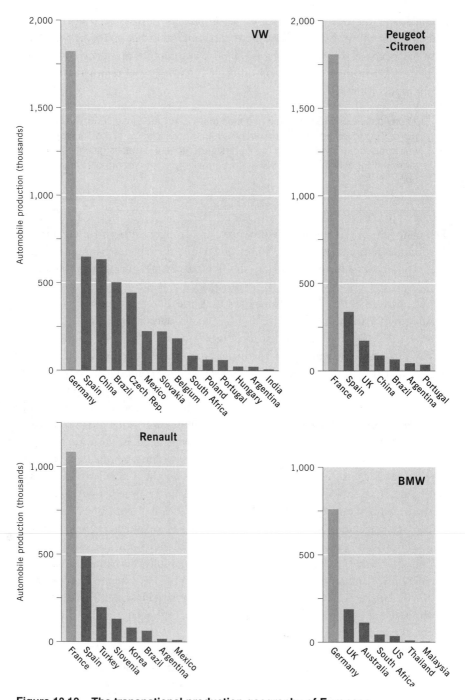

Figure 10.10   The transnational production geography of European producers: VW, Peugeot-Citroën, Renault and BMW

Source: calculated from SMMT, 2005, *World Automotive Statistics*

constrain its activities. State control has been reduced to 15 per cent and Renault has been involved in major restructuring, including its major coup in acquiring a large equity stake in Nissan. This has become the central pillar of Renault's strategy. For Renault itself, France remains the dominant production location with 52 per cent of its total world production. A further 24 per cent is located in Spain.

While VW was expanding its European production base to incorporate Spain in the 1980s, the Italian automobile firm Fiat initially moved in the opposite direction and reconcentrated production in its home market. A key element in Fiat's recent strategy has been to create an extensive production network in the former Soviet Union and Eastern Europe where it has the longest-established links of any automobile producer in the world. Fiat's 'grand European design' was to build a manufacturing network extending from the Mediterranean to the Urals. At the same time, Fiat strengthened its position in Brazil. Both Brazil and Poland became the major bases for Fiat's small 'world car', the Palio, which began production in Brazil early in 1996. Fiat aimed to produce more than 50 per cent of its cars outside Italy by the year 2000 but failed to reach that target.

Korean producers – and now there is one

The US, Japanese and European automobile companies exert such market dominance that there have been virtually no new entrants to the industry during the past 30 years. The major exception is in South Korea, which has emerged in the space of a just a few years as a significant international force directly as a result of state involvement.[37] 'Automobiles were identified as one of the priority industries in the Heavy and Chemical Industry Plan of 1973. In 1974 an industry-specific plan for automobiles was published covering the next ten years. The objectives were to achieve a 90 per cent domestic content for small passenger cars by the end of the 1970s and to turn the industry into a major exporter by the early 1980s'.[38]

Although the Korean government effectively made Hyundai the leading producer in the industry, giving it an enormous relative advantage, by 1997 there were still five Korean automobile producers. Today, there is just one – Hyundai – the others being victims, directly or indirectly, of the East Asian financial crisis. The second biggest Korean producer, Daewoo, was acquired by GM. Daewoo, in fact, had originally developed out of a joint venture with General Motors (hence its original name, GM Korea).

Both Hyundai and Daewoo depended for their early development on close technological and marketing relationships with US and Japanese firms. Hyundai's strategy was to compete with the Japanese in a very narrow product range and entirely on price. Its initial export success was remarkable. In 1986, 300,000 cars were exported to North America, a level more or less maintained for the following two years. On the strength of this success, Hyundai built a second new plant in Korea. More ambitiously still it built a plant in Canada, with the capacity to produce 120,000 cars and to employ 1200 workers directly, but this plant was closed in 1991 because of quality problems.

As a result of the 1997 crisis, Hyundai was forced to undertake major restructuring and refocusing of its operations. First, it acquired Kia, one of the smaller Korean automobile firms. Second, in 2000, it separated from its parent company, the Hyundai *chaebol*. Third, it began to move away from being merely a low-price regional producer of cars primarily for the Asian market to one with much wider ambitions. Geographically, however, Hyundai remains overwhelmingly oriented to Korea where 70 per cent of its production is concentrated. But this is changing rapidly. Hyundai sees itself as a major global producer of high-quality vehicles (operating a two-brand strategy, with Kia providing the lower-cost cars). It is shedding its cheap-car image and focusing, like Toyota, on superior quality control (to the extent of offering 10-year warranties on its products). It has rapidly expanded geographically to operate plants in China, India, Turkey, the United States (a new plant in Alabama opened in 2005) and Slovakia (operated by its Kia affiliate), and is to build a new plant in the Czech Republic. It has a remarkable record of rapid growth and has become the world's fifth largest automobile producer (see Table 10.3).

Regionalizing production networks in the automobile industry

Although, in one sense, the automobile industry is one of the most globalized of industries, it is also an industry in which the *regionalization* of production and distribution is especially marked.[39] Figure 10.11 shows that three-quarters of automobile trade in both North America and Europe is intra-regional, while the intra-regional share in East Asia has increased from 19 to 23 per cent in just two years. Geographically, rather than attempting to organize (and reorganize) operations on a truly global scale, the tendency of most of the leading automobile producers is towards the creation of distinctive production and marketing networks within each of the three major regions of the global triad.

Europe

Figure 10.12 maps the current pattern of automobile production within Europe. Europe has the most complex automobile production networks in the world, a reflection of the legacy of formerly nationally oriented automobile industries, the particular ways in which the political environment has evolved in the past four decades, and the responses of automobile firms to these changes in the context of increasingly intensive global competition.

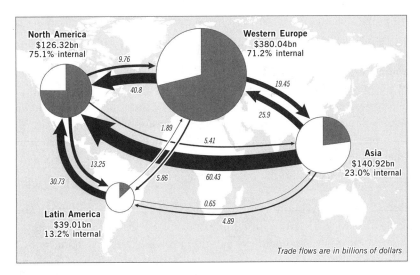

Figure 10.11   Global trade network in automobiles

Source: calculated from WTO, 2004, *World Trade Report*: Table A10

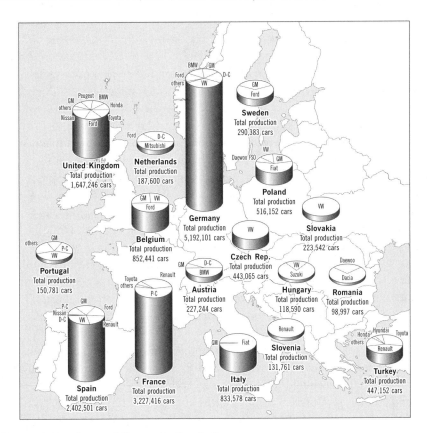

Figure 10.12   Automobile production in Europe

Source: calculated from SMMT, 2005, *World Automotive Statistics*

Two political-economic transformations during the 1990s dramatically reshaped the region's automobile industry.

- Completion of the Single Market in 1992. This removed the remaining technical and physical barriers to the flows of vehicles and components, whilst individual national policies (for example, on import quotas and local content levels) were abolished with the introduction of EU automobile industry policy.
- The opening up – and subsequent political integration – of Eastern Europe. This created both a low-cost production location for sourcing components and finished vehicles and also the potential of a growing consumer market.

The actual geographical configuration of automobile production within Europe (Figure 10.12) bears the very strong imprint of each firm's national origins and the history of its development within this evolving political framework. For example, Ford and GM have been in Europe for almost 100 years, building up multilocational, initially nationally oriented, production networks. Currently, as we have seen, both GM and Ford are in crisis and engaged in major reorganization and rationalization. For example, Ford has ceased to assemble cars at Dagenham in the UK (when built in 1931 this was its first fully integrated plant outside the United States) and transformed it into a centre for the design and production of diesel engines. In fact, the UK's position within Ford is becoming primarily that of engine and transmission production rather than car assembly. GM has followed a similar path. Just as Ford closed its assembly plant at Dagenham, GM has closed its Luton plant. Much to the consternation of the Swedes, the next generation of Saab models will be built in Germany, not Sweden.

Whereas the geographical reconfiguration of the US automobile manufacturers has evolved from a long-established presence in Europe, the position of the Japanese producers is very different. With no history of European car production and no inherited structure, the Japanese were able to treat Europe as a 'clean sheet'. Beginning in the early 1980s, Japanese firms established production facilities in Europe. All three leading Japanese firms – Toyota, Nissan, Honda – built their plants in the UK. Although the UK is a major market in itself, the Japanese plants in the United Kingdom were specifically oriented towards the European market. This led to political friction within the EU during the 1980s and 1990s. Significantly, Toyota's second European plant was built in northern France. However, Japanese firms now produce more than half of all the cars manufactured in the UK.

The geographical configuration of the indigenous European automobile producers is, of course, much more embedded in their national contexts. Only VW has anything approaching a pan-European production network, focused around the three nodes of Germany, Spain, and its acquired Eastern European plants in the Czech Republic and Slovakia. Prior to the opening up of Eastern Europe, VW had developed a clear strategy of spatial segmentation. High-value, technologically advanced cars were produced in the former West Germany; low-cost, small cars were produced in Spain. After 1990, VW moved very rapidly to establish production of small cars in Eastern Germany and to take a controlling stake in the Czech firm Skoda.

Indeed, it is the developments in Eastern Europe that are now the primary focus of change in European automobile production networks.

> In just a few years all the major Czech, Hungarian and Polish automotive firms were privatised and sold to foreign investors, the existing car assembly and components manufacturing capacity was substantially modernised and extended, and new firms were established ... The region has become integrated into the Western European automotive space through ownership links, production, procurement and sales networks.[40]

In addition to joining with (or taking over) existing local automobile firms, automobile producers are building new plants in Eastern Europe. For example, Toyota has established a joint venture with Peugeot-Citroën to develop and assemble small cars for the European market at Kolin in the Czech Republic; Peugeot-Citroën itself has built a plant in Slovakia and is closing its UK plant; the Hyundai affiliate Kia has established an assembly plant in Slovakia, while Hyundai itself is to build a large assembly plant in the Czech Republic; Renault and Nissan are considering buying the former Daewoo plant in Romania; and the Chinese firm Chery plans to build an assembly plant in Eastern Europe.

At the same time, substantial components production is also shifting towards the East. On the one hand, there are the affiliates of foreign companies set up primarily to follow the assemblers. While some of these investments may be genuinely 'new' (in the sense that they did not formerly exist elsewhere), or are acquisitions of local companies by foreign firms, others are, in effect, locational transfers from elsewhere in Europe. On the other hand, there are the indigenous suppliers, many of them the successors of formerly state-owned enterprises prior to the onset of privatization. In many cases, these have been restricted to the production of low-value components and are rather peripherally connected into transnational production networks.[41]

In sum, three major trends have recently occurred within the European automobile production network:

- First, large-scale geographical rationalization of core country operations as major firms grapple with problems of over-capacity and outdated physical plant.
- Second, substantial development of automobile production in the south-western periphery of Europe, notably in Spain and, to a lesser extent, Portugal, as US, French and German manufacturers establish production in lower-cost locations capable of serving the entire European market.
- Third, and most recently, significant development of automobile production (and, especially, of component production) facilities in Eastern Europe. Here,

the major developments have been in the Czech Republic, Slovakia and Hungary. Almost 20 per cent of EU automobile production is now located in Eastern Europe.[42]

North America

Although political-economic integration in North America is much shallower than in the EU, in the case of the automobile industry political agreements have had profound repercussions on its geographical structure. The 1965 Automobile Pact between the United States and Canada reshaped the industry, producing an integrated structure in which production by the major manufacturers was rationalized and reorganized on a continental, rather than a national, basis.[43] By the early 1970s, in fact, the US–Canadian automobile industry was fully integrated. The 1988 Canada–United States Free Trade Agreement (CUSFTA) redefined the level of 'North American content' necessary for a firm to be able to claim duty-free movement within the North American market. The 1994 North American Free Trade Agreement (NAFTA) had even more far-reaching implications for the automobile industry because it incorporated the Mexican auto industry with its vastly lower production costs.

The NAFTA regulations for the automobile industry included the following important provisions:[44]

- Tariffs on trade in automobiles between the United States and Canada on the one hand and Mexico on the other were to be phased out completely by 2004.
- By the end of the transition period at least 62.5 per cent of the vehicle's value added had to derive from NAFTA countries.
- Until 2004, existing limitations on passenger car imports into Mexico remained in force. This meant that only the five foreign producers with existing production facilities in Mexico were allowed to import (at levels linked to their volume of local production).

Figure 10.13 shows the broad geographical structure of the North American production system. Prior to the 1980s, it was totally dominated by US producers; attempts by Volkswagen and Renault to produce cars successfully in the United States had failed. But from the mid 1980s onwards the position changed dramatically. Three major waves of foreign involvement have occurred.

First, each of the major Japanese firms established large-scale production facilities in the United States and Canada during the 1980s. Today, there are around 15 Japanese automobile assembly plants, with plans for more (for example, Toyota is to build a new assembly plant in Ontario, Honda will build a large new assembly plant in Indiana). The pioneer was Honda, which established a manufacturing plant at Marysville in Ohio in 1982. This was followed, in 1983, by the Nissan plant at Smyrna, Tennessee. In contrast, Toyota entered North America very cautiously

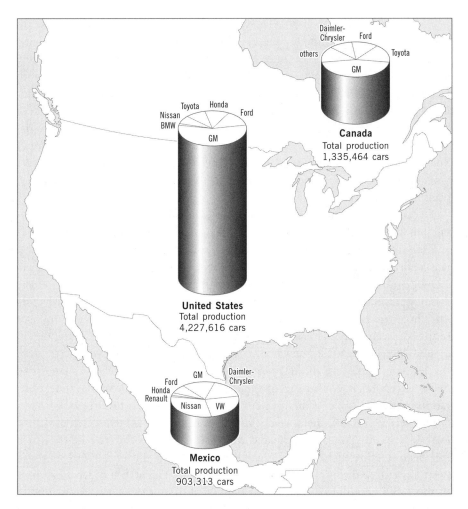

Figure 10.13 Automobile production in North America

Source: calculated from SMMT, 2005, *World Automotive Statistics*

in 1984, through a joint venture (NUMMI) based at GM's Fremont, California, plant. Since then, each of the major Japanese firms has continued to increase its capacity and to make major investments in engine, transmission and components plants. As the Japanese plants progressively increased their North American content they were followed by a wave of Japanese component manufacturers. In other words, during a period of less than a decade an entirely new Japanese-controlled automobile industry was created in North America in fierce, direct competition with domestic manufacturers. This 'new' automobile industry had a very different geography from that of the traditional one, simply because the Japanese had no existing plants or allegiance to specific areas. With few exceptions, the old-established automobile industry centres were not favoured.

The second, very much smaller, wave of foreign investment in the North American automobile industry began in the mid 1990s in the form of German luxury car manufacturers, Daimler-Benz and BMW. Daimler-Benz built a new plant at Tuscaloosa, Alabama in 1993; BMW built a plant at Spartanburg, South Carolina in 1994. Subsequently, of course, Daimler-Benz created a major shake-up of the North American automobile industry when it acquired Chrysler in the late 1990s. These incursions by the two upmarket German producers were the first major European involvement in North America after the failed ventures of Volkswagen and Renault in the 1970s. In 2005, the Korean firm Hyundai opened a major plant in Montgomery, Alabama and its subsidiary Kia is to build a new plant in Georgia.

The third development has been the large-scale investment by foreign automobile producers in Mexico.[45] Prior to the Mexican economic reforms of the 1980s and later the NAFTA in 1994, most of the automobile investment in Mexico was market oriented. Subsequently, these investments were replaced by 'strategic asset-seeking and cost-reducing FDI'.[46] Most of these newer investments in vehicle assembly and major component plants have rather different locational characteristics from the earlier import-substituting investments. The latter, being oriented to the Mexican domestic market, tended to be concentrated in the central region around Mexico City. The newer investments, oriented to the North American market as a whole, are located nearer the border with the United States. The major exception is VW's large integrated facility in Puebla. At the same time, rationalization of some of the former core-region plants has occurred.

By the early 2000s, then, a very different regional production network had evolved in North America compared with that existing before the 1980s, which had been dominated by the US manufacturers. From the mid 1960s they had been integrating their Canadian and US operations. The arrival of Japanese firms in the 1980s consolidated that pattern, while also creating a new geography of production away from the old-established automobile concentration in the US Midwest. The NAFTA, together with earlier reforms within Mexico, transformed that system by incorporating into the North American regional production network a production location with very low costs (and a potentially fast-growing domestic market).

East Asia

The development of distinctive regional automobile production networks in both Europe and North America reflects the combination of two forces: the size and affluence of the market and the political-economic integration of the market through the EU and the NAFTA respectively. In these circumstances, the development of a high level of intra-regional integration of supply, production and distribution becomes possible. The situation in East Asia is rather different: the region remains primarily a series of individual national markets, some of them very heavily protected against automobile imports. On the other hand, the undoubted potential of the East Asian market, set against the saturation of most Western markets, makes it an absolutely necessary focus for the leading automobile manufacturers. It is against this background that the current automobile production network in East Asia needs to be set.

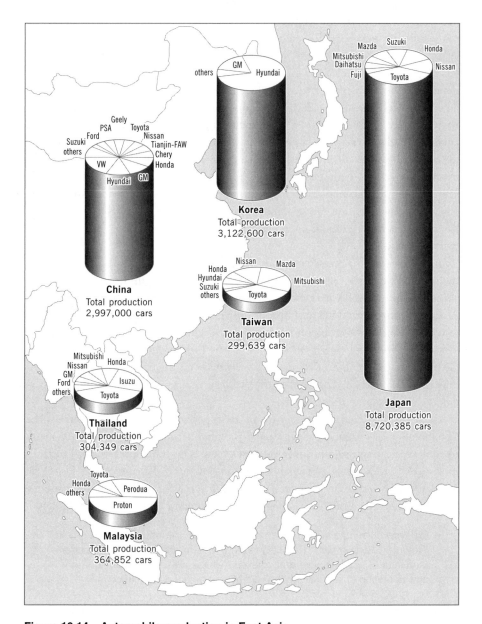

Figure 10.14   Automobile production in East Asia

Source: calculated from SMMT, 2005, *World Automotive Statistics*; Liu, 2006

As Figure 10.14 shows, automobile production in East Asia is dominated by Japan and, to a lesser extent, Korea and China. Not surprisingly, therefore, the region's automobile production is dominated by Japanese firms.[47] Through a network of assembly plants and joint ventures with domestic firms, Japanese cars are assembled in Thailand, Malaysia, the Philippines, Indonesia, Taiwan and China. In several of these countries, Japanese manufacturers totally dominate the automobile market. In Thailand, for example, they have a market share of more than 90 per cent; Toyota alone controls almost 30 per cent of the Thai vehicle market. Most of these are assembled locally in individual countries to serve the local market. This is less out of choice on the part of the Japanese manufacturers than out of the necessity created by high levels of import protection in virtually all the East Asian countries, particularly those in South East Asia (notably Malaysia). Faced with increasingly difficult circumstances in the Japanese market itself (for example, the problems created by the high value of the yen, the slowdown in demand), Japanese firms have placed increased emphasis on raising their penetration of the Asian market by developing cars specifically tailored to that market and not just versions of existing models. In the late 1990s, for example, both Toyota and Honda introduced completely new models based upon a very different approach to producing cost-efficient cars for a low-income market.

In comparison, Korean firms have preferred to serve East Asian markets from their domestic bases, although Hyundai has operations in Indonesia. Western automobile companies have only recently taken a really serious interest in East Asia. Of course, several US and European firms have had small kit-assembly plants in different parts of the region for many years, while GM and Ford have had significant equity involvement in Japanese firms (Isuzu and Suzuki in the case of GM; Mazda in the case of Ford). Today, virtually all the major Western automobile companies are in the process of establishing operations in the region, particularly in Thailand and in China.

The continued restrictive nature of trade policy within ASEAN makes the development of a genuinely regionally integrated automobile industry difficult to achieve, although significant regulatory changes have been introduced to facilitate cross-border flows of vehicles and components.[48] Within South East Asia, Thailand has become the preferred production focus for many major automobile assemblers and component manufacturers, a deliberate outcome of state policy.

> Malaysia pursued a strategy of creating a national car industry [based on the Proton] ... By contrast, Thailand's strategy was to attract car manufacturers (and later on multinational parts-makers) whose presence would lead to an agglomeration of suppliers that would benefit from the infrastructure.[49]

Thailand, therefore, is currently regarded as the 'car capital' of South East Asia, with virtually all the major foreign producers having a presence there. Not only does it have a major concentration of Japanese automobile and components production but also it is the favoured point of entry for Western car manufacturers, notably GM and Ford through their Japanese partners (Isuzu/Suzuki and Mazda respectively). BMW has also opened an assembly plant for its 3 Series model in Thailand. As a consequence, Thailand has become the third largest exporter of automobile products in East Asia, after Japan and Korea, especially of pickup trucks and basic cars and components.

Whereas Western automobile firms see Thailand as potentially a base for serving the whole of South East Asia, their reasons for wishing to establish operations in China are rather different: access to potentially the world's biggest market. Because all the major automobile manufacturers are extremely anxious to establish themselves in China, the Chinese government has been able to retain the bargaining power to impose specific entry restrictions (see Chapter 8).[50]

The Chinese automobile industry consists of a small number of state corporation groups together with a number of joint ventures between members of these groups and foreign firms (Figure 10.15). VW was one of the earliest Western automobile firms to establish a joint venture in China, first with the Shanghai Automotive Industry Corporation (SAIC) in 1985, and then with First Auto Works (FAW). For more than 10 years, VW was virtually unchallenged in the Chinese market and it still has the biggest market share, although its position has been heavily eroded by new entrants. US firms have found entry to China rather more difficult. GM also established a joint venture with SAIC in Shanghai. Ford took considerably longer and agreed a joint venture with the CAIC Group only in 2001. Japanese firms continue to strengthen their presence in China: Toyota has established a new joint venture with Guangzhou Automotive. Hyundai is also expanding in China through its Kia affiliate, which already had Chinese operations before it was acquired by Hyundai.

The prospect of gaining access to what is seen as the world's largest and fastest-growing consumer market has led to a scramble by automobile producers to enter China. But the Chinese government has exerted virtually complete control over such entry and has adopted a policy of limited access for foreign firms, including the form that their involvement can take. Here, therefore, we have the obverse of the usual situation. Whereas, in many cases, TNCs are able to play off one country against another to achieve the best deal, in the Chinese case it is the state whose unique bargaining position has enabled it to play off one TNC against another.[51]

Although access to the Chinese market has been the major motivation for foreign automobile firms, some are now making serious plans to export cars from their Chinese plants. Honda, as so often, is in the forefront, exporting its Jazz model to Europe from its Guangzhou plant. At the same time, the Chinese company SAIC plans to export cars based on the MG Rover designs to Europe.[52]

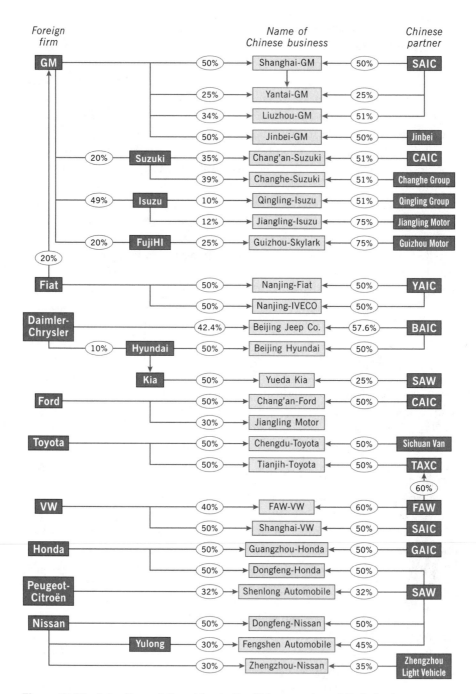

Figure 10.15   Inter-firm relationships in the Chinese automobile industry

Source: based on Liu and Dicken, 2006: Figure 5

Conclusion

In summary, the strategies of the major automobile producers are more diverse than is often realized, a fact not unrelated to their national origins. They are also in a condition of flux as major technological and organizational changes sweep through the industry. There is, undoubtedly, a global battle raging in the automobile industry in which there will certainly be more casualties.

The strategy of the leading Japanese companies is to establish a major integrated production system in each of the three global regions: Asia, North America and Western Europe. GM and Ford are moving in the same direction and have the advantage of already operating a sophisticated integrated network within Europe. In comparison, the European manufacturers remain far more limited geographically. Only VW has much of an internationally integrated system. However, the European automobile production network is being transformed by the new developments in Eastern Europe.

However, whilst intense competition continues in both Europe and North America (where NAFTA is changing the regional production map) the competitive focus is shifting to Asia. Although Japanese producers currently have a dominant position, they are being threatened by the Koreans and by other low-cost regional producers, as well as by the intensified efforts being made by GM and Ford to increase their regional market penetration. Most significantly, however, is the growth of China as a major centre of automobile production but one whose characteristics are fundamentally shaped by the ability of a 'strong' state to influence the decisions of the transnational automobile producers – at least up to now. The automobile industry remains one of the most 'politicized' of industries.

NOTES

1 Drucker (1946: 149).
2 *The Economist* (2004) (4 September 2004).
3 ABN-AMRO (2000: 11), Freyssenet and Lung (2000: 83).
4 UNIDO (2003: 22).
5 *Financial Times* (6 December 2005).
6 See 'Detroit's gas-guzzling business model runs out of road', *Financial Times* (19 May 2005).
7 *Financial Times* (9 November 2001).
8 Veloso and Kumar (2002: 3).
9 *Financial Times* (15 April 2003).
10 Veloso and Kumar (2002: 6).
11 Delphi Automotive Systems, quoted in EIU (2000: 1).

12 Berger (2005: 86).

13 EIU (2000: 7).

14 *Financial Times* (16 June 2004).

15 Reich (1989).

16 Humphrey and Oeter (2000: 42).

17 Humphrey and Oeter (2000).

18 Humphrey and Oeter (2000: 46).

19 Huang (2002).

20 Humphrey and Oeter (2000: 55–6).

21 The *Economist* (10 September 2005: 73).

22 Kang and Sakai (2000: 4–5), Mockler (2000: Figure 1.5).

23 *Financial Times* (8 September 2005).

24 *Financial Times* (4 March 2003).

25 Frigant and Lung (2002: 745–7).

26 *Financial Times* (29 March 1996).

27 *Financial Times* (27 July 2000).

28 *Financial Times* (10 May 2002).

29 Frigant and Lung (2002: 746).

30 *Financial Times* (24 January 2006).

31 *Financial Times* (4 January 1999).

32 *Financial Times* (22 November 2005).

33 *Financial Times* (24 January 2006).

34 *The Economist* (29 January 2005: 73).

35 *Financial Times* (27 November 2001).

36 *Financial Times* (11 February 2006).

37 Amsden (1989), Kim and Lee (1994), Lee and Cason (1994), Wade (1990; 2004a).

38 Wade (1990: 310).

39 See Dicken (2003a), Freyssenet et al. (2003), Rugman and Collinson (2004), UNIDO (2003: 9–16).

40 Havas (2000: 239).

41 Hudson (2002: 271–2).

42 *Financial Times* (8 March 2005).

43 See Eden and Molot (1993), Holmes (1992), Sturgeon and Florida (2004).

44 UNCTAD (2000: 131).

45 Carillo (2004), Mortimore (1998: 411–17), UNCTAD (2000: 130–1).

46 Mortimore (1998: 417).

47 Horaguchi and Shimokawa (2002).

48 Guiher and Lecler (2000).

49 Leclair (2002: 800).

50 Liu and Dicken (2006).

51 Liu and Dicken (2006).

52 *Financial Times* (11 April 2006).

Eleven
'Chips with Everything': The Semiconductor Industry

The semiconductor has emerged as the dominant influence of the last four decades, extending its transformative effects into all branches of life, often without our being aware of its existence. It is, without doubt, 'the engine of the digital age' – today's 'industry of industries'. Yet it is a very young industry, dating back only to the late 1940s and 1950s, when first the *transistor* and then the *integrated circuit* (a number of transistors connected together on a single piece or 'chip' of silicon) were developed in the United States. By the early 1970s a number of very sophisticated solid-state circuits could be placed onto a single chip of silicon the size of a fingernail to create a *microprocessor,* able to perform functions which, only two decades earlier, had taken a whole roomful of valve (vacuum tube) computers.

The progressive refinement of these basic innovations over a very short period made it possible to pack millions of individual circuits onto a single chip of silicon less than one centimetre square. In fact, increased *miniaturization* has been *the* fundamental development, for it permits the incorporation of electronic components into a vast range of products, some of which could not otherwise exist, others which have been transformed by their incorporation. Indeed, it is this *increasingly pervasive application* of semiconductors which makes the industry so very significant.

It is an industry in a perpetual state of flux: an industry of boom and slump, of continuous technological and organizational innovation, and of geographical reconfiguration. Significantly, the semiconductor industry was the first to which the label 'global factory' could be applied. It was here that a *spatial hierarchy of production* at the global scale first became apparent, with clear geographical separation between different stages of the production process.

The semiconductor production circuit

The semiconductor production circuit is highly complex, not least because of the wide diversity of semiconductor products, ranging from mass produced standard devices to highly specific custom devices, and the vast range of end uses. Production of standard devices demands large-scale mass production; manufacture of custom chips demands much smaller runs. However, the development of semi-custom and application-specific (ASIC) devices has extended the use of mass production techniques into new areas.

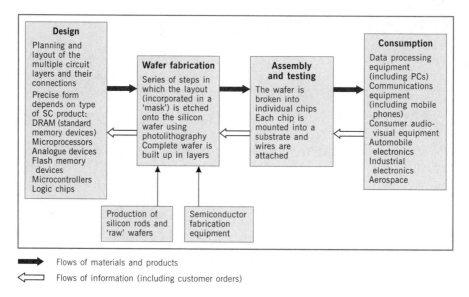

Flows of materials and products

Flows of information (including customer orders)

Figure 11.1   The semiconductor production circuit

The circuit of semiconductor production is shown in Figure 11.1. The differing characteristics of each stage have very important implications for the global geography of the industry. A particularly important distinction exists between the design and wafer fabrication stages on the one hand and the assembly and testing stage on the other. Each has different production characteristics and neither needs to be located in close geographical proximity to the other. Design and fabrication require high-level scientific, technical and engineering personnel while the fabrication stage itself requires an extremely pure production environment and the availability of suitable utilities (pure water supplies, waste disposal facilities for noxious chemical wastes).

In contrast, the assembly of semiconductors is carried out using low-skill labour, notably female, and although there is still a need for a 'clean' production environment this is less critical than for wafer fabrication. The low-weight/high-value characteristics of semiconductors permit their transportation over virtually any geographical distance. Hence, the assembly stage of the production sequence has been most susceptible to relocation to lower-labour-cost areas of the world. Overall, therefore, the manufacture of semiconductors is a highly capital- and research-intensive industry, but one in which there are distinct 'breaks' in the production sequence. As we shall see, semiconductor producers use such breaks to create a complex and dynamic global map of production.

Global shifts in the semiconductor industry

The growth of production of semiconductors has been truly phenomenal since their commercial introduction in the 1950s. Output virtually doubled every year throughout the 1970s and, as Figure 11.2 shows, this vertiginous growth rate has continued. Since 1980, in fact, there has been 'a twenty-fold expansion … making semiconductors by far the fastest growing industry in the world'.[1] But, clearly, it is also an industry subject to both booms and slumps: set against the peaks of 1994, 1997, 2000 and 2004 are the troughs of 1996, 1998 and 2001.

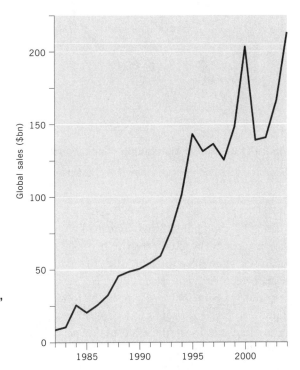

Figure 11.2   Growth in global semiconductor sales, 1982–2004

Source: based on US Semiconductor Industry Association data

Unlike clothing and automobiles, the geographical distribution of semiconductor production is much less widely spread. It is also highly segmented according to stage in the production circuit (Figure 11.1). Figure 11.3 shows the extent to which semiconductor fabrication capacity has become increasingly dominated by producers in the Asia–Pacific region, a major global shift indeed.[2] Geographically, the commercial production of semiconductors began in the United States during the 1950s and for nearly two decades the United States completely dominated world production. During the 1980s, however, Japan overtook the United States to become the world's leading semiconductor producer, creating major political tensions between the United States and Japan.

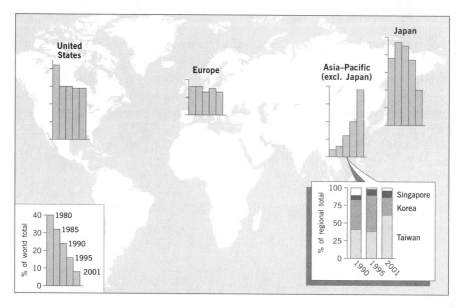

Figure 11.3   Global distribution of semiconductor fabrication capacity

Source: calculated from Leachman and Leachman, 2004: Tables 8.2, 8.3

In 1980, the United States contained 42 per cent of world semiconductor fabrication capacity; by 2001 its share had fallen to 29 per cent. Conversely, by 1985, Japan was responsible for almost half the world total. But the United States' industry staged a recovery while Japan's share fell back to 20 per cent. At the same time – and very much a contributory factor in the relative decline in Japan's share – Korea and Taiwan emerged as major producers. Meanwhile, Europe's position remained more or less stagnant. Today, therefore, the ownership of fabrication capacity (as opposed to its physical location) is dominated by Asia–Pacific and United States' firms, each with almost 40 per cent of the total. Japanese firms own just over 20 per cent, European firms a mere 8 per cent.[3]

However, in interpreting these trends, we need to take into account changes in the product composition of the industry – especially the distinction between memory chips (dynamic random access memory, DRAM) and more sophisticated microprocessors. The basis of the United States' recovery as a major semiconductor producer was its shift out of memory chips and into more design-intensive devices.[4] Conversely, Japanese producers made far less of a product shift, concentrating on 'next-generation' DRAMs. However, in such essentially 'commodity' products, low production costs become especially important and this is where Korea and Taiwan were first able to enter the semiconductor market.[5] In fact, Korea and Taiwan emerged as significant semiconductor producers in the late 1980s and early 1990s.

> The Korean semiconductor industry has grown with an overwhelming focus on exports. In 1995, exports accounted for 91 per cent of production, making semiconductors the largest item in Korea's exports. Exports of chips as a proportion of total output have actually been increasing, from 81 per cent of production in 1991, to 87 per cent in 1993, and 90 per cent in 1994.[6]

Taiwan's growth has been similarly impressive. By the mid 1990s, Taiwan had become the fourth largest semiconductor producer in the world, overtaking Germany. Elsewhere in East Asia, other important centres of semiconductor production have emerged, although none is anywhere near as important as either South Korea or Taiwan. The most significant so far are Singapore, China and Malaysia.

Changing patterns of consumption

Unlike automobiles or garments, most people don't go shopping for semiconductors. Despite Intel's advertising slogan 'Intel Inside', attached to other manufacturers' PCs, semiconductors are invisible. Demand for semiconductors is a derived demand: it depends upon the growth of demand for products and processes in which they are incorporated. In the United States, most of the initial stimulus came from the defence-aerospace sector, and this explains much of the early growth of the semiconductor industry in the United States. In Europe, and especially in Japan, defence-related demand was far less important than industrial and, particularly, consumer electronics applications. Nevertheless, governments have often been dominant customers, either directly or indirectly, throughout the semiconductor industry's short history.

As Figure 11.2 shows, global semiconductor sales in 2004 were more than $213 billion. On average, the global market for semiconductors has been trebling every five years. Until recently, by far the biggest end user was the computer industry and, especially, the personal computer industry. In the mid 1990s, around 60 per cent of semiconductor revenues were derived from sales to computer manufacturers. Indeed, the two industries – PCs and semiconductors – had become symbiotically locked together. On the one hand, demand by PC manufacturers drove much of the demand for semiconductors; on the other hand, the leading semiconductor manufacturers (notably Intel and AMD) created demand for more powerful PCs by introducing new generations of increasingly powerful chips. Such a cumulative process was reinforced by the dominant position of Microsoft, the world's leading software company, whose successive new generations of software have required increasingly powerful microprocessors.

Recently, however, the market for semiconductors has broadened significantly beyond the PC sector. By the turn of the millennium, the PC sector consumed only 40 per cent of semiconductor production whilst other markets, primarily communications equipment (including Internet infrastructure, mobile telephones) and sophisticated digital consumer electronic products (such as digital TVs, digital cameras, digital music players, handheld PDAs, video game consoles) had become increasingly important. Much of this relates to what has been termed the 'Net World Order (NWO)'.[7] The increasing centrality of the Internet, and of communication in general, not only stimulates a shift towards products capable of transmitting as well as receiving information but also means that semiconductor producers have to relate to

> the carrier, who controls the infrastructure that connects users to networks. Carriers have a regional focus, since they must operate under regional standards and usually have a regionally situated infrastructure for delivery (satellites being the exception). For these reasons, carriers play the regional gatekeepers to the rest of the NWO value chain ... Network dependence tends to increase the bargaining power of the network service provider.[8]

A major reason for the spectacular growth in demand for semiconductors has been the vertiginous decline in their selling price. Partly because of technological developments (discussed in the next section) and partly because of fierce price competition between producers, the price of semiconductors is a tiny fraction of that prevailing only a short time ago. This process of continuously falling prices per unit of capacity has shown no sign of abating. Indeed, higher-capacity DRAMs now appears so quickly that each generation affects the price of its predecessors. In the early 1990s, for example, the price per megabyte (MB) of DRAM was $25; by the late 1990s, the price per megabyte was down to a little over one dollar.[9] Today, it is just a few cents.

The semiconductor industry is subject to spectacular fluctuations in demand: gluts and shortages are endemic. Supply gluts further intensify price competition; at other times, severe shortages of chips create major problems and raise prices for end users. The basic problem results from a combination of factors: the massive fixed costs of production facilities (see next section), the fact that they take at least a year to build, and the fact that demand for chips can change rapidly. In addition, there are many different types of semiconductor demand: it is a highly *segmented* market.

Geographically, the demand for semiconductors is overwhelmingly concentrated in the three major regions of North America, Europe and East Asia. But, as in so many aspects of the global economy, the fastest potential growth is expected to be in China – possibly 20 per cent per year for the foreseeable future. Much of this projected growth is likely to be in the expansion of the mobile phone market and in the fast-expanding local consumer electronics and related industries (for example, cars, smart toys).

Production costs and technology

In the fiercely competitive environment of the global semiconductor industry the drive to push down production costs is unavoidable. The industry has become increasingly capital intensive, while retaining a considerable degree of labour intensity in certain parts of its production circuit. The urge to minimize production costs helps to drive both the rapid rate of technological change and the changing geography of production.

Basic parameters of semiconductor production

Two considerations are especially important in the manufacture of semiconductors:

- *The reliability of the product*: a great deal of effort and expenditure is devoted to improving the 'yield' of the manufacturing process and reducing the number of rejected circuits. A fundamental need is a pure environment for wafer fabrication. Such pure environments are one of the reasons for the very high capital costs of semiconductor fabrication plants. One approach is to create 'mini-environments' which are, in effect, sealed containers to ensure that as the wafer travels through the factory from one processing step to another during its production cycle, it never comes into contact with the 'dirty' factory environment.
- *The need to pack as many circuits as possible on to a single chip*: by the mid 1990s, a microprocessor could contain as many as 5.5 million individual transistors, while the Pentium 4 of the late 1990s contained 42 million transistors. Compare these chip capacities with the mere 3500 on Intel's first commercial microprocessor produced in 1971. In 2000, Intel predicted it would produce a microprocessor containing 400 million transistors within four years. In fact, the prediction by Charles Moore (co-founder of Intel) in 1965 that the number of transistors per chip would double every 18 months has been borne out (hence the description 'Moore's Law'). This 'law' is shown on the left-hand axis of Figure 11.4.

It is the combination of these two requirements – to increase both the yield and the number of circuits per chip – that accounts for the steep increase in the cost of setting up a new semiconductor plant (see the right-hand axis of Figure 11.4). Every major increase in the number of components per chip demands far more sophisticated and expensive manufacturing equipment which, in turn, becomes obsolete more quickly. At the same time, the trend is towards larger, more capital-intensive and research-intensive production. The pace of technological change has been so rapid and far-reaching in semiconductor production that it is now necessary for production equipment to be replaced or updated every three to five years.

The cost of establishing a semiconductor plant has escalated accordingly: each new generation of semiconductors has involved a doubling of plant construction

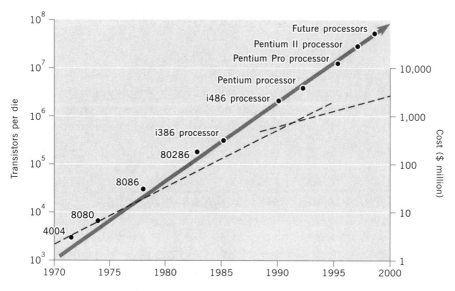

Figure 11.4   Exponential increase in the capacity of semiconductors and the cost of new manufacturing plants

Source: based on material in the *Financial Times*, 4 February 1998

costs. Today between $2 billion and $3 billion is needed to create a state-of-the-art chip plant; as a result, the *capital barriers to entry* have increased very substantially. Not surprisingly, therefore, the industry is one of the most capital and research intensive of all manufacturing industries.

Developments in the technology of semiconductor production – new lithographic techniques, new methods of wafer fabrication, and the introduction of automation at all stages of the process – have created substantial changes in the organization and structure of the industry. In particular, the relative importance of labour costs has been greatly reduced, especially for higher-value products. At the same time, a wafer fabrication plant, to be profitable, needs to produce at very high volumes.

Apart from continuing attempts to introduce new materials to supplement or to replace silicon, the most important current developments in the technology of the semiconductor are related to the shift from 200 mm to 300 mm diameter wafers. This 1.5-fold increase in diameter will increase by 2.6 times the number of identically sized chips on a single wafer.[10] If the size of each chip is also reduced, say from 0.20 microns to 0.13 microns, then the number of chips per wafer can be increased further. At the same time, such smaller chips are more efficient because the electrons have a smaller distance to travel. On the other hand, smaller chip geometries increase the risk of 'interference' between the circuits. Packing more on to a single chip creates major problems of heat generation and increased energy consumption.

The transition to 300 mm wafers and smaller chip sizes is both difficult and expensive. It also depends upon the availability of new-generation scanners – the equipment that forms the images of the chips which are then etched onto the silicon.[11] Firms are seeking new ways of solving the problems of increasing miniaturization and more sophisticated semiconductor devices – 'systems-on-a-chip'.[12] New three-dimensional chips are being developed that are still within an individual sliver of material only 0.3 mm thick.[13]

But whatever the precise nature of these new technological developments, they serve to reinforce the knowledge and capital intensity of the semiconductor industry and the problems of generating sufficient investment capital and scientific and technological expertise to keep up with the ever-quickening game. Indeed, speed is really of the essence in this industry, not just the 'clock speed' of the devices themselves but also the speed with which a new design can be marketed. As we have seen, the price of a new generation of chips falls very quickly after introduction.

Characteristics of the labour force in semiconductor production

Two characteristics are especially important. First, the particular configuration of the semiconductor production process creates a clear *polarization* of skills between highly trained professional and technical workers on the one hand and low-skilled production workers on the other. Increased automation has led to a much steeper relative decline in the number of production workers, particularly the less skilled. But there is also a clear *geographical* pattern of skills in the semiconductor industry. Overall, professional and technical occupations are far more important in the developed countries than in the developing countries, a reflection of the fact that it is mainly the more routine assembly tasks which have been located in lower-cost developing countries while the more sophisticated design, R&D and wafer fabrication stages remain more localized. Hence most, though by no means all, of the semiconductor employment in Silicon Valley and the other production clusters in the United States and Japan is in the higher-skill categories; conversely, virtually all the semiconductor employment in the EPZs of the East Asian countries and in Mexico is of production workers. However, the increasingly complex intra-regional division of labour within the semiconductor industry in East Asia is removing some of this simplicity, as we shall see later in this chapter.

The second characteristic of employment in the semiconductor industry is the predominance of female workers in the assembly stages. In this respect, there are clear parallels with the clothing industries. Female workers are the norm in the semiconductor assembly process wherever the plants are located. Although labour regulations are generally more stringent in developed countries, some undesirable characteristics may still apply. The gleaming plate-glass- and metal-clad plants of the modern semiconductor industry look, at first sight, to be the complete opposite of the dark, satanic mills of the nineteenth century and the squalid sweatshops of the garment industry. In some senses, of course, the contrast is undoubtedly what it seems. For many of the workers employed in the semiconductor industry the wages are high, the work is stimulating and the conditions are superb. But for others the story is rather different. At worst, conditions are just as bad, with long working hours, an unpleasant and noxious working environment and little or no job security. It is indeed paradoxical that an industry epitomizing all that is new and up-to-date at the same time harbours some of the oldest and least desirable attributes of work in manufacturing industry.

The role of the state

It is easy to understand why governments have attempted to intervene in the development of the semiconductor industry. Semiconductor production is a key technology with enormous ramifications throughout the economy; access to what is an expensive and rapidly changing technology is critical. Such access may be achieved in several ways: building an indigenous production capacity based upon domestically owned firms; attracting foreign semiconductor firms to establish production units; purchasing semiconductors on the open market; and concentrating on developing the end uses. There are problems with all three options. Setting up a viable domestic industry is beyond the means of many countries. Relying on foreign investment or the open market may lead to problems of dependency of supplies on foreign sources. The particular policies pursued by governments reflect their specific national circumstances, including the country's relative position in the global semiconductor industry.

United States

Although the United States does not have a formal industry policy it has been heavily involved in the development of the semiconductor industry from its earliest days, when the dominant formative influences were the federal defence and aerospace sectors. In effect, these set the direction and nature of the industry's development because they were its dominant customers. Although defence-related forces are now less important to the US semiconductor industry their influence persists. In the 1980s, as the United States' lead in semiconductor production began to be eroded, strong political and industry lobbying urged government intervention. This came in the form of trade policy and the facilitation and encouragement of industry consortia.[14]

The prevailing view in the United States was that Japan was engaging in 'unfair' trade practices: 'dumping' chips at unreasonably low prices in the United States (and therefore undercutting US producers) and unfairly restricting access to their domestic market. In retaliation, the US government 'persuaded' the Japanese government to sign the US–Japan Semiconductor Trade Agreement in 1986. The initial pact expired in 1991 and was renegotiated for a further five years, with a provision on access for American semiconductor manufacturers to the Japanese market. When the pact finally expired in 1996, the US and Japan agreed to set up two new international industry bodies. The effects of the semiconductor pact were limited, although not insignificant for the US semiconductor industry.[15] More recently, the United States has been involved in confrontation with Korea over the alleged illegal subsidization of Hynix semiconductors, and with China over problems of access to the Chinese market for US semiconductor firms.

The second policy strand has been the encouragement and facilitation of inter-firm collaboration through industry consortia. In 1987, SEMATECH was founded by the 14 major US semiconductor manufacturers with funding from the firms themselves and, initially, from the federal government. The focus was to be on medium-term research. The fact that much of the consortium's funding was defence related, together with the political priorities underlying its establishment, meant that non-US firms were excluded. More recently, the SEMATECH consortium launched a new collaborative initiative – the International 300 mm Initiative (I300I) – involving both US and non-Japanese foreign firms (European, South Korean and Taiwanese).[16]

Whereas the role of the state in the development of the semiconductor industry in the United States has been important it has not been, in a direct sense, central to that development. In contrast, in the case of those East Asian countries that have emerged as major players in the semiconductor industry, the state's role has been absolutely central. However, each state has developed its semiconductor industry in rather different ways. [17]

Japan

The initial aim of Japanese policy was to avert technological dominance by the United States by discouraging direct foreign participation in the Japanese semiconductor industry and acquiring foreign technology through other means. Protection of the industry was extremely tight until the end of the 1970s, using import controls and restrictions on inward investment. A major government initiative was the Very Large Scale Integration (VLSI) Project begun in 1976 on the initiative of MITI, involving research collaboration between the five major Japanese companies – NEC, Fujitsu, Hitachi, Toshiba and Mitsubishi. The objective was to develop highly sophisticated integrated circuits for the next generation of computers. The Japanese government funded some 40 per cent of the total cost between 1976 and 1979. Indeed, it was the example of the Japanese VLSI Project that helped fuel the debate within the United States and eventually led to the creation of SEMATECH.

Japanese direct financial support of the VLSI Project ceased after 1979 but the more general context of supportive policies continued in line with overall Japanese industrial policy. In 1996, a new initiative, Selete – encouraged, though not directly driven, by the government – was established. Like the I300I programme in the United States, Selete was a response to the problems of moving to the 300 mm wafer standard.[18] Selete consists only of Japanese firms, unlike SEMATECH which involves both US and non-Japanese foreign firms.

Korea

Whereas Japan's objective in stimulating its semiconductor industry was to emulate the United States, the objective of South Korea's state initiatives was to emulate both. The Korean government used a variety of devices – including its control of the telecommunications industry – to foster the entry of large Korean firms into the semiconductor industry. Indeed, the government explicitly used the major *chaebols* as the vehicles for developing a national semiconductor industry in a more or less direct imitation of the Japanese system.

Table 11.1   Stages in the evolution of the Korean semiconductor industry

Stage I	Stage II	Stage III	Stage IV
Pre-1974	**1974–81**	**1982–88**	**1989–98**
Preparation	**Seeding/implantation**	**Propagation**	**Roots of sustainability**
Assembly and test operations established by foreign companies	Indigenous industry established and beginnings of IC fabrication in Korea	Incursion into mass memory chip production	Fully-fledged memory chip production and diversification/ consolidation of a semiconductor industry
KIST (Korea Institute of Science and Technology) established	KIET (Korea Institute of Electronics Technology) leads technology leverage	VLSI take-off	Ancillary industries established
Expansion of technical education		Joint development of 4 MB DRAMs	Collaborative R&D programmes Science and technology strategy as a political priority

Source: Mathews and Cho, 2000: Table 3.1, reprinted with permission of Cambridge University Press

    Table 11.1 shows the developmental stages through which the Korean semiconductor industry has evolved. In the first, 'preparation' stage, an important catalytic influence was the establishment of semiconductor assembly and test operations in Korea by some of the leading US companies during the 1960s. By 1974, there were nine such US-owned facilities in South Korea.[19] Domestically, major investments were made in technical education through the Korea Institute of Science and Technology (KIST), set up in 1966. The Korea Institute of Electronics Technology

(KIET), established in 1976, was responsible for 'planning and coordinating semiconductor R&D, importing, assimilating, and disseminating foreign technologies, providing technical assistance to Korean firms, and undertaking market research'.[20] KIET played a key role in establishing an indigenous semiconductor industry in South Korea during the second half of the 1970s.

It was during the 1980s, then, that Korea began to emerge as a major semiconductor producer. In 1982, the Long-Term Semiconductor Industry Promotion Plan was announced, making available substantial fiscal and financial incentives to the four leading semiconductor *chaebols*. Within this supportive framework,

> the government … designated R&D on the 4M DRAM as a national project in October 1986. The Electronics and Telecommunications Research Institute (ETRI), a government research institute (GRI), served as a coordinator in the consortium of three *chaebol* semiconductor makers – Samsung, LG, and Hyundai – along with six universities. The objective was to develop and mass-produce the 4M DRAM by 1989 and completely close the technology gap with Japanese firms. The consortium spent $110 million for R&D over three years (1986–1989); the government shared 57 per cent of the total R&D expenditure, a disproportionately large share in comparison to other national projects.[21]

Taiwan

In both Japan and Korea, the result of state–private collaboration was the creation of large semiconductor operations within massive electronics groups. The Taiwanese semiconductor industry evolved rather differently.

> Taiwan's semiconductor industry is a flourishing market-driven industry. It has no 'nationalized' firms within it: all are privately owned and managed. It is strongly export-oriented. It has never imposed any protective tariffs on its semiconductor products. There are no government 'handouts' to any of the firms involved in the industry for any of their current activities. And yet its *creation, its nurturing and its guidance have been entirely the product of government and public sector institutions.*[22]

Table 11.2 summarizes the developmental stages of the Taiwanese semiconductor industry

As in Korea, the initial development of semiconductor production derived from the decisions of United States' firms to locate assembly operations in low-cost East Asian areas in the 1960s. The Taiwanese government immediately capitalized on this trend by building the world's first export processing zone (EPZ) for semiconductors in 1965. In the early 1970s, the Industrial Technology Research Institute (ITRI) was set up to promote technological leverage. ITRI was a specialist semiconductor and electronics laboratory which eventually evolved into the Electronics Research Service Organization (ERSO).

Unlike Korea at a similar stage, there were no Taiwanese firms willing to enter such a capital-intensive and risky industry as semiconductors. The solution

Table 11.2   Stages in the evolution of the Taiwanese semiconductor industry

Stage I Pre-1976 Preparation	Stage II 1976–79 Seeding	Stage III 1980–88 Technology absorption and propagation	Stage IV 1989–98 Sustainability
Labour-intensive semiconductor back-end operations (assembly) and testing	Licensing of IC fabrication technology, and its adoption by public sector R&D institute	Technology absorption and enterprise diffusion	Entry of firms to cover all phases of semiconductor manufacturing and full product range, including DRAMs
Dominated by foreign multinationals	Phase I of Electronics Industry Development Program	Establishment of secure infrastructure in the form of the Hsinchu Science-Based Industry Park	From VLSI to ULSI technology
Establishment of ITRI (Industrial Technology Research Institute) and ERSO (Electronics Research Service Organization)		ERSO acquires skills covering all phases of semiconductor manufacturing moving from LSI to VLSI	Submicron stage of public-sector-led R&D
		Spin-off of private companies and entry of private sector	Cooperative R&D system of innovation established

Source: Mathews and Cho, 2000: Table 4.1, reprinted with permission of Cambridge University Press

adopted by the Taiwanese government was to launch a publicly financed firm United Microelectronics Corporation (UMC) through ITRI/ERSO. At the same time, the Hsinchu Science-Based Industry Park was opened near Taipei. This has become the focus of the entire Taiwanese semiconductor industry, creating what has been called a 'Silicon Valley of the East'.[23] But even as late as 1985, when the competition in semiconductors was intensified by the entry of Korean firms, there was only the one Taiwanese-owned fabrication plant operated by UMC.

The government's response to this serious problem was to facilitate the creation of a new joint venture company, Taiwan Semiconductor Manufacturing Corporation (TSMC), between a spin-off from ITRI and the European firm Philips. TSMC took over the existing VLSI pilot operation built by ERSO and also constructed its own advanced VLSI fabrication plant.

> The immediate effect of TSMC's launching and its success was to take Taiwan to a new level of technical sophistication and to spark the formation of dozens of small IC design houses in the Hsinchu region. Its demonstration effect also attracted serious Taiwan capital to invest in the semiconductor industry. Thus, the object of the exercise – to launch UMC and then TSMC as 'demonstration' vehicles – was achieved, as skilled engineers were voting with their feet, and capitalists were voting with their finance, to enter the new industry.[24]

Today, Taiwan is the world's third largest semiconductor producer. It is difficult to believe that it could have achieved such a position without the focused involvement of the state. 'The semiconductor industry in Taiwan has been created and established through a process of *resource leverage* that has been *accelerated* by a judiciously constructed institutional framework'.[25]

Singapore

Apart from the very early stages of their development, the semiconductor industries in Japan, Korea and Taiwan have been predominantly indigenous industries. In Singapore, on the other hand, government industrialization policies have relied overwhelmingly on foreign firms to develop their semiconductor industries (Table 11.3). The focus of Singapore's strategy, implemented largely through the Economic Development Board, has been continuously to upgrade its semiconductor sector through leveraging the technologies and resources of foreign TNCs. In addition, however, the Singapore government has used one of its state-owned companies, Singapore Technologies Group (STG), to develop an indigenous presence in semiconductors. Like Taiwan, Singapore used a technology agreement with a foreign firm – in this case a United States firm, Sierra Semiconductor – to create a state-owned IC foundry: Chartered Semiconductor Manufacturing (CSM). Subsequently, STG bought out the US partner and modelled CSM's activities on the Taiwanese foundry firm TSMC.

> CSM expanded its production facilities rapidly in the 1990s to become one of the world's largest IC foundries. It built a second and third fabrication facility in the Woodlands semiconductor park and has entered into joint ventures with two of the world's leading semiconductor firms, HP and Lucent, to build a further two. Recently, the Singapore Technologies Group has launched a third semiconductor business, in the form of Singapore Technologies Assembly and Test Services (STATS), offering 'back-end' test and assembly contract services to complement its 'front-end' IC design and IC fabrication services. By the late 1990s, STG was becoming an integrated contract IC producer.[26]

The Singapore government's direct involvement in semiconductor production through CSM is impressive but is only a small component of the country's activities. Most semiconductor production in Singapore is performed by foreign TNCs. But their continued presence and development testify to the kind of supportive environment provided by the state.

Table 11.3   Stages in the evolution of the Singaporean semiconductor industry

Stage I Pre-1976 Preparation	Stage II 1976–85 Seeding/ implantation	Stage III 1986–present Diffusion	Stage IV 1991–present Roots of sustainability
Preparation by Economic Development Board	Upgrading of TNC activities and expansion of their scope	Further upgrading and expansion of TNC activities	Stock of both TNC and indigenous firms engaged in all phases of IC and wafer fabrication
Attraction of TNCs	Internal and external leverage via TNCs	Beginnings of TNC-based wafer fabrication	Framework of R&D support to upgrade activities
Assembly and test operations established by foreign companies	Beginnings of indigenous industry established as contract service providers to TNCs	Beginnings of local wafer fabrication	Establishment of institutional R&D support
Expansion of technical education		Expansion of local service industries	Development of Woodlands wafer fabrication 'park'

Source: Mathews and Cho, 2000: Table 5.1, reprinted with permission of Cambridge University Press

Europe

Not only does the European semiconductor industry lag way behind those of the United States, Japan, Korea and Taiwan but also much of its production capacity is in US-, Japanese- and, more recently, Korean-owned plants. In general, European governments have not only welcomed such investments but also assiduously courted the foreign semiconductor firms. Since the early 1980s European governments have also been involved in supporting major collaborative ventures in information technologies in general (the ESPRIT programme) and specifically in semiconductors. The $4 billion JESSI programme – the Joint European Submicron Silicon Initiative – was concerned with developing advanced microchip technology. There has also been increasing pressure on non-European semiconductor manufacturers to locate more of their design and fabrication activities in Europe.

The aims of these European initiatives were threefold: to protect European capacity in the core technologies of microelectronics; to accelerate innovation; and to encourage cross-border links between national electronics firms within Europe. A major problem, however, has been whether or not to allow non-European firms with major European operations to participate in the collaborative programmes. At the individual state level the French have been most directly involved in attempting to establish and nurture a 'national champion' semiconductor firm.

The European Commission has also been active in trade protection. In 1990 it secured a voluntary agreement with the Japanese on the minimum prices at which standard memory chips would be sold in the EC. Subsequently, the Commission initiated anti-dumping measures against Korean semiconductor manufacturers who were alleged to be dumping chips at below acceptable prices. In effect, this was the price agreed with the Japanese, who were regarded as the lowest-cost producers. These minimum prices were reimposed in 1997 on the biggest-selling semiconductor chips from 14 Japanese and Korean manufacturers.[27] Nevertheless, despite these national and EU-wide initiatives, Europe has not succeeded in sustaining a strong indigenous semiconductor industry.[28]

Corporate strategies in the semiconductor industry

Consolidation and concentration

Although the semiconductor industry is dominated by very large TNCs (reflecting the necessity of huge capital investment in new plants and technologies), it is less concentrated than the automobile industry. This is largely because new niches within the semiconductor production system are continuously being created. Nevertheless, the industry is far less fragmented than in the past, when small and medium firms were the norm. This is seen most clearly in the United States where, in the pioneering late 1950s and 1960s, entry into the semiconductor industry was relatively easy. However, the proliferation of new small semiconductor firms slowed down as the barriers to entry increased during the 1970s and 1980s. The dominance of large firms became especially marked in the manufacture of standard semiconductor devices, which depend on mass production technology. Even so, very high levels of new firm start-ups continued in the US semiconductor industry through the 1980s, especially among firms engaged in application-specific products.[29]

Rank by market share	1978	Market share (%)	1989		1995		1999		2004	
1.	Texas Instruments, U.S.	12.4	NEC	8.9	Intel	8.9	Intel	15.9	Intel	15.0
2.	Motorola, U.S.	7.5	Toshiba	8.8	NEC	7.3	NEC	5.5	Samsung	7.7
3.	NEC, Japan	5.6	Hitachi	7.0	Toshiba	6.6	Toshiba	4.5	Texas Instruments	5.2
4.	Philips, Netherlands	4.8	Motorola	5.9	Hitachi	6.1	Samsung	4.2	Infineon	4.5
5.	Hitachi, Japan	4.3	Fujitsu, Japan	5.3	Motorola	5.9	Texas Instruments	4.2	Renesas, Japan	4.4
6.	Toshiba, Japan	3.7	Texas Instruments	5.0	Samsung, Korea	5.4	Motorola	3.8	STM	4.3
7.	Fairchild, U.S.	3.7	Mitsubishi, Japan	4.7	Texas Instruments	5.2	Hitachi	3.3	Toshiba	4.1
8.	National Semiconductor, U.S.	3.5	Intel	4.4	Fujitsu	3.6	Infineon, Germany	3.1	TSMC, Taiwan	3.7
9.	Siemens, Germany	2.9	Philips	4.2	Mitsubishi	3.3	STM, Italy/France	3.0	NEC	3.2
10.	Intel, U.S.	2.9	National Semiconductor	2.8	Hyundai, Korea	2.8	Philips	3.0	Philips	2.8

Figure 11.5   Changes among the world's leading semiconductor producers

Figure 11.5 shows how the composition of the world's top 10 semiconductor producers has changed since the late 1970s. The five time slices plot the fluctuating fortunes of US, Japanese and European firms, and the rise of Korean and Taiwanese producers. In the late 1970s, five US companies were in the top 10 including – in 10th position – the subsequent industry leader, Intel. The five US firms accounted for 30 per cent of global semiconductor sales. Three of the top 10 firms were Japanese and two were European. A decade later, the number of US firms in the top 10 had fallen to four and their share had fallen to 18 per cent. On the other hand, Japanese companies had surged up the league table, clearly shattering US dominance. By the late 1980s there were five Japanese firms in the top 10, including the three top positions. The combined Japanese share had grown to 35 per cent while the US share had fallen from 30 to 18 per cent.

By the mid 1990s a US recovery could be discerned. Although the number of US firms in the top 10 had fallen to three, their share of global semiconductor sales had improved from 18 to 20 per cent. More significant, in the light of subsequent events, was the dramatic rise of Intel from eighth position in 1989 to first in 1995. Every leading Japanese company in the top 10 lost ground while their combined share of sales fell from 35 to 27 per cent. The other striking feature of the mid 1990s position was the appearance of two Korean semiconductor producers in the top 10, displacing the European firms. Indeed, in the mid 1990s it looked as though the European semiconductor industry was in terminal decline.

However, by 2004 the position had changed again, although Intel continues to dominate. The most significant aspects of the current position are: the rise of the Korean company Samsung to second place; the decline of several Japanese firms (notably NEC and Toshiba); the appearance of a Taiwanese firm, TSMC; and the improved ranking of a European firm, Infineon. Indeed, the three European firms in the top 10 had the same market share as the Japanese firms.

In striving to compete in global markets, semiconductor firms have become increasingly interconnected through processes of acquisition, merger and strategic alliances. By 1980, only seven of 36 post-1966 start-up companies in the United States were still independent. Within Europe the biggest merger was that between the Italian firm SGS and the French firm Thomson to form SGS-Thomson, which subsequently became STMicroelectronics. Most recently, there has been a further upsurge in acquisitions in the aftermath of the East Asian financial crisis of the late 1990s. For example, Hyundai acquired LG Semicon in 1999 and then transformed itself into Hynix. In Japan, a number of mergers have occurred, most notably that between Hitachi and Mitsubishi's semiconductor business to form Renesas which became the fifth-ranking producer in 2004.

The use of strategic alliances has been even more pervasive than acquisition. The massive costs of R&D, the incredibly rapid pace of technological change, and the escalating costs of installing new capacity all contribute towards the attractiveness of forming strategic alliances. Table 11.4 shows some of the more significant recent

Table 11.4 International strategic alliances in the semiconductor industry

Alliance partners	Purposes of the alliance
Motorola (US), IBM (US), Siemens (Germany), Toshiba (Japan)	To develop next generation of memory chips, including a 1 GB DRAM device. Builds upon existing alliance between IBM, Siemens and Toshiba
Siemens (Germany), Motorola (US)	To build a new advanced memory chip plant in the United States
Mitsubishi (Japan), Umax Data Systems (Taiwan), Kanematsu (Japan)	To build a new advanced memory chip in Taiwan
NEC (Japan), Samsung (South Korea)	To collaborate in the production of memory chips for the European market
IBM (US), Toshiba (Japan)	To build a semiconductor facility in the United States to produce next-generation memory chips
UMC (Taiwan), Kawasaki Steel (Japan)	To develop ASICs
Fujitsu (Japan), AMD (US)	To develop production of flash memory devices
Hitachi (Japan), NEC (Japan)	To create a joint venture – Elpida – to produce DRAM
IBM (US), AMD (US)	To develop new technology for high-end microprocessors
NEC (Japan), state-owned firm (China)	To produce DRAM in China
Intel (US), BT (UK)	To develop wireless applications
STMicroelectronics (Fr/It), Philips (Neth.), Motorola (US)	To create R&D alliance based near Grenoble, France
AMD (US), UMC (Taiwan)	To build $5bn fabrication plant in Singapore
Sony (Japan), Toshiba (Japan), IBM (US)	To develop a new microprocessor (Cell)
Toshiba (Japan), Xilinx (US)	To supply high-end programmable microchips
Intel (US), Micron Technologies (US)	To form IM Flash Technologies to produce flash memory chips

strategic alliances in the semiconductor industry. In addition, there has been a considerable development of multilateral alliances within consortia, often stimulated or facilitated by national governments. Figure 11.6 shows the company membership of the two most recent consortia in the United States and Japan. In Europe an alliance of producers from different European companies formed European Silicon Structures (ES2) to manufacture custom chips.

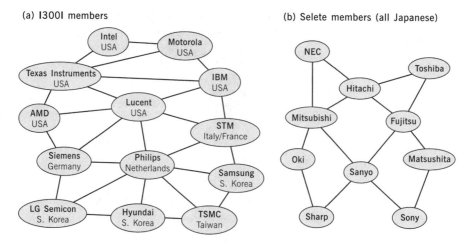

Figure 11.6   Company alliances within semiconductor consortia

Source: based on Ham et al., 1998: Table 1

Changing strategies of semiconductor producers

In such a volatile technological and competitive industry as semiconductors, firms inevitably employ a whole variety of strategies to ensure their survival and in pursuit of growth.[30] Some firms pursue a 'niche' strategy; others have preferred to increase their vertical integration. Cutting across these strategies are those of increasing transnationalization, rationalization and reorganization of production on a global scale. A major influence is, undoubtedly, *a firm's geographical origins*: the domestic context in which it has developed. There are substantial differences in behaviour between US, East Asian and European semiconductor firms.

Semiconductor firms can be classified into several broad types:

- *Vertically integrated captive producers* manufacture semiconductors entirely for their own in-house use. This has been a common strategy in Japan, where virtually all of the major semiconductor producers were part of diversified electronics groups. It was also the model used by the US company IBM for many years, and by European producers.
- *Merchant producers* manufacture semiconductors for sale to other firms. This strategy developed especially in the United States in such firms as Texas Instruments and Motorola.
- *Vertically integrated captive–merchant producers* manufacture semiconductors partly for their own use and partly for sale to others. This hybrid strategy was adopted by a number of the vertically integrated captive producers, notably IBM, to expand their market.

- *'Fabless' semiconductor firms* design semiconductors but do not manufacture them. These emerged especially in the United States during the 1980s. They 'serve a variety of fast-growing industries, especially personal computers and communications, by offering more innovative designs and shorter delivery times'.[31]
- *Foundries* manufacture semiconductors to customer specifications but do not design them. They include both 'pure-play' (i.e. dedicated) foundries as well as the excess capacity of more traditional semiconductor firms. They have become especially important in East Asia. The foundry is, in effect, the obverse of the fabless firm.

> The pure-play foundry was an organizational innovation that transformed the electronics industry, for it allows the sequence of functions in the making of components to fragment. Manufacturing at lead firms can be broken off and transferred to chip-manufacturing specialists at these foundries. New firms can be created (and old ones transformed) to carry out only one, two, or several functions in placed of the formerly integrated companies.[32]

- *Specialized providers of semiconductor designs:* a recent development of 'Electronic Design Automation (EDA) suppliers and systems houses that compete in the provision of intellectual property (IP) design blocks and system-on-a-chip (SOC) technology'.[33]

United States firms

The commercial production of semiconductors began in the United States in the 1950s and, for almost three decades, US firms dominated before being overtaken by the Japanese. Subsequently, US firms staged a major recovery, primarily by moving out of the commodity DRAM market into more sophisticated semiconductor devices. In the view of Michael Borrus, US firms developed a new form of strategic competition that he calls 'Wintelism'.[34] The term derives from a combination of the dominant software system (*Win*dows) and the dominant microprocessor producer (In*tel*). It materialized in the increasing use of a 'merchant producer' strategy, which enabled US firms to specialize in developing and selling semiconductor devices to a wide variety of end users rather than to captive parts of a vertically integrated organization.

A key element of this strategic reorientation by US semiconductor firms was its manipulation of the *geography* of production. From the beginning, the geographical pattern of the semiconductor industry within the United States was highly concentrated around two foci: the Route 128 corridor near Boston, Massachusetts and the Santa Clara Valley in California – later to be dubbed 'Silicon Valley'.[35] Subsequently, Silicon Valley became emblematic of the new semiconductor industry.[36]

Offshore production by US semiconductor firms first occurred in the early 1960s when a number of American firms began to seek out low-labour-cost locations for their more routine assembly operations. The first offshore assembly plant was set up by Fairchild Semiconductor in Hong Kong in 1962. In 1964 General Instruments transferred some of its microelectronics assembly to Taiwan. In 1966 Fairchild opened a plant in Korea. Around the same time, several US manufacturers set up semiconductor assembly plants in the Mexican border zone. In the later 1960s US firms moved into Singapore and subsequently into Malaysia. During the following decade semiconductor assembly grew very rapidly in these Asian locations and spread to Indonesia and the Philippines. Despite some developments elsewhere (for example in Central America and the Caribbean) most of the growth of US-owned semiconductor assembly remained in East Asia.

By the early 1970s, every major US semiconductor producer had established off-shore assembly facilities, a tendency greatly encouraged by the offshore assembly provisions operated by the US government. However, the structure of US semiconductor operations within East Asia has changed substantially since the early focus on low-labour-cost assembly operations. We will look at this in some detail later in the chapter. Within Europe there is a long-established US semiconductor presence, initially related to trade barriers, with a particularly heavy concentration in Scotland and, to a much lesser extent, Ireland. However, after the deep recession in the IT industries in 2001, most of the US semiconductor firms in Scotland cut back on their production and laid off large numbers of workers.

The dominant US semiconductor manufacturer is, of course, Intel.[37] Although Intel has operations throughout the world, it retains a very strong US focus:

- headquartered in Silicon Valley, California
- design and development of components and other products performed in the United States (Arizona, California, Oregon, Washington) and in Israel and Malaysia
- 70 per cent of its wafer production located within the United States (Arizona, California, Massachusetts, New Mexico, Oregon)
- major fabrication operations in Israel and Ireland
- microprocessor- and network-related board-level products manufactured in Malaysia, Oregon and Washington
- components assembly and testing in China, Costa Rica, Malaysia and the Philippines, and at a new plant to be built in Vietnam.

Despite its dominance, Intel faces intensifying competition, especially as the PC market has become relatively less important. In 2005, Intel announced a radical reorganization away from a dual structure based on chip architecture and communications groups to a structure of

> five new business units to bring its product groups into line with a strategy of developing complete technology platforms. Platformization began with the successful Centrino mobile platform, where functions including processing and wireless communications were bundled together … three groups were being created to lead its efforts in platforms for mobility, the digital enterprise and digital home. 'The platform-based organizations … *reflect the ongoing convergence of computing and communications* by incorporating both capabilities across the new groups'.[38]

Japanese firms

The Japanese dominance of the global semiconductor industry in the 1980s and early 1990s rested on three basic elements:

- targeted state involvement
- concentration on memory chips (DRAMs)
- production for in-house use, within large diversified electronics business groups.

> Japanese semiconductor firms' dominance of DRAM markets during the 1980s rested on low prices and high quality ... in 1980, leading Japanese memory producers averaged 160 defective parts per million (PPM) while U.S. producers averaged 780 PPM for the same devices. Their skills in managing the development and introduction of new process technologies also enabled Japanese producers to 'ramp' output of new products more rapidly than their U.S. counterparts. Faster achievement of high production volumes gave Japanese firms advantages in defining product standards for leading-edge memory devices, strengthening their market position.[39]

The organization of many Japanese firms into complex groups (*keiretsu*) has meant that most Japanese semiconductor production has been vertically integrated: driven by the needs of intra-group businesses, notably consumer electronics and telecommunications. There are some obvious advantages in this. Equally, as the success of the US merchant producers showed during the 1990s, there are dangers of insularity. The revival of the US semiconductor industry on the one hand and the emergence of Korean and Taiwanese producers on the other have put a severe squeeze on Japanese semiconductor firms. In contrast to the US producers, Japanese firms have tended to remain in the DRAM market and, therefore, have been especially heavily affected by Korean and Taiwanese competition.

Although Japanese semiconductor producers have concentrated most of their activities within Japan, they have also developed a considerable degree of offshore production. There is an especially heavy emphasis on East Asia at around 70 per cent, with 17 per cent in North America (including Mexico) and 10 per cent in Europe. Of course, proximity to the Japanese domestic production base facilitated such developments in East Asia and helped to create a complex intra-regional division of labour.

The establishment of semiconductor production facilities by Japanese firms in both Europe and North America is much more recent than that of American firms overseas. Until the 1970s there was virtually no direct Japanese investment in electronics in the developed countries. There are now substantial numbers of Japanese electronic component firms operating plants in both the United States and Europe (more than two-thirds of these are in the United Kingdom and Germany).

The completion of the Single European Market in 1992 and the revalued yen – but especially the former – stimulated a new, and highly significant, wave of Japanese semiconductor investment in Europe. This new wave was led by Fujitsu's decision to build a $100 million chip fabrication plant in north-east England in 1989. The other leading Japanese companies followed this lead. Japanese semiconductor production in the United States is much more firmly established than in Europe. In fact 60 per cent of all Japanese electronic component plants outside Asia are located in North America.

Faced with intensifying competition in the DRAM market, Japanese firms adopted a number of strategies. One has been to diversify into the flash memory market, where demand has grown very rapidly with the development of internet-related mobile phones and related products. A second strategy has been to outsource more production. For example, in 2000 Toshiba announced its intention to double its outsourcing of semiconductors over three years while Hitachi planned to almost quadruple its outsourcing of semiconductors to reduce costs.[40] Such moves go very much against the traditional preference of Japanese electronics firms to keep their semiconductor production in-house to control quality. As a consequence of these large-scale reductions in DRAM production a number of plants have been closed and thousands of jobs lost.

A third strategy has been to pull out of DRAM production altogether. The biggest Japanese semiconductor producer, NEC, transferred its DRAM production to Elpida, its joint venture with Hitachi (see Table 11.4).[41] Similarly, Toshiba cut more than 18,000 jobs in 2001, sold its only US semiconductor plant to Micron Technologies, and began negotiations with Infineon of Germany about merging their memory chip operations.[42] Fujitsu announced a huge restructuring programme to take three semiconductor plants out of production (including its plant in north-east England), to 'consolidate its 12 chip lines into nine and bring a flash memory factory in Oregon into a Japanese joint venture with Advanced Micro Devices'.[43]

European firms

Although Europe has not produced a strong indigenous semiconductor sector (much semiconductor production within Europe is either American or East Asian owned), three of the world's leading firms are European (see Figure 11.5). Having been out-competed by US, Japanese and other East Asian firms, Infineon, STMicroelectronics and Philips have staged something of a revival since the late 1990s, primarily by vastly improving their production efficiency and by diversifying into the broader communications markets.

Historically, European semiconductors producers have not engaged in international production to the same extent as US and Japanese firms, although as a diversified electronics company, Philips always had an extensive geographical network.

> The big European companies have opted for different routes. Philips has been the most willing to outsource manufacturing, through its relationship with TSMC in Taiwan. It has also decided not to build any further wholly-owned fabs.[44]

But Philips still seems uncertain about whether or not to retain its four fully owned semiconductor factories. On the other hand, Philips has recently created a new R&D centre in Shanghai 'to develop chip sets for "ultra low cost" mobile phone handsets that can be manufactured for less than $20'.[45] Both Infineon and STMicroelectronics are more aggressively committed to expanding semiconductor production and sustaining leading positions in the industry:

- The German company Infineon employs around 36,000 worldwide. It operates R&D at around 12 sites in Europe, six in North America and four in Asia, and has nine manufacturing facilities in Europe, four in East Asia and one in the United States.
- STMicroelectronics, a Franco-Italian company headquartered in Geneva, Switzerland, employs around 50,000 worldwide. The firm operates 16 R&D sites and 16 manufacturing facilities (primarily in France, Italy, the United States and Singapore).

Korean firms

The speed with which the leading Korean conglomerates – Samsung, Goldstar (subsequently LG) and Hyundai – developed as semiconductor producers was astonishing, occurring within just a single decade. None of the Korean electronics firms was involved in semiconductors until Samsung entered the business in the mid 1970s. Goldstar followed in 1979, and Hyundai and Daewoo (which were not then electronics companies) entered in 1983. Initially the Korean firms were totally dependent on technology acquired from the United States and Japan. They focused on simple types of semiconductor of low capacity – the products that were being abandoned by the US and Japanese companies. In the mid 1970s, the technology gap between Samsung and the industry standard was around 30 years; today it has long since disappeared. Indeed by 1995, as Figure 11.5 shows, Samsung had become the sixth biggest semiconductor firm in the world; today it is second.

Such a dramatic, almost overnight, emergence of Korean firms as global players was based upon 'a relentless concentration on commodity memory chips'.[46] In particular, their aims have been to drive Japanese firms out of the memory chip market (US firms, with a few exceptions, having largely exited that market some years ago). Korean firms entered the semiconductor markets as imitators rather than innovators.[47] Samsung, for example, at various critical stages in its development, sought out and purchased small firms (including US firms) with the necessary technology but which were in financial difficulties. One of its major advantages was its 'ready access to funds siphoned from cash-cow industries within the *chaebol*'.[48] In that regard, Korean firms followed the Japanese model of developing a semiconductor capability within user electronics firms.

Each of the Korean firms established production facilities in the United States, primarily to absorb state-of-the-art technology. They then planned to move into

Europe, mainly to circumvent protectionist barriers as well as to be close to their growing markets. In the mid 1990s, Samsung, Hyundai and LG all announced plans to build massive semiconductor plants in the UK. However, the financial crisis of 1997 put an end to these. In addition, LG was taken over by Samsung, and Hynix (formerly Hyundai) has had problems, not least over allegations of its receiving illegal state subsidies (strenuously denied). Samsung, on the other hand, is the big success story.

> Samsung's success stems from a radical restructuring in the late 1990s, when the company was a struggling maker of cheap electronics ... 'Our history was based on manufacturing volume and market share, not profitability and technology' ... today the company sits near the top end of every market in which it competes ... [*a major*] *advantage of Samsung's business model is the reliable supply of chips ... it guarantees to the company's consumer goods businesses.* Motorola, which spun off its semiconductor division ... suffered shortage of chips ... In semiconductors, Samsung is rapidly increasing the size of its silicon wafers from 200 mm to 300 mm, reducing production costs by one-third.[49]

Samsung operates three semiconductor plants in Korea and one in the United States (with a possible second in prospect), with other, more limited, semiconductor facilities in East Asia and Europe. Samsung has an extremely ambitious investment programme: in 2005 it announced plans to invest $33bn in new chip fabrication and R&D facilities, as well as contemplating entering the semiconductor foundry business (to supply chips to other companies).[50] Samsung now has more than half of the rapidly growing Nand flash memory chip market.[51]

Taiwanese firms

While Korean firms have followed the Japanese model of development in their semiconductor businesses – focusing on DRAM production within the boundaries of huge electronics conglomerates – the approach adopted by Taiwanese firms has been very different. Again, as we saw earlier, the role of the state has been absolutely central to the industry's development but its mode of operation was different. Whereas Korean semiconductor firms developed out of the private sector *chaebol*, the leading Taiwanese firms TSMC and UMC were the product of government initiatives (though not as state-owned companies). They are especially distinctive because they are semiconductor *foundries*, producing chips to order from third-party companies (including the fabless design houses). TSMC and UMC are the two biggest semiconductor foundry companies in the world. The basis of their success has been the development of massive economies of scale which also accelerate technological learning. However, TSMC is now investing more heavily in its own design capabilities, rather than simply building to customer specifications. It is to provide 'design for manufacturing' chips as well as continuing its pure-play foundry business.[52]

So far, Taiwanese semiconductor firms have remained strongly embedded geographically within their domestic environment, although they have long had a presence in Silicon Valley in order to be at the perceived centre of the semiconductor industry. Overseas investment, otherwise, has been confined mainly to East Asia – including Japan, Singapore and, most recently, China. In 2002, the Taiwanese government ended its ban on Taiwanese firms building semiconductor plants in China but restricted such plants to 200 mm wafer technology. In order to qualify, a Taiwanese firm must already have begun 300 mm production in Taiwan; only then can it ship 200 mm equipment to China.[53] So far, only TSMC has actually established a semiconductor plant in China, located in Shanghai, although there is speculation that other Taiwanese firms may have done so illegally, through third parties. The sensitivity of relationships between the PRC and Taiwan is clearly reflected in this industry, where there are strong commercial pressures to combine Taiwanese technological superiority with low-cost production in China.

Regionalizing production networks in the semiconductor industry: the case of East Asia

As in the other industry cases discussed in this book, the pattern of production has a strong regional dimension. In the semiconductor industry, however, one particular region – East Asia – is especially important. East Asian production networks have been particularly critical in the development and redevelopment of the US semiconductor industry, both in the early stages of the industry (in the early 1960s) and also in the period of the resurgence of the US industry since the late 1980s.[54] In the 1960s, as we saw earlier, US semiconductor firms shifted the less skilled, labour-intensive stages of semiconductor production to East Asia. This was a very simple geographical division of labour, albeit a critical one for the competitiveness of US firms at that time. But it was not enough to offset the immensely efficient Japanese producers of memory chips.

The subsequent US recovery in the semiconductor market was based, as we have seen, on the shift out of memory chips and into far more sophisticated devices within the framework of 'Wintelism'. The development of an East Asian production network was absolutely central to this.

> The unique heterogeneity of Asia's regional economy, with different tiers of nations (Japan, Four Tigers, ASEAN, and coastal China, interior China, and India) at different stages of development provided the fertile ground for technical and production specialization that enabled the creation of [cross-border production units] … e.g. software in Bangalore, process engineering in Singapore, component assembly in Malaysia, printed circuit board (PCB) assembly in coastal China, semiconductor memory in Korea, digital design and final assembly in Taiwan.[55]

Three stages in the development of such intricate production networks can be identified:[56]

- *Stage I (1960s to late 1970s)*: US firms sought low-cost production locations through Asian affiliates as part of their export-oriented transnational production networks.
- *Stage II (1980 to 1985)*: US-owned assembly platforms were technically upgraded to encompass a wider range of production stages (for example, from simple assembly to the more complex stage of testing). Their Asian affiliates developed extensive local relationships, particularly through increased local sourcing of components. 'The result, by the end of the 1980s, was burgeoning indigenous electronics production throughout the region, with most of it, outside Korea, under the control of Overseas Chinese (OC) capital'.[57]
- *Stage III (1985 to early 1990s)*: US firms shifted their focus to new product definition and design and software development, further upgrading their Asian affiliates. Local firms gained much greater manufacturing responsibilities and greater autonomy in sourcing key components within the region.

As a result,

> the strongest indigenous Asian producers began to control their own production networks ... In sum, by the early 1990s, the division of labor between the United States and Asia, and within Asia between affiliates and local producers, deepened significantly, and U.S. firms effectively exploited increased technical specialization in Asia.[58]

More recently, apart from the increasing focus on China as a semiconductor location, one of the most significant regional developments has been the rise of East Asia as a location for semiconductor *design*.[59] UNCTAD calculates that the region's share of semiconductor design grew from virtually zero 10 years ago to around 30 per cent today. Several related factors explain this significant development: the lower cost and the higher availability of Asian design engineers; the increasing size and sophistication of the local market (especially China); proactive state policies; and changes in the nature of chip design (especially greater complexity and miniaturization which increase design costs).

In particular, design networks have become more complex, multilayered structures whose precise form varies in accordance with specific projects.

> A possible network might be comprised of the following players: a Chinese system company for the definition of the system architecture; an electronic manufacturing supplier from Taiwan Province of China; a United States integrated device manufacturer; a European 'silicon intellectual property' firm; design houses from the United States and Taiwan Province of China; foundries from Taiwan Province of China, Singapore, and China; chip packaging companies from China; tool vendors for design automation and testing from the United States and India; and design support service providers from various Asian locations.[60]

Conclusion

In a number of ways, the global pattern of semiconductor production has shifted geographically and changed organizationally as a result of interactions between the evolving corporate strategies of the major firms and the actions of national governments. Fierce competition has forced semiconductor firms to increase their degree of functional integration, to diversify into new product lines, to relocate production to more favourable locations in terms of markets or costs, and generally to rationalize their operations on a global basis. There has been considerable geographical shift to some developing countries, mostly in East Asia. Initially this was just the assembly stage – a clear example of the global combination of highly capital-intensive technology with low-cost, labour-intensive production. But the simple global division of labour characteristic of the 1960s and early 1970s no longer applies.

Today's global map of semiconductor production is far more complex. So, too, is the organization of the industry. The emergence of fabless design houses and semiconductor foundries has transformed the industry, creating new opportunities for producers lacking either the captive markets of the vertically integrated producers or the established customer links of the major merchant producers. One of the most striking developments has been the rise, fall and rise again of the semiconductor industry of the United States, which is, once again, the dominant player in the more advanced semiconductor product markets.

NOTES

1 Mathews and Cho (2000: 32).
2 These data are derived from Leachman and Leachman (2004: 208–13).
3 Leachman and Leachman (2004: Table 8.4).
4 Macher et al. (1998: 112).
5 Mathews and Cho (2000) provide an excellent analysis of the development of the semiconductor industry in East Asia.
6 Mathews and Cho (2000: 44).
7 Linden et al. (2004).
8 Linden et al. (2004: 242–3).
9 *Financial Times* (16 July 1999).
10 Ham et al. (1998: 157).
11 *Financial Times* (6 December 2000).
12 Linden and Somaya (2003).
13 *Financial Times* (30 November 2005).
14 See Angel (1994: 156–86), Ham et al. (1998), Macher et al. (1998).
15 Macher et al. (1998: 125–6).
16 Ham et al. (1998) provide a detailed discussion of I300I.
17 Mathews and Cho (2000) provide the most recent and most comprehensive account of the role of the state in the development of the semiconductor industry in East Asia. Mathews

(1997) and Fuller et al. (2003) describe the case of Taiwan. Mathews (1999) deals with Singapore.

18 Ham et al. (1998: 151-2).
19 Mathews and Cho (2000: 112).
20 Wade (1990: 313).
21 Kim (1997: 94–5).
22 Mathews (1997: 28).
23 Mathews (1997).
24 Mathews (1997: 36).
25 Mathews (1997: 40).
26 Mathews (1999: 66–7).
27 *Financial Times* (1 April 1997).
28 Owen (1999: 275).
29 Angel (1994: 38–9).
30 See Linden and Somaya (2003), Macher et al. (2002) for useful reviews.
31 Macher et al. (2002: 157).
32 Berger (2005: 79).
33 Macher et al. (2002: 157–8).
34 Borrus (2000).
35 See Saxenian (1994) for a detailed comparison of the evolution of the electronics industries along Route 128 and in Silicon Valley.
36 In addition to Saxenian (1994), see also Angel (1994) and various chapters in Kenney (2000).
37 See Curry and Kenney (2004), Mortimore and Vergara (2004: 505–12).
38 *Financial Times* (18 January 2005).
39 Macher et al. (1998: 113–14).
40 *Financial Times* (21 August 2000; 28 September 2000).
41 *Financial Times* (1 August 2001).
42 *Financial Times* (28 August 2001).
43 *Financial Times* (21 August 2001).
44 *Financial Times* (12 April 2002).
45 *Financial Times* (10 November 2005).
46 Mathews and Cho (2000: 45).
47 Kim (1997) provides a detailed account of Samsung's development as a leading semi-conductor producer.
48 Kim (1997: 88).
49 *Financial Times* (6 September 2004, emphasis added).
50 *Financial Times* (30 September 2005).
51 *Financial Times* (15 December 2005).
52 *Financial Times* (4 January 2006).
53 *Financial Times* (30 March 2002).
54 See Gereffi (1996), Borrus (2000).
55 Borrus (2000: 58–9).
56 Borrus (2000: 68-74).
57 Borrus (2000: 70).
58 Borrus (2000: 73).
59 UNCTAD (2005: Chapter V Annex).
60 UNCTAD: (2005: 175).

Twelve
'We Are What We Eat': The Agro-Food Industries

Transformation of the food economy: the 'local' becomes 'global'

The production, distribution and consumption of food – *the* most basic of human needs – have been transformed during the past four decades.[1] Although for millions of people, basic subsistence is still the norm and starvation is always imminent, for millions of others food has become as much a statement about lifestyle as about survival.

> The food economy is on the one hand increasingly differentiated in new sorts of ways at the level of *consumption* – some within LDCs are eating better at a time when others in Africa are descending into a universe of ever-greater food insecurity, millions in California go hungry while others consume 'designer' organic vegetables shuttled around the world in a sophisticated … 'cool chain' …

> On the other hand, at the level of *production and distribution* the food economy is being restructured in radically new ways … increasingly driven by global demand and internationalization of the agro-food industry. The giant food companies and the large retailers are aggressively transforming the world agro-food economy.

> Classical export commodities (coffee, tea, sugar, tobacco, cocoa and so on) have been increasingly displaced by so-called 'high value foods' (HVF), such as fruits and vegetables, poultry, dairy products, shell fish.[2]

In some respects, therefore, the modern agro-food industry may seem little different from clothing, automobiles or electronics. Indeed, one writer predicted the development of the 'world steer (cow)' as the direct parallel of the 'world car'.[3] But, despite the industrialization of much food production (including 'global sourcing'), this greatly oversimplifies what are highly complex and geographically differentiated industries. The basic fact remains that food production is fundamentally different from other industries in one particular way: its, quite literal, *grounding* in biophysical processes.

> Agriculture is a unique branch of industry in that it is constrained by natural processes which act to limit the productivity of labor and restrict capital investment. Here *the role of biology in plant and animal growth is key* … on a farm – unlike a factory – it is the biological time necessary for plant and animal growth that dictates the work schedule … In addition, the land-based character of farm production poses severe constraints to industrialization … because land is a fixed and limited resource, and because land markets are deeply colored by localized social conditions, farmers cannot easily or quickly adjust their investment in land.[4]

Food *production*, therefore, remains an intensely *local* process, bound to specific climatic, soil and often socio-cultural conditions. At the same time, certain kinds of local production, notably high-value foods, have become increasingly *global* in terms of their *distribution* and *consumption*. For the affluent consumer, with access to the overflowing cornucopias of supermarket shelves, the seasons have been displaced by 'permanent global summertime' (PGST).[5] But such apparently idyllic circumstances for affluent consumers have a dark and contentious side.

Producing food for a global market requires huge capital investment and gives immense power to the transnational food producers and the big retailers. It creates serious problems – as well as opportunities – for food suppliers as they become increasingly locked into (or out of) transnational agro-food production networks. Global food production and distribution create huge environmental disturbances, in terms of excessive exploitation of sensitive natural ecosystems, the application of chemical fertilizers and pest-controlling agents, the increasing attempts to genetically modify seeds, plants and even animals and to 'patent life', and the transportation of HVFs over vast geographical distances. These processes make agro-food an intensely sensitive industry. It is at the centre of the continuing acrimonious arguments within the WTO to agree a 'development dimension' to the Doha Round of trade negotiations (see Chapter 19).

Cutting across trade issues are those relating to food safety and to the ethics of genetic modification (GM) of seeds, plants and animals. In the past few years, for example, there have been several serious food safety scares: BSE ('mad cow disease'), foot (hoof) and mouth disease, and avian flu. Such outbreaks have a huge impact on agro-food trade and, therefore, on the livelihoods of farmers, growers and distributors. They create massive fluctuations in consumer buying patterns, often out of ignorance. At the same time, there is widespread scepticism – and considerable fear – of genetic modification. Both food safety and GM help to stimulate consumer resistance to the products of the global agro-food industries and to reinforce demands for a return to local sourcing of organically grown products. Without doubt, the agro-food industry has become a battleground with several 'fronts': between producers and producers, between producers and consumers, between producers and governments (not least because agro-food is one of the most heavily regulated industries), and between governments.

Agro-food production circuits

In the case of clothing, automobiles and semiconductors it is possible to identify a basic production circuit that applies to virtually all products in each sector. In contrast, production circuits in the agro-food industries are immensely varied. In the case of traditional commodities, like grains, the circuit is relatively simple (though more intricate than in the past). In the case of high-value foods (HVFs), however – the focus of this chapter – the situation is far more complex. For that reason, we provide several examples here.

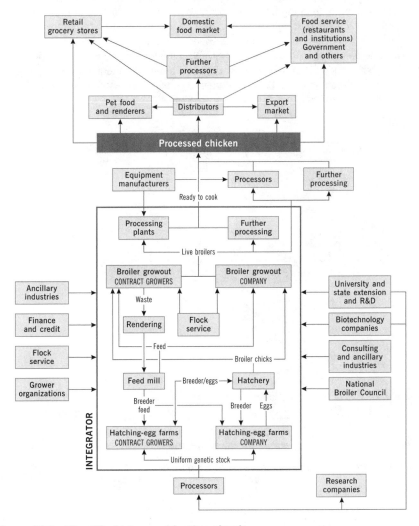

Figure 12.1   The US chicken production circuit

Source: based on Boyd and Watts, 1997: Figure 8.4

Figure 12.1 shows the highly complex structure of the US chicken (broiler) production circuit. This is an industry which has become increasingly dominated by very large integrated producers. From a producer's perspective, a major advantage of integrated chicken production is that it facilitates the coordination of chicken raising processes which are subject to intrinsic biological lags. It isn't possible to speed up the 'assembly line' as can be done in automobiles. It is, however, as much a 'just-in-time' system as that in automobile production. At the same time, integration gives closer control over product quality and food safety.

> Integrated operations are supported by a host of technology and input suppliers such as primary breeders, equipment suppliers, pharmaceutical and chemical firms, feed ingredient suppliers, as well as a panoply of research and technical support entities ... Not only do integrators have to coordinate their own activities but relations between integrators and suppliers also have to be closely coordinated. Effective vaccine development, for example, depends on long-term cooperation between breeders, integrators, vaccine manufacturers and research laboratories ... *In effect, the system as a whole functions as an industrial network.*[6]

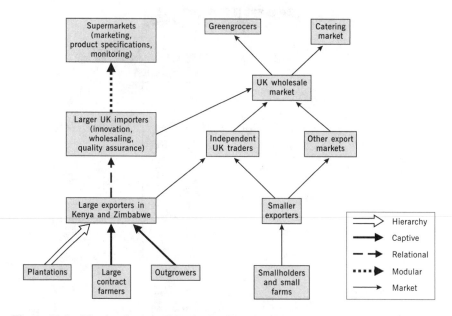

Figure 12.2   The fresh vegetable production circuit

Source: based on Dolan and Humphrey, 2002: Figure 3

Figure 12.2 displays the fresh fruit and vegetable production circuit between the producing countries of Kenya and Zimbabwe and the consumer markets of Europe, particularly the UK. Its focus is more on the distribution and marketing functions of these agro-food production circuits and their coordination and governance. In particular, it sets out the different ways of coordinating transnational production networks discussed in Chapter 5 (Figure 5.9 and associated text). The key point to make about the fruit and vegetable production circuit is that it is driven by the large supermarket chains, rather than by the producers of the crops themselves.

Figures 12.1 and 12.2 both depict conventional agro-food production circuits. However, there are other, 'alternative', circuits which involve the production of organic food and/or the involvement of various kinds of non-economic actors, notably 'fair trade organizations'.[7] The driving forces underlying the development of some alternative food networks are the increasing concerns with food quality and food safety and with fairer treatment of farmers/growers in developing countries (see later in this chapter). Such networks 'redistribute value through the network against the logic of bulk commodity production ... reconvene "trust" between food producers and consumers ... and ... rearticulate new forms of political association and market governance'.[8] Figure 12.3 provides one example of an alternative agro-food production circuit: fair trade coffee.[9]

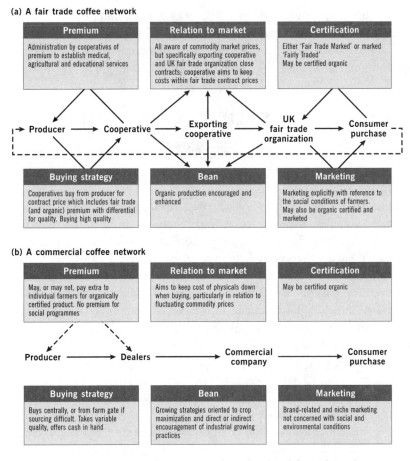

Figure 12.3 'Alternative' agro-food production circuits: fair trade and commercial coffee

Source: based on Whatmore and Thorne, 1997: Figure 11.2

The kind of food circuit shown in Figure 12.3 is just one of a number of alternatives to the tightly controlled, highly integrated, industrially based agro-food circuits that have become so dominant in recent years. Currently, there is also a growing (re-)emergence of explicitly *territorially based* food production networks.

> A key characteristic of the new supply networks is their capacity to re-socialize or re-spatialize food, which comes to be defined by its locale ... by drawing upon an image of the farm or the region as a source of 'quality', alternative food networks 're-localize' food.[10]

Although these developments need to be kept in perspective – the vast bulk of modern food production and distribution is contained within the big producer- and buyer-dominated networks – they do suggest that we need to adopt a more nuanced position. The agro-food industries seem to be bifurcating into two main sets of processes:[11]

- standardized, specialized production processes responding to economic standards of efficiency and competitiveness
- localized, specialized production processes attempting to trade on the basis of environmental, nutritional or health qualities.

Global shifts in the agro-food industries

In this chapter, our concern is with the *high-value* segments of agricultural production and trade. We will focus, in this section, on three examples: chicken, fresh fruit and vegetables, and coffee.

As our discussion of Figure 12.1 revealed, chicken production has become an immensely complex, highly integrated agro-food industry. At the global scale, chicken production is dominated by three countries: the United States, China and Brazil, with Mexico some way behind in fourth place (Figure 12.4). Until very recently, the United States was also the world's leading exporter of chickens but it has been overtaken by Brazil. 'In 1997, Brazil's chicken exports were less than a third those of the US. But production has taken off since 2000 – growth averaged 230 per cent the past two years – and today Brazil exports to 127 different countries and controls 36 per cent of the world share'.[12]

Fresh fruits and vegetables production is also heavily concentrated at the global scale (Figure 12.5). China (36 per cent of the world total) is by far the world's biggest producer (although much of this production is consumed domestically). India is far behind at 10 per cent, followed by the United States (5 per cent) and Brazil (3 per cent). However, the composition and pattern of trade in fruits and vegetables have changed markedly during the past two decades.[13] In 1977–81, fruits and vegetables constituted almost 12 per cent of the total value of

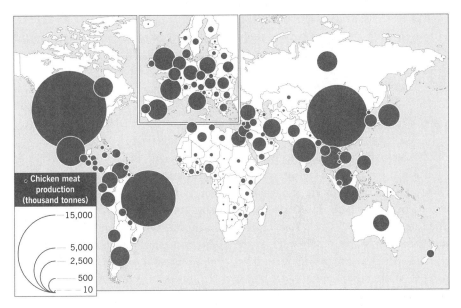

Figure 12.4   Global production of chickens

Source: FAO *Statistical Yearbook*, 2004: Table B8

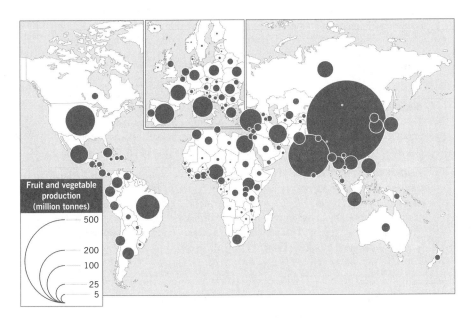

Figure 12.5   Global production of fruits and vegetables

Source: FAO *Statistical Yearbook*, 2004: Table B3

agricultural exports. By 1997–2001, this share had increased to almost 17 per cent. However, these aggregate figures obscure significant differences between various products. Export growth rates of traditional products (e.g. oranges, canned pine-apples, canned mushrooms, concentrated orange and apple juices) were very low. Non-traditional products grew fastest: 'Some commodities – mangoes, frozen potatoes, single-strength orange and apple juices, fresh mushrooms, garlic, sweet corn (prepared or preserved), and avocado – achieved, or were close to, double-digit growth rate in their exports during 1989–2001'.[14]

The geography of global trade in fruits and vegetables is strongly regionalized. Not only are Europe and North America the leading importers of such products (along with Japan), they are also substantial exporters. Both regions contain a variety of climatic conditions conducive to certain kinds of fruit and vegetable production: the Mediterranean rim in the case of Europe; Mexico and the Caribbean in the case of North America. Figure 12.6 shows the origins of imports of fruits and vegetables to the world's 30 leading importers. Quite apart from the intra-regional trade flows to the highest-income countries of North America, Europe and Japan, the role of the Southern Hemisphere countries is especially significant. Unlike the banana-producing countries, for which this single product accounts for almost 90 per cent of their fresh fruit exports, the Southern Hemisphere countries produce and export an increasing variety of products for consumption in the affluent markets of the Northern Hemisphere.

> With a crop production cycle opposite to that of the Northern Hemisphere, the Southern Hemisphere exporters … play a vital role in making the year-round supply of fresh fruits possible. These countries have taken advantage of the seasonal differences to expand their exports, particularly for many temperate-climate fruits …
>
> During 1999–2001, more than half of the fresh fruits exported by the Southern Hemisphere countries were temperate-climate fruits such as grapes, apples, and, to a much lesser extent, pears. About two-thirds of apples exported by the Southern Hemisphere countries came from Chile and New Zealand, while Chile and Argentina were the dominant suppliers for grapes and pears …
>
> In addition to fresh fruits, the group of Southern Hemisphere countries is a major supplier for fruit juices, accounting for nearly one-third of the import value for juices purchased by the world's top 30 importers during 1999–2001 … Brazil accounted for nearly three-fourths of the region's juice exports, while Argentina (shipping mainly apple and grape juices) was the second largest exporter in the region (11 per cent of the exports).[15]

Finally, Figure 12.7 maps global exports of coffee. As coffee *aficionados* will know, there are two major types of coffee bean: arabica beans, grown at higher altitudes and more difficult to grow, and robusta beans, grown on the low lands in the humid tropics. In general, arabica beans are regarded as being of higher quality though, as always, it is not quite as simple as this. Five countries generate 67 per cent of total coffee exports: Brazil (29.1 per cent, of which 91 per cent is arabica), Vietnam

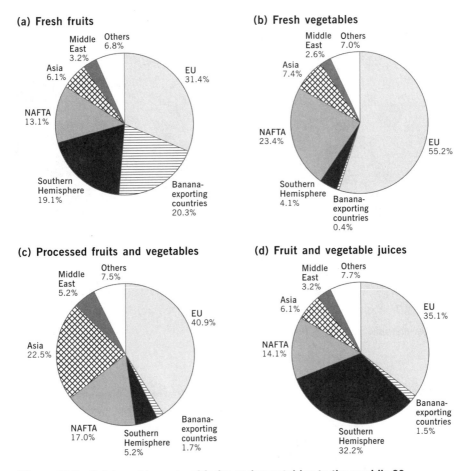

(a) Fresh fruits

Others 6.8%
Middle East 3.2%
Asia 6.1%
NAFTA 13.1%
Southern Hemisphere 19.1%
Banana-exporting countries 20.3%
EU 31.4%

(b) Fresh vegetables

Middle East 2.6%
Others 7.0%
Asia 7.4%
NAFTA 23.4%
Southern Hemisphere 4.1%
Banana-exporting countries 0.4%
EU 55.2%

(c) Processed fruits and vegetables

Others 7.5%
Middle East 5.2%
Asia 22.5%
NAFTA 17.0%
Southern Hemisphere 5.2%
Banana-exporting countries 1.7%
EU 40.9%

(d) Fruit and vegetable juices

Middle East 3.2%
Others 7.7%
Asia 6.1%
NAFTA 14.1%
Southern Hemisphere 32.2%
Banana-exporting countries 1.5%
EU 35.1%

Figure 12.6 Origins of imports of fruits and vegetables to the world's 30 leading importers

Source: based on Huang, 2004: Figure 2.2

(16.4 per cent, all robusta), Colombia (11.2 per cent, all arabica), Indonesia (6 per cent, 84 per cent robusta) and India (4 per cent, 40 per cent arabica).

The pattern of production and trade in high-value foods, therefore, combines elements of global, regional and local scales. Globally, the emergence of Southern Hemisphere producers, basing their advantage on their seasonal complementarity with the temperate markets of the Northern Hemisphere, generates massive flows of long-distance trade. Regionally, the existence of areas of more exotic production within the major regional markets of North America, Europe and East Asia has led to strong intra-regional trade flows of high-value foods. Locally, the increasing interest in alternative food networks, especially those which focus on local (often organic) production, has created much shorter movements of agro-food products.

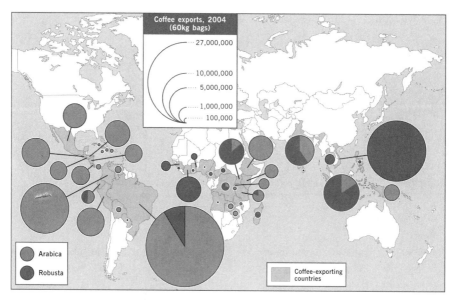

Figure 12.7   Exporters of coffee

Source: International Coffee Organization data

Specialized agro-food production spaces

Cutting across these scales, we can identify large, highly specialized agro-food 'production spaces', within developed countries as well as within the NACs (see Chapter 2). In both cases, the catalyst is the existence of highly favourable *biological* conditions (of soil and climate), which have been exploited by the big integrated food companies able to organize both highly efficient, localized production networks and geographically extensive distribution and marketing networks. But such developments also reflect *specific local geographies, histories* and *socio-cultural institutions and practices*.

For example, in the United States, the integrated production of chickens (Figure 12.1) has become highly concentrated in distinctive territorial complexes in the southern states.[16] Here, distinctive attributes of chicken production (notably breeding practices, biological time-lags in production, and chicken microbiology) have developed in historically developed, locally specific conditions:

> The first was the existence of a class of small marginal farmers, located primarily on the periphery of the cotton belt, who confronted an agricultural crisis and a need for alternative forms of rural livelihood ... Second, the role of merchants and feed dealers in extending credit to small farmers, proved to be indispensable to the origins of the industry. By providing the institutional imperative for the contract grow-out arrangement that spread so prolifically in parts of the South during the 1950s, these feed dealers were instrumental in

establishing the production base for the integrated system. Finally, the existence of a pool of surplus rural labour available to work in the processing facilities was also critical.[17]

Examples of 'new agrarian regions' within the developing world are especially evident in Brazil, particularly away from the traditional agricultural areas close to the coast. Large areas of the Amazon rainforest are being cleared for new agricultural production. As one commentator put it: 'You cannot see the wood for the beans in an ever-widening expanse of the Amazon'.[18] Away from the Amazon, the São Francisco Valley of north-east Brazil is 'the largest irrigated agricultural development in Latin America. It is based upon the development of export agriculture, particularly mangoes, grapes and tomatoes'.[19] An intricate agro-food production complex has developed, focused on the main local service centre of Petrolina.

Climatic and biological conditions in this area allow grapes to ripen in only 120 days, compared with the norm of 180, giving two harvests per year. But, as in the case of southern US chicken production, it is the particular combination of biological and socio-economic factors that explains the specific nature of the São Francisco Valley agro-food territorial complex. The rate of growth of production and export of grapes and mangoes from the region has been spectacular, as Figure 12.8 shows.

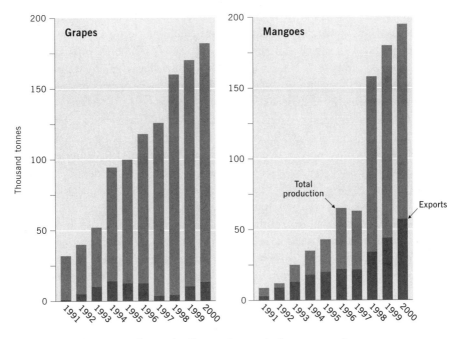

Figure 12.8   Increases in production and export of grapes and mangoes from the São Francisco Valley, Brazil

Source: based on Marsden and Cavalcanti 2001: Table 1

Consumer choices – and consumer resistances

Food, as we have already observed, is the most basic of human needs. For most of human history, people have had to struggle to obtain enough food to survive. Only a very tiny proportion of the population could afford to obtain the more exotic foods from distant places. That is, of course, still the case today for millions of people in the poorest countries and for some people in affluent countries. But as incomes have risen for many through economic growth, and with the associated urbanization of the population, demand for food has changed dramatically. In developed economies, consumers now spend only around one-tenth of their income on food, compared with one-third 50 or 60 years ago. However,

> food differs fundamentally from other commodities … Food's unique nature has
> made it uniquely central to human social life … and therefore a carrier of histor-
> ically constructed meanings, both intimate and political … These meanings in
> turn enter into food markets at a variety of levels. Different groups' and societies'
> ideas of food purity and danger, of the 'proper meal' and the proper treatment of
> farmland and livestock, of government's responsibility to protect producers and
> consumers from food risks … [mean] that food is never free of the meanings that
> make it the subject of bread riots, trade wars, and media scares.[20]

These factors make the relationship between food production and consumption more problematic than is often assumed. What we choose to eat has become a far more complex process: a mix of taste, culture, religion, health concerns, ethical position and lifestyle as well as of disposable income. On the one hand, food producers strive to produce and market foods that will attract the largest number of consumers (and enhance profits) whilst, on the other hand, consumers themselves have widely varying 'food agendas'.

In the affluent consumer markets of North America, Europe, and parts of East Asia, it is the *changing patterns* of demand and consumption, rather than the overall level of food consumption, that are especially important. Increasing affluence stimulates a desire for greater choice in food products. As a result – but also, of course, driven by the marketing strategies of the transnational food producers – the market for food has become highly segmented.

At one level, this is reflected in the huge diversity of products sold through the major supermarkets and, especially, their provision of all-year-round perishable foods from across the globe. It is reflected in the rapid growth of new food products: for example, the chilled convenience food market in the UK has doubled in size in the past 10 years.[21] It is reflected in the ever-changing dietary fashions of the affluent in search for the route to beauty and long life. It is reflected, too, in the development of the specialist 'lifestyle' drinks markets, for example, the 'latte revolution' driven by Starbucks' colonization of much of the world.[22]

At the same time, however, there is increasing *consumer resistance* to many of the food products being sold through the big supermarkets, as well as to the more traditional providers of fast food. In early 2006, for example, McDonald's announced that it was actually going to close a significant number of its outlets in the UK. McDonald's, of course, has long been the focus of much criticism for its allegedly unhealthy products, culminating in the movie *Supersize Me*. In some countries, though not all, there is widespread opposition to GM foods and to the use of non-organic production methods. In the case of GM foods, there is considerable difference in consumer attitudes between the US and Europe (and even within Europe itself). A European Commission public opinion survey in 2001 found that 80 per cent of those surveyed did not want GM food and 95 per cent wanted the right to choose.[23]

There are also pressures to 'relocalize' food production: both to rely more on local sources and also to stimulate and protect areas of local production of key products. Such resistances derive from a combination of concern over environmental damage and fears about the safety of foods grown using what are increasingly regarded as suspect or ethically unacceptable methods. For example, 'fresh' supermarket food

> is predicated on a new nature-defying order where every conceivable fruit and vegetable grown anywhere is available all the time ... PGST [permanent global summertime] may look good, but in the name of consumer choice and public health the irregularity and diversity that is part of the natural order has been eliminated, not to benefit consumers but to fit the way our big food retailers like to do business. In essence, this means sourcing vast quantities of easy-to-retail, long shelf-life standard varieties, grown to rigid size and cosmetic specifications, that can be supplied 365 days a year ... 'Hi-tech, low-taste, odour-free produce is the norm.'[24]

There has been significant recent growth in the ethical consumer movement in the agro-food industries.[25] For example, some 5 million farmers and workers are now covered by the 'Fairtrade' charitable scheme, which pays a guaranteed price covering basic costs and a surplus to reinvest in further development. Fairtrade is especially active in such foods as coffee (see Figure 12.3), tea, bananas and chocolate, although the proportion of total world trade in these product covered by Fairtrade agreements remains small: for example, around 5 per cent of the UK banana market.[26]

Set against these kinds of consumer resistance, we have to recognize that such movements are, at least in part, facilitated by the choices of the affluent consumer. While there is no doubt that demand is growing from consumers for food whose quality and geographical provenance are regarded as being superior to food from the large-scale sources, for most people the overwhelming need is still for enough food to survive. For every 'enlightened' consumer pursuing her organic food, or life styler drinking his designer coffee, there are many for whom such foods are out of reach. For people working long hours or the elderly, the availability of convenience foods is a major benefit. The fact that such foods may not be especially healthy is another issue. Clearly, therefore, demand for and consumption of food represent an extremely complex set of processes. As we shall see in subsequent sections of this chapter, this has major implications for the changing technologies of food production, for state regulatory policies, and for the strategies of the transnational food producers.

Transforming technologies in agro-food production

Global cool chains

Traditionally, food production was a relatively simple process. Of course, technological innovations in crop growing and animal husbandry have been significant throughout human history. For example, without the important agricultural innovations of the eighteenth and nineteenth centuries, the Industrial Revolution simply could not have occurred. Indeed, it has sometimes been argued that the Industrial Revolution was really an *agricultural* revolution. Equally, without the developments in transportation and communications discussed in Chapter 3, long-distance movement of agricultural products would not have been possible. These, together with innovations in refrigeration and food-freezing technologies – the development of what have been called 'global cool chains' – transformed the availability of a much wider range of agricultural products over vast geographical distances.[27] One of the most widely quoted indicators of globalization is the distance over which the food in our shopping basket or on our dinner table has travelled. For example, a basket of 20 fresh foods bought from major UK retailers was found to have clocked up a total of 100,943 miles.[28] Not surprisingly, one critic writes of the 'barminess of the long distance breakfast'.[29]

The rapid growth in world trade in fresh foods depends critically on the ability to move fragile and perishable products over long distances without destroying their 'freshness'.

> In particular, advances in controlled atmosphere (CA) technologies have extended the shelf life of perishable products … With CA, products hold up better during transportation. CA technologies allow operators to lower the respiration rate of produce by monitoring and adjusting oxygen, carbon dioxide, and nitrogen levels within a refrigerated container. In this way, CA can slow ripening, retard discoloration, and maintain freshness of perishables like lettuce, asparagus, peaches, mangoes, and avocados, that would not remain fresh during ordinary refrigerated ocean transport. Some sophisticated CA systems are combined with systems that maintain relative humidity – a crucial factor for some produce such as grapes, fruits with pits, and broccoli – and that control levels of ethylene, a naturally occurring gas that accelerates the ripening of fresh fruits and vegetables.[30]

Such technologies of fresh food preservation are, of course, greatly enhanced by the use of air freight to transport low-weight/high-value exotic foods to distant, affluent markets.

Industrialization of food production and the shift towards biotechnology

The technologies of agro-food production have been transformed by their *industrialization* and, most recently, through the introduction of *biotechnologies*.[31] Such developments are intimately related to the increasing role of very large agro-food corporations in all aspects of food production. We will look specifically at the corporate dimension in a later section of this chapter. Here, our concern is with the increasing importance of biotechnologies in food production.

The application of industrially produced chemicals to agricultural production (fertilizers to stimulate higher crop yields, pesticides to inhibit disease and insect damage) has been common for many decades. The development of newer varieties of crops has also been a continuing process. The so-called 'Green Revolution' of the 1960s and 1970s was the most significant combination of such practices: an attempt to solve the food problems of poor countries through the development of new varieties of basic crops such as wheat, rice and maize. The Green Revolution 'depended on applications of fertilizers, pesticides and irrigation to create conditions in which high-yielding modern varieties could thrive'.[32] The Green Revolution had an enormous though far from beneficial effect on developing country farmers and on the environment, resulting in massive increases in the use of pesticides to cope with the susceptibility of large-scale monoculture to disease and pests. The result was widespread contamination of soil and water and massive water depletion.

The Green Revolution was, in many ways, a precursor of what has become the most controversial aspect of agro-food production: *genetic modification* (GM). As before, the objective is to improve plants' resistance to disease and to herbicides, to increase yields, and to improve nutritional value. This is done by changing basic genetic structures and producing new varieties of seeds.

> Biotechnological plant breeding 'recombines' the DNA of the target plant by altering its genetic sequence, or in the case of transgenic plants, by adding one or more genes from a donor organism ... The recombinant (rDNA) process involves three key steps. The isolation of the coding sequence for the gene(s) associated with the desired trait (identification); the replication and transfer of this gene to plant cells (T-DNA vector construction); and the regeneration and developmental regulation of the gene in the target plant (propagation and expression control) using conventional tissue culture techniques.[33]

Such GM techniques are immensely complex and costly. They involve massive levels of capital expenditure of a scale that can be afforded only by the big biotechnology and agro-food companies. Not least, they encourage the patenting of what had hitherto been regarded as 'public' goods: the seeds needed to produce the next generation of crops. This is the patenting of life itself. Traditionally, a farmer would set aside some seeds from one year for use in the following year. GM seeds, on the other hand, 'belong' to the seed company, which produces 'terminator' seeds which cannot be reproduced by the user, who has to purchase next year's seeds from the seed company.

Whereas the application of biotechnologies is relatively recent, and mainly applied in the early stages of the agro-food production circuit, the use of chemical additives in food products themselves has been common for some time. One of the results of the increasing industrialization of food production was a loss of some of the desirable qualities of taste, texture, colour and so on. To counteract these changes, and to enhance the attractiveness of food products, producers have developed a bewildering variety of food additives: preservatives, antioxidants, emulsifiers, flavourings, colourings. One calculation is that some 4500 different flavouring compounds are available to food manufacturers and that 90 per cent of additives are purely cosmetic.[34]

What about the workers?

The impacts of these technological transformations in how food is produced, and how far and how quickly it can be transported, are immense. In addition to their effect on what people eat – and the potential effects on their health – they also impact greatly on those people who work in agriculture. The proportion of the labour force working on the land has fallen markedly, especially in developed countries. The industrialization of agro-food processes has, in effect, shifted the locus of much of the work from the field to the factory or the packaging plant. The seasonal rhythms of agricultural work have been displaced for many by the rhythms of the food processing and packaging assembly lines. To that extent, many workers in the agro-food industries are more like workers in automobile or electronics production than farmers.

Because all governments are heavily involved in regulating their food industries for health and safety reasons (see next section), the working conditions in processing and packaging plants are more tightly monitored than is the case in some other industries (such as clothing). The work itself may be mind-numbingly boring and repetitive but so, too, are many other jobs in today's society (and not just in manufacturing: think of telephone call centres). Of course, wide variations in working conditions exist, especially between developed and developing countries, despite the ubiquitous involvement of the big supermarket chains in sourcing from such plants.

But not all jobs in food processing and packaging are permanent or full-time. Agro-food is probably the largest user of casual labour of all modern industries. Indeed, these industries depend fundamentally on a huge floating labour force of workers who are employed only when the producer needs them and who are often organized by subcontractors or 'gangmasters'. Since the supply of such labour invariably exceeds demand, wages are extremely low and working hours very long. The majority of such workers are migrants, with virtually no bargaining power and often very little protection from abuse. The seasonality of agricultural processes creates vast periodic movements of migrant workers within and across borders. In the United States, the majority of these workers are Hispanic (especially Mexican); in Europe, they come predominantly from Eastern Europe or from North Africa.

A recent Oxfam report on American agriculture, *Like Machines in the Fields*, provides graphic details of what the report terms 'sweatshops in the fields'.

> Farmworkers are among the poorest – if not *the* poorest – laborers in the United States ... farm labor is also one of the most dangerous jobs in America. At work, farmworkers suffer higher rates of toxic chemical injuries than workers in any other sector of the US economy, with an estimated 300,000 suffering pesticide poisonings each year. They also suffer extremely high rates of workplace accidents ...
>
> Farmworkers are much more likely to have temporary jobs ... Just 14% of all workers in crop agriculture are employed full time in year-round positions, while fully 83% work on a seasonal basis ... 56% of farmworkers in crop agriculture are migrant workers, travelling more than 75 miles to get a job ...
>
> Thirty percent of migrant workers (or 17% of all crop workers) are characterized as 'follow-the-crop' migrants, moving year-round like those portrayed in John Steinbeck's *The Grapes of Wrath* ... These migrant farmworkers generally follow one of three migration streams: the eastern stream originates in Florida and extends up the East Coast; the Midwestern stream originates in Texas and extends to the Great Lakes and Great Plains states; the western stream originates in California and extends along the West Coast as far as Washington ...
>
> The vast majority of crop workers in the US are foreign-born. While most (95%) are from Mexico ... others come from Central America (primarily Guatemala and El Salvador) or the Caribbean (primarily Haiti and Jamaica) ...
>
> Farmworkers in general and immigrant farmworkers in particular, have low levels of education ... Their literacy and communication skills in English are especially limited ...
>
> Finally, yet perhaps most significantly, these immigrant workers typically lack work authorization ... Given the vulnerabilities of their legal status, US farmworkers tend to face widespread workplace and human rights abuses, and are rarely able to take the risk of challenging abuses when they occur.[35]

While some of these characteristics of the agro-food workforce are far from new, there is little doubt that they have intensified as the industry's production circuits have become more tightly controlled by larger and larger producers and buyers.

The role of the state

The state plays an immensely important role in the agro-food industries, which are among the most highly regulated, heavily subsidized, and vigorously protected of all economic activities.

Regulating agro-food industries

A vast array of government agencies and departments operates to oversee various parts of the agro-food industries, for example, the UK Department of Environment, Food and Rural Affairs and the Food Standards Agency, and the US Department of Agriculture and its Food and Drug Administration. A primary focus of such regulatory activity is the issue of food safety, a problem greatly exacerbated by the growth of international trade in food. Before the 1970s, as much as 90 per cent of world food production was consumed in the country in which it was produced.[36] That situation has changed dramatically. As a result, national food regulatory measures have become increasingly embedded in international codes, such as the Codex Alimentarius, set within the Food and Agricultural Organization and the World Health Organization. This consists of 'over 200 standards, forty codes and guidelines for food production and processing, maximum levels for about 500 food additives, and 2700 maximum-residue limits for pesticide residues in foods and food crops'.[37]

A striking feature of regulatory policies in these industries is the extent to which they are deeply intertwined with the strategies of the major food producers.

> The biggest funder of the establishment of the Codex Alimentarius Commission was not the US state but the US food industry … Indeed, the Codex has become one of the more industry-dominated international organizations. More corporations have members of delegations to Codex committees (140) than nations (105) … However, we can assign too much significance to the raw numbers of industry participants in the Codex process. Typically, the most strategic bargaining in Codex expert committees is done by government representatives of the key states.[38]

In other words, there is a substantial amount of 'private' regulation in the agro-food industries sanctioned by national governments.

Even within the EU, there are national variations in regulatory practices. This is seen most clearly in the context of the food safety scares that have occurred from time to time – for example, salmonella in eggs, 'mad cow disease' (BSE), foot (hoof) and mouth disease – and also in the GM food controversy. Such cases invariably create conflict between the state within which the disease originates and its export markets, leading to trade boycotts and disputes over the efficacy of safety measures. The case of GM food is potentially the biggest source of difference between states and one that spills over into trade disputes, especially between the US and the EU. The US position is that GM foods are not only safe but also vital to increasing food supply in poor countries. Driven by consumer resistance (see earlier), the EU has taken a more restrictive position. Although it lifted its six-year moratorium on GM food in 2004 and allowed limited approvals of GM products, several EU states continue to ban them.

A major problem facing food safety regulators is the continuing proliferation of new products that cross the boundaries between food and medicine: the development of so-called functional foods or 'nutriceuticals', which claim to improve various aspects of health.

> The regulatory framework for functional foods is still in the process of being constructed. In Japan, which has led the functional foods strategy ... legislation was put into place in the early 1990s. In the US, the nutritional labelling legislation has permitted generic health claims and ... opened the way for positioning functional foods as dietary supplements. In Europe, national regulations interpret community directives each in their own way leading to widely differing applications.[39]

There is also a proliferation of nationally specific regulations governing the extent to which foreign food retailers are permitted to operate. For example, it has been quite common for countries to restrict entry of foreign food retailers. Although this practice has declined in recent years, the rules of operation in national retail markets continue to differ substantially. The United States has the most lenient regulatory system towards retailing. The Japanese retail market is far more heavily regulated, although Japan has revised the Large-Scale Retail Store law, which had been a major obstacle to the growth of large retail stores. Certainly the law had discouraged major foreign retailers from trying to enter the Japanese market. Following the financial crisis of 1997, several East Asian countries have relaxed the restrictions on foreign ownership of property. India, on the other hand, continues to operate a highly restrictive policy, although it announced in 2005 that it was considering relaxing some restrictions on the entry of foreign retailers. In Eastern Europe, although retailing has been opened up, there is considerable opposition to the spread of foreign-owned supermarkets.

Subsidizing and protecting agro-food industries: the major focus of trade conflict

For reasons that lie deeply embedded in national emotions, as well as in the need to guarantee a secure food supply for its citizens, most countries have adopted policies to nurture, sustain and, where felt necessary, protect their agro-food industries from external competition. For example, both Japan and Korea adopt a highly protectionist attitude towards their rice industries, which have deep cultural as well as dietary meanings. In Europe the French, in particular, regard the rural economy as sacrosanct. In the United States, farming remains a national obsession, a reflection of the country's desire for food security as well as the emotional connotations of the development of the national space in the nineteenth century. As a result, financial subsidization of some, or all, agricultural production is common in many countries. Despite the general reduction in tariffs, many agricultural products remain heavily protected to the detriment, especially, of poor countries.

In the original European Community, established in 1957, agriculture played a central role. The EU's Common Agricultural Policy (CAP) has long absorbed the largest single share of the EU's total budget and has, as a result, become a source of dissatisfaction for several member states (including the UK). The CAP has become increasingly controversial, not only within the EU itself but also in the context of the WTO trade negotiations. The CAP was reformed most recently in 2003, when the level of subsidy to farmers was separated from production, a practice which had led to notorious cases of over-production in the past. Now, subsidy is linked to 'compliance with environmental, food safety, and animal welfare standards' and part of the process of 'transforming the CAP from a sectoral policy of farm community support to an integrated policy for rural development'.[40] To some member states (Austria, Denmark, Finland, the Netherlands, Sweden, and the UK) further radical reform of the CAP is essential.

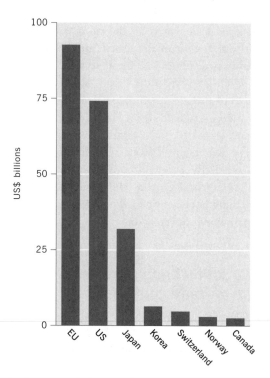

Figure 12.9   Agricultural subsidies: the EU, US and Japan
Source: WTO

However, large-scale agricultural subsidies are not confined to the EU. Far from it, as Figure 12.9 shows. The subsidies to US farmers began in the 1930s under the New Deal programme. In fact, 'farm subsidies quadrupled in the USA and doubled in the EC, in the early 1980s. The resulting surpluses substantially depressed world agricultural prices – from a mean of 100 in 1975 down to 61 in 1989 (a decline of 39 per cent). In this instability lay the origins of the Cairns Group of agro-exporting nations that have sought to abolish subsidized agriculture through GATT'.[41] The issue of agricultural subsidies, therefore, has become possibly the biggest bone of contention in the current WTO negotiations, especially in the context of the Doha 'development round'. It is often pointed out, for example, that the average subsidy per cow in the EU is more than the $2 per day on which half the world's population has to live, whilst US farm subsidies allow 'farmers to export wheat at 28 per cent less than it costs to produce, corn at 10 per cent less and rice at more than a quarter less than cost price'.[42] We will discuss the WTO negotiations more fully in Chapter 19.

Corporate strategies in the agro-food industries

Concentration and consolidation

The massive transformation of the agro-food industries that has occurred during the past few decades is inexorably bound up with the increasing dominance of very large transnational firms. This is apparent at all stages in the production circuit, from seeds through growing, processing and retailing. What was historically a highly fragmented set of industries – although some parts were always more concentrated than others – has become one in which a relatively small number of giant, transnational firms shapes what food is produced, how it is produced, who produces it, and how it is marketed and distributed to final consumers.

Top 10 seed companies		$m sales 2004
1	DuPont (Pioneer) USA	2,600
2	Monsanto USA	2,277
3	Syngenta Switzerland	1,239
4	Groupe Limagrain France	1,044
5	KWS Germany	622
6	Seminis USA	526
7	Land O'Lakes USA	416
8	Bayer Germany	387
9	Taikii Japan	366
10	DLF-Trifolium Denmark	320

Top 10 pesticide companies		$m sales 2004
1	Bayer Germany	6,120
2	Syngenta Switzerland	6,030
3	BASF Germany	4,141
4	DOW USA	3,368
5	Monsanto USA	3,180
6	DuPont USA	2,211
7	Koor Israel	1,358
8	Sumitomo Japan	1,308
9	Nufarm Australia	1,060
10	Arysta Japan	790

Top 10 food and beverage companies		$m sales 2004
1	Nestlé Switzerland	63,575
2	ADM USA	35,944
3	Altria Group USA	32,168
4	Pepsico USA	29,261
5	Unilever UK/Netherlands	29,205
6	Tyson USA	26,441
7	Cargill USA	24,000
8	Coca-Cola USA	21,962
9	Mars USA	18,000
10	Groupe Danone France	17,040

Top 10 food retailers		$m sales 2004
1	Wal-Mart USA	287,989
2	Carrefour France	99,119
3	Metro Germany	76,942
4	Ahold Netherlands	70,439
5	Tesco UK	65,175
6	Kroger USA	56,434
7	Costco USA	52,935
8	ITM France	51,800
9	Albertson's USA	39,897
10	Edeka Zentrale Germany	39,100

Figure 12.10 Dominant firms in the global agro-food industries

Source: ETC Group, 2005, Communiqué 91: www.etcgroup.org

Figure 12.10 lists the 10 leading companies in the world in four agro-food industries: seeds, pesticides, food and beverage manufacture, and food retailing. Global seed production is dominated by European (five) and US (four) firms. In fact, US dominance increased in 2005 with Monsanto's acquisition of Seminis, to create the world's largest seed company. US firms also dominate food and beverage production, although the world's biggest food manufacturer, Nestlé, comes from one of the smallest European countries, Switzerland, while the fifth largest is the UK/Netherlands company Unilever. In global food retailing, six of the top 10 are European, and four are American, including by far the largest, Wal-Mart. According to one agro-food industry observer:[43]

- Almost half of the world seed market is controlled by the leading 10 companies.
- Four-fifths of the world pesticide market is controlled by the leading 10 firms.
- One-quarter of the world packaged food market is controlled by the leading 10 firms.
- One-quarter of the global food market is controlled by the top 10 global retailers.

As in the other industries we have been discussing in this book, much of this increased concentration in the agro-food industries is the result of *merger and acquisition*. Indeed, these have been among the most takeover-intensive industries in recent years, as firms have striven not only to acquire a wider portfolio of brands (as well as to drive out competition for their own existing brands) but also to extend their reach into new geographical markets. Among the more important examples in recent years has been that of the US tobacco company Philip Morris, which has transformed itself through a whole series of acquisitions to become the Altria Group. The sequence was as follows:[44]

- 1985: Philip Morris acquires General Foods, one of the largest US food companies and owner of a wide portfolio of brands.
- 1988: acquires Kraft Foods of the United States.
- 1989: combines General Foods and Kraft to form Kraft General Foods and Kraft General Foods International. Constitutes the largest food company in the United States.
- 2000: acquires Nabisco Holdings of the United States; integrates Nabisco brands into Kraft Foods worldwide.
- 2003: changes company name to Altria Group. Becomes the world's third largest food company (see Figure 12.10).

Virtually all the other leading food companies have followed a similar trajectory varying, of course, according to their specific company history. Among the diversified food companies, Unilever acquired Brooke Bond in 1984 to make it the world's leading tea company. In 2000, Unilever acquired the US food company

Bestfoods, as well as Ben & Jerry's ice cream. The more narrowly specialized food companies have also grown through acquisition as well as through organic growth (no pun intended). Tyson Foods, for example, the world's biggest poultry company, began its 'expand or expire' strategy in 1963 by acquiring the Garrett Poultry Company of Arkansas and then made 19 further acquisitions between 1966 and 1989. In 1995, Tyson purchased Cargill's US broiler operations and has subsequently made acquisitions in other food companies outside poultry. By 2001, Tyson was 'the world's largest processor and marketer of not only chicken but also red meat with the acquisition of beef and pork powerhouse, IBP, Inc.'.[45]

Acquisition and merger have also been important factors in the growth of the major transnational food retailers. One of the biggest was Wal-Mart's acquisition of the British supermarket chain Asda for almost $11 billion, and its acquisition of the German chain Wertkauf and the Japanese company Seiyu. A second example was Ahold's acquisition of retailers in the United States, Scandinavia, Argentina, Brazil and Chile. Because of national regulatory restrictions, the major food retailers have often had to enter foreign markets through joint ventures with local partners. Examples include Tesco's alliance with Samsung in Korea and its planned link-up with the Indian firm Bharti.

Strategies of combining 'global' brands with 'local jewels'

The agro-food industries are dominated by the drive to introduce, develop and sustain *branded products*.[46] It is through branding that firms strive to convince consumers that there is something special (in terms of quality, reliability, safety and so on) about the foods they are purchasing. The degree of product differentiation through branding is probably greater in the agro-food industries than in most other industries. Each of the leading agro-food companies has a vast portfolio of brands serving different market segments. Nestlé, for example, has around 8000 brands in up to 20,000 variants.[47] Such brand portfolios are excessive; most have evolved through acquisition and merger, as we have seen. One important strategy being pursued by all the leading food companies, therefore, is to rationalize their brand portfolio, usually by selling some of them to other firms. Unilever, for example, announced a 'Path to Growth' strategy in 1999, part of which was to reduce the company's brands from 1600 to 400 (and the food brands from 70 to 60). In 2006, for example, Unilever announced its intention to sell its European Frozen Food business.

Obviously, the aim has to be to sell each brand to the largest number of consumers. The ideal, from a producer viewpoint, would be brands that sell everywhere without any need for modification. But agro-food markets are not like that, as we have seen. A major problem for the big agro-food producers, therefore, is in creating *global* brands in circumstances where much food consumption is still very strongly influenced by *local* tastes and preferences. A distinction must, of course, be made between the manufacture of a product for a global market (based on large-scale production plants serving geographically extensive markets) and the way that product is actually sold to the local consumer. A product may be sold overtly as a global brand (like Coke or Pepsi) but it may also be sold under a more local label and packaging, even if the product itself is the same everywhere.

While some food companies do market their products as global brands, others are less inclined to do so. Nestlé, for example, dismisses the idea of 'global brands':

> There is a trade-off between efficiency and effectiveness in global brands ... Operational efficiency comes from our strategic umbrella brands. But we believe there is no such thing as a global consumer, especially in a sector as psychologically and culturally loaded as food. As a result, Nestlé retains its brand strength by using ... very strong local brands.[48]

The increased consumer interest in food health and safety has important implications for food producers' strategies. Capitalizing on the enhanced interest in local and organic foods becomes increasingly important: hence the idea expressed in the heading to this section – the development of 'local jewels' (where 'local' invariably means an individual country). All the big companies are having to deal with these market changes. They are doing so in various ways: for example by acquiring local companies and by retaining their brand identities rather than rebranding them with the corporate identity. Nestlé, for example, recently announced its intention to 'accelerate the evolution of Nestlé from a respected, trustworthy Food and Beverage Company to a respected, trustworthy *Food, Nutrition, Health and Wellness* Company'.[49] Note the very significant change of emphasis.

Changes in organizational and geographical architectures

Traditionally, the major food manufacturers expanded overseas by setting up (or acquiring) operations in each of their major geographical markets. The existence of highly protected domestic food markets, together with the idiosyncrasies of local consumer tastes, made each national market distinctive. As a result, the leading transnational food producers established organizational structures that were strongly *multi*national, with all the characteristics shown in Figure 4.6.[50]

> In the current context of constant product innovation, leading global players can still rely on their portfolios developed in the home country, but the demands of adaptation and monitoring of the evolution of the local market become much greater. The result is a permanent global–local dilemma which has led major firms ... to successive reorganizations at all levels – production, management, and marketing – in an effort to establish a workable balance between the respective benefits of centralization/decentralization.[51]

The agro-food industries, therefore, are the clearest example of the 'global–local tension' discussed in Chapter 4. Because the traditional organizational-geographical structures are less and less effective, all the major food producers are engaged in large-scale reorganization programmes. Two examples help to illustrate these processes.

Nestlé currently has operations in 80 countries and employs 250,000. In announcing its intention to transform itself into a 'Food, Nutrition, Health and Wellness Company', the company recognized that this has implications that go beyond products and brands.

> Nestlé is changing from a decentralized multinational company to a global, and ultimately, a global multifocal company.[52]

In fact, Nestlé has been involved in substantial geographical reorganization for some time, as its actions within South East Asia reveal.[53] With the increasing liberalization of agro-food trade within ASEAN, Nestlé progressively rationalized its multidomestic operations there (in the early 1990s it had more than 40 factories in the region). Under the 'centres of excellence' programme, the company established such centres for production of breakfast cereals in the Philippines, chocolate and confectionery in Malaysia, non-dairy creamer in Thailand, soya sauce in Singapore, and instant coffee in Indonesia.

Such developments formed part of the company's broader reorganization programme at the global scale.

> In the 1990s, the company dramatically expanded its intra-firm trade of final products ... Because many Nestlé products sell under globally recognised brand names and are manufactured according to globally consistent recipes, there is a certain footlooseness in the intra-firm trade patterns of the company.[54]

Like Nestlé, *Unilever* has long operated a decentralized multinational strategy but it, too, has made strenuous efforts to create a more efficient and responsive global structure especially in its much publicized 'Path to Growth' strategy of the late 1990s.[55] At that time, Unilever operated around 300 food factories, with a presence in virtually every country in the world. The acquisition of Bestfoods in 2000 brought in a further 70 factories in 60 countries.

In its two major divisions, Food and Home and Personal Care Products, Unilever's plan has been to focus on a much smaller number of brands (see earlier). This involves closing a large number of plants in favour of concentrating production on around 150 key sites, together with a further 200 sites for manufacturing local market-leading or nationally profitable products.

> The aim is to reach annual savings of about €1.5 billion per year ... Unilever expects to save another €1.6 billion per year by switching from local/national to global buying or sourcing. This means a switch to fewer suppliers and they in turn are urged to deliver higher volumes for lower costs. This cost-cutting strategy cascades down through the entire supply chain.[56]

These kinds of organizational and geographical restructuring are typical of all the major multibrand food producers, although with variations in detail from one company to another. The pattern of development is rather different in less diverse companies such as the world's biggest poultry producer, Tyson Foods. Whereas firms like Nestlé, Unilever, Altria or PepsiCo operate in many countries across the world, Tyson Foods remains overwhelmingly a US-based company. More than 170 of its 193 operations are in the United States, and 63 per cent of these are in the southern states (see Figure 12.1). Of the 22 overseas operations, 14 are engaged in sales or service.[57]

'Big Food' and 'Big Retail': two sides of the same coin

These developments in the strategies of the major transnational food producers have to be seen within the context of the retailing systems through which their products are sold. One of the most significant developments in the agro-food industries, in fact, has been the evolving, symbiotic relationship between the big food producers and the big supermarket chains.

> 'Big Food' and 'Big Retail' are really two sides of the same coin. Big global food manufacturers need big supermarket chains to get their products on to the shelves and our big supermarkets need big food processors … Mass-produced food can be churned out over and over again in vast, uniform quantities, made by a handful of big manufacturers who jump to the big retailers' tune, processed food lends itself to supermarket retailing: it gives them the ability to put a standard, regular product into every store nationwide, a product that does not require any specialist handling … Industrial food lends itself to the supermarkets' heavily centralised, highly mechanical distribution systems.[58]

This is an arena of continous power struggles in which, as the quotation suggests, power lies increasingly with the big transnational food retailers. Of course, these companies sell many products other than food: clothing, footwear, toys, consumer electronics and many others. But food retailing is at the core of their activities. And there is no doubt that the biggest food retailers have become increasingly *transnational* after being essentially domestically oriented for most of their histories. Figure 12.11 maps the transnational operations of four of the leading transnational retailers. There are some significant differences between them.

Wal-Mart is overwhelmingly a North American company in terms of its store geography: 70 per cent of its stores are in the United States and Canada. If its Mexican stores are included, the regional share rises to 83 per cent. However, Wal-Mart has been pursuing an aggressive transnationalization strategy, although at present it operates in only 10 countries. In East Asia it already has more than 40 stores in China (with plans for more) and 16 in Korea, and has acquired an equity stake in the Japanese retailer Seiyu. However, Wal-Mart has now sold its

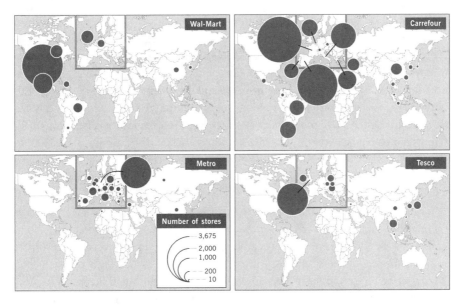

Figure 12.11   The global geographies of leading transnational food retailers

Source: company reports

Korean stores to a domestic Korean firm. In Europe, it is focusing on Eastern European countries in particular to add to its 282 stores in the UK and its 91 stores in Germany. In Central America, Wal-Mart has acquired a substantial equity stake in the region's largest retailer from Ahold.[59]

The European retailers Carrefour, Metro and Tesco, whilst also strongly home country oriented, are far less so than Wal-Mart. The French retailer *Carrefour* has a strong presence in Spain and Italy, in addition to France (which has 34 per cent of its world total). Overall, 89 per cent of Carrefour's stores are in Europe. Elsewhere, 8 per cent are in Latin America (primarily in Argentina and Brazil) and 3 per cent in Asia, where it has more than 200 stores in China. However, Carrefour has a policy of getting out of countries in which it cannot become one of the top three retailers.[60] In this context, it has recently withdrawn from Japan, Korea and Mexico, and sold its stores in the Czech Republic and Slovakia to Tesco. At the same time, it has bought Tesco's Taiwan operations. The German retailer *Metro* has almost three-quarters of its stores in Germany and only a few outside Europe (the largest number, 23, being in China). It has a strong and growing presence in Eastern Europe (8 per cent of its total stores). Its current transnationalization strategy is to invest heavily in Eastern Europe and Asia (especially in China and India).

The UK retailer *Tesco* has grown extremely rapidly in recent years. It now totally dominates the UK grocery market, where 75 per cent of its stores are located, making it, like Wal-Mart in the United States, a target for strong opposition from different interest groups. Tesco is pursuing a very aggressive – but geographically focused – transnationalization strategy, based on expansion in East Asia (12 per cent of its stores) and Eastern Europe (9 per cent). Interestingly, Tesco has no stores in Western Europe, apart from Ireland. Its recent buying and selling deals with Carrefour are part of this strategy, strengthening Tesco's position in Eastern Europe. In East Asia, Tesco's major store concentrations are in Thailand, Japan, China and Korea (where it operates as a joint venture with Samsung[61]). In 2005–6, Tesco planned to open a further 150 stores in East Asia and around 50 in Eastern Europe.[62] Having stayed out of the United States, Tesco recently announced it would set up convenience stores, initially on the West Coast, planning to open 100 stores in southern California during its first years of operations.[63] It also plans to enter the highly protected Indian market through a local collaboration.

Thus, there has been very considerable growth in the transnational operations of some of the leading retail chains. But such expansion has not been problem-free. Carrefour failed to transfer its hypermarket model to the United States, while Wal-Mart has had major difficulties with its acquired German affiliates, particularly by failing to understand the fundamental differences between the German and the US retail food distribution system. Tesco faces opposition in Poland to its hypermarket expansion there. The use of local partners within a joint venture often helps to avoid the problems of misunderstanding local market conditions. But even joint ventures are not without their difficulties, especially if the foreign partner fails to learn from the knowledge embedded in the local partner. It is also the case that, while the strength of most of the leading retailers is based on their high levels of profitability in their home market, their returns on international operations are often far lower.

So, the transnationalization of food retailing is far from being a straightforward or unproblematic process. Competing head-to-head with local firms is particularly difficult in this sector. A major problem is that of identity. Because food retailing has traditionally been very much a domestic activity, there is little knowledge of foreign retail store brands (as opposed to product brands). For many customers outside the United States, for example, Wal-Mart is a totally unknown quantity. The same applies to non-French residents' knowledge of Carrefour, or non-UK residents' awareness of Tesco. Yet building up a respected and trusted brand identity takes a long time. Meanwhile, local competition remains, in most cases, a very serious problem for transnational food retailers.

Geographical expansion of the store network is one dimension of food retailers' strategies. A second dimension is from whom, and from where, its products are sourced. The big retail chains have vastly increased the geographical extensiveness of their sourcing systems as well as exerted increasing power and influence over their suppliers. We saw in the case of clothing that the major retailers have come to dominate their supply networks, forcing suppliers to meet their increasingly stringent demands on price, delivery and quality. Precisely the same applies to supplier relationships in the food industry – perhaps even more so. For example,

> by the end of the 1990s, the UK supermarkets had restructured the fresh vegetable value chain, moving away from their initial reliance on wholesale markets to tightly knit supply chains ... there had been a move away from arm's-length relationships. The information requirements for the value chain had become much more complex, and resources were being invested in specifications of systems and monitoring of performance. Each retailer developed its own value chain control system. The supermarket–importer relationship moved towards the relational pattern, whereas the importer–exporter relationship displayed the characteristics of a captive linkage ... The exporters were transactionally dependent upon their UK importers and subject to rigorous control.[64]

There is a great deal of criticism of the treatment of suppliers by the big supermarkets, although suppliers are often afraid to object out of fear of losing their contracts. A recent investigation of the accounts of transnational food retailers claims that they gain huge financial benefits simply by delaying payments to their suppliers:

> stock is turned into cash at the check-out counters long before suppliers have to be paid ... In effect, suppliers have acted as surrogate bankers ... [however] the burden is not shared equally ... the most powerful manufacturers are able to shunt the burden of increased trade debt down the supply chain ... life is very much tougher for smaller suppliers who do not have the luxury of their burden down the line.[65]

It is also increasingly common practice for the big supermarket chains to ask the major food producers to pay for 'preferred status'.[66]

As the big food retailers have increased their direct presence in foreign countries (especially in the emerging market economies) they have also drastically changed the geography and organization of the sourcing networks, both for their local stores and for their entire network.[67] Typically, the degree of centralization of procurement has greatly increased. When a transnational retailer establishes operations in a specific country, one of its first actions is to replace 'a per store procurement system with the distribution centre (DC) model used in established markets. Each DC may have responsibility for a particular range of products or a particular territory'.[68]

The second aspect of the changing procurement practices of transnational retailers is the changing balance between global and local sourcing.

> On the one hand, transnational retailers have increased levels of global sourcing for their home markets ... On the other hand ... there are the supply chain impacts that result from the retailers establishing store operations *within* the various markets ... The foreign subsidiaries of retailers such as Tesco, Ahold, and Carrefour commonly source over 90% of products from within the country ... *contra* accounts of the continuing rise of global sourcing, local sourcing may actually *increase* over time as the supply base develops and retailers therefore import fewer products.[69]

Conclusion

The geographies of agro-food production, distribution and consumption have been transformed dramatically during the span of only a few decades. This is especially true of the high-value foods which have been the main focus of this chapter. In the case of fresh fruits and vegetables, for example, the counter-seasonality of production in the Southern Hemisphere for Northern Hemisphere consumers has created what has been termed 'permanent global summertime'. Significant territorial production complexes have developed in agro-food production both in developed countries and also in the newly agriculturalizing countries.

Underlying these developments is a complex interplay between the major producers, retailers, states and consumer groups. Agro-food production and distribution have become increasingly industrialized, in terms of both the technologies of food production and how and where such production and distribution are organized. Increasingly complex production circuits have developed in which the degree of integration implemented by the major producers and buyers has deepened dramatically. The agro-food industries are also vigorously contested activities, imbued with high levels of ethical, safety and environmental concerns. They are industries in which consumer pressures are becoming increasingly important. They are industries in which states (at both national and international scales) have always been heavily involved: they are among the most highly regulated, heavily subsidized and vigorously protected of all economic activities. Above all, they are industries which have become dominated by very large transnational firms, both producers and retailers. 'Big Food' and 'Big Retail' are, indeed, two sides of the same coin.

NOTES

1 Bonanno et al. (1994), Ward and Almås (1997), Watts and Goodman (1997), Wilkinson (2002).
2 Watts and Goodman (1997: 2–3, 10–11).
3 Sanderson (1986).
4 Page (2000: 245, emphasis added).
5 Blythman (2004: Chapter 11).
6 Boyd and Watts (1997: 204, emphasis added).
7 Raynolds (2004), Whatmore and Thorne (1997), Whatmore et al. (2003).
8 Whatmore et al. (2003: 389).
9 Ponte (2002) examines the conventional coffee production circuit at a global scale.
10 Sonnino and Marsden (2006: 183).
11 Sonnino and Marsden (2006: 183).
12 www.newfarm.org/international/news/120104/121604/br_chicken/html.

13 This section draws heavily on Huang (2004). See also Barrett et al. (1997) for a study of Kenya's horticultural production and trade.

14 Huang (2004: 3).

15 Huang (2004: 7–8, 10, 11).

16 Boyd and Watts (1997: 207–14).

17 Boyd and Watts (1997: 209).

18 *The Guardian* (10 November 2005).

19 Marsden (1997: 174). See also Marsden and Cavalcanti (2001) and *The Economist* (5 November 2005, Special Report on Brazilian Agriculture).

20 Freidberg (2004: 10–11).

21 *Financial Times* (8 August 2005).

22 See Ponte (2002).

23 Quoted in Paul and Steinbrecher (2003: 173).

24 Blythman (2004: 76, 77).

25 See Blowfield (1999), Browne et al. (2000), Hughes (2001), Leclair (2002).

26 *The Guardian* (8 March 2006).

27 Friedland (1994: 223).

28 *The Guardian* (10 May 2003).

29 *London Evening Standard* (12 December 2001).

30 Huang (2004: 19).

31 Goodman et al. (1987), Sorj and Wilkinson (1994).

32 FAO report in 1996, quoted in Paul and Steinbrecher (2003: 4).

33 Whatmore (2002: 131).

34 Millstone and Lang (2003).

35 Oxfam (2004: 2–3, 7, 8).

36 Grigg (1993: 236).

37 Braithwaite and Drahos (2000: 401).

38 Braithwaite and Drahos (2000: 401).

39 Wilkinson (2002: 338).

40 Sonnino and Marsden (2006: 192).

41 McMichael (1997: 641).

42 US Institute for Agriculture and Trade Policy, quoted in *The Observer* (3 July 2005).

43 ETC Group (2005): www.etcgroup.org

44 Altria Group website: www.altria.com

45 Tyson Foods Inc. website: www.tysonfoodsinc.com

46 For 'local jewels' see Capgemini News (2002): www.capgemini.com

47 *Financial Times* (22 February 2005).

48 Nestlé CEO, quoted in the *Financial Times* (22 February 2005).

49 Nestlé company website: www.nestle.com, emphasis added.

50 For cases in the agro-food industries, see Pritchard (2000a; 2000b; 2000c).

51 Wilkinson (2002: 335).

52 Nestlé company website: www.nestle.com

53 Pritchard (2000a: 252–4).

54 Pritchard (2000a: 253).

55 Company reports; FNV Mondiaal *Unilever Corporate Profile 2003*.

56 FNV Mondiaal *Unilever Corporate Profile 2003*: 12.

57 Tyson Foods Inc. website: www.tysonfoodsinc.com
58 Blythman (2004: xvii, 73).
59 *Financial Times* (21 September 2005).
60 *The Economist* (16 April 2005).
61 Coe and Lee (2006).
62 *Tesco Annual Report* (2005).
63 *Financial Times* (29 June 2006)
64 Dolan and Humphrey (2002: 501).
65 *Financial Times* (7 December 2005).
66 *Financial Times* (30 March 2006).
67 Coe and Hess (2005) provide an excellent analysis of the situation in Eastern Europe and East Asia.
68 Coe and Hess (2005: 464).
69 Coe and Hess (2005: 463).

Thirteen
'Making the World Go Round': Financial Services

Money counts

Financial services are fundamental to the operation of all production circuits and networks (Figure 1.4). Every economic activity (whether a material product or a service) has to be financed at all stages of its production. Without the parallel development of systems of money- and credit-based exchange there could have been no development of economies beyond the most primitive organizational forms and the most geographically restricted scales.

> The geographical circuits of money and finance are the 'wiring' of the socio-economy … along which the 'currents' of wealth creation, consumption and economic power are transmitted … money allows for the deferment of payment over time–space that is the essence of credit. Equally, money allows propinquity without the need for proximity in conducting transactions over space. These complex time–space webs of monetary flows and obligations underpin our daily social existence.[1]

Finance is also one of the most controversial of economic activities because of its historical relationship with state 'sovereignty'. Ever since the earliest states emerged on the scene, the creation and control of money have been regarded as central to their legitimacy and survival.[2] Today, in the context of a globalizing world economy, this tension has become especially acute as national states see their traditional control of their monetary affairs being threatened by external market forces. The financial crises that periodically engulf some or all parts of the world economy intensify these tensions and also create legitimate fears over the accountability and responsibility of financial institutions.

In 1986, Susan Strange coined the graphic term 'casino capitalism' to describe the international financial system:

every day games are played in this casino that involve sums of money so large that they cannot be imagined. At night the games go on at the other side of the world ... [the players] are just like the gamblers in casinos watching the clicking spin of a silver ball on a roulette wheel and putting their chips on red or black, odd numbers or even ones.[3]

Twelve years later, she used the term 'mad money' to reflect the increased volatility of the financial system and the uncertainty it generates throughout the world economy.

International financial flows and foreign currency transactions have reached unprecedented levels. They totally dwarf the value of international trade in manufactured goods and in other services and have done so at an increasing rate over the past three decades, as Figure 13.1 shows. Of course, some of those financial transactions are directly related to international trade (and production) and are essential for that purpose. But only around 10 per cent of international financial transactions are of this kind. The remainder take the form of what are, essentially, *speculative* dealings – aimed at making short- or long-term profits as ends in themselves – through a bewildering variety of financial instruments. Of course, it is often difficult to draw a clear line between speculative and productively essential financial transactions.[4]

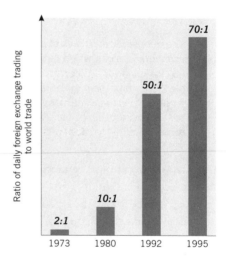

Figure 13.1   The growing disparity between foreign exchange trading and world trade

Source: based on Eatwell and Taylor, 2000: 3–4

Clearly, a distinct financial system has come into being in which changes in the level and in the geography of the flows of international portfolio investments (investments in stocks and bonds) and of foreign currencies have an enormous impact on the financial well-being of entire countries. For example, a heavy dependence on such 'hot' money was a major factor in the Mexican peso crisis of 1994 and in the East Asian financial crisis of 1997–8.

Financial services have a dual character:

- They are *circulation* services, fundamental to the operation of every aspect of the economic system.
- They are *commodities* or *products* in their own right, in the sense that they are produced and traded in the same way as more tangible manufactured goods are traded.

They are also intrinsically *geographical* in their production and consumption. Indeed, one of the paradoxes of the global economy is that the one sector which would appear to be the most geographically 'footloose' is, in fact, very highly concentrated into a small number of international financial centres.

The structure of the financial services industries

> The function of the financial system as a whole is financial *intermediation*, the pooling of financial resources among those with surplus funds to be lent out to those who choose to be in deficit, that is to borrow … With financial intermediation, investors in new productive activities do not themselves have to generate a surplus to finance their projects; instead the projects can be financed by surpluses generated elsewhere within the economy.[5]

The process of intermediation has constituted the basic function of banks from the very beginning. Figure 13.2 shows the sequence of development of the banking system. The first two stages of credit provision to borrowers depend greatly upon the nature of geographically specific knowledge to legitimize lending and borrowing. The geographical scope of knowledge and trust grows through such developments as inter-bank lending (that is, lending outside the local area) in stage 3 and, eventually, of a central bank that ultimately acts as the lender of last resort to the banking system as a whole (stage 4). Subsequent developments, especially in securitization, create a totally different scale and complexity of financial activity.

Such increased complexity of the financial system has resulted in the development of a huge variety of different types of financial institution, each of which has a specific set of core functions (Figure 13.3). In fact, the boundaries between these individual activities and institutions have become increasingly blurred. The first three types of financial institution shown in Figure 13.3 are concerned with the creation and distribution of credit in various forms. The two remaining categories

The stages of banking development	Banks and space	Credit and space
Stage 1: Pure financial intermediation Banks lend out savings Payment in commodity money No bank multiplier Saving precedes investment	Serving local communities Wealth based, providing foundation for future financial centres	Intermediation only
Stage 2: Bank deposits used as money Convenient to use paper money as means of payment Reduced drain on bank reserves Multiplier process possible Bank credit creation with fractional reserves Investment can now precede saving	Market dependent on extent of confidence held in banker	Credit creation focused on local community because total credit constrained by redeposit ratio

Stage 3: Inter-bank lending
Credit creation still constrained by reserves
Risk of reserves loss offset by development of
  inter-bank lending
Multiplier process works more quickly
Multiplier larger because banks can hold lower
  reserves
	Banking system develops
	at national level
		Redeposit constraint relaxed
		somewhat, so can lend wider
		afield

Stage 4: Lender-of-last-resort facility
Central bank perceives need to promote
  confidence in banking system
Lender-of-last-resort facility provided if
  inter-bank lending inadequate
Reserves now respond to demand
Credit creation freed from reserves constraint
	Central bank oversees
	national system, but
	limited power to
	constrain credit
		Banks freer to respond to
		credit demand as reserves
		constraint not binding and
		they can determine volume
		and distribution of credit
		within national economy

Stage 5: Liability management
Competition from non-bank financial
  intermediaries drives struggle for market share
Banks actively supply credit and seek deposits
Credit expansion diverges from real economic
  activity
	Banks compete at
	national level with
	non-bank financial
	institutions
		Credit creation determined by
		struggle over market share
		and opportunities in
		speculative markets
		Total credit uncontrolled

Stage 6: Securitization
Capital adequacy ratios introduced to curtail
  credit
Banks have an increasing proportion of bad
  loans because of over-lending in stage 5
Securitization of bank assets
Increase in off-balance-sheet activity
Drive to liquidity
	Deregulation opens up
	international competition,
	eventually causing
	concentration in financial
	centres
		Shift to liquidity by emphasis
		being put on services rather
		than credit; credit decisions
		concentrated in financial
		centres; total credit determined
		by availability of capital,
		i.e. by central capital markets

Figure 13.2   The sequence of development of the banking system

Source: based on Dow, 1999: Tables 1 and 2

Type	Primary functions
Commercial bank	Administers financial transactions for clients (e.g. making payments, clearing cheques). Takes in deposits and makes commercial loans, acting as intermediary between lender and borrower
Investment bank/ securities house	Buys and sells securities (i.e. stocks, bonds) on behalf of corporate or individual investors. Arranges flotation of new securities issues
Credit card company	Operates international network of credit card facilities in conjunction with banks and other financial institutions
Insurance company	Indemnifies a whole range of risks, on payment of a premium, in association with other insurers/reinsurers
Accountancy firm	Certifies the accuracy of financial accounts, particularly via the corporate audit

Figure 13.3   Major types of financial service activity

are concerned with different forms of risk indemnification (insurance companies) and with certifying the accuracy of financial accounts (accountancy firms).

The dynamics of the market for financial services

From being fairly simple and predictable, the markets for financial services have become increasingly diverse and far less predictable. The intensity of competition for consumers (whether corporate or individual) has increased enormously. Four processes are especially important:

- *Market saturation:* by the late 1970s, traditional financial services markets were reaching saturation. There were fewer and fewer new clients to add to the list; most were already being served, particularly in the commercial banking sector but also in the retail sector in the more affluent economies.
- *Disintermediation:* this is the process whereby corporate borrowers make their investments or raise their needed capital without going through the 'intermediary' channels of the traditional financial institutions, particularly the bank. Instead, they have increasingly sought capital from non-bank institutions, for example through securities, investment trusts, and mutual funds. However, the process is extremely dynamic. Disintermediation may well be challenged by reintermediation, as new forms of supplier–customer relationship emerge (as, for example, through the Internet).[6]
- *Deregulation of financial markets:* financial services markets have traditionally been extremely closely regulated by national governments. One of the most important developments of the past few years has been the increasing deregulation or liberalization of financial markets. Deregulation has been aimed primarily at three areas:
  o the opening of new geographical markets
  o the provision of new financial products
  o changes in the way in which prices of financial services are set.

- *Internationalization of financial markets:* demand for financial services is no longer restricted to the domestic context: financial markets have become increasingly global. Three types of demand are especially significant:
  o the massive growth in international trade
  o the global spread of transnational corporations
  o the vastly increased institutionalization of savings.

Each of these forces for change in the demand for financial services is interrelated; together they create a *new competitive environment*.

Technological innovation and the financial services industries

> Travelling at the speed of light, as nothing but assemblages of zeros and ones, global money dances through the world's fiber-optic networks in astonishing volumes … National boundaries mean little in this context: it is much easier to move $41 billion from London to New York than a truckload of grapes from California to Nevada.[7]

Technological developments in communications technologies and in the broader sphere of information technology are absolutely fundamental to the financial services industries. *Information is both the process and the product of financial services.* Their raw materials are information: about markets, risks, exchange rates, returns on investment, creditworthiness. Their products are also information: the result of adding value to these informational inputs. In the words of one financial services executive:

> We don't have warehouses full of cash. We have *information* about cash – *that* is our product.[8]

The *speed* with which financial service firms can perform transactions and the *global extent* over which they can be made are especially important. Not surprisingly, therefore, all the major financial services firms invest huge sums in information technologies. In the mid 1990s, for example, the top 10 US banks spent more than $10 billion on information technology; the top 10 European banks spent almost $12 billion in the same year.[9] In particular, computers

> have changed the prevailing system of financial settlement out of all recognition. For hundreds of years, current payments and capital transfers were completed in negotiable cash – coins and notes – and in written promises to pay. They were recorded by handwriting in ledgers. Coins and notes gave way progressively to cheques and other promises to pay, such as letters of credit. Since the early 1970s, these too have become obsolete. *Computers have made money electronic* … by the mid 1990s, computers had not only transformed the physical form in which money worked as a medium of exchange, they were also in the process of transforming the systems by which payments of money were exchanged and recorded.[10]

Together with the introduction of microprocessors into credit and debit cards, the enhanced ability instantly to transmit information globally, and the revolutionary implications of the Internet, IT developments have

- vastly increased productivity in financial services
- altered the patterns of relationships both within financial firms and also between financial firms and their clients

- greatly increased the velocity of turnover of investment capital (for example, the ability to transfer funds electronically – and, therefore, instantaneously – has saved billions of dollars in interest payments formerly incurred by the delay in making transfers)
- enabled financial institutions both to increase their loan activities and also to respond immediately to fluctuations in exchange rates in international currency markets.

From a technological viewpoint, therefore, it is now possible for financial services firms to engage in global 24-hours-a-day trading – to 'follow the sun' – whether this be in securities, foreign exchange, financial and commodities futures or any other financial service.[11] As Figure 13.4 shows, the trading hours of the world's major financial centres overlap.

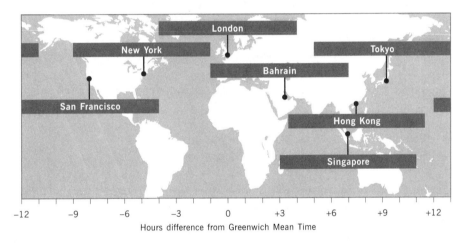

Figure 13.4   The potential for 24-hour financial trading

Source: based on Warf, 1989: Figure 5

   In fact, genuine 24-hour trading is currently limited to certain kinds of transaction partly because, although the technology is available, either the organizational structure or the national regulatory environment remain an obstacle. In addition, the nature of the product can be a limiting factor.

> Global trading is encouraged by products that are homogeneous, familiar and tradable in large quantities, as, for example, trading in foreign exchange and US government bonds. A fully internationalised banking and finance system would operate on a 24-hour basis with information flowing between different countries with minimal institutional and regulatory barriers. Clearly, this situation has not been reached and there are numerous barriers still in existence.[12]

To the extent that such electronic transactions do not require direct physical proximity between seller and buyer they are a form of 'invisible' international trade. In that sense, therefore, financial services are one form of service activity that is *tradable*. Electronic communications have also contributed greatly to the bypassing of the commercial banks and the trend towards the greater *securitization* of financial transactions: the conversion of all kinds of loans and borrowings into 'paper' securities which can be bought and sold on the market. Such transactions may be performed directly by buyers and sellers without necessarily going through the intermediary channels of the commercial banks.

The global integration of financial markets brings many benefits to its participants in the speed and accuracy of information flows and the rapidity and directness of transactions, even though the participants may be separated by many thousands of miles and by several time zones. But such global integration and instantaneous financial trading also create costs. 'Shocks' occurring in one geographical market now spread instantaneously around the globe, creating the potential for global financial instability. Financial 'contagion' is endemic in the structure and operation of the contemporary financial system.

Innovations in telecommunications and in process technologies, therefore, have helped to transform the operations of financial services firms. In addition, a whole new array of financial instruments (product innovations) has appeared on the scene, notably:

- those providing new methods of lending and borrowing
- those facilitating greater spreading of risk.

Figure 13.5 gives some examples. The most important product innovation since the mid 1980s has been the phenomenal growth of the *derivatives* markets. Derivatives are

> financial tools derived from other financial products, such as equities and currencies. The most common of these are futures, swaps, and options ... The derivatives market aims to enable participants to manage their exposure to the risk of movements in interest rates, equities, and currencies.[13]

Without question, therefore, technological developments in telecommunications and information technology, as well as in product innovations, have transformed the financial services industries. The global integration of financial markets has become possible, collapsing space and time and creating the potential for virtually instantaneous financial transactions in loans, securities and a whole variety of financial instruments. However, completely borderless financial trading does not actually exist, for the simple reason that most financial services remain very heavily supervised and regulated by individual national governments. Let us now see how the regulatory system operates and how it is changing.

Type of financial instrument	Basic characteristics
Floating-rate notes (FRNs)	Medium- to long-term securities with interest rates adjusted from time to time in accordance with an agreed reference rate, e.g. the London Inter-Bank Offered Rate (LIBOR) or the New York banks' prime rate
Note issuance facilities (NIFs)	Short- to medium-term issues of paper which allow borrowers to raise loans on a revolving basis directly on the securities markets or with a group of underwriting banks
Eurocommercial paper	Non-underwritten notes sold in London for same-day settlement in US dollars in New York. More flexible than longer-term Euronotes of 1, 3 or 6 months' duration
Loan sales	The sale of a loan to a third party with or without the knowledge of the original borrower
Interest rate swaps	A contract between two borrowers to exchange interest rate liabilities on a particular loan, e.g. the exchange of fixed-rate and floating-rate interest liabilities
Currency swaps	Financial transactions in which the principal denominations are in different currencies

Figure 13.5 Examples of product innovations in financial markets

The role of the state: regulation and deregulation in financial services

A tightly regulated system

> Financial systems are also *regulatory spaces* ... Money has a habit of seeking out geographical discontinuities and gaps in these regulatory spaces, escaping to places where the movement of financial assets is less constrained, where official scrutiny into financial dealing and affairs is minimal, where taxes are lower and potential profits higher.[14]

Before the 1960s a 'world' financial market did not exist. The IMF, together with the leading industrialized nations, operated a broadly efficient global mechanism for monetary management based, initially, on the post-war Bretton Woods agreement (see Chapter 19). At the national level, financial markets and institutions were very closely supervised, primarily because of concerns over the vulnerability of the financial system to periodic crisis and because of the centrality of finance to the operation of a country's economy.

Financial services, therefore, have been the most tightly regulated of all economic activities, certainly more so than manufacturing industries. Such regulation can be divided into two major types:[15]

- Those governing the *relationships* between different financial activities. National financial services markets have generally been *segmented* by regulation: banks performed specified activities; securities houses performed other activities. Neither was allowed to perform the functions of the other.

- Those governing the *entry* of firms (whether domestic or foreign) into the financial sector. Restricted entry into the different financial services markets has been virtually universal, although the precise regulations differ from country to country. Governments have been especially wary of a too-ready expansion of the *branches* of foreign banks and insurance companies. This is because, unlike separately incorporated subsidiaries, branches are far more difficult to supervise: they form an integral part of a foreign company's activities. In almost every case, there are limits on the degree of foreign ownership permitted in financial services.

'The crumbling of the walls'

Although many of these restrictions still exist, the regulatory walls have been crumbling – even collapsing altogether in some cases. The process was relatively slow at first but accelerated rapidly after the late 1980s. Deregulatory pressures have come from several sources, most notably the increasing abilities of transnational firms to take advantage of 'gaps' in the regulatory system and to operate outside national regulatory boundaries.

The starting point was the emergence of the Eurodollar (that is, offshore) markets in the 1960s. Initially, Eurodollars were simply dollars held outside the US banking system largely by countries, like the USSR and China, anxious to prevent their dollar holdings being subject to US political control. 'Once it was appreciated that Eurodollars ... were free of US political control, it did not take bankers long to recognise that the dollar balances were also free of US banking laws governing the holding of required reserves and controls upon the payment of interest'.[16]

The rapid growth of this new currency market outside national regulatory control was reinforced by pressure from banks and other financial services firms to operate in a less constrained and segmented manner, both domestically and internationally. The internationalization of financial services and the deregulation of national financial services markets, therefore, are virtually two sides of the same coin. Forces of internationalization were one of the pressures stimulating deregulation; deregulation is a necessary process to facilitate further internationalization.

Major deregulation has occurred in all the major developed economies. In the United States, a series of changes since the 1970s has both eased the entry of foreign banks into the domestic market and facilitated the expansion of US banks overseas. In 1981, the United States allowed the establishment of international banking facilities (IBFs), which created 'onshore offshore' centres able to offer specific facilities to foreign customers. Later, the federal government introduced a major reform of the US financial system allowing banks to become involved in a whole variety of financial services and to operate nationwide branching networks.

In the United Kingdom, the so-called 'Big Bang' of October 1986 removed the existing barriers between banks and securities houses and allowed the entry of foreign firms into the Stock Exchange. In France, the 'Little Bang' of 1987 gradually opened up the French Stock Exchange to outsiders and to foreign and domestic banks. In Germany, foreign-owned banks were allowed to lead-manage foreign issues, subject to reciprocity agreements. In Japan the restrictions on the entry of foreign securities houses were relaxed (though not removed) and Japanese banks can open international banking facilities. But the Japanese financial system remained tightly regulated. In 1996, the Japanese government announced its intention to undertake a wide-ranging deregulation of the country's financial system by 2001. That process is still ongoing in the context of Japan's attempt to recover from its deep financial crisis. Even the Singaporean government has progressively loosened the restrictions on the financial sector in order to maintain the country's position as a major Asian financial centre. Most recently, China has allowed foreign participation in big state-owned banks and has listed them on overseas stock exchanges. However, other countries (such as India) retain very strong protectionist barriers around their national financial markets.

Increasing deregulation of financial services is an important component of the major regional economic blocs. A financial services agreement was part of the Canada–United States Free Trade Agreement whilst, under the NAFTA, Mexico must open up its financial sector to North American firms by 2007. EU reforms aim at removing the individual national financial regulatory structures inhibiting the creation of an EU-wide financial system so that financial services flow freely throughout the EU and financial services firms can establish a presence anywhere within the Single Market. The creation of a single European currency transformed wholesale financial markets in the EU but retail financial markets remain fragmented on national lines.

> The European Union ... is aiming for financial-market integration but is being held back by its own diversity and by political disagreements about the purpose of national banking systems. For the past couple of years, Italy has been the pariah, seemingly resisting one of the EU's fundamental principles: the free flow of capital among member countries ... But Italy's is not the only European banking system to resist foreign buyers. Most German banks are still publicly or mutually owned and therefore cannot be sold. The same goes for half of the French banking system ... Even in Britain, a foreign purchase of one of the dominant clearing banks would probably be self-defeating.[17]

In addition, negotiations within the WTO to complete a multilateral regulatory framework for financial services continue. An interim financial services agreement, involving some 30 countries, was reached by the WTO in 1995, which aimed to guarantee access to banking, securities and insurance markets. But the United States did not fully accept the agreement and reserved the right to restrict access to its own financial markets on the basis of reciprocity (although existing arrangements stayed in place). We will return to the contentious issue of a 'new financial architecture' for the world economy in Chapter 19.

The accelerating deregulation of financial services is the most important current development in the globalization of the financial system. Even so, there remain substantial differences in the extent and nature of financial services regulation in different countries. The international regulatory environment is highly asymmetrical. Such differences, of course, have a powerful influence on the locational strategies of international financial services firms.

Corporate strategies in financial services

The environment in which financial services companies must operate consists of the following interacting elements:

- shifting patterns in the demand for financial services
- technological innovations which affect how these services can be delivered
- a changing, but geographically diverse, regulatory framework.

Of course, financial services firms are not simply passive respondents to these environmental conditions. There is no simple cause–effect relationship; rather, there is a complex interplay in which the major financial services firms – the banks, securities houses, insurance companies and the like – have themselves been highly influential in changing that environment.

In this section we focus on the specialist financial services companies and exclude the in-house financial activities of TNCs in other industries. Of course, all TNCs have large-scale, sophisticated financial operations. The corporate treasury departments of major companies are significant entities in their own right and are larger than many specialist financial services companies. However, the primary functions of these in-house operations are to contribute towards the efficiency of their own corporate systems. They do this by, for example, optimizing internal flows of funds, investing surplus capital effectively on the financial markets, and hedging against the foreign currency fluctuations endemic in the operations of any TNC, whatever its line of business.

As far as the specialist financial services firms are concerned, we will explore four major strategic trends:

- *concentration and consolidation* through merger and acquisition
- *transnationalization* of operations
- *diversification* into new product markets
- *outsourcing* of business functions.

Concentration and consolidation

The history of the financial services industries – particularly of banking – is a continuous trend towards greater concentration into a smaller number of companies. As Figure 13.2 suggests, this process is intrinsically geographical: smaller financial

Table 13.1   The world's top 25 banks

	Bank	Home country	Assets ($ million)	Number of host countries	Foreign affiliates as % total
1	UBS	Switzerland	1,533,036	48	83.9
2	Citigroup	United States	1,484,101	77	53.2
3	Mizuho Financial Group	Japan	1,295,942	15	47.1
4	HSBC Holdings	United Kingdom	1,276,778	48	59.0
5	Crédit Agricole Group	France	1,243,047	41	43.8
6	BNP Paribas	France	1,233,912	48	54.8
7	J.P. Morgan Chase & Co	United States	1,157,248	27	50.8
8	Deutsche Bank	Germany	1,144,195	40	69.1
9	Royal Bank of Scotland	United Kingdom	1,119,480	26	17.1
10	Bank of America Corp.	United States	1,110,457	14	14.6
11	Barclays Bank	United Kingdom	992,103	37	23.1
12	MTFG	Japan	980,285	37	59.8
13	Crédit Suisse	Switzerland	962,783	37	83.1
14	Sumitomo Mitsui Financial Group	Japan	896,909	14	45.8
15	ING Bank	Netherlands	828,961	34	39.1
16	ABN AMRO Bank	Netherlands	828,961	48	42.8
17	Société Générale	France	818,699	47	64.3
18	Santander Central Hispano	Spain	783,707	27	71.1
19	HBOS	United Kingdom	759,594	10	38.4
20	Groupe Caisse d'Epargne	France	740,821	–	–
21	UFJ Holdings	Japan	730,394	11	39.7
22	Dresdner Bank	Germany	710,000	–	–
23	Industrial & Commercial Bank of China	China	680,000	–	–
24	Fortis Bank	Belgium	650,000	15	13.9
25	Rabobank Group	Netherlands	640,000	22	31.8

Source: based on data in *The Banker*, July 2005; UNCTAD, 2005

institutions, set up to serve geographically localized markets, have become swallowed up (or have themselves swallowed up others). Within this long-term trend towards concentration, periods of frenetic merger and acquisition activity – often associated with significant regulatory changes – have greatly accelerated concentration and consolidation.

In 1989, no fewer than 17 of the top 25 banks, including the top seven, were Japanese, a reflection of Japan's spectacular economic growth during the 1980s. However, Japanese banks suffered very badly in the collapse of the Japanese 'bubble economy' in the early 1990s. Having lent heavily (and, with hindsight, misguidedly) during the boom years of the 1980s, many of the major Japanese banks had to make large provisions in their balance sheets to cover losses on these big loans. As a result, only four of the leading 25 banks in 2004 were Japanese (Table 13.1).

More generally, a number of very large bank mergers have occurred since the mid 1990s so that, by 1996, six of the world's biggest banks had been reduced to three through merger. Some of the biggest mergers were:

- Fuji Bank, Dai-Ichi Kangyo Bank, and the Industrial Bank of Japan merged to form the Mizuho Financial Group.
- Bank of Tokyo, Mitsubishi Bank and Mitsubishi Trust merged to form MTFG.
- Sanwa Bank and Tokai Bank merged in 2002 to form the UFJ Bank.
- Citicorp merged with Travelers Group to form Citigroup.
- Chase Manhattan Bank merged with J.P. Morgan to form J.P. Morgan Chase.
- HSBC acquired the French bank CCF.
- MTFG merged with UFJ in 2005 to form the world's largest bank.

However, not all planned mergers and acquisitions came to fruition. The most notable was the planned merger of the two major German banks, Dresdner Bank and Deutsche Bank, which was called off before completion.

While such giant mergers have reconfigured the world banking system in spectacular ways, there is considerable doubt as to their real value. According to the Federal Reserve Bank of New York, there is no evidence that in-market mergers lead to significant improvements in efficiency.[18] One potential problem is the nature of financial services products themselves:

> Mergers between car manufacturers or consumer goods companies allow production to be centralized because the products are essentially the same … The same is not true for many financial services … 'A bottle of beer is a real thing but financial products are intangible constructs of regulation, culture and behaviour. A current account is a different product in every country, while life insurance policies are tax-driven products and tax systems are not harmonised. Where will the synergies come from?'[19]

Nevertheless, the merger waves in the financial sector show no sign of disappearing.

Transnationalization of banking operations

Although banks have long engaged in international business – for example, through foreign exchange dealing or providing credit for trade – historically, this kind of business was carried out from their domestic locations. Any business that could not be carried out by mail or using telecommunications was handled by local correspondent banks; there was no need for a direct physical presence abroad. A small number of banks certainly set up a few overseas operations towards the end of the nineteenth century. But even in the early part of the twentieth century, the international banking network was very limited indeed. Almost all international banking operations were 'colonial' – part of the imperial spread of British, Dutch, French and German business activities. In 1913 the four major US banks had only six overseas branches between them. By 1920 the number of branches had grown to roughly 100 but there was little further change until the 1960s. During that period, Citibank had by far the most extensive international network of any US bank.

As with TNCs in manufacturing industries, the most spectacular expansion of transnational banking occurred in the 1960s and 1970s. In both cases, the initial surge was led by US firms, a reflection of both the focal role of the United States in the post-war international financial and trading system and also the rapid proliferation of US manufacturing TNCs. But European, and later Japanese, banks increasingly transnationalized. Figure 13.6 shows that the number of foreign affiliates of banks increased from 202 in 1960 to 1928 in 1985. Growth in this transnational network was especially rapid in the 1960s and 1970s. At the same time, the *geographical composition* of the international banking network changed. US banks became less dominant than in the 1960s and 1970s, whilst European and Japanese banks increased the size of their international branch network.

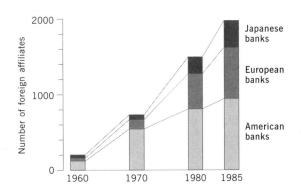

Figure 13.6   Growth in the number of overseas bank affiliates, 1960–85

Source: based on *OECD Observer*, 1989: 36

Although size and transnationality are related, it is not necessarily the biggest banks in terms of total assets that are the most transnationalized, as the right-hand column of Table 13.1 shows. Only 10 of the world's 25 largest banks have more than 50 per cent of their affiliates outside their home country, whilst only a further four of the top banks have more than 40 per cent of their affiliates abroad. Some, however, are seeking to become genuinely global banks. For example,

Citigroup plans to generate more than 50 per cent of its income from outside the United States.[20] The specific reasons for, and the pace of, the transnationalization of banks and other financial services may well vary from one case to another, but there is an overall general pattern.

- In the immediate post-1945 period, the major overseas function of the small number of US transnational banks was to provide finance for trade. But as the overseas operations of US TNCs in manufacturing and other activities accelerated in the early 1960s, the functions of the US transnational banks evolved to meet their particular demands. Thus, the staple business of US banks in London became that of servicing the financial needs of their major industrial clients who were rapidly extending their operations outside the United States.
- In the second half of the 1960s, transnational banking functions took on a significant additional dimension with the growth of the Eurodollar market. This, as we have seen, was a market outside the regulatory control of the US government. As a consequence, US banks could raise money there for relending domestically as well as overseas.
- The attractions of transnational banking widened progressively both as the global capital market evolved and as local capital markets overseas developed. As banks from the other industrialized economies also transnationalized, the process became *self-reinforcing*. All large banks had to operate transnationally; they had to have a presence in all the leading markets.
- The transnationalization of financial markets received fresh impetus in the 1970s from two sources: first, the 'windfall' capital acquired by the OPEC oil exporting countries which they sought to lend abroad; and, second, the progressive dismantling of exchange controls on capital movements by the Federal Republic of Germany, the United States, Japan and the United Kingdom.
- The major trend towards greater deregulation of national financial markets from the late 1980s and through the 1990s gave a fresh impetus to transnationalization, especially to transnational mergers and acquisitions in the banking and financial services sector in general.

Figure 13.7 summarizes the major phases of development of transnational banking. As with all such models it should be seen as a broad general framework within which the precise detail varies from case to case.

These broad developments have pulled more and more securities firms into international operations.

> Up to 1979/80, the US multinational investment bank had little more than a large office in London and perhaps some much smaller ones in other European countries, perhaps an Arab country and possibly (though less likely) Japan. From 1980 onward, the development of the US investment bank as a multinational changed qualitatively.[21]

	Phase I National banking	Phase II Transnational banking	Phase III Transnational full-service banking	Phase IV World full-service banking
Transnationalization of customer companies	Export–import	Active direct overseas investment	Transnational corporation	
Transnational operations in banking	Mainly foreign exchange operations connected with foreign trade Capital transactions are mainly short-term ones	Overseas loans and investments become important, as do medium- and longer-run capital transactions	Non-banking fringe activities such as merchant banking, leasing, consulting and others are conducted Retail banking	
Methods of transnationalization	Correspondence contracts with foreign bank	To strengthen own overseas branches and offices	By strengthening own branches and offices, capital participation, affiliation in business, establishing non-bank fringe business firms, the most profitable ways of fund-raising and lending are sought on a global basis	
Customers of transnational operations	Mainly domestic customers	Mainly domestic customers	Customers are of various nationalities	

Figure 13.7   Major phases in the development of international banking

Source: based on Fujita and Ishigaki, 1986: Table 7.6

The same process applied to the leading Japanese securities houses, such as Nomura, Daiwa, Nikko and Yamaichi, although with differences of detail. Finally, as both industrial and service companies spread globally a further boost was given to the increased internationalization of accountancy companies.

As a result, all of the transnational financial services firms have been basing their strategy on a direct presence in each of the major geographical markets and on providing a local service based on global resources. They are, in fact, selling a *global brand image,* with the clear message that a global company can cope most easily and effectively with every possible financial problem that can possibly face any customer wherever they are located. But again, as in the case of mergers, there is considerable doubt that such a 'one size fits all' strategy really works in a world that continues to be highly differentiated.

Diversification into new product markets

It is a short, and supposedly logical, step from this kind of global strategic orientation to the notion that global financial services should not only operate globally in their own core area of expertise but also supply a *complete package of related services.* The financial conglomerate or the financial supermarket has arrived, greatly stimulated by increasing deregulation. It is increasingly possible for banks to act as securities houses, for securities houses to act as banks, and for both to offer a bewildering array of financial services way beyond their original operations. A leading bank's current portfolio of offerings now includes: clearing banking, corporate finance, insurance broking, commercial lending, life assurance, mortgages, unit trusts, travellers' cheques, treasury services, credit cards, stockbroking, fund management, development capital, personal pensions, and merchant banking. At the same time, entirely new non-bank financial services companies have emerged.

Although some of this diversification has taken place via new greenfield investment, most has occurred through acquisition and merger. This has been particularly the case in the rush of the securities houses to gain a foothold in newly deregulated markets and in the creation of totally new financial services conglomerates. The rationale for diversification, both into new products and into new geographical markets, is the familiar one of economies of scale and scope. The argument for a strategy of globally diversified financial service operations, then, is that it offers a complete package of services − a 'one-stop shop' − to customers.

Their supply by a large, globally recognized, brand-name firm − backed up, of course, by lavish advertising expenditure − is supposed to give a reassurance to potential customers that they will receive the highest-quality service. Whether or not economies of scale and scope really do exist to a significant extent is a matter over which there is considerable disagreement. The large financial companies themselves certainly seem to think so. Although there have been divestments as some financial services firms have shed parts of their diversified portfolio of activities, the diversification and transnationalization trend continues. On the other hand, there are distinct benefits in geographical specialization:[22]

- The administrative benefits of size can be exaggerated.
- Local knowledge may be a major advantage in assessing local risks and opportunities.
- The benefits of recognition through size may be less in national markets where names have already established themselves.

In other words, 'localization might be a marketing advantage'. This, of course, has been recognized, at least in principle and in their advertising blurbs, by the major financial services companies in their emphasis on providing a locally sensitive service within a global organization. HSBC, for example, ran a series of advertisements showing examples of different national meanings given to the same behavioural symbol under the heading: 'Never underestimate the importance of local knowledge.'

In fact, there is considerable variation in the attitudes of TNCs towards the banks they use. A survey of more than 2000 foreign affiliates of TNCs operating in 20 European countries found that two-thirds chose local (host country) banks for cash management services rather than home country or global banks.

> That is, affiliates of [trans]national corporations often use banks that know the local market, culture, language, and regulatory conditions, rather than banks that are more familiar with the conditions in the corporation's home market, or have direct ties to the corporation in the home nation.[23]

The major exceptions to this were TNCs operating in the transitional economies of Eastern Europe, presumably because of their less developed financial systems.

Outsourcing

Like all business firms, financial services companies face the internalization/externalization decision: which functions should they perform for themselves, and which should they procure from outside suppliers? Given that the basic functions of financial services companies are to do with collecting, interpreting and manipulating *information*, then it is essentially certain kinds of information function that tend to be outsourced. Traditionally, these have been the more routine data processing functions, such as the processing of cheques, credit card transactions, insurance claims and the like. More recently, some of the customer-related functions have also become increasingly outsourced with the development of *call centres*.

Until fairly recently, the outsourcing of financial services functions was geographically quite limited. However, as in the case of manufacturing – though rather later in the day – financial services firms have begun to outsource transnationally on an increasing scale. Indeed, when 'outsourcing' is discussed today it almost always means 'offshoring': the relocation of functions to other countries. We will look at this in more detail later in this chapter.

Geographical structures of financial services activities

At first sight, IT developments would appear to release financial services companies from geographical constraints. Such firms appear to be especially foot-loose: they are not tied to specific raw material locations, whilst at least some of their transactions can be carried out over vast geographical distances, using telecommunications facilities. Such considerations have led many to write geography and distance out of the script as far as financial services are concerned.[24]

There is no doubt that the revolutionary developments in the time–space shrinking technologies permit financial transactions to whizz around the world electronically, and that deregulation has reduced the resistance of national boundaries to financial flows. But, far from heralding the 'end of geography', this has in fact made geography *more* – not less – important. Indeed, we find that, at global, national and local scales, the major financial services activities continue to be extremely *strongly concentrated geographically*. They are, in fact, more highly concentrated than virtually any other kind of economic activity except those based on highly localized raw materials. However, there are some subtle variations according to the particular function involved: there is a division of labour within financial services firms, parts of which may show a greater degree of geographical decentralization. In this section, two aspects of this issue are explored: first, the relationship between geography and the nature of financial products; and, second, the geographical structure of financial services.

Geography and the nature of financial products

The informational content of financial products themselves is inextricably related to the geographical structure of the global finance industry.[25]

> The geography of places shapes the operation of finance markets and the global economy primarily because of the geography of information that is embedded in the provision of specific financial products. There is, in effect, a robust territoriality to the global financial industry ... financial products often have a distinct spatial configuration of information embedded in their design.[26]

Financial product type	Probable market scope and type of financial centre	Information intensity	Specialist expertise required	Perceived risk-adjusted return
Transparent	Global	Ubiquitous	Low, e.g. equity products like stock market indices. Traded internationally	Low
Translucent	National	Third party Market specific	Significant. Require detailed knowledge of national/local companies. May be traded internationally	Medium
Opaque	Local	Transaction specific	Vital. Applies to products based upon trust and long-term relationships	High

Figure 13.8   Financial products and their geographical characteristics

Source: based on Clark and O'Connor, 1997: Table 4.1, pp. 99–104

Figure 13.8 identifies three types of financial product, each of which has distinctive, geographically derived, properties. The specific informational nature of these three types of financial product, and the fact that such information is locationally specific, are reflected in a particular 'topography' of financial markets:

The global network of financial centres

Financial services activities of all kinds tend to be very strongly concentrated in key metropolitan centres. These form a complex network spanning national boundaries and connecting major cities across the world (see Figure 2.27). Figure 13.9 shows one representation of this global network of financial centres, based on such indicators as:

- the volume of international currency clearings
- the size of the Eurocurrency market
- the volume of foreign financial asset
- the number of headquarters of the large international banks
- the degree of connectivity of financial centres.

This network of global financial centres has three major levels, with London and New York being in the top level. In this scheme, Tokyo appears in the second tier.

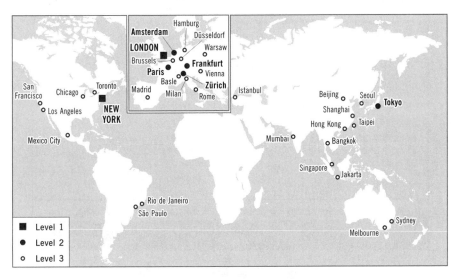

Figure 13.9 The global network of financial centres

Source: based on Reid, 1989: Figure 1; data in Taylor, 2004

Although this global network has certain stable features – for example, the long-sustained dominance of London, New York and, to a much more limited extent, Tokyo – it is by no means static. Financial centres may change their status in the network in response to changing conditions. Much may also depend on the measure used.[27] Figure 13.10 illustrates both of these circumstances for the top 12 banking centres. Measured in terms of aggregate income (Figure 13.10a), the effect of the Japanese financial crisis is abundantly clear. However, Tokyo's position in terms of assets (Figure 13.10b) has remained stable. Figure 13.10 also shows the emergence of several new centres, notably Beijing and Charlotte, North Carolina. Clearly, what we have is a global network of financial centres that is highly dynamic whilst, at the same time, containing elements of relative stability. Such stability reflects the operation of the path dependent forces outlined in Chapter 1.[28]

The widespread deregulation of financial services during the 1980s and 1990s had an especially significant impact on the structure of the global network of financial centres. Such developments

> made it possible for footloose financial firms to set up directly in the world's leading metropolitan centres rather than serving them from a distance ... International firms want to locate in these different metropolitan financial centres because each has a different financial specialization, and each is the hub of a different (although of course overlapping and inter-connected) continental global region. Foreign banks and related institutions have moved into these

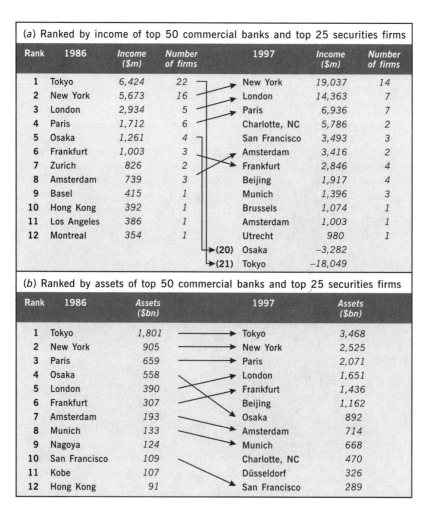

Figure 13.10   The changing composition of the leading banking centres

Source: based on Sassen, 2001: Table 7.5

centres precisely because of geography, that is to expand their presence or gain access to specific markets, to capitalise on the economies of specialization, agglomeration and localization (skilled labour, expertise, contact, business networks, etc.) available in these centres, or to specialize their own operations and activities geographically.[29]

The locational attractions of international financial centres consist of four interlocking processes:[30]

- The characteristics of the *business organizations* involved in international financial centres: (a) much of the production of financial products and services occurs at the boundaries of firms and there is a strong reliance on repeat business; (b) the firms tend to be 'flattened and non-hierarchical' and based around 'small teams of relationship and product specialists'; (c) firms need to cooperate as well as to compete, as in the case of syndicated lending; (d) firms need to compare themselves with one another to judge their performance; (e) there is a need for a constant search for new business and for rapid response.

> These shared characteristics point to two important correlates. First, in general, these firms must be sociable. Contacts are crucially important in generating and maintaining a flow of business and information about business. 'Who you know' is, in this sense, part of what you know ... 'relationship management' is a vital task for both employees and firms. Second, this hunger for contacts is easier to satisfy if contacts are concentrated, are proximate. When contacts are bunched together they are easier to gain access to, and swift access at that.[31]

- The *diversity of markets* in international financial centres: (a) their large size which makes them both flexible in terms of entry and exit and also socially differentiated: 'more likely to consist of social "micro-networks" of buyers and sellers, whose effect on price-setting can sometimes be marked'; (b) their basis in rapid dissemination of information which may lead to major market movements; (c) their speculative and highly volatile nature. 'Again, as with the case of organizations, there are the two obvious corollaries to these characteristics: the twin needs for sociability and proximity'.[32]

- The *culture* of international financial centres: (a) such centres receive, send and interpret increasing amounts of information; (b) they are the focus of increasing amounts of expertise which arise from a complex division of labour involving workforce skills and machinery; (c) they depend on contacts and such contacts have become increasingly reflexive because of their basis in trust founded on relationships. Such cultural aspects of international financial centres are actually increasing in importance.

- The *dynamic external economies of scale* which arise from the sheer size and concentration of financial and related services firms in such centres. Such economies include: (a) the sharing of the fixed costs of operating financial markets (for example, settlement systems, document transport systems) between a large number of firms; (b) the attraction of greater information turnover and liquidity; (c) the enhanced probability of product innovations in such clusters (the 'sparking of mind against mind'); (d) the increased probability of making contacts which rises with the number of possible contacts; (e) the attraction of linked services such as accounting, legal and computer services, which reduces the cost to the firm of acquiring such services; (f) the development of a pool of skilled labour; (g) the enhanced reputation of a centre which, in a cumulative way, increases that reputation and attracts new firms. In other words, the constellation of traded and untraded interdependencies, described in Chapter 1, tends to be especially strongly developed in international financial centres.

In effect, therefore, a small number of cities control almost all the world's financial transactions. It is a remarkable level of geographical concentration. But such cities are more than just financial centres. There is clearly a close relationship with the distribution of the corporate and regional headquarters of transnational corporations. These global cities are, indeed, the *control points of the global economic system*. As far as financial services firms are concerned, the global financial centres – especially those at or near the top of the hierarchy – are their 'natural habitat'. Financial companies agglomerate together in these centres for the reasons outlined above.

But this does not mean that all global financial centres – even those at the top of the hierarchy – are identical. Far from it. Each has distinctive characteristics reflecting its specific history and geography. On criteria measuring both the breadth and depth of global financial activity, New York and London stand at the apex of the global financial hierarchy, significantly more important globally than Tokyo. London is the more broadly based international financial centre.

> London's capacity as an international financial centre rests on three core activities: foreign exchange, international equities and exchange-traded derivatives ... London has increased its share of international foreign exchange in the last five years. It remains the largest international centre.[33]

The daily turnover on the London foreign exchange market is almost as large as the turnover in New York and Tokyo put together. London's current significance as a global financial centre can be attributed to the following factors:[34]

- The historical evolution of the City as a world centre has created both a large pool of relevant skills and also an almost unparalleled concentration of linked institutions within a very small geographical area.
- Its geographical position locates it in a time zone between Asia and New York.
- The regulatory environment has encouraged the growth of transnational banking. Foreign banks in London can operate as 'universal' banks, a particularly important feature for US and Japanese banks. They can combine both banking and securities businesses there in ways that are prohibited so far in their domestic operations.
- Its key role as a 'capital switching centre': 'The London market brings together in one place great diversity of market participants and, consequently, great diversity of risk preferences and profiles'.[35]

But, given the recent development of the Eurozone, and the fact that the UK remains outside, how secure is London's position? So far, the evidence is that London's status as a global financial centre has not been adversely affected and that its standing as the leading European financial centre is still not seriously threatened. London has the largest concentration of foreign banks in the world, with more than 470 foreign branches, subsidiaries and representative offices located there in 2001. In comparison, Paris had 214 and Frankfurt 320. In 2001,

London had 20 per cent of all cross-border bank lending worldwide, 52 per cent of foreign equity trading, and 31 per cent of all foreign exchange dealing: as much as its three closest rivals (New York, Tokyo, and Singapore) put together.[36]

In an attempt to capitalize on the rapid growth of Islamic financial activities across the world, the current UK government is aiming to establish London as:

a global centre for Islamic finance by offering regulatory and tax-regime measures to support the creation of products targeted at devout Muslims ... this market ... [is] ... estimated to be worth $300bn-$400bn globally ... The defining feature of the Islamic finance industry is that it offers products without interest payment ... The business initially emerged in the Middle East and Asia a couple of decades ago, when local banks started to offer Islamic-compliant consumer financial products. However, in recent years it has spread to the UK as well, with some institutions offering consumer Islamic products to Muslims and others using London to develop Islamic investment banking products for global distribution. London is widely perceived to be the main centre for Islamic investment banking outside the Muslim world.[37]

London's strength as a financial centre, therefore, rests on the scale of its foreign exchange business, its deregulated securities markets, and its wide diversity of financial activities. New York, in comparison, is the world's largest securities market in addition to having a huge concentration of transnational banks and other financial activities. London and New York stand apart from Tokyo as truly global financial centres. The international significance of Tokyo has rested primarily on the strength of the Japanese economy itself. Tokyo's financial community has been far more domestically oriented. Japanese banks, as we saw earlier (Figure 13.6), were later entrants to international banking. The ongoing financial problems of the Japanese economy have not helped Tokyo's position, although there are clear signs of recovery.

Geographical decentralization and offshoring of business functions

All of the discussion of global financial centres suggests that the potential for other cities, outside the favoured few shown in Figure 13.9, to develop as significant centres of finance and related activities is very limited. In the United Kingdom, for example, the sheer overwhelming dominance of London makes it extremely difficult for provincial cities to develop more than a very restricted financial function.

It is, of course, the 'higher-order' financial and service functions which are especially heavily concentrated in the major global financial centres. So-called 'front-office' functions, by definition, must be close to the customer: hence the huge branch networks of the retail banks and other financial services supplying final demand. However, the advent of Internet and telephone banking is allowing retail banks to rethink their locational strategies in this regard.

As we have seen, the essence of financial services activities is the transformation of massive volumes of *information*. Much of that activity is routine data processing performed by clerical workers. Such 'back-office' activity can be separated from the front-office functions and performed in different locations. The early adoption of large-scale computing by banks, insurance companies and the like from the late 1950s initially led many of them to set up huge *centralized* data processing units. To escape the high costs (both land and labour) in the major financial centres, such units were often relocated to less expensive centres or in the suburbs. Access to large pools of appropriate (often female) labour was a key requirement.

The introduction of dispersed computer networks made such centralized processing units unnecessary and the tendency in recent years has been to *decentralize* back-office functions more widely. At the same time, the distinction between back-office and front-office functions has become less clear. In fact, it is not only routine back-office activities that have been decentralized. It has become increasingly common for some of the higher-skilled functions to be relocated away from head office into dispersed locations, both nationally and, in some cases, transnationally. In fact, centralization of back-office functions is by no means obsolete. For example, Citicorp brought back all its back offices from around the world and recentralized them in large and more efficient centres to achieve economies of scale.

> Across Europe, banks have been making use of new technology to overhaul back-office operations by bringing processing out of their branches and into centralised units. For example, Lloyds TSB, the UK bank, announced an efficiency drive ... which involved the loss of 3,000 jobs, stemming mainly from the centralisation of computer operations and consolidation of large-scale processing operations, including the complete removal of all back-office processing from branches ... ABN-Amro bank, the Dutch bank, said it was in negotiation with at least three European banks over a possible alliance of their back offices.[38]

Even in the case of back offices, therefore, geographical centralization is far from dead.

The terrorist attack on lower Manhattan on 11 September 2001 inevitably raised the question of whether very high levels of geographical concentration of higher-level financial activities are prudent. After all, similar concentrations exist in all the other prominent financial centres. Some degree of relocation of financial services had been happening already in New York for some years, as some firms relocated within the city or outwards to Connecticut or New Jersey. However, there are major perceived barriers to major relocation of key financial functions from the leading centres, as the following statements of leading bankers imply:

> If you disperse it you lose that level of ingenuity and intellectual talent. We are committed to Manhattan. We are committed to Wall Street. But we will have facilities outside Manhattan.[39]

> At the end of the day, global firms need to have a presence in lower Manhattan. Prestige is still important.[40]

Beyond the continued tendency to retain their major presence in key financial centres, the major financial services firms have become increasingly involved in relocating some of their functions overseas, that is *offshoring*. (Note that the term 'offshoring' in this context does not have the same meaning as the 'offshore financial centres' discussed in the next section.)

> Last year [2003], approximately 25 per cent of financial services companies utilized offshoring to some degree; today [2004], more than two thirds do ... The emerging global operating model for financial services companies is unleashing a new competitive dynamic that is changing the rules of the game for everyone ...
>
> As the pace of offshoring accelerates, companies are broadening the functions and processes they are relocating. Financial institutions are steadily increasing operations conducted offshore to include more back-office and customer-facing services.[41]

Deutsche Bank, for example, expects to have relocated more than half of all its back-office jobs in its sales and trading activities to India by the end of 2007.[42]

	Vendor direct	**Captive direct**	**Vendor indirect**
	Place contract with specialist firm in specific country	Set up own directly-owned operation(s) overseas	Place contract with specialist firm with operations in several countries
Potential benefits	Cost reduction Use of specialist expertise Speed	Increased control Reduced risk Greater security	Lower costs Use of specialist expertise Vendor reputation
Potential costs	Loss of control Security risks	High cost of establishment and control	Control issues Security risks

Figure 13.11 Alternative modes of offshoring

Source: based, in part, on material in the *Financial Times* (18 February 2004)

As the extent of offshoring has increased, its particular organizational form has become more varied, as Figure 13.11 shows. Initially, most offshoring by financial services firms was *vendor direct outsourcing* to a foreign firm located overseas. By far the most popular location was India, where IT-related outsourcing has grown at a phenomenal rate. As a result, the region around Bangalore has become one of the biggest geographical concentrations of IT-related activity. As major financial services firms became increasingly involved in offshoring they began to establish systems of *captive direct offshoring*: setting up their own subsidiary operations in other countries. The third and most recent arrangement, *vendor indirect offshoring*, reflects the tendency for specialist outsourcing companies to establish their own transnational networks to serve more diverse customers. So, for example, one of the leading Indian IT outsourcing companies, Tata Consultancy Services (TCS), has recently established its own operations in Hungary, as part of its increasingly global network.

> TCS first opened a software centre in Hungary in 2001 … it is now building a *global delivery network* so that projects can be completed closer to the customer where necessary … TCS chose Hungary because, as well as English, many Hungarians can speak other European languages, including German, French, and Italian. 'We want to provide a European front and [point of contact] for our European customers' … TCS preferred eastern Europe over western Europe because it was much cheaper … Eastern Europe is where India was a decade ago.[43]

Offshore financial centres

> Scattered across the globe, a series of little places – islands and micro-states – have been transformed by exploiting niches in the circuits of fictitious capital. These places have set themselves up as offshore financial centres; as places where the circuits of fictitious capital meet the circuits of 'furtive money' in a murky concoction of risk and opportunity. Furtive money is 'hot' money that seeks to avoid regulatory attention and taxes.[44]

The speculative nature of financial transactions and flows and the desire to evade regulatory systems have led to the development of a number of offshore financial centres (OFCs).[45] With few exceptions, the sole rationale for such centres is to operate outside the regulatory reach of national jurisdictions. They attract investors through their low tax levels and 'light' regulatory regimes.[46] For example, although more than 500 banks are 'located' in the Cayman Islands – with more than $500 billion in their accounts – only around 70 of these actually have a physical presence there. The vast majority are no more than 'a brass or plastic name plate in the lobby of another bank, as a folder in a filing cabinet or an entry in a computer system'.[47] Similarly, there are roughly 300,000 companies registered in the Virgin Islands, although 'only 9,000 of them show any signs of activity locally'.[48]

Figure 13.12 shows the geographical distribution of offshore financial centres. Each tends to fill a specific niche which it exploits in competition with other centres in the same geographical cluster and with similar niche centres elsewhere in the world. Much of their growth occurred in the 1970s, in places that were already operating as tax havens, to act as banks' 'booking centres' for their Eurocurrency transactions.

> By operating offshore booking centres international banks could act free of reserve requirements and other regulations. Offshore branches could also be used as profit centres (from which profits may be repatriated at the most suitable moment for tax minimization) and as bases from which to serve the needs of multinational corporate clients.[49]

The location of these offshore centres, and especially their geographical clustering, is partly determined by time zones and the need for 24-hour financial trading.

The OECD has initiated a drive to make OFCs more accountable by enforcing greater disclosure and ending preferential treatment for foreigners. Not

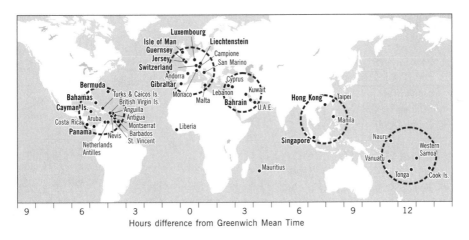

Figure 13.12   Offshore financial centres

Source: based on Roberts, 1994: Figure 5.1

surprisingly, the 35 OFCs identified by the OECD resist tighter regulation. The problem is exacerbated by the fact that many OFCs are located in small, poor countries. The pressure to regulate OFCs intensified after 9/11, as part of US-led efforts to improve anti-money-laundering controls.

Conclusion

The financial services industries are central to the operation of all aspects of the economy. They are inherently embedded within all production networks and are essential to production and trade. At the same time, financial services firms generate 'products' in their own right, including those on which speculative profits can be made. Much of the rapid growth in financial services has been facilitated by the progressive deregulation of financial markets across the world, by innovations in information technology, and by the ingenuity of firms in creating novel financial products, such as derivatives. As a consequence, the world financial system has become increasingly volatile and unpredictable, with shocks originating in one part of the world spreading to other parts of the world at exceptional speed through the processes of 'financial contagion'.

A particularly significant feature of these industries is the fact that, although potentially highly flexible geographically, they continue to display an extremely high level of geographical concentration. In particular, the global financial system is articulated through a selective network of major cities. Without question, the 'end of geography' has not happened – even in such a 'footloose' activity as financial services, whose structure is geographically extremely complex.

NOTES

1 Martin (1999: 6, 11).
2 See Braithwaite and Drahos (2000) for a discussion of the history of money and its regulation.
3 Strange (1986: 1).
4 Pauly (2000: 120).
5 Dow (1999: 33).
6 French and Leyshon (2004).
7 Warf and Purcell (2001: 227).
8 *Financial Times* (16 March 1994).
9 *The Economist* (26 October 1996).
10 Strange (1998: 24).
11 See Langdale (2000).
12 Langdale (2000: 94).
13 Kelly (1995: 229). See also Strange (1998: Chapter 2).
14 Martin (1999: 8, 9).
15 This discussion of regulation and deregulation draws on Braithwaite and Drahos (2000). Eatwell and Taylor (2000).
16 Lewis and Davis (1987: 225, 228).
17 *The Economist* (2005a: 4)
18 *Financial Times* (9 February 2001).
19 *Financial Times* (26 May 2000).
20 *Financial Times* (17 April 2006).
21 Scott-Quinn (1990: 281).
22 Davis and Smailes (1989).
23 Berger et al. (2003: 412).
24 O'Brien's (1992) book on financial services, subtitled *The End of Geography,* is widely cited in this respect, although his position was more subtle than his critics often allowed.
25 Clark and O'Connor (1997).
26 Clark and O'Connor (1997: 90, 95).
27 See the examples given in Sassen (2001).
28 See Porteous (1999) for a discussion of this process in the development of financial centres.
29 Martin (1999: 19–20).
30 Thrift (1994).
31 Thrift (1994: 333).
32 Thrift (1994: 334).
33 Budd (1999: 126, 127).
34 Lewis and Davis (1987).
35 Clark (2002: 448).
36 *Financial Times* (8 February 2002).
37 *Financial Times* (13 June 2003).
38 *Financial Times* (26 May 2000).

39 Quoted in the *Financial Times* (21 September 2001).

40 Quoted in the *Financial Times* (14 September 2001).

41 Lowes et al. (2004: 24).

42 *Financial Times* (27 March 2006).

43 *Financial Times* (3 August 2005, emphasis added).

44 Roberts (1994: 92).

45 See Hudson (2000), Palan (2003), Roberts (1994).

46 Eatwell and Taylor (2000: 189).

47 Roberts (1994: 92).

48 *Financial Times* (7 December 2001).

49 Roberts (1994: 99).

Fourteen
'Making the Connections, Moving the Goods': The Logistics and Distribution Industries

'Whatever happened to distribution in the globalization debate?'

Good question.[1] Whereas there is a huge literature and a continuous, often frenzied, debate on the role of financial services in the processes of globalization, distribution rarely makes an appearance on the stage. It remains hidden, mainly confined to the specialist fields of supply chain management and transportation. The distribution processes get taken for granted. It is more or less assumed that, as transportation and communication systems have allegedly shrunk geographical distance, the problems of getting products from points of production to points of consumption have been solved.

This is far from being the case. Indeed, the *circulation processes* that *connect* together all the different components of the production network are absolutely fundamental. They have become central to the abilities of transnational companies to operate their complex, geographically dispersed operations. The logistics industries themselves are huge, worth around $4 billion per year. They have become especially significant in the light of the broader forces of change discussed in earlier chapters, notably:

- new production methods, involving increased flexibility
- changing relationships between customers and suppliers
- increasing use of just-in-time procurement and delivery systems
- increasing geographical complexity and extent of production networks
- changing consumer preferences.

In particular, *time* has come to be seen as the essential basis of successful competition.[2] In such a context, the nature and efficiency of distribution systems become central.

Time- and quality-based competition depends on eliminating waste in the form of time, effort, defective units, and inventory in manufacturing-distribution systems ... [requiring] firms to practice such logistical strategies as just-in-time management, lean logistics, vendor-managed inventory, direct delivery, and outsourcing of logistics services so that they become more flexible and fast, to better satisfy customer requirements.[3]

The structure of the logistics and distribution industries

The essential function of the distribution services is to *intermediate* between buyers and sellers at all stages of the production circuit (Figure 14.1). This involves not only the *physical movement* of materials and goods but also the transmission and manipulation of *information* relating to such movements. They involve, above all, the organization and coordination of complex flows across increasingly extended geographical distances. In that respect, these services have been revolutionized by the kinds of technological developments in transportation and communication discussed in Chapter 3. They have also been transformed by the increased outsourcing of logistics and distribution services by manufacturing firms, by the intensifying pressures from the big retailers, and by the emergence of new forms of logistics service providers.

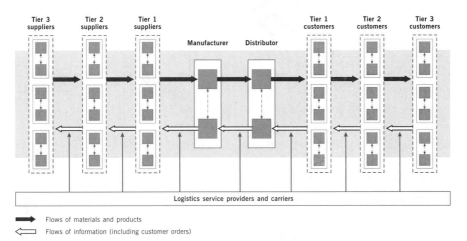

Figure 14.1   Logistics and distribution in the production circuit

Source: based, in part, on Schary and Skjøtt-Larsen, 2001: Figure 1.6.

In Figure 14.1 there are no political, or other, boundaries or obstacles to complicate the basic system. In reality, of course, their existence greatly affects the structure and operation of logistics and distribution processes. Two kinds of 'barrier' to movement are especially significant:

- Physical conditions that necessitate the transfer from one transportation mode to another – for example, land/water interfaces.
- Political boundaries that create complications of customs clearance, tariffs, duties, administration and the like. Such barriers have become increasingly significant as economic activity has become more globalized.

Hence, as Figure 14.2 shows, there are many stages in the process of getting products to their final market.

> A typical door-to-door journey for containerised international shipments involves the interaction of approximately 25 different stakeholders, generates 30–40 documents, uses two to three different transport modes and is handled in 12–15 physical locations.[4]

However, the more extended the production circuit becomes – both organizationally and geographically – the greater the potential problems.

> Companies deciding to source in China rather than producing in-house may think they are adding a single link to their supply chain ... In fact, they are probably adding at least five: the production agent in China, a logistics company in China, China customs, the freight shipper, customs and transport in the domestic market.[5]

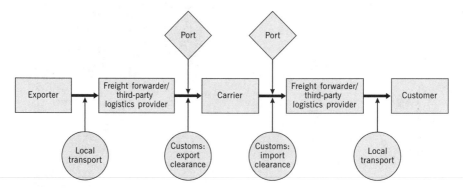

Figure 14.2   Logistics processes in a transnational context

Source: adapted from Schary and Skjøtt-Larsen, 2001: Figure 11.7

The logistics and distribution processes shown in Figures 14.1 and 14.2 can be performed in a variety of ways and by a variety of organizational forms. At one extreme, each individual transaction may be performed by a separate firm; at the other extreme, the entire process may be carried out by a single integrated firm or related group of firms. The major types of organization involved in logistics and distribution include:

- transportation companies
- logistics service providers

- wholesalers
- trading companies
- retailers
- e-tailers.

Transportation companies (rail, road, shipping, airlines), wholesalers and retailers perform fairly clearly defined and restricted roles in the production circuit. On the other hand, trading companies and the more recent logistics service providers perform a far broader range of activities. Of course, the categories are not clear-cut. Not only are the boundaries between them often blurred, but also one form of organization may mutate into another, as in the case of some transportation and trading companies that have evolved into more comprehensive logistics service providers (see later in this chapter). At the same time, the significance of some types of intermediary has changed.

For example, traditionally the wholesaler played a major role in collecting materials or products from a range of individual producers and then distributing them to the next stage of the production process, or to the retailer in the case of final demand. However, the importance of the wholesaler as the key intermediary has changed substantially as the major retailers have bypassed the wholesaler to deal directly with the manufacturer, or as other forms of logistics and distribution services have developed. In a similar way, the development of e-commerce makes it possible to bypass the traditional retailer as the key intermediary between producer and final consumer and to create a new type of retailer: the *e-tailer*. Figure 14.3 shows just one possible way in which the production/supply circuit may be organized internationally, although great diversity remains.

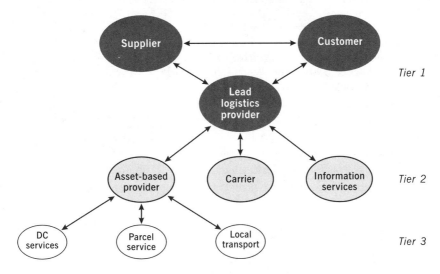

Figure 14.3   A potential way of organizing logistics services

Source: adapted from Schary and Skjøtt-Larsen, 2001: Figure 7.4

The dynamics of the market

In aggregate terms, the growth of the market for logistics and distribution services is closely related to growth in the economy as a whole. But the market for logistics and distribution services is highly heterogeneous and we would not expect all parts to grow at similar rates. Demand for certain kinds of distribution services has been growing more rapidly than that for others, driven by the different characteristics of individual services.

Ultimately, as Figure 14.1 shows, the system is driven by the demands of the final consumer, although this influence becomes increasingly indirect the further back up the production circuit we go. Although the primary driver of final consumer demand is the level of disposable income, consumption, as we have discussed at various points in this book, is an immensely complex socio-cultural process. The intensely competitive retail markets have major repercussions on the demand for distribution and logistics services further up the production circuit. As we saw in the case of the clothing and agro-food industries, the major retailers and buyers exert very heavy pressure on their suppliers to deliver more rapidly, more cheaply, and in greater variety. This in turn creates opportunities (and challenges) for the suppliers of logistics and distribution services to provide a faster and more integrated supply system between the different components of the production circuit.

The enhanced power of the major retailers greatly affects the logistics firms responsible for getting the products to the retail stores.

> A shifting power structure in the retail trade not only changes market shares but also the structure of the distribution network. The major international retail customers ask for customized logistics solutions across borders. Apart from negotiating frame [i.e. bulk] orders with significant price advantages with suppliers, the most powerful retailers also require information sharing services, such as electronic data interchange, advance shipping notices via the Internet and track and trace capabilities. They typically prefer delivery to their own distribution centers where goods are consolidated with other products for delivery to their retail stores.[6]

Such demands intensify the search for new logistical systems, in terms of both their technology and their organization. In such a context, the role of electronic systems has become increasingly important. Hence, there is a link between the changing demand pressures on the suppliers of logistics and distribution services and changing technologies. Let's now look at these technologies.

Technological innovation and the logistics and distribution industries

Three criteria dominate the logistics and distribution industries: speed, flexibility and reliability. Technological innovations have transformed each of them. At a general level, technological developments in transportation and communication have been immensely important in transforming the basic time–space infrastructure of the logistics and distribution industries. Likewise, these industries have also been revolutionized by the shift in process technologies from mass production to more flexible and customized production systems. The mass production systems of the late nineteenth and first two-thirds of the twentieth century were facilitated by mass distribution systems, based on the rapidly developing rail, road and ocean shipping networks (see Chapter 3).

Such mass distribution systems depended heavily on large warehouses which were used to store components and products and from which deliveries were made to customers on an infrequent basis. This was an immensely expensive system in terms of the capital tied up in large inventories. It was also, very often, a source of waste in terms of faulty products that were not discovered until they were actually used. Together with a major shift towards lean, just-in-time, systems of production there has also been a parallel shift towards *lean systems of distribution*, whose purpose is to minimize the time and cost involved in moving products between suppliers and customers, including the holding of inventory.

Three key elements form the core of such lean distribution systems:[7]

- *Electronic data interchange (EDI)*: this enables the rapid transmission of large quantities of data electronically (rather than using paper documents). Such data can encompass all aspects of the logistics and distribution system throughout the production circuit, including the retailer. Information on product specifications, purchase orders, invoices, status of the transaction, location of the shipments, delivery schedules and so on can be exchanged instantly. EDI requires a common software platform to enable data to be read by all participants in the chain.
- *Bar code systems and radio frequency identification technology*: bar codes were first developed in the 1970s by grocery manufacturers and food chain stores to enable each item to be given a unique, electronically readable identity. They have now become ubiquitous throughout the production circuit. 'Bar codes permit organizations to handle effectively the kind of vast product differentiation that would have been prohibitively expensive in an earlier era. They also facilitate instantaneous information at the point of sale, with significant effects on inventory management and logistics'.[8] Now, radio frequency identification (RFID) technology is beginning to be applied to increase the sophistication and flexibility of the bar code principle. Bar codes can be difficult to read (they require a clear line of sight). RFID gets around this by using small radio tags that allow goods to be instantaneously tracked throughout their journey through a production circuit. By combining the tag with a unique electronic product code (EPC) it becomes possible to incorporate a large quantity of information about each object. One observer (admittedly with a vested interest in RFID) described it as 'a bar code on steroids'.[9]
- *Distribution centres*: modern distribution centres hold inventory for much shorter periods of time and turn it over very frequently. 'Four technologies have made the modern distribution centre possible: (1) bar codes and associated software

systems; (2) high-speed conveyors with advanced routing and switching controls; (3) increased reliability and accuracy of laser scanning of incoming containers; and (4) increased computing capacities.'[10] In the most advanced distribution centres - such as the system used by the US retailer Wal-Mart – a method known as 'cross-docking' is used. This is a 'largely invisible logistics technique ... [in which] goods are continuously delivered to Wal-Mart's warehouses, where they are selected, repacked, and then dispatched to stores, often without ever sitting in inventory. Instead of spending valuable time in the warehouse, goods just cross from one loading dock to another in 48 hours or less. Cross-docking enables Wal-Mart to achieve economies that come with purchasing full truckloads of goods while avoiding the usual inventory and handling costs. Wal-Mart runs a full 85% of goods through its warehouse system – as opposed to only 50% for Kmart.'[11] Of course, few firms can afford the scale of investment required to implement such a sophisticated system.

E-commerce: a new logistics revolution

Computer-based electronic information systems are at the heart of all these tech-nological developments in logistics and distribution systems. They have evolved over a period of 30 years, often incrementally rather than as a spectacular 'revolution'. However, the later 1990s brought into existence a new set of distribution methods based upon the Internet: *e-commerce*. In essence, e-commerce has developed out of the convergence of several technological strands: electronic data interchange, the Internet, e-mail, and the medium of the World Wide Web (WWW).[12] Few devel-opments in recent years have been so heavily hyped as the e-commerce revolution but there is no doubt that the emergence of the Internet has huge implications for logistics and distribution systems. The main reason, once again, is *speed*.

Although four types of e-commerce are shown in Figure 14.4, two dominate: B2B and B2C. The majority of e-commerce is B2B, although the balance between the two is changing.

- *B2B (business to business)*: this encompasses potentially the whole range of trans-actions between businesses, notably procurement of products and services and logistics. B2B websites are electronic 'marketplaces' where firms come together to buy or sell products and services. They may be 'vertical', that is, industry specific (for example, the B2B procurement system Covisint established in 2000 by GM and Ford, together with some other automobile manufacturers, to increase the efficiency of component purchasing), or they may be 'horizontal', organized around the products and services provided rather than the industry. Connecting together large numbers of buyers and sellers through electronically automated transactions has a number of potential benefits, notably vastly increas-ing choice to both sellers and buyers, saving costs on transactions, and increas-ing the transparency of the entire supply chain.

Types of e-commerce

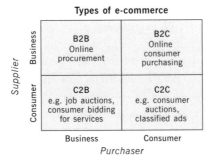

The major components of e-commerce organization

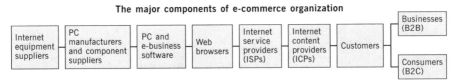

Figure 14.4　**Types of e-commerce**

Source: based, in part, on Gereffi, 2001: Figure 2

- *B2C (business to consumer)*: B2C business is simply the selling of consumer products and services directly over the Internet by a web-based firm. The most often-quoted examples are Amazon, eBay and Dell but, of course, the dotcom revolution has created many thousands of 'e-tailers', some with a very brief life. E-tailing has also been adopted by the traditional retailers. Indeed, contrary to predictions that traditional retailers would be adversely affected by Internet shopping, the opposite has occurred.

> Far from wrecking retailers' businesses, the web plays to their strengths. Shopping-comparison sites, including America's Shopzilla and Ciao in Europe, are among the fastest-growing destinations on the web. These sites allow users to compare products, read reviews – and most important of all – see who is offering the lowest prices. They make money from advertising or charging retailers when users click on a link to the retailer's website.[13]

As in the case of B2B, the potential benefits of B2C transactions are, for the consumer, greater choice, ease of comparison of prices, and instant (or very fast) delivery to the home; and, for the seller, direct access to a massive potential market without the need for physical space in the form of retail outlets and the associated inventory and staffing costs. As a result, online shopping is now growing extremely rapidly (by more than 35 per cent in 2005 in the UK).[14]

E-commerce (both B2B and B2C) has enormous potential implications for the traditional intermediaries of business and retail transactions (for example, wholesalers, shippers, travel agents and the like). The early predictions were that many would disappear as their functions were displaced by direct online transactions. In fact, this has not happened to the extent predicted. On the contrary, e-commerce has actually enhanced opportunities for such roles. Some traditional intermediaries have adapted and found new ways of adding value as providers of logistics, information and financial services; new intermediaries have emerged.[15] Some of these are what are sometimes termed *infomediaries*, notably the Internet service and content providers shown in Figure 14.4. To some extent, these new infomediaries are increasing their relative bargaining power *vis-à-vis* producers and buyers. More generally, Figure 14.5 shows that in the case of both physical goods and electronic goods and services, either old intermediaries transform themselves or new ones appear.

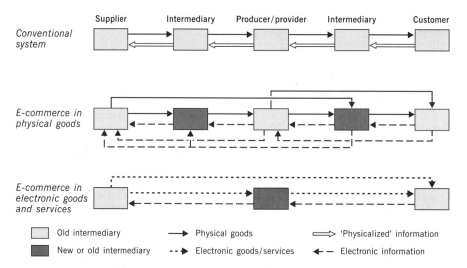

Figure 14.5   The continuing role of intermediaries in an e-commerce world

Source: based on Kenney and Curry, 2001: Figure 3.2

Related to the perception that e-commerce spells the end for the traditional intermediary organization is the idea that the traditional physical infrastructures will also be displaced by the new technological forms. Again, this is an illusion. As Figure 14.6 shows, there are several ways of fulfilling e-commerce orders.[16]

- *'Dell' model*: the supplier receives orders for specific products, which are then integrated directly into production. 'The customer order activates the supply chain. Customers can "design" their products from a list of options to be incorporated into a production schedule. The order then initiates a flow of component parts from suppliers to be assembled into a final product, turned over to a logistics service provider, merged with a monitor from another source and delivered to a final customer. The system avoids holding finished product inventory, providing both lower cost and more product variety'.[17]

- *Drop-shipment model*: the e-commerce firm receives an order and passes the order on to a manufacturer for production and delivery direct to the customer.
- *'Amazon' model*: this is the electronic version of an old mode of direct retailing where a 'catalogue' of products is held electronically and accessed via the Internet. Customer orders are either fulfilled by the seller from its own distribution centre or 'drop-shipped' direct from the manufacturer.
- *Bricks-and-mortar model*: combines conventional retail stores, fed by distribution centres, with an Internet website that channels orders to the same distribution centres. A problem with this system is that whereas retail orders generally require large orders, individual Internet-based orders require individual units.
- *Inventory pooling model*: enables customers in a specific industry to acquire common components from an inventory pool controlled by a web-based provider.
- *Home delivery model:* customers requiring regular deliveries of, say, groceries can place their order with a web-based service which will deliver on a routine basis to the customer's home address.

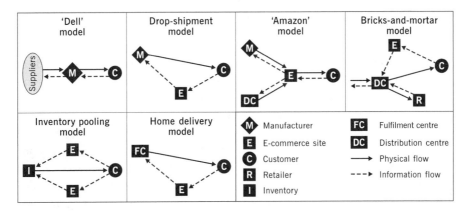

Figure 14.6   Methods of fulfilling e-commerce orders

Source: based, in part, on Schary and Skjøtt-Larsen, 2001: Figure 4.5

Inherent in each of these different ways of fulfilling e-commerce business is the question of physical space which, paradoxically, becomes more, not less, important. In addition to warehousing space the Internet-based system also requires another kind of dedicated physical space: the so-called *Internet 'hotels'*,

> centres where companies with big Internet operations, frequently telecoms carriers who look after the web traffic of many of their customers, can house their web servers ... With ultra high-speed telecoms connections, massive power supplies, arctic air conditioning and squads of trained engineers, the facilities are capable of housing thousands of computers connected to the Internet in optimal conditions ... Prime locations are essential; the hotels should be near the customers they serve, such as City banks, as the

high-capacity fibre-optic communications pipes they require are expensive to run far. Some of the sites are cooled by the kind of water-circulation systems normally associated with nuclear power plants, because the computers generate so much heat.[18]

Thus, a whole variety of technological developments has helped to transform the nature and operations of the distribution industries. Initially, in the early twentieth century, they were the technologies that enabled mass distribution systems to facilitate the output of the mass production systems of the day. In the last two decades of the twentieth century, the technologies were predominantly electronic. Such technologies facilitate the operations of the more flexible production systems of today through their ability to transmit and process vast quantities of information on all aspects of the supply circuit.

The role of the state: regulation and deregulation in the logistics and distribution industries

There is a significant obstacle to the smooth, seamless operation of the kinds of logistics and distribution systems made possible by these technological innovations: state *regulatory* regimes. By definition, transnational production and distribution involve crossing political boundaries. All national governments regulate, in various ways, the cross-border movement of goods and services. We have discussed one aspect of this in Chapter 6 and in the other case study chapters: the variety of trade measures (both tariff and non-tariff barriers) that states use. The existence of such regulations, for example customs requirements and procedures, creates a major discontinuity in the geographical surface over which distribution services operate (see Figure 14.2).

In this section, we are concerned with the regulatory structures affecting the basic functions of the logistics and distribution industries themselves, especially transportation and communication systems. Such regulations are devised and implemented at various political-geographical scales: international, regional and national. They are also in a continuous state of flux. In the past three decades, in particular, there have been major waves of deregulation.

Regulation and deregulation of transportation and communications systems

Regulation of transportation and of communications has a very long history.[19] It has involved a varying mix of national- and international-level systems in the public sphere, as well as private organizations in the form of the operators themselves.

Much of the regulatory system in air and sea transportation relates to issues of safety and security, whilst in communications a key issue is harmonization of standards to enable communications originating in one place to be received and understood in other places. Such international bodies as the IMO (International Maritime Organization), the ICAO (International Civil Aviation Organization) and the ITU (International Telecommunication Union) constitute the primary elements of such an international regulatory framework.

Within such an international framework, individual national states have also regulated telecommunications and air transportation. They are both sectors in which states believe (or have believed until recently) that 'natural monopolies' exist. For example,

> telecommunications' regulation contains one of the earliest examples of inter-national regulatory cooperation between states, with the creation of the International Telegraphic Union (ITgU) in 1865. But in other respects the regulation of telecommunications is a story of territorial containment. Much of the early regulatory development in the first half of the twentieth century was influenced by the economic view that telecommunications is a 'natural mono-poly'. But no state thinks that there should be one world monopolist. Instead, the contours of this natural monopoly correspond with state boundaries.[20]

In most cases this involved state ownership, although in the United States it was a private monopoly, AT&T.

Similarly in the air transportation industry,

> just as almost every nation has its telecommunications carrier (and rarely more than one), almost every nation has a flagship airline (and rarely more than one). The state controls landing rights (just as it tends to control the telecommunications infrastructure) and rations those rights, usually in ways that favour the national flag-carrier.[21]

In both cases, the operation of the regulatory framework has involved a tension between

- the desire on the part of most states for control over their own national spaces
- the drive (primarily by business organizations) for the least possible regulation consistent with safety and efficiency.

As globalizing processes have intensified, however, the balance has shifted decisively towards greater deregulation of the nationally based systems and the privatization of state-owned companies. In the case of telecommunications, the initial impetus came in the United States, with the enforced break-up of the AT&T monopoly in the early 1980s, when the company was forced to divest itself of its local Bell Telephone operating companies. The US example stimulated a wave of European deregulation in telecommunications during the 1980s, led by the UK's Thatcher government.

The air transportation industry has experienced a similar wave of deregulation. Again, the early moves began in the United States. In the late 1970s, a bilateral agreement was signed between the United States and the United Kingdom, which helped to undermine the cartelization of the airline industry within IATA. In 1978, the US domestic airline industry was deregulated. Subsequently,

> both the US and the UK then set about reshaping their bilateral agreements towards more liberal policies. For example, France has been the most vigorous opponent of liberalization, so the US worked at isolating France by negotiating open-skies agreements with Belgium and other countries around France...In short, the process in the 1990s is US-led liberalization that is seeing the world become gradually and chaotically more competitive. The process is chaotic because even the most liberal states, such as the US and UK ... are 'liberal mercantilists' ... Another chaotic element is that many European, African, and South American states support liberalization within their continents but want protection from competition outside the continent (especially from the US).[22]

The continuing tensions between states in terms of their own air spaces (and often their 'national' airlines) has important implications for the logistics and distribution industries. Two examples illustrate this. First, the continuing disagreement between the United States and the United Kingdom over mutual access on the North Atlantic route means that, on the one hand, US airlines have restricted access to London Heathrow whilst, on the other, British airlines are not allowed to fly routes onwards within the United States beyond their initial point of entry. This also means that the US company FedEx cannot operate a fully fledged operation from its UK base at Stansted; it is allowed to fly from the US to Stansted but only to a small number of destinations from there. Instead, it has to charter planes to fly from Britain to Paris to connect with its European hub.[23]

A second example is the recently resolved dispute between the United States and Hong Kong, which was especially important for the large express couriers (FedEx, DHL and UPS). FedEx was allowed only five flights a week from Hong Kong to destinations outside the United States. This was because Hong Kong wanted its airline, Cathay Pacific, to be able to fly within, as well as to, the United States – which the United States refused to allow. The agreement signed in 2002 increased the daily flights for all-cargo carriers from eight to 64, phased in over three years, although Cathay Pacific insisted that the agreement over-benefited United States carriers.[24]

Some of the biggest changes in the regulatory environment have resulted from the emergence of regional economic blocs, such as the EU and the NAFTA. The completion of the Single European Market in 1992 removed virtually all obstacles to internal movement of goods and services within the EU. Liberalization of trucking within and between the United States, Canada and Mexico was also a part of the NAFTA. Under this agreement, the United States and Mexican border states were to be opened to international trucking by the end of 1995. By January 1997, Mexican trucking companies would be allowed to operate as domestic carriers in the border states and for international cargo in the rest of the United States. By January 2000, Mexican truckers would be able to file to operate in the entire United States. In fact this did not happen. Not until 2004 did the US Supreme Court overthrow the opposition of the House of Representatives to Mexican and US truckers operating in each other's domestic markets.

Regulating transportation and communication systems is not easy, given the number of conflicting interests involved. But it is far easier than regulating the Internet. As a medium that 'knows no boundaries' and that is allegedly (although, as we have seen, not actually) 'placeless', it involves some intractable issues as to who regulates it. The answer is far from clear, not least because of the very newness of the Internet and e-commerce and its phenomenally rapid growth. The key issue is 'whose laws apply?' when e-commerce transactions transcend different national jurisdictions.

The EU has ruled that 'companies which "direct their activities" to consumers in other European countries can be sued in those overseas countries'. In other words,

> the law of the consumer's home jurisdiction would apply in any case where a purchase was made through a web site. When a Christmas tree catches fire in a US home, can the store where the tree was bought sue the Hong Kong company if it has a presence in cyberspace? But that raises many problems for US companies, because European privacy and consumer protection laws are much tougher. Companies are forced either to set up separate web sites to comply with local laws, or one megasite which meets every conceivable national and local legal requirement … Both options are costly and defeat the supposed efficiency gains of globalized commerce.[25]

Corporate strategies in the logistics and distribution industries

The logistics and distribution industries cover an immensely wide range of activities and consist of a mix of traditional shipping and carriage of goods through to the highly complex and sophisticated *logistics service providers* (LSPs), from trading companies to large transnational retail chains. In this section we outline the major trends in corporate strategies as firms respond to market, technological and regulatory forces. Although there are many niche areas within the distribution sector there is a broad tendency in most activities for the size of firms to be increasing and for higher degrees of concentration in a smaller number of large firms. As in the other industries we have been examining, mergers and acquisitions have been important mechanisms in the process of consolidation. Table 14.1 lists the largest TNC logistics firms, based on their degree of transnationalization.

Table 14.1   Leading transnational logistics firms

Company	Country	Sales 2003 ($m)	Employment	Foreign affiliates	Number of countries
Deutsche Post World Net	Germany	45,267	341,572	307	99
UPS Inc.	USA	33,485	357,000	97	34
FedEx Corp.	USA	22,487	190,918	59	32
Nippon Express	Japan	13,951	40,081	3,020	
TPG NV	Netherlands	13,423	163,028	241	29
Neptune Orient Lines	Singapore	5,386	12,218	57	14
CNF Inc.	USA	5,104	26,000	21	12
Geodis	France	4,020	22,519	59	20
DSV	Denmark	2,987	9,249	47	9
EIW	USA	2,625	8,600	66	44

Source: based on UNCTAD, 2004: Table A.III.5

Global logistics: from transportation companies to integrated logistics service providers

As customer demands have become more complex (and more global), the providers of logistics and distribution services have responded in a number of ways.[26] Some have diversified into complete 'one-stop shops'; others have remained more narrowly focused on providing a limited range of functions. In both cases, the trend has been towards greater consolidation and concentration through acquisition and merger. Some examples illustrate the trend.

In the shipping sector, Maersk acquired Nedlloyd in 2005 to become the largest container operator in the world. In the logistics field, acquisitions and mergers have accelerated since 2000, when Exel was formed from a merger of a shipping company, Ocean, and a contract logistics supplier, NFC. In 2004, Exel purchased the second largest UK logistics company, Tibbett and Britton, to become the sector leader with 111,000 employees in more than 135 countries.

> The group is the largest stand-alone provider of supply chain management and freight forwarding services in the world. It moves freight in 14 of the 15 largest economies and its customers include 75 per cent of the world's largest non-financial companies.[27]

In turn, in 2005, Deutsche Post World Net (owner of DHL, which it had acquired in 2003) acquired Exel. This created by far the world's largest logistics service company, providing air freight, ocean freight and contract logistics services. The new group, which will use the DHL brand name, will have two divisions: DHL Exel Supply Chain and DHL Global Forwarding. It will employ around 500,000 people worldwide.[28]

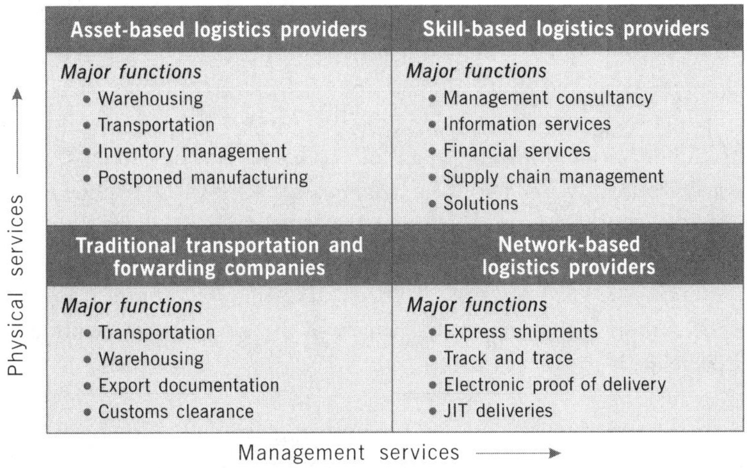

Asset-based logistics providers	**Skill-based logistics providers**
Major functions • Warehousing • Transportation • Inventory management • Postponed manufacturing	*Major functions* • Management consultancy • Information services • Financial services • Supply chain management • Solutions
Traditional transportation and forwarding companies	**Network-based logistics providers**
Major functions • Transportation • Warehousing • Export documentation • Customs clearance	*Major functions* • Express shipments • Track and trace • Electronic proof of delivery • JIT deliveries

Physical services →

Management services →

Figure 14.7 **Types of logistics service providers**

Source: based, in part, on Schary and Skjøtt-Larsen, 2001: Figure 7.3

As a result of such developments, together with the movement of other service companies into logistics provision, we can identify four major types of logistics service firm, classified according to the kinds of physical and management services provided (Figure 14.7).

The simplest functions are provided by the *traditional transportation and forwarding companies*. The *asset-based logistics providers* first emerged during the 1980s, developing primarily from the diversification of some traditional transportation companies into more complex logistics service providers. Several of the world's leading container shipping companies, such as Maersk-Sealand and Nedlloyd/P&O, moved in this direction. For example, in 1992 Nedlloyd took on responsibility for all of IBM's distribution activities as part of its strategy to become a worldwide logistics provider.

During the early 1990s, a number of *network-based logistics providers* appeared on the international scene, notably DHL, FedEx, UPS and TNT.

> These third-party logistics providers started as couriers and express parcel companies and built up global transportation and communication networks to be able to expedite express shipments fast and reliably. Supplemental information services typically include electronic proof-of-delivery and track-and-trace options from sender to receiver ... Recently, these players have moved into the time-sensitive and high-value-density third-party logistics market, such as electronics, spare parts, fashion goods and pharmaceuticals, and are competing with the traditional asset-based logistics providers in these high margin markets.[29]

These companies have built up massive networks of operations across the globe, as their annual reports make clear.

- Prior to its acquisition of Exel, DHL had operations in more than 630 cities in 230 countries (Figure 14.8), employed 64,000 people, operated thousands of vehicles (both air and road) and had one of the largest private telecommunications networks in the world linking the company's 3000 stations and 35 hubs.
- FedEx offers services in more than 220 countries and has 250,000 employees. These are seriously big TNCs by any measure. Unlike DHL, however, FedEx remains focused on transporting small packages and the light freight market, resisting the temptation to diversify into a full-service logistics company.
- TNT, a Dutch-owned company, has decided to sell off its contracts logistics business, even though it is the second largest in the world, to concentrate on post and parcel delivery.[30]

Each of these companies operates on the basis of global hub-and-spoke transportation networks, owned either by themselves or with partners. Exel, for example, has preferred to use third-party carriers whereas DHL and FedEx have tended to develop their own transportation fleets.

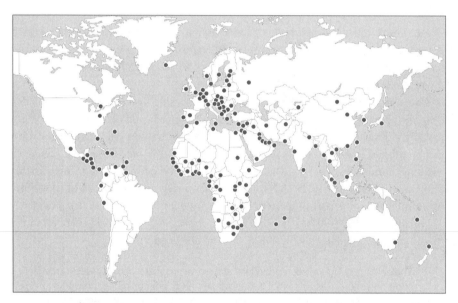

Figure 14.8   DHL's global network
Source: DHL

The fourth type of logistics service firm shown in Figure 14.7 – the *skill-based logistics providers* – became increasingly significant in the later 1990s. These are firms that do not own any major physical logistics assets but provide a range of primarily information-based logistics services. These encompass consultancy services (including supply chain configuration), financial services, information technology services, and a range of management expertise. Examples of such skill-based logistics service providers include GeoLogistics, a firm created in 1996 through the merger of three existing logistics companies (Bekins, LEP, Matrix).

Global trading companies

Trading companies have a history which goes back many hundreds of years. From the earliest days of long-distance trade they played an especially important role in facilitating trade in materials and products. Here we look at two important contemporary examples, both taken from East Asia.

The first example is the Japanese *sogo shosha*. The common translation of the term *sogo shosha* is 'general trading company'; but they are very much more than this, having been central to the development of the Japanese economy since the late nineteenth century.[31] The six leading *sogo shosha* — Mitsubishi, Mitsui, Itochu, Marubeni, Sumitomo and Nissho-Iwai — are gargantuan commercial, financial and industrial conglomerates. They operate a massive network of subsidiaries and thousands of related companies across the globe (Figure 14.9) and handle tens of thousands of different products. In the early 1990s, they were responsible for roughly 70 per cent of total Japanese imports and for 40 per cent of Japanese exports. This is the true Japanese general trading oligopoly, each member of which has a major coordinating role within one of the Japanese *keiretsu* (see Figure 4.11).

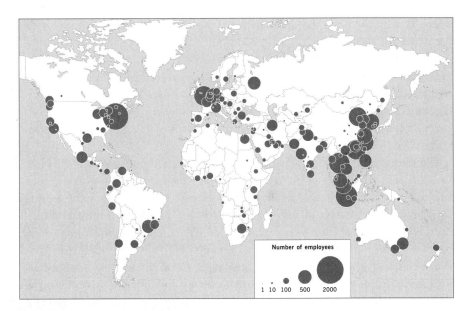

Figure 14.9   The global distribution of Japanese *sogo shosha* offices

Source: company reports

Historically, the *sogo shosha* developed initially to organize exchange and distribution within the Japanese domestic market. Subsequently, they became the first Japanese companies to invest on a large scale outside Japan. These foreign investments were primarily designed to organize the flow of imports of much-needed primary materials for the resource-poor Japanese economy and to channel Japanese exports of manufactures to overseas markets. It was the particular demands of these Japanese-focused trading activities that necessitated the development of the globally extensive networks of *sogo shosha* offices around the world. In other words, they were to set up a global marketing and economic intelligence network. Once in place, this network, with all its supporting facilities, not only facilitated the growth of Japanese trade but also enabled a whole range of Japanese firms to venture overseas. Indeed, a good deal of the early overseas investment by Japanese manufacturing firms was organized by the *sogo shosha*.

They perform four specific functions:

- *trading and transactional intermediation*: matching buyers and sellers in a long-term contractual relationship
- *financial intermediation*: serving as a risk buffer between suppliers and purchasers
- *information gathering*: collecting and collating information on market conditions throughout the world
- *organization and coordination of complex business systems*: for example, major infra-structural projects.

As the position of the Japanese economy in the global system has changed – especially with the deep and continuing recession of the 1990s – the role of the *sogo shosha* has also had to change. In addition to their traditional roles of import–export business between Japan and other parts of the world and their activities as traders in commodities markets, the *sogo shosha* have increasingly become engaged in 'third-country trade'. In this context, they act as intermediaries between firms in countries other than Japan in their trade with other parts of the world. In fact, this has been the fastest-growing area of transactions of the *sogo shosha* in recent years.

The second example of a trading company, also taken from East Asia, is the Hong Kong based firm Li & Fung.[32] This firm is not only the biggest export trading company in Hong Kong but also – and more importantly – a highly innovative logistics company, with offices spread across 40 countries (see Figure 14.10). Established in Canton in 1906, Li & Fung was originally a simple commodity broker, connecting together buyers and sellers for a fee. Today, although still a Chinese family firm, it has been transformed from the simple brokerage to an immensely sophisticated organizer of geographically dispersed manufacturing and distribution operations in a whole variety of consumer goods, but with a strong specialization in garments (see Chapter 9). As Figure 14.11 shows, Li & Fung controls and coordinates all stages of the supply chain, from design and production planning, through finding suppliers of materials and manufacturers of products, to the final stages of quality control, testing and the logistics of distribution.

These two examples show that traditional trading companies have carved out new roles for themselves, both responding to, and creating, new demands for distribution and logistics services.

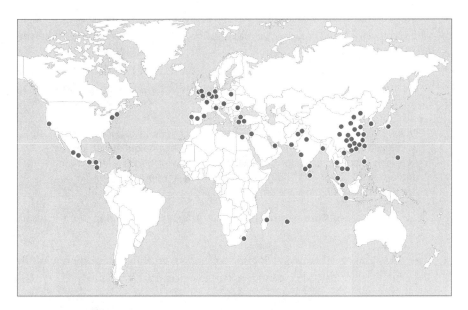

Figure 14.10   The global spread of the offices of Li & Fung

Source: Li & Fung

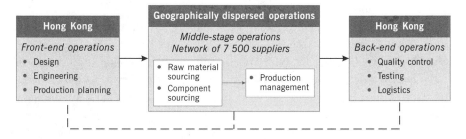

Figure 14.11   Organization of Li & Fung's logistics operations

Source: based on Magretta, 1998: 111

Logistics 'places': key geographical nodes on the global logistics map

These developments in the global logistics industries have a highly distinctive geography, which both helps to shape their activities and is also shaped by them. Amongst the thousands of seaports and airports across the world, a few key nodes have become increasingly important. They reflect three sets of forces:

- Their position in the twenty-first century global economy, especially in the light of the global shifts in economic activities we have been discussing throughout this book.
- The strategies of states in investing in port, airport and IT facilities. An example is the highly focused investment strategies of the Singaporean government to create a 'globally integrated logistics hub':

  > an integrated IT platform that manages the flow of trade-related information … will enable exchange of information between shippers, freight forwarders, carriers and financial institutions to facilitate the flow of goods within, through and out of Singapore … The government will invest up to S$50 million over five years to develop the platform.[33]

  Other states are pursuing broadly similar strategies to develop their logistics capabilities.

- The strategies of the major logistics firms in creating their own globally dispersed operations and choosing certain key places as their 'hubs'.

Figures 14.12 and 14.13 show two examples of these trends. The emergence of East Asia as the most dynamic economic region in the world is clearly reflected in Figure 14.12. No fewer than 11 of the 20 leading container ports are located in East Asia, four of them in China. Neither Shanghai nor Shenzhen were in the top 10 in 1995; now they rank third and fourth respectively. Conversely, the leading European container ports – Rotterdam and Hamburg, as well as some US ports – have slipped down the rankings.

The global map of leading cargo airports (Figure 14.13) has some features in common with the container ports map. Certain key cities, notably Hong Kong, Singapore, Shanghai, Tokyo, Los Angeles, New York and Dubai, have developed as leading air cargo hubs as well as major container ports. They are, indeed, key geographical nodes on the global logistics map. But there are other leading cargo airports that are quite different. They are the strategic hubs of the leading specialist freight and logistics firms. The place of Memphis, Tennessee as the world's biggest cargo airport is entirely due to its role as FedEx's main hub. In similar vein, Louisville, Kentucky is the UPS main hub. The importance of Anchorage airport in Alaska, likewise, is attributable to its role as a key hub for FedEx and UPS in their links with China.

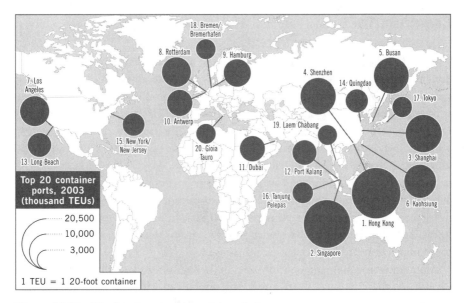

Figure 14.12   The leading world container hubs

Source: based on US Bureau of Transportation statistics, 2006.

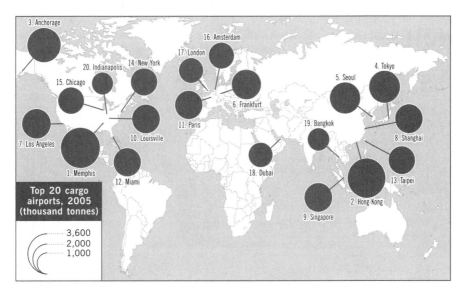

Figure 14.13   The leading world cargo airports

Source: based on Airports Council International statistics, 2006

Conclusion

Like the financial services industries discussed in the previous chapter, the distribution industries are circulation activities and, as such, are absolutely central to the operation of all economic activities. Their basic function is to act as intermediaries, connecting together all stages of the production circuit in fast, flexible and reliable ways by overcoming the friction of geographical distance. As manufacturers and other service firms have become increasingly obsessed with time-based competition, the demands on the distribution services to 'deliver' have intensified. As in financial services, too, deregulation has been an important factor – especially deregulation of transportation and telecommunications systems. In recent years, the distribution industries have been transformed, first by such developments as electronic data interchange and bar codes and, second, by the emergence of Internet-based transactions in the form of e-commerce.

Alongside some quite dramatic changes in the logistics and distribution industries we can see the continuation of some long-established forms of organization. For example, although the emergence of highly sophisticated third- and fourth-party logistics service providers represents a major development, there continues to be a wide variety of firms engaged in providing logistics and distribution services. These developments have created a distinctive geography of logistics services, which reflects the significant global shifts in economic activity of recent years. Certain key nodes have emerged as the major hubs of the global logistics industries.

NOTES

1 The question is from Wrigley (2000).
2 Stalk and Hout (1990), Schoenberger (2000).
3 Min and Keebler (2001: 265).
4 OECD (2004: 178).
5 *Financial Times* (25 August 2005).
6 Schary and Skjøtt-Larsen (2001: 129).
7 Abernathy et al. (1999: Chapter 4).
8 Abernathy et al. (1999: 61).
9 Quoted in the *Financial Times* (20 April 2005).
10 Abernathy et al. (1999: 66).
11 Stalk et al. (1998: 58).
12 Leinbach (2001: 15).
13 *The Economist* (3 December 2005).
14 *Financial Times* (6 April 2005).
15 US Department of Commerce (2000: 18).
16 Schary and Skjøtt-Larsen (2001: 132–6).
17 Schary and Skjøtt-Larsen (2001: 132).

18 *Financial Times* (1 August 2000).
19 See Braithwaite and Drahos (2000).
20 Braithwaite and Drahos (2000: 322).
21 Braithwaite and Drahos (2000: 454).
22 Braithwaite and Drahos (2000: 456–7).
23 *The Sunday Times* (23 January 2005).
24 *Commercial Aviation Today* (21 October 2002).
25 *Financial Times* (23 December 1999).
26 This section is based primarily on Schary and Skjøtt-Larsen (2001: 230–41). See also Beukema and Coenen (1999).
27 *Financial Times* (2 September 2005).
28 Company sources (December 2005).
29 Schary and Skjøtt-Larsen (2001: 231).
30 *Financial Times* (7 December 2005).
31 See Dicken and Miyamachi (1998).
32 See *The Economist* (2 June 2001), Magretta (1998), Schary and Skjøtt-Larsen (2001: 383–5).
33 Singapore Economic Development Board, June 2004.

PART FOUR
WINNING AND LOSING IN THE GLOBAL ECONOMY

Fifteen
Winning and Losing: An Introduction

From processes to impacts

Globalizing processes revisited

Our focus so far has been on the *patterns* and *processes* of global shift: on the *forms* being produced by the globalizing of economic activities and on the *forces* producing those forms. During the past 50 years, the world economy has experienced enormous *cyclical* variation in economic activity: the unparalleled growth of the long boom lasting from the early 1950s to the mid 1970s being followed by periods of rapid growth interspersed with recession, stagnation and, often, unanticipated crises. Cyclical volatility is the norm.

Underlying these global cyclical trends are *structural* changes associated with the *geographical rescaling* – local, national, regional, global – of economic activities. Geographically, the global economy has become increasingly *multipolar*. New centres of production – new geographical divisions of labour – have emerged in parts of what had been, historically, the periphery and semi-periphery of the world economy. Most significant of all has been the resurgence of Asia (especially East Asia, and most recently China) to a position of global significance commensurate with its importance before 'the West' overtook it in the nineteenth century. At the same time, many parts of the world remain, to a greater or lesser degree, disarticulated from the engines of economic growth. The geography of the world economy is, indeed, a 'mosaic of unevenness in a continual state of flux'.[1]

These transformations of the geo-economy are the outcome of extremely complex processes, involving major changes in the nature of production, distribution and consumption. Within such complexity, my argument has been that three closely interconnected processes are especially important in reshaping the global economic map:

- *Transnational corporations* are the *primary movers and shapers* of the global economy because of their potential ability to control or coordinate production networks across several countries; to take advantage of geographical differences in factor distributions; and to switch and reswitch resources globally. TNCs are both intricate organizational networks in their own right and also deeply embedded within dynamic networks of inter-firm relationships and alliances. They continuously – and unevenly – reshape the geography of the global economy. But TNCs are not converging to a single 'global' organizational form. Diversity – much of it derived from the geographical origins of TNCs – continues to exist.

- *States* continue to be a major influence in the global economy through their continuing attempts to *regulate* economic transactions within and across their territorial boundaries. Two sets of political forces have been especially significant in the past few years. One is the spread of 'deregulatory' forces as access to national markets has been opened up, initially to trade and more recently to foreign direct investment and to financial flows in general. The other is the proliferation of regional trade agreements that, in effect, shift the regulatory processes to a different geographical and institutional scale.

- *Technology* is a fundamental *enabling* force in the globalizing of economic activities. However, technology is not a determining force. It is a socially embedded process in which choices have to be made. Most significant from the viewpoint of the global shift of economic activities have been innovations in the time–space shrinking technologies and, especially, the information (digital) technologies which have transformed the processes and organization of production (especially in terms of increased flexibility and modularity).

The precise manner in which these three sets of processes occur and interact varies greatly between different types of economic activity, as the case study chapters of Part Three have shown. There certainly is not a unidirectional trajectory of change. In the case of consumption, for example, the notion of a global consumer market driven by TNCs, wiping out local tastes and preferences, is a myth – notwithstanding some claims to the contrary. Globalizing processes do not lead to the erasure of local differences. On the contrary: geographical differentiation has become even more important in a world in which distances are alleged no longer to matter. Contrary to the assertion of Thomas Friedman the world is emphatically not 'flat'.[2]

Relatedly, the real *effects* of globalizing processes are felt not at the global or the national level but at the *local* scale: the communities within which real people struggle to live out their daily lives. It is at this scale that physical investments in economic activities are actually put in place, restructured or closed down. It is at this scale that most people make their living and create their own family, household and social communities. In fact, much of the 'energy' driving economic change is generated at a relatively localized geographical scale – what Scott calls 'regional motors of the global economy' (Figure 15.1).

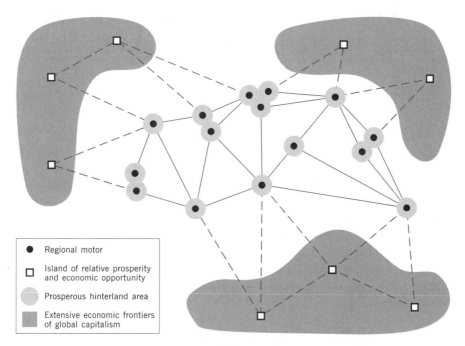

Figure 15.1 **A schematic representation of the geography of the global economy**

Source: based on Scott, 1998: Figure 4.2

Winning and losing: the impacts of globalizing processes

What do globalizing processes mean for people? Are they beneficial or are they detrimental? What can or should be done to make things better? Views are strongly polarized (see Chapter 1). To its proponents, globalization is a savage process but it also a beneficial one, in which the winners far outnumber the losers.[3] It is certainly true that the growth of the global economy during the past few decades has dramatically increased the material well-being of many people. This should not be forgotten amidst the clamour of anti-globalization rhetoric. However, the benign view must be challenged. For many, many people, the global economy has not yet brought either material gifts or the hope of a better life.[4]

In this brief chapter, which acts as an introduction to subsequent chapters in Part Four, we examine the broad 'contours' of economic development. This provides a framework within which we can explore the problems facing people in developed and developing countries alike as they struggle to cope with the processes of global shift. In Chapter 16, we look at the potential impacts (positive as well as negative) of TNCs on national and local economies, whilst in Chapters 17 and 18 we focus on the specific impact of globalizing processes on jobs and incomes in developed and developing countries respectively. There is a very good reason for adopting such a perspective. *Income* is the key to an individual's or a family's material well-being. However, income – or lack of it in the form of *poverty* – is not an end in its own right but, rather, a means towards what Amartya Sen[5] calls 'development as freedom'. In that sense, poverty is an 'unfreedom':

> There are good reasons for seeing poverty as a deprivation of basic capabilities, rather than merely as low income. Deprivation of elementary capabilities can be reflected in premature mortality, significant undernourishment (especially of children), persistent morbidity, widespread illiteracy and other failures.[6]

The major source of income (for all but the exceptionally wealthy) is *employment* or *self-employment*. Hence, the question '*Where will the jobs come from?*' is a crucial one throughout the world.

In attempting to unravel this question in terms of the processes discussed throughout this book we find a very complex picture.

> Establishing a link between globalization and inequality is fraught with difficulty, not only because of how globalization is defined and how inequality is measured, but also because the entanglements between globalization forces and 'domestic' trends are not that easy to separate out ... However, there is sufficient evidence to conclude that contemporary processes of globalization have been accompanied by a rise in global inequality and vulnerability.[7]

The major employment changes that have been occurring in both developed and developing countries are the result of an intricate interaction of processes. Job losses in the developed market economies, for example, cannot be attributed simply to the relocation of production to low-cost developing countries. There is far more to it than this. What is clear, however, is that the industrialized economies face major problems of adjusting to the decline in manufacturing jobs in particular but also, increasingly, in some service jobs. The problems facing developing countries are infinitely more acute, despite the spectacular success of a small number of newly industrializing economies.

The contours of economic development

The contours of the development map show a landscape of staggeringly high peaks of affluence and deep troughs of deprivation interspersed with plains of greater or lesser degrees of prosperity. In the words of the 2005 UN *Human Development Report,*

> The era of globalization has been marked by dramatic increases in technology, trade and investment – and an impressive increase in prosperity. Gains in

> human development have been less impressive. Large parts of the developing world are being left behind. Human development gaps between rich and poor countries, already large, are widening ... The scale of the human development gains registered over the past decade should not be underestimated – nor should it be exaggerated. Part of the problem of global snapshots is that they obscure large variations across and within regions ... Progress towards human development has been uneven across and within regions and across different dimensions.[8]

At the global scale, the development gap is stunningly wide. The 20 per cent of the world's population living in the highest-income countries control well over 80 per cent of world income, trade, investment and communications technology. The 20 per cent of the world's population living in the poorest countries controlled around 1 per cent.

The contours of world poverty

Before the beginning of the nineteenth century, the differences in levels of income between different parts of the world were relatively small:

> At the dawn of the first industrial revolution, the gap in per capita income between Western Europe and India, Africa, or China was probably no more than 30 per cent. All this changed abruptly with the industrial revolution.[9]

Figure 15.2 shows how dramatically the gap between the richest and poorest countries widened progressively through to the 1990s, although the gap between the richest and poorest countries has not widened further since.

A rather different way of measuring poverty is to calculate the number of people living on less than $1 per day. Despite very considerable advances in some parts of the world, one in five people (around 1.2 billion) live on less than $1 per day. Nearly 70 per cent of these utterly impoverished people live in South Asia and sub-Saharan Africa (Figure 15.3). The extent to which the income gap is widening, narrowing or staying about the same is controversial and depends on how it is measured.[10] But whichever measure is used, the fact remains that the gap between rich and poor countries is enormous and that any improvements have been relatively small compared with the sheer scale of the problem. Of course, that doesn't mean that no country has improved its position. Some – especially in East Asia – certainly have. As Figure 15.3 shows, the number of people in East Asia and the Pacific living on less than $1 per day fell by around 140 million between 1987 and 1998.

In general, however, the winners – and the losers – have been the usual suspects. The already affluent developed countries have sustained – even increased – their affluence, and some developing countries have made very significant progress, but there is a hard core of exceptionally poor countries that remain stranded. Despite the generally rising tide associated with overall world economic growth, it has not

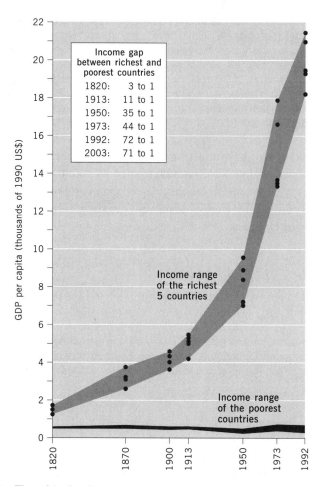

Figure 15.2   The widening income gap between countries

Source: based on UNDP *Human Development Report*, 1999, Figure 1.6; World Bank, 2005b: Table 1

lifted all boats. Most strikingly, the countries of sub-Saharan Africa and parts of South Asia have benefited least: in both cases, the number of people living on less than $1 per day increased very substantially, as Figure 15.3 shows.

Focusing the analytical lens at the country level provides a useful first cut at mapping the contours of development. But, of course, such a focus obscures the detail of the economic landscape both at smaller geographical scales and in terms of non-geographical criteria (for example, gender, social class and so on). We will look at such criteria in more detail in Chapters 17 and 18. Here, however, we can note just two broad groups of winners and losers, which cut across the broad development divide.

The clear winners are the elite *transnational capitalist class*[11] whose members are predominantly, although not exclusively, drawn from developed countries. The

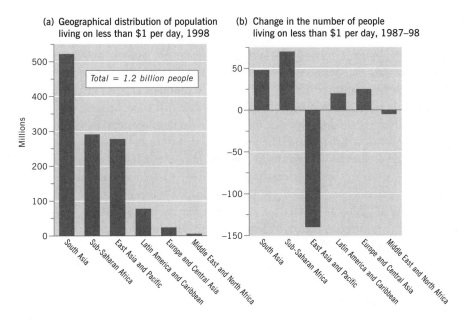

(a) Geographical distribution of population living on less than $1 per day, 1998

(b) Change in the number of people living on less than $1 per day, 1987–98

Total = 1.2 billion people

Figure 15.3   Distribution of the world's poorest population

Source: based on World Bank, 2001: Figures 1 and 2

dominant group is made up of the owners and controllers of the major corporations – the globe-trotting, jet-setting TNC executives. To these we can add globalizing bureaucrats and politicians, globalizing professionals (with particular technical expertise – even including some academics), merchants and media people. Without question, these are winners in the global economy and are highly influential in global policy discourses.[12]

This transnational capitalist class (TCC) displays a number of significant characteristics:[13]

- Economic interests increasingly globally linked rather than exclusively local and national in origin.
- Behaviour based on specific forms of global competitive and consumerist rhetoric and practice.
- Outward-oriented global, rather than inward-oriented local, perspectives on most economic, political and culture ideology issues.
- Similar lifestyles, especially patterns of higher education (for example, in business schools) and consumption of luxury goods and services. 'Integral to this process are exclusive clubs and restaurants, ultra-expensive resorts in all continents, private as opposed to mass forms of travel and entertainment and, ominously, increasing residential segregation of the very rich secured by armed guards and electronic surveillance.'
- Self-projection as citizens of the world as well as of place of birth.

While transnational elites are clear winners, *women* – at least in many parts of the world – tend to be losers in the global economy. A staggering two-thirds of the world's population living on less than $1 per day are women, surviving 'on the margins of existence without adequate food, clean water, sanitation or health care, and without education'.[14] In Sen's terms of 'development as freedom', women are significantly more disadvantaged than men. At the same time, because of their key role in nurturing children, women hold the key to development, especially in the poorest countries of the world. The problem is that in many developing countries (as opposed to developed countries), women have a much higher mortality rate and lower survival rate than men. As a result, the female/male ratio is lower than in developed countries, implying a phenomenon of 'missing women'. Where this occurs – as in China and India, for example – the main explanation would seem to be 'the comparative neglect of female health and nutrition, especially, but not exclusively, during childhood.'[15]

The contours of world population

Geographical variations in population growth rates, as well as in the age composition of the population, have an extremely important influence on how globalizing processes are worked out in different places. They also relate, very clearly, to issues of poverty, to the ability of people in different places to make a living through employment, and to issues of environmental impact'.[15]

Population growth

At the beginning of the twenty-first century the world's population reached a total of 6.1 billion. One hundred years earlier, world population was less than 2 billion. Not unreasonably, then, has the twentieth century been called 'the century of population' and the 'explosion of population … [been seen as] one of its defining characteristics'.[16]

> This is an absolute increase that far exceeds that which has occurred in any other period of human experience. It took until 1825 to reach one billion humans *in toto*; it took only the next 100 years to double; and the next 50 years to double again, to 4 billion in 1975. A quarter of a century later, as we were celebrating the millennium, the total jumped to six billion. True, the pace of increase has been slowing in the last decade or so but, like a large oil tanker decelerating at sea, that slowdown is a protracted process.[17]

The UN's latest medium-term projection is that world population in 2050 will be around 9.1 billion (rather lower than previously predicted), although it could be as high as 10.7 billion or as low as 7.7 billion, depending on what happens to fertility rates.[18]

The most striking feature of world population growth is that it now occurs overwhelmingly in developing countries. In 2005, 81 per cent of the world's 6.5 billion population was in the developing countries. Figure 15.4 shows this massive – and accelerating – divergence in population growth between developed and developing countries. The year 1950 was an especially significant turning point. That year marked the beginnings of the 'population explosion' brought about by the rapid fall in death rates in Africa, Asia and Latin America coupled with continuing high fertility rates in those areas. Since then, the contrast between the very low population growth rates of the developed countries and the very high rates in many developing countries has become even more marked.

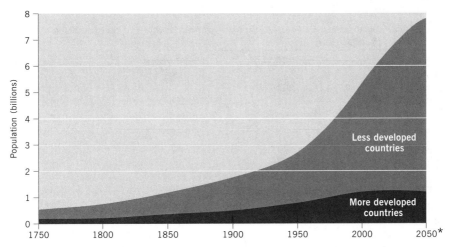

*based on medium fertility variant

Figure 15.4   World population growth

Source: based on data in UN Population Division, 2005

Just to replace an existing population requires a fertility rate of 2.1 children per woman. In most developed countries, fertility rates are now well below the replacement level – at 1.56 and declining – although with an expected rise in the mid twenty-first century to just below replacement levels. In contrast, fertility rates in the developing world as a whole are currently at 2.9. But although they are projected to decline a little, and are well below that in one or two cases (for example, China is now actually below replacement level), in the very poorest countries of the world fertility rates are exceptionally high: 5.02 in 2000–5, although the UN estimates a possible decline to 2.57 by 2045–50. Despite high mortality rates through HIV/AIDS in many of these poorest countries, their population is expected to grow from 658 million to 1.73 billion by 2050. For developing countries as a whole, the 2050 population is predicted to be about 7.8 billion (that compares with a *total* world population in 2005 of 6.5 billion).

'Old' and 'young' populations

Persistent unevenness in fertility rates between developed and developing countries creates significant differentials in the *age composition* of the population. Put in a nutshell, developed countries are ageing while most developing countries continue to be youthful in population terms. Table 15.1 shows the marked geographical variations in the relative importance of different age groups. Europe, North America and Japan all have relatively old populations; 'young' countries (in population terms) are overwhelmingly in the developing world, particularly in Africa, which is the youngest region in the world. Such wide variations in age structure are enormously important in terms of economic and social development; we will have more to say about them in Chapters 17 and 18.

Table 15.1   Geographical variations in the age composition of the population (% in region)

Region	Under 15 years	15–64 years	Over 65 years
World	31	62	7
Europe:	18	68	14
Northern	19	66	15
Western	17	68	15
Eastern	20	67	13
Southern	17	67	16
North America	21	66	13
Japan	15	69	16
Oceania	26	64	10
Africa:	43	54	3
Northern	38	58	4
Western	45	52	3
Eastern	46	51	3
Middle	47	50	3
Southern	35	60	5
Latin America and Caribbean:	33	62	5
Central America	36	60	4
Caribbean	31	62	7
South America	32	62	6
Asia:	32	62	6
Western	37	59	4
South Central	37	59	4
South East	34	62	4
East	25	67	8

Source: based on Population Reference Bureau, 1999: 2–9

An urban explosion

Not only is most of the world's population, and population growth, located in developing countries but also that population is increasingly concentrated in cities (see Figure 2.26). In complete contrast to the older industrialized countries, therefore, where a growing *counter-urbanization* trend has been evident for some years, urban growth in most developing countries has continued to accelerate. The highest rates of urban growth are now in developing countries, where the number of very large cities has increased enormously. In some cases, notably the cities within the newly industrializing economies of East Asia, such growth is driven, and sustained, by the forces of economic dynamism. However, in most cases of rapid urbanization within developing countries, the link between economic growth and urban growth is less clear, and owes more to high rates of population fertility coupled with rural poverty (see Chapter 18).

> The urban population of the world is estimated to increase from 2.86 billion in 2000 to 4.98 billion by 2030 … By comparison, the size of the rural population is expected to grow only very marginally, going from 3.19 billion in 2000 to 2.29 billion in 2030.[19]

By 2007, more than half of the world's population will be living in urban areas. The extent to which populations are urbanized varies significantly from one part of the world to another, as Table 15.2 shows.

Table 15.2 Geographical variations in levels of urbanization

| | Level of urbanization (%) | | | Annual growth (%) | | | |
| | | | | Urban | | Rural | |
Region	2000	2015	2030	2000–15	2015–30	2000–15	2015–30
Africa	37.9	46.5	54.5	3.5	2.8	1.1	0.7
Asia	36.7	44.7	53.4	2.4	2.0	0.2	−0.4
Europe	74.8	78.6	82.6	0.3	0.1	−1.2	−1.6
Latin America	75.3	79.9	83.2	1.7	1.2	−0.1	−
North America	77.2	80.9	84.4	1.0	0.8	−0.5	−0.8
Oceania	70.2	71.3	74.4	1.2	1.2	0.9	0.1
World	47.0	53.4	60.3	2.0	1.7	0.3	−0.2

Source: based on UN Centre for Human Settlements, 2001: Table A.2

International migration

Of course, neither rates of population growth nor changes in age composition are caused solely by differences in fertility rates. That would be the case only in an entirely closed system where neither *in-migration* nor *out-migration* occurred.[20] The number of people migrating between different parts of the world is both huge and growing, despite the fact that political barriers to migration are significantly higher than they were several decades ago. In 2005, there were 185 million documented migrants in the world. In absolute terms this is the largest recorded figure though, in relative terms, international migration is significantly lower than in the nineteenth century, when it accounted for 10 per cent of the world population compared with 2.9 per cent today.

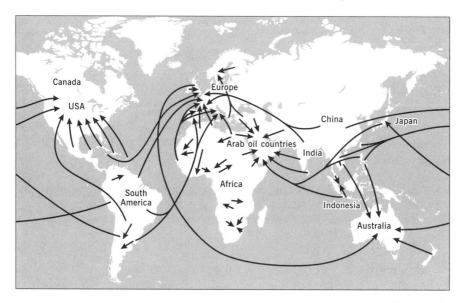

Figure 15.5 Major migration movements

Source: based on Castles and Miller, 1993: Map 1.1

The geographical distances over which international migration occurs are enormously varied, as Figure 15.5 shows. A large proportion of migrant flows are to countries close to the place of origin – for fairly obvious reasons, including cost, greater knowledge of closer opportunities, possibly greater cultural compatibility. But over and above such short-distance migrations are the long-distance, often intercontinental, flows. Certain migration paths are especially important. For example, there are massive movements across the Mexican–United States border and from parts of Asia to the United States. Australia has become an important focus of migration from East Asia. In the recent past, there were huge migration flows from the Caribbean and South Asia to the United Kingdom and from countries around the Mediterranean basin to Germany.

In general, developing countries are the major sources of out-migration:

- Within Africa, the biggest outflows are from Eastern Africa, while there are significant inflows to Northern Africa.
- Within Asia, the biggest outflows are from South Central and South East Asia.
- Within Latin America and the Caribbean, the biggest outflows are from Central America. Mexico is the biggest single source of international migrants in the world, overwhelmingly to the United States.

Conversely, the biggest in-migrations are to developed regions, notably North America and Europe (Table 15.3).

- North America is by far the largest net recipient of migrants.
- The pattern of migration within Europe is complex. There are huge inflows into Western Europe and large outflows from Eastern Europe.
- Oceania, notably Australia and New Zealand, continue to be important destinations for international migration.

Table 15.3 Major migrant destinations, 2000

Host country	Migrants (millions)	% of world migrant stock
United States	35.0	20.0
Russian Federation	13.3	7.6
Germany	7.3	4.2
Ukraine	6.9	4.0
France	6.3	3.6
India	6.3	3.6
Canada	5.8	3.3
Saudi Arabia	5.3	3.0
Australia	4.7	2.7
Pakistan	4.2	2.4
United Kingdom	4.0	2.3
Kazakhstan	3.0	1.7
China, Hong Kong SAR	2.7	1.5

Source: based on International Organization for Migration, 2005: Table 23.3

One of the most important results of international migration is the creation of geographically dispersed *transnational migrant communities*, particularly in cities in developed countries.[21] These complex networks created by migrants – especially labour migrants – between their places of origin and their places of settlement constitute particular kinds of transnational social spaces held together by financial remittances and social networks derived from ethnic ties. Such transnational communities play an extremely significant role not only in channelling subsequent migrant flows but also in investment patterns and in the creation of distinctive forms of entrepreneurship.

In Chapters 17 and 18 we will explore the implications of international migration for both sending and receiving countries. But it is worth noting at this stage that international migration has a significant effect on population growth (and its composition) in the developed countries. The UN estimated that

> without migration, the population of more developed regions as a whole would start declining in 2003 rather than in 2025, and by 2050 it would be 126 million less than the 1.18 billion projected under the assumption of continued migration.[22]

Making a living in the global economy

> We are facing a global jobs crisis of mammoth proportions, and a deficit in decent work that isn't going to go away by itself. (Juan Somavia, ILO Director-General, 2006)[23]

People strive to make a living in a whole variety of ways: for example, exchanging self-grown crops or basic handcrafted products; providing personal services in the big cities; working on the land, in factories or in offices as paid employees; running their own businesses as self-employed entrepreneurs; and so on. For the overwhelming majority of people, *employment* (full- or part-time or as self-employment) is the most important source of income and, therefore, one of the keys to 'development as freedom'. However, not only are there not enough jobs 'to go round' – one estimate suggested that 400 million new jobs needed to be created over 10 years just to absorb newcomers to the labour market[24] – but also the volatility of employment opportunities appears to be increasing.

> Unemployment is the global problem of our times, and more than that: it is a protracted tragedy at the personal level, and destabilizing at the social level.[25]

The total global labour force (which includes both employed and unemployed) increased by almost 440 million between 1995 and 2005, a growth rate of almost 17 per cent. In 2005, approximately 192 million people were unemployed in the world economy, 'an increase of 2.2 million since 2004 and 34.4 million since 1995' (and this figure refers only to 'open' unemployment; it does not include the millions of people suffering from 'hidden' unemployment who are not measured in the official figures).[26] And, as always, the pattern of unemployment is extremely uneven between different parts of the world, between different parts of the same country, and between different population groups.

Serious as the unemployment position is in the industrialized nations, it pales into insignificance compared with the problems of most developing countries, particularly the least industrialized countries. At least in older industrialized countries the growth of the labour force is now easing. Only 1 per cent of the projected growth of the global labour force between 1995 and 2025 will be in the high-income countries, while more than two-thirds of the projected growth will occur in developing countries. As Figure 15.6 shows, the low- and middle-income countries already account for 85 per cent of the global labour force. In many of these countries, extremely high rates of population growth mean that the number of young people seeking jobs will continue to accelerate for the foreseeable future.

The 'double exposure' problem

In discussing outcomes of globalization processes we have focused primarily on one set of issues: those to do with economic well-being. Of course, there are many others and these may interact in such a way as to change the identity of winners and losers. One example is the dual, and related, effects of economic globalization and climate change:

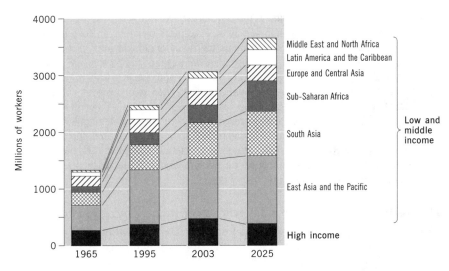

Figure 15.6   Distribution of the global labour force

Source: calculated from World Bank, 1995, *World Development Report*. Table 1; World Bank, 2005a: Table 2

> Both climate change and economic globalization are ongoing processes with uneven impacts, and both include implicit winners and losers … Double exposure refers to cases where a particular region, sector, ecosystem or social group is confronted by the impacts of both climate change and economic globalization. It recognizes that climate impacts are influenced not only by current socioeconomic trends, but also by structural economic changes that are reorganizing economic activities at the global scale … different outcomes emerge when the two processes are considered together.[27]

Regionally, although Africa is a 'double loser' overall, the situation is more varied than this generalization suggests. There are sectoral effects, as in the case of Mexican agriculture: 'farmers who are trying to compete in … international markets as well as agricultural wage labourers in Mexico are … likely to be double losers in terms of climate change and globalization'.[28] Among the most vulnerable social groups affected by double exposure are the poor residents of cities in both the developed and the developing worlds:

> At the same time that globalization is contributing to the economic vulnerability of disadvantaged residents of US cities, climate change may increase the physical vulnerability of these groups to weather-related events. Climate change may increase mean summer temperatures in Northern cities, and may also increase the frequency and magnitude of summer heat waves, consequently increasing heat-related illnesses and deaths … Residents of poor, inner-city communities are among the most vulnerable to heat waves due to lack of resources to pay for air conditioning or to leave stifling central city areas … Globalization and climate change thus represent a dual threat to these groups.

> For poor residents of cities in the developing world, the double impacts of globalization and climate change may be even more severe ... In conjunction with increased financial vulnerability as the result of globalization, poor residents of developing world cities are also among the groups that are most vulnerable to climatic change. Many of the urban poor live in shantytowns and squatter settlements located in precarious areas such as on hillsides ... Such areas are especially vulnerable to mudslides or flooding as the result of severe storms, events that may increase in both frequency and magnitude as the result of climate change. In addition to direct physical hazards, the urban poor are also vulnerable to climate change related health-hazards, particularly outbreaks of diseases such as cholera and malaria, both of which increase with warm spells and heavy rains.[29]

The argument, therefore, is that different sets of winners and losers from globalization may emerge when the effects of the two sets of global processes are superimposed on both those who are vulnerable and those who may benefit.

Conclusion

All parts of the world face major adjustment to the problem of making a living. But the nature of the problem, and certainly its perception, vary according to each country's position in the global system. In this respect, the view from the older industrialized countries is very different from the view from the newly industrializing economies and different again from the least industrialized countries. But position in the global economy is only part of the picture. It is far too simplistic to 'read off' a country's or a region's problems (and solutions) solely from its place in a global division of labour. Internal circumstances – cultural, social, political as well as economic – are of enormous importance. Nevertheless, there are problems affecting older industrialized countries, newly industrializing countries, and least developed countries in different ways as *groups* of countries as they grapple with the repercussions of global economic change and attempt to adjust to its employment impact. In the following two chapters we look, first, at the problems facing the developed or industrialized countries and, second, at the infinitely more intractable problems facing developing countries.

NOTES

1 Storper and Walker (1984: 37).
2 Friedman (2005).
3 Micklethwait and Wooldridge (2000: ix).
4 Kapstein (1999: 16).
5 Sen (1999).

6 Sen (1999: 20).

7 Amin (2004: 218).

8 UNDP (2005: 19, 21).

9 Bairoch quoted in Cohen (1998: 17).

10 See Kaplinsky (2001: 48–50), Wade (2004b), Wolf (2004: Chapter 9), World Bank (2001: Chapter 1).

11 Sklair (2001).

12 Carroll and Carson (2003).

13 Sklair (2001: 18–23).

14 Department of International Development (2000: 12).

15 Sen (1999: 106).

16 Population Reference Bureau (1999: 1).

17 Kennedy (2002: 3).

18 UN Population Division (2005: Table 1).

19 Cohen (2004: 27).

20 See Castles and Miller (1993), International Organization for Migration (2005).

21 Coe et al. (2003a).

22 UN Population Division (2001: vii).

23 ILO Press Release (24 January 2006).

24 FIET (1996: 16).

25 Luttwak (1999: 102).

26 ILO (2006a: 2).

27 O'Brien and Leichenko (2000: 227).

28 O'Brien and Leichenko (2000: 229).

29 O'Brien and Leichenko (2000: 229).

Sixteen
Good or Bad?: Evaluating the Impact of TNCs on Home and Host Economies

A counterfactual dilemma

In Chapter 8, we explored the complex interrelationships between TNCs and states, noting that such relationships are rarely straightforward: it is impossible to assert unequivocally that the advantage always lies with one or the other. This ambiguity also applies when we try to evaluate the impact of TNCs' activities on home countries (i.e. the effects of *outward* investment) or host countries (i.e. the effects of *inward* investment). Indeed, virtually every aspect of TNC operations – economic, political, cultural – has been judged in diametrically opposed ways by the proponents and opponents of TNCs. Depending on one's ideological viewpoint, TNCs may

- expand national or local economies or exploit them
- act as a dynamic force in economic development or as a distorting influence
- create jobs or destroy them
- spread new technology or pre-empt its wider use.

The major problem in trying to evaluate the impact of TNCs is that it is a *counterfactual* problem: we cannot fully measure what would have happened if the TNC activity did not exist. In other words, we generalize about the impact of TNCs at our peril. A realistic assessment must be based on a careful evaluation of specific cases. What is true in one set of circumstances may not be true in others. We need to avoid 'kneejerk' reactions, whether positive or negative.

The problem is made more intractable by the increasingly diverse and intricate modes of TNC operation. As we have seen, TNCs operate through a complex mix of intra-organizational and inter-organizational networks: the internalized networks of TNCs themselves, varying from centrally controlled hierarchies to flatter 'heterarchies'; and the externalized captive, relational and modular networks created through strategic alliances and various kinds of subcontracting and

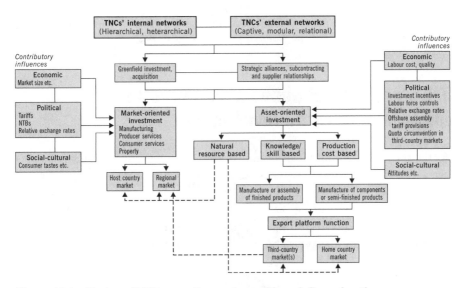

Figure 16.1   Modes of TNC operation and conditions influencing them

supplier relationships. Figure 16.1 provides a diagrammatic summary. These complex, geographically extensive TNC networks are grounded in specific places: organizational networks connect into geographical networks. As a result, firms in specific places – and, therefore, the places themselves and their inhabitants – are increasingly connected into transnational networks. The precise roles played by firms – and 'parts' of firms – in these networks have very significant implications for the communities in which they are based.

Four highly interconnected sets of relationships are especially relevant:

- *intra-firm relationships*: between different parts of a transnational business network, as each part strives to maintain or to enhance its position *vis-à-vis* other parts of the organization
- *inter-firm relationships*: between firms belonging to separate, but overlapping, business networks as part of customer–supplier transactions and other inter-firm interactions
- *firm–place relationships*: as firms attempt to extract the maximum benefits from the communities in which they are embedded and as communities attempt to derive the maximum benefits from the firms' local operations
- *place–place relationships*: between places, as each community attempts to capture and retain the investments (and especially the jobs) of the component parts of transnational corporations.

Each of these sets of relationships is embedded within and across *national/state* political and regulatory systems that help to determine the parameters within which firms and places interact. In this chapter, we look first at the 'home' country/community impact of TNC activities, and then at the other side of the coin: the impact of TNCs on 'host' countries/communities.

TNCs and 'home' economies: potential impacts of outward investment

As the number of countries of origin of TNCs has increased (Chapter 2), concern over the implications of firms' transferring some of their activities out of their home countries/communities has become more widespread. In the past, these were concerns largely restricted to the dominant developed economies (notably the United States, European countries and Japan) but as firms from some developing countries become TNCs this is no longer the case.

Questions and assumptions

What are the implications for the firm's home country of increased transnational production? Does it adversely affect the country's economic welfare by, for example, drawing away investment capital, displacing exports or destroying jobs? Or is it an inevitable feature of today's highly competitive global economy that forces firms to expand overseas in order to remain competitive?

- *Proponents* of overseas investment argue that the overall effects on the domestic economy will be positive, raising the level of exports and of domestic activity to a level above that which would prevail if overseas investment did not occur. Profits from overseas operations will flow back to the home country, enhancing the firm's competitive position and making funds available for investment in appropriate activities in the home country, as well as overseas. This will have a positive impact on home country employment.
- *Opponents* of overseas investment argue that the major effect will be to divert capital that could have been invested at home and to displace domestic exports. Profits earned overseas will be reinvested in other overseas ventures, rather than in creating new job opportunities at home.

A critical issue is the extent to which domestic investment could realistically be *substituted* for overseas investment. This raises a whole series of questions:

- What would have happened if the investment had not been made abroad?
- Would that investment have been made at home?
- Would the resources that went into the foreign investment have been used in higher levels of consumption and/or public services?
- What would have been the effect of foreign investment on domestic exports?

- Would the foreign sales of the product of the investment have been filled by exports from the home economy in the absence of the investment?
- Or would they have been taken over by foreign competitors?

Is there a choice?

It is virtually impossible to say with any certainty that overseas investment could equally as well have been made in the firm's home country. We can make various assumptions about what might have happened, but that is all. Ultimately, the key lies with the *motivations* that underlie specific investment decisions. As we have seen, firms invest abroad for a whole variety of reasons, for example:

- to gain access to new markets
- to defend positions in existing markets
- to circumvent trade barriers
- to diversify the firm's production base
- to reduce production costs
- to gain access to specific assets and resources.

It might be argued that foreign investment undertaken for *defensive* reasons – to protect a firm's existing markets, for example – is less open to criticism than *aggressive* overseas investment. The argument in the case of defensive investment would be that, in its absence, the firm would lose its markets and that domestic jobs would be lost anyway. Such investment might be made necessary by the erection of trade barriers by national governments, by their insistence on local production, or by the appearance of competitors in the firm's international markets. But, presumably, defensive investment might also include the relocation of production to low-wage countries in order to remain competitive. Here, the alternative might be the introduction of automated technology in the domestic plant that would lead to a loss of jobs anyway without any outward investment.

Although there may well be some clear-cut cases – particularly where access to markets is obviously threatened or where proximity to a localized material is mandatory – there will inevitably be many instances where there is substantial disagreement over the need to locate overseas rather than at home. The various interest groups will have different perceptions of the situation.

Potential employment impacts of outward investment

Establishing an overseas operation will have implications for the home country's balance of payments, through its influence on capital and financial flows and its effects on trade. But the most obvious implication for the average citizen is the effect on employment. The possible direct employment effects of outward investment fall into four categories:[1]

- *Export stimulus effect* (XE): employment gains from the production of goods for export created by the foreign investment which would not have occurred in the absence of such investment.
- *Home office effect* (HE): employment gains in non-production categories at the company's headquarters made necessary by the expansion of overseas activities.
- *Supporting firm effect* (SE): employment gains in other domestic firms supplying goods and services to the investing firm in connection with its overseas activities.
- *Production displacement effect* (DE): employment losses arising from the diversion of production to overseas locations and the serving of foreign markets by these overseas plants rather than by home country plants, that is, the displacement of exports.

Thus, the *net employment effect* (NE) of overseas investment on the home economy is:

$$NE = - XE + HE + SE + DE$$

Unfortunately, the data needed to disaggregate employment change into these components are rarely available so that, once again, large assumptions have to be made. That is why the estimates of the numbers of jobs either created or destroyed vary so widely – often by hundreds of thousands.

Area of impact	DIRECT		INDIRECT	
	Positive	Negative	Positive	Negative
Quantity	Creates or preserves jobs in home location, e.g. those serving the needs of affiliates abroad	Relocation or 'job export' if foreign affiliates substitute for production at home	Creates or preserves jobs in supplier/service industries at home that cater to foreign affiliates	Loss of jobs in firms/ industries linked to production/activities that are relocated
Quality	Skills are upgraded with higher-value production as industry restructures	'Give backs' or lower wages to keep jobs at home	Boosts sophisticated industries	Downward pressure on wages and standards flows on to suppliers
Location	Some jobs may depart from the community, but may be replaced by higher-skilled positions, upgrading local labour market conditions	The 'export' of jobs can aggravate regional/local labour market conditions	The loss of 'blue-collar' jobs can be offset by greater demand in local labour markets for high-value-added jobs relating to exports or international production	Demand spiral in local labour market triggered by layoffs can lead to employment reduction in home country plant locations

Figure 16.2   The potential effects of outward investment on home country employment

Source: based on UNCTAD *World Investment Report*, 1994: Table IV.1

Figure 16.2 summarizes the potential positive and negative aspects of both direct and indirect employment effects of outward investment in terms of three attributes: quantity of jobs, quality of jobs and location of jobs. In interpreting Figure 16.2 we need to bear in mind that the precise effects of outward investment on home country employment are highly contingent on the specific circumstances involved. But even though we cannot put precise numbers on jobs gained or lost through FDI, one thing is clear: *the winners and the losers are rarely the same.* At one time, it could be said with some accuracy that the dominant losers were production workers while the major gainers were white-collar managerial workers. But such a simple distinction no longer holds, as the recent rapid increase in the outsourcing/offshoring of white-collar jobs has shown. Also, given the complex geography of TNC networks (Chapter 5), it is almost inevitable that jobs created in the home country through outward investment effects (XE, HE, SE in the above formula) will be in *different places* from those where jobs are lost.

TNCs and 'host' economies: potential impacts of inward investment

Today, few parts of the world do not have a TNC presence – although, as we saw in Chapter 2, the distribution of FDI is geographically very uneven. Hence, we can think of places, at whatever geographical scale, as having an *organizational ecology* (Figure 16.3): a mix of firms and parts of firms, foreign and domestically owned, connected together through geographically extensive production circuits and networks. In other words, national and local economies are connected into much larger organizational and geographical structures. In the case of branches and affiliates they are obviously part of a specific corporate structure and will be constrained in their autonomy by parent company policy. The extent to which they are functionally connected into the local economy will be enormously variable. Even the 'independent' firms in a local economy may, in fact, be rather less independent than they appear at first sight. Some, at least, will be integrated into the supply networks of larger firms, again including TNCs, whose decision-making functions are very distant. Other local firms may be linked together through strategic alliances or they may be a part of the flexible business networks coordinated by key 'broker' firms.

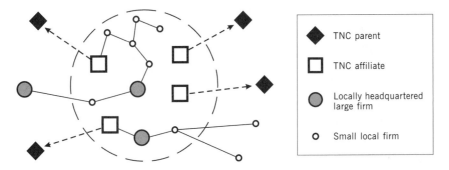

Figure 16.3   A place's 'organizational ecology'

The key issue, then, is to what extent the presence of the affiliates of foreign-owned/controlled TNCs creates net benefits or net costs for a local economy acting as a 'host' for such activities. There is no doubt that foreign direct investment has made major contributions to economic development as a whole, and this should not be underestimated. Indeed, without the capital, technology and access to markets that may be provided through involvement in transnational production networks, many places would be, in a material sense, far poorer. *Not being connected into these bigger structures is a major developmental problem*; hence the continuing scramble by states (and areas within states) to attract TNC activities. Nevertheless, the picture is not entirely positive. There are potential costs as well as benefits although, again, there are major conceptual and, especially, empirical problems in evaluating these.

In this section we concentrate on the *economic* impact of TNCs on host economies. However, they are extremely significant transmitters not only of economic change but also of *social, cultural and political* change. This is particularly the case for developing countries. A person employed in a TNC factory or office acquires not just work experience but also possibly a new set of attitudes and expectations. The effect of the employment of women is often to transform, not always favourably, family structures and practices. TNCs introduce patterns of consumption that reflect the preferences of industrialized country consumers. In this respect, the transnational advertising agencies are especially important. They 'are not simply trying to sell specific products in the Third World, but are engineering social, political and cultural change in order to ensure a level of consumption that is "the material basis for the promotion of a standardized global culture"'.[2]

In this discussion, we explore some of the broad areas of potential economic impact using Figures 16.4 and 16.5 as a basis. These should be referred to as we focus on specific issues. Figure 16.4 sets out the major dimensions of potential TNC impact and provides the logic for the way the discussion is organized. Figure 16.5 shows how complex the impact of a TNC presence may be and emphasizes the importance of tracing both direct and indirect effects. Such potential impacts are contingent upon the interactions between the nature of the foreign-owned operation – its mode of entry, functions and basic attributes – and the nature and characteristics of the host economy itself.

Injecting capital?

The inflow of capital is the most obvious impact of foreign investment, especially for those countries suffering from capital shortage. TNCs have certainly been responsible for injecting capital into host economies, both developed and developing. But not all new overseas ventures undertaken by TNCs involve the actual transfer of capital into the host economy. One estimate for the 1990s was that around 50 per cent of US foreign direct 'investment' was actually raised on host country capital markets and not imported.[3] Thus, local firms may be bought with

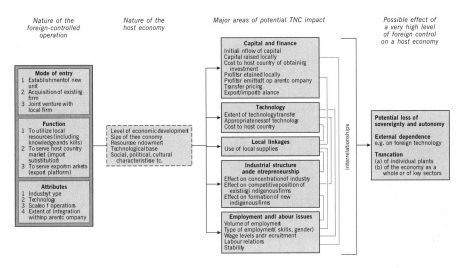

Figure 16.4 Major dimensions of potential TNC impact on host economies

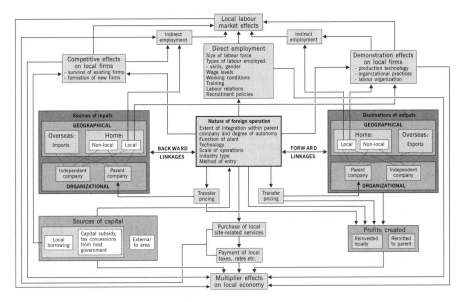

Figure 16.5 Making the connections: tracing the direct and indirect effects of a TNC on a host economy

local money. Local firms may even be squeezed out of local capital markets by the perceived greater attractiveness of TNCs as a use for local savings.

Even where capital inflow does occur there will, eventually, be a reverse flow as the foreign operation remits earnings and profits back to its parent company. This outflow may, in time, exceed the inflow. Any net financial gain to the host country also depends on the trading practices of the TNC. A host country's balance of payments will be improved to the extent that the foreign plant exports its output and reduced by its propensity to import. A vital issue, therefore, is the extent to which financial 'leakage' occurs from host economies through the channel of the TNC. This raises the question of the ability of host country governments to obtain a 'fair' tax yield from foreign firms, many of which are capable of manipulating the terms of their intra-corporate transactions through transfer pricing (see Chapter 8).

Transferring technologies?

As we saw in Chapter 3, technological change through innovation is the primary engine of economic growth and development. TNCs are major generators of technological change through their large-scale investments in R&D. In some respects, therefore, transferring technologies to host countries is a more important mechanism than transferring financial capital, especially given the comments above.[4] Hence, simply by locating some of its operations outside its home country the TNC engages in the geographical transfer of technologies. In this respect, technologies are, indeed, spread more widely. In so far as a foreign affiliate employs local labour there will be a degree of technology transfer to elements of the local population through training in specific skills and techniques. But the mere existence of a particular technology within a foreign-controlled operation does not guarantee that its benefits will be widely diffused through a host economy. The critical factor here is the extent to which the technology is made available to potential users outside the firm either directly through linkages with indigenous firms, or indirectly via 'demonstration effects'.

In fact, the very nature of the TNC inhibits the spread of its proprietary technologies beyond its own organizational boundaries. Such technologies are not lightly handed over to other firms. Control over their use is jealously guarded: the terms under which technologies are transferred are dictated primarily by the TNC itself in the light of its own overall interests. They tend to transfer the results of innovation but not the innovative capabilities – the 'know-how' rather than the 'know-why'. A major tendency, as we saw in Chapter 5, is for TNCs to locate most of their technology-creating activities either in their home country or in the more advanced industrialized, and some of the more advanced newly industrializing, countries. So far, relatively little R&D has been relocated to developing countries other than lower-level support laboratories. In some cases this is a direct result of host government pressure on TNCs to establish R&D facilities in return for entry. Such leverage is greatest where the TNC wishes to establish a branch plant to serve the host country market itself.

The evidence for TNCs transferring technologies beyond a fairly basic level to developing countries is very mixed. A study of the electronics industry in Malaysia, the Philippines and Thailand was fairly positive:

> Their [TNCs'] participation has at least produced latent technological capabilities for absorption by local firms ... foreign firms' participation and the high levels of ... [human resource and process technology] capabilities generated have at least transformed the local environment to facilitate export manufacturing in these countries involving a high-tech industry.[5]

On the other hand, the conclusions of a study of two Caribbean countries (Trinidad and Tobago and Costa Rica) were more pessimistic:

> Despite the attractiveness of both countries to foreign investors, foreign investment has made only a minimal contribution to strengthening local innovation systems in these countries.[6]

In addition to the extent to which TNCs actually transfer technologies to host countries there are two other important questions. One is whether or not the technologies being transferred are *appropriate* to local conditions and needs. This applies both to process technologies (are they labour or capital intensive?) and to products (are they relevant and accessible to local people?). In the case of process technologies, are they *environmentally* sound? TNCs are often accused of exporting technologies to developing countries that are environmentally objectionable or that are less safe than they should be. There have been claims that TNCs systematically shift some of their environmentally noxious or more hazardous operations to developing countries with less stringent environmental and safety standards. In other words, they make use of 'pollution havens'. Although this may indeed occur in specific cases, there is no evidence to suggest that this is the general practice.[7]

A rather different aspect of the problem relates to safety and environmental management. Industrial disasters, such as the one at the Union Carbide plant at Bhopal, India, in 1994, focused attention on the safety practices of TNCs. A frequent claim was that TNCs tend to adopt less stringent safety practices in their developing country plants than in their home plants. The more recent conflict in Nigeria involving Shell's environmental practices in Ogoniland raised both environmental and political issues. The now-collapsed – and infamous – US company Enron clearly rode roughshod over environmental regulations in many parts of the world with some devastating effects. These cases, and others, are serious in the extreme. But it is dangerous to generalize from them to produce universal statements on the environmental behaviour of *all* TNCs.

The second question relates to the *cost* to a host country of acquiring TNC technologies. The problem here is that technology is only one part of the overall package of attributes that the TNC brings to a host country; it is difficult to separate it out. Assessing the cost involved assumes that it can be measured against alternative ways of acquiring the same technologies. The two major alternatives are:

- to buy or license the technologies alone from its owner (the TNC) – that is, to 'unbundle' the TNC package
- to produce the technologies domestically.

The parallel usually drawn is with Japan, which rebuilt its post-war economy without the introduction of direct foreign investment, mainly by *licensing* technologies from Western firms. Although a great deal of technology is licensed by developing (and developed) countries from TNCs, it is not always a feasible alternative. A TNC may be unwilling to license the technology or it may charge an exorbitant price. It may be a question of the host country accepting the entire TNC package or getting nothing at all. The possibility of producing the technology domestically may be feasible for some of the more advanced industrial nations but is rarely so for developing countries.

Creating local linkages?

Inter-firm linkages are the most important channels through which technological change is transmitted. By placing orders with domestic suppliers for materials or components that must meet stringent specifications, technical expertise is raised. The experience gained in new technologies by local firms enables them to compete more effectively in broader markets, provided, of course, that they are not tied exclusively to a specific customer. The sourcing of materials locally may lead to the emergence of new domestic firms to meet the demand created, thus increasing the pool of local entrepreneurs. The expanded activities of supplying firms, and of ancillary firms involved in such activities as transportation and distribution, will result in the creation of additional employment. But such beneficial spin-off effects will occur only *if* the foreign affiliates of TNCs *do* become linked to local firms. Where TNCs do not create such linkages they remain essentially foreign enclaves within a host economy, contributing little other than some direct employment.

As far as local linkages are concerned, the most significant are *backward* or *supply* linkages (see Figure 16.5). Here, the crucial issue is the extent to which TNCs either import materials and components or procure them from local suppliers. The actual incidence of local linkage formation by foreign-controlled plants depends upon three major influences:

- *The particular strategy followed by the TNC and the role played by the foreign operation in that strategy*: TNCs that are strongly vertically integrated at a global scale are less likely to develop local supply linkages than firms with a lower degree of corporate integration. But even where vertical integration is low, the existence of strong links with independent suppliers in the TNC's home country or elsewhere in the corporate network may inhibit the development of local linkages in the host economy. Familiarity with existing supply relationships may well discourage the development of new ones, particularly where the latter are perceived to be potentially less reliable or of lower quality. Foreign plants that serve the host market are more likely to develop local supply linkages than export platform plants.

- *The characteristics of the host economy*: in general, we would expect to find denser and more extensive networks of linkages between TNCs and domestic enterprises in developed, compared with developing, economies. Within developing countries such linkages are likely to be greatest in the larger and more industrialized countries than in others. In many developing countries, the existing supplier base is simply not sufficiently developed to meet TNC criteria, that is, the *absorptive capacity* is too low. However, host country governments may well play a very important role in stimulating local linkages, both by implementing policies to upgrade local suppliers and through local content policies. But much depends on the relative strength of the host country's bargaining power *vis-à-vis* the TNC. Again, it tends to be in the larger and the more industrialized developing countries that such local content policies have the greatest impact, and also in those TNC activities serving the local market. Indeed, it could be that the export-oriented industrialization strategies of developing countries actually inhibit the development of local supply linkages.
- *Time*: local supply capabilities do not develop overnight. Particularly in view of the closer relationships between firms and their suppliers which have been emerging (see Chapter 5) it should not be expected that a foreign plant, newly established in a particular host country, would immediately develop local supplier linkages. Not only do appropriate suppliers have to be identified but also it takes time for supplier firms to 'tune in' to a new customer's needs.

The linkage question is not only a quantitative one. More important than the number of TNC linkages is their *quality* and the degree to which they involve a beneficial transfer of technology (either production or organizational) to supplier firms. A common criticism is that many TNCs tend to procure only 'low-level' inputs from local sources, for example cleaning services and the like. This may be because of deliberate company policy to keep to established suppliers of higher-level inputs or because such inputs are simply not available locally (or are perceived not to be so). Where development of higher-level supply linkages occurs there does seem to be a positive effect on supplier firms. Figure 16.6 summarizes the differences between 'dependent' and 'developmental' linkages. Clearly, from a host country perspective, the aim must be to achieve a linkage structure that is developmental. In this respect, much will depend upon its bargaining power (see Chapter 8).

Empirical evidence of local linkage formation by TNCs presents a very uneven picture.[8] Studies within smaller developing countries, particularly those with a short history of industrial development, tell a fairly uniform story of shallow and poorly developed supply linkages between local firms and foreign-controlled plants. A common observation is that foreign plants located in export processing zones (EPZs) are particularly unlikely to develop supplier linkages with the wider economy. In the case of the Mexican *maquiladora* plants, for example, less than 5 per cent of the inputs used are sourced from within Mexico. Additionally, most

Attribute	Dependent structure	Developmental structure
Form of local linkages	Unequal trading relationships Conventional subcontracting Emphasis on cost-saving	Collaborative, mutual learning Basis in technology and trust Emphasis on added value
Duration and nature of local linkages	Short-term contracts	Long-term partnerships
Degree of local embeddedness of inward investors	Weakly embedded Branch plants restricted to final assembly operations	Deeply embedded High level of investment in decentralized multifunctional operations
Benefits to local firms	Markets for local firms to make standard, low-technology components Subcontracting restricts independent growth	Markets for local firms to develop and produce their own products Transfer of technology and expertise from inward investor strengthens local firms
Prospects for the local economy	Vulnerable to external forces and 'distant' corporate decision-making	Self-sustaining growth through cumulative expansion of linked firms
Quality of jobs created	Predominantly low-skilled, low-paid jobs May be high level of temporary and casual employment	Diverse, including high-skilled, high-income employment

Figure 16.6   Dependent and developmental linkage structures

Source: based on Turok, 1993: Table 1

of those inputs are low-value and low-technology products whose production does little to upgrade the local technological and skill base.

In some cases, there may be a considerable amount of local sourcing but with relatively little involvement of genuinely local firms. For example, although the new foreign manufacturing plants established in the Johor region of southern Malaysia 'are sourcing a large part of their inputs in Johor ... the regional effect is confined to foreign, mainly Japanese and Singaporean, suppliers. As a result, the linkages of the new manufacturing plants are only in part beneficial to the local economy'.[9] Overall, Japanese firms in the Malaysian electronics industry tend 'to rely more heavily on relocated suppliers from their home country, supporting the general belief about the effect of Japanese business ties'.[10]

> In the electronics industry, sourcing patterns appear to differ significantly by host country. For example, in 2001, foreign affiliates in the colour TV industry in Tijuana, Mexico, sourced about 28 per cent of their inputs locally, of which only a very small proportion (3 per cent) was supplied by Mexican-owned firms ... in Malaysia, locally-procured components by foreign affiliates in the electronics and electrical industries comprised 62 per cent of exports in 1994; the corresponding figure for Thailand was 40 per cent. However, in both countries, *the most strategic parts and components were supplied mainly by foreign-owned companies rather than domestic ones*.[11]

This latter point is confirmed by a study of 227 Japanese electronics companies operating in 24 countries[12] which found that although local procurement was widespread, such increase in local content did not necessarily involve local suppliers:

- Local content was achieved primarily through the use of group (*keiretsu*) suppliers rather than through the use of local firms, despite the intentions of state local content policies.
- The mode of entry of the TNC was highly significant. Acquired firms and joint ventures displayed much higher linkage with local suppliers than green-field investments.
- Operations established to serve the domestic market had higher local linkages than export-oriented operations.

Of course, these are anecdotal examples and many influences are involved. Nevertheless, they do indicate that the creation of genuinely local linkages by TNCs is far from certain. Of course, attracting tied suppliers may well be beneficial to the host economy, but this does not directly enhance the local supply base.

Squeezing or stimulating local firms?

Although not all TNCs are giant corporations, it is often the case that foreign plants are larger than their domestic competitors, especially in developing countries. Hence, the entry of a foreign plant into a host economy may have a number of repercussions on the structure of domestic industry, particularly on the competitiveness, survival and birth of domestic enterprises. But, as in other aspects of TNC impact, there is no inevitability about such structural effects. Much will depend upon the specific domestic context itself and on the relative size and market power of the TNC affiliate in that context.

In general, the difference between a foreign enterprise and a domestic enterprise in developed market economies, especially those that are themselves sources of TNCs, will be far less than that in developing countries. The industrial structure of most developing economies tends to be much more strongly *dualistic,* with a small, technologically advanced sector (relatively speaking) which is oriented to the more modern urban market and a technologically less advanced sector characterized by traditional production and attitudes. TNCs in developing countries are, by definition, part of the technologically advanced sector of the host economy.

A major long-term effect of the entry of TNCs into a host economy – both developed and developing – is likely to be an increase in the level of *firm concentration*. The number of firms is likely to be reduced and the dominance of very large firms increased. Thus, two important *negative* effects of TNCs on host economies are:

- the possible squeezing out of existing domestic firms
- the suppression of new domestic enterprises.

Both of these fears have been voiced especially strongly in the case of developing countries but there is no reason to believe that they do not also apply to particular sectors in developed economies if high TNC penetration has occurred.[13] But, clearly, the less developed the domestic sector, the greater is the likelihood of its being swamped by foreign entry and of local entrepreneurship being suppressed.

However, we should beware of assuming that the involvement of TNCs in a particular host economy will inevitably destroy or suppress domestic enterprise. There may well be *positive* effects:

- Where substantial local linkages are forged by foreign plants, particularly on the supply side, opportunities for local businesses may well be enhanced. Existing firms may receive a boost to their fortunes or new firms may be created in response to the stimulus of demand for materials or components.
- The formation of new enterprises may be stimulated through the 'spin-off' of managerial staff setting up their own businesses on the basis of experience and skills gained in employment with the foreign firm.

Creating good jobs?

For most ordinary people, as well as for many governments, the most important issue in the debate over the TNC is its effect on jobs:

- Does the entry of a foreign-controlled plant create new jobs?
- What kinds of jobs are they?
- Do TNCs pay higher or lower wages than domestic firms?
- Do TNCs operate an acceptable system of labour relations?

Number of jobs

The number of jobs created (or displaced) by a TNC operation in a host economy consists of both those created in the operation itself and those created/displaced elsewhere in the domestic economy.

The number of *direct jobs* created in a particular TNC operation depends upon two factors:

- the *scale* of its activities
- the *technological* nature of the operation, particularly on whether it is capital intensive or labour intensive.

The number of *indirect* jobs created also depends upon two major factors:

- the extent of *local linkages* forged by the TNC with domestic firms
- the *amount of income generated* by the TNC and *retained* within the host economy. In particular, the wages and salaries of TNC employees and of those in linked firms will, if spent on locally produced goods and services, increase employment elsewhere in the domestic economy (Figure 16.5).

Against the number of jobs *created* in, and by, TNCs we need to set the number of jobs *displaced* by any possible adverse effects of TNCs on domestic enterprises (see the previous section). Hence, the overall employment effect of TNCs in host economies depends upon the balance between job-creating and job-displacing forces. The *net* employment contribution of a TNC to a host economy, or the net jobs (NJ), can therefore be expressed as:

$$NJ = DJ + IJ - JD$$

where DJ is the number of direct jobs created in the TNC, IJ is the number of indirect jobs created in firms linked to the TNC, and JD is the number of jobs displaced in other firms.

Quality of jobs

The number of jobs created by TNCs in host economies is only part of the story. What kind of jobs are they? Do they provide employment opportunities that are appropriate for the skills and needs of the local labour force? The answer to these questions depends very much on the attributes of the foreign operation (see Figure 16.5). In particular, where the operation 'fits' into the TNC's overall structure and how much decision-making autonomy it has are key factors. In general, the fact that TNCs tend to concentrate their higher-order decision-making functions and their R&D facilities in the developed economies produces a major geographical bias in the pattern of types of employment at the global scale.

In developing countries, the overwhelming majority of jobs in TNC plants are *production* jobs. In export processing zones, of course, low-level production jobs, especially for young females, are the rule, although this partly reflects the types of industry that dominate in EPZs. Overall, the proportion of higher-skilled workers employed by TNCs in developing countries has tended to increase over time, as has the proportion of local professional and managerial staff. Such changes have progressed furthest in the more advanced industrializing countries of Asia. For example, the shift of IT activities to India involves more than low-level call centre jobs:

> Anyone who assumes J.P. Morgan will simply be doing low-level 'back office' tasks in the country – a bit of data entry and paper-shuffling – would be flat wrong. One task for the new recruits is to settle complex structured-finance and derivative deals, what one insider calls 'some of the most sophisticated transactions in the world'.[14]

Even so, the TNC labour force in developing countries remains concentrated in low-skill production and assembly occupations. The experience of individual developing countries varies in the extent of TNC-induced labour upgrading, as a study of the TV industry in East Asia shows.[15] In each case, the extent of human capital formation in the industry was very limited prior to the mid 1990s:

- In Malaysia, specialized staff were still foreign but there was significant training by leading firms and for their partners in their regional TNC networks. There was evidence of rising skill levels and increasing numbers of specialized technical and managerial staff.
- In Mexico, the first signs of upgrading were apparent as a result of significant training efforts and linking with local education institutions, rising labour skill levels, and increasing numbers of specialized technical and managerial staff.
- In Thailand, the skill levels of the labour force were low but rising with increased emphasis on labour training. But there was not much evidence of an increasing involvement of more highly educated specialist staff.

Wages and salaries

In so far as TNCs take advantage of geographical differences in prevailing wage rates between countries they do, in fact, 'exploit' certain groups of workers. The exploitation of cheap labour in developing countries at the expense of workers in developed countries is one of the major charges levelled at the TNC by labour unions in Western countries (an issue we will return to in Chapter 17). The major problem here seems to be in relation to conditions in subcontracting companies rather than in the TNCs themselves, and there is much controversy over this issue. The general response of TNCs facing such allegations is that they do not have complete control over what goes on in independent factories although, in the face of these criticisms, many TNCs are now implementing codes of practice to which their subcontractors must conform (see Chapter 19).

However, as far as their *directly owned affiliates* are concerned, the general consensus seems to be that TNCs generally pay either at or above the 'going rate' in the host economy. Figure 16.7 shows two aspects of the wage comparison. First, the relative height of the columns shows that TNCs pay very much more to workers in high-income countries than to those in middle- and low-income countries. This differential reflects a number of factors, including the composition of economic activity, educational and skill levels, cost of living and so on. Second, Figure 16.7 shows that although TNCs certainly pay higher wages overall than domestic firms in the same country group (a ratio of 1.5), the pattern varies between country groups: 1.4 in high-income countries, 1.8 in middle-income countries and 2.0 in low-income countries.

TNCs that do pay above the local rate may well 'cream off' workers from domestic firms and possibly threaten their survival. This point relates to the kinds of *recruitment policies* used by TNCs. Invariably, TNCs tend to operate very careful screening procedures when hiring workers. This may well mean that employees for a newly established foreign plant are drawn from existing firms rather than from the ranks of the unemployed. Another aspect of recruitment, at least in some industries, is the extent to which TNCs recruit particular types of workers to keep labour costs low. In the textiles, garments and electronics industries, for example,

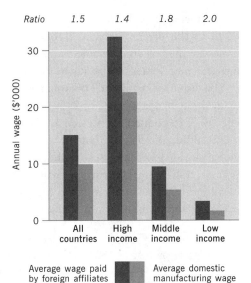

Figure 16.7   Differences in average wages paid by foreign affiliates of TNCs and domestic firms

Source: based on Crook, 2001: 15

there is a very strong tendency to prefer females to males in assembly processes and, in some cases, to employ members of minority groups as a means of holding down wage costs and for ease of dismissal. But such practices appear to be specific to particular industries and should not necessarily be regarded as universally applicable to all TNCs in all industries.

Labour relations

In many developing countries, either labour is weakly organized or labour unions are strictly controlled (or even banned) by the state. Even in developed economies some major TNCs simply do not recognize labour unions in their operations, while deregulation of labour markets has become widespread. But most TNCs, however reluctantly, do accept labour unions where national or local circumstances make this difficult to avoid. Whether labour unions are involved or not, the question of the nature of labour relations within TNCs focuses on whether they are 'good' or 'bad', that is, harmonious or discordant. Some studies suggest that TNCs tend to have better labour relations in their plants than domestic firms; others point to a higher incidence of strikes and internal disputes in TNCs. But it is often difficult to separate out the 'transnational' element. In the case of strikes, for example, it may be plant or firm size that is the most important influence rather than nationality of ownership.

One of the most acute concerns of organized labour is that decision-making within TNCs is too remote: that decisions affecting work practices and work conditions, pay and other labour issues are made in some far-distant corporate headquarters which has little understanding or even awareness of local circumstances. Some labour relations decisions are far more centralized than others, either being made at corporate headquarters or requiring its approval. These areas tend to relate to the parent company's concern to control financial and labour costs. However, there is considerable variation between TNCs in their degree of head-quarters' involvement in labour relations.

The dispersed nature of TNC operations and the tendency towards remoteness in decision-making have made it very difficult for labour unions to organize effectively to counter such issues as plant closure or retrenchment. Two develop-ments, although relatively limited so far, are significant.[16] One is the initiation by global union federations (such the International Confederation of Free Trade Unions, ICFTU) of networks of workers within specific TNCs in an attempt to move industrial relations issues to the global level. The second development has occurred within the European Union:

> As part of the social protocol of the 1993 Maastricht Treaty, at least 15 million employees in some 1500 [TNCs] operating in Europe now have rights to infor-mation and consultation on all matters that affect more than one member state ... Each company that employs more than 1000 people, of whom at least 150 are located in two (or more) member states, has to meet the representation, transportation, accommodation, and translation costs of bringing employee representatives from across the European Union on an annual basis ... they provide a new opportunity for workplace representatives to meet their counterparts from every division of their companies' operations in Europe.[17]

Despite such developments, labour unions remain primarily contained within national state boundaries while TNCs are not. This structural difference creates inevitable tensions.

Dangers of over-dependence on TNCs

It should now be clear that no unambiguous evaluation of TNC impact on host economies can be made. Whether a foreign plant creates net costs or net benefits will depend on the *specific context*: the interaction between the attributes and func-tions of the plant itself within its corporate system and the nature and characteris-tics of the host economy. It also depends, critically, on the alternatives realistically available. But what if a host economy develops a very high level of foreign TNC involvement? As Table 2.4 shows, there are some countries which do, indeed, have a huge dependence on inward FDI. Does the presence of such a large foreign-controlled sector tip the balance of evaluation?

In a long-term sense, the answer would appear to be 'yes'. Whereas the involve-ment of some foreign activities in a host economy will have beneficial effects – not only in creating employment but also in introducing new technologies and business practices – overall dominance by foreign firms is almost certainly undesirable from a host country viewpoint. There are real dangers in acquiring the status of a *branch plant economy*. Precisely what constitutes an undesirable level of foreign penetration is open to debate. Indeed, a country may be dominated by, and dependent upon, external forces even where there is very little direct foreign investment in the economy. This may occur, for example, where a large segment of an economy is engaged in subcontracting work for foreign customers. Here, however, our concern is with the effects of a large *direct* foreign presence. Until recently, most of the debate was focused upon developing countries and formed part of the broader dependency debate. But such concern is no longer confined to developing countries. Most obviously, those developed economies that have a very high level of foreign direct investment also share the same kinds of problem.

A high level of dependence on foreign enterprises *potentially* reduces the host country's sovereignty and autonomy: its ability to make its own decisions and to implement them. At the heart of this issue are the different – often conflictual – goals pursued by nation-states on the one hand and TNCs on the other (see Chapter 8). Each is concerned to maximize its own 'welfare' (in the broadest sense). Where much of a host country's economic activity is effectively controlled by foreign firms, non-national goals may well become dominant. It may be extremely difficult for the host government to pursue a particular economic policy if it has insufficient leverage over the dominant firms.

The tighter the degree of control exercised by TNCs within their own corporate hierarchies, and the lower the degree of autonomy of individual plants, the greater this loss of host country sovereignty is likely to be. In the *individual* case this may not matter greatly, but where such firms *collectively* dominate a host economy or a key economic sector it most certainly does matter. The most significant aspect of dependence upon a high level of foreign direct investment is *technological*: the inability to generate the knowledge, inventions and innovations necessary to generate self-sustaining growth. However, this is not to argue that foreign investment should be avoided completely. On the contrary. What should be avoided by host economies is an *excessive* degree of foreign penetration.

Ownership and control matter in other ways.

> What happens at headquarters matters. In any organization, the ablest people gravitate to power. Head office is the place where political influence is exerted, and on which political pressure is exerted. Modern business is multinational but the companies that take part in it retain an identifiable nationality.[18]

Conclusion

The purpose of this chapter has been to explore the potential implications of TNC activities for those economies in which they operate. In the cases of both home and host country impacts, the analytical problem is the same: it is a counterfactual situation. We cannot unambiguously establish what would have happened *if* the TNC were *not* involved in a particular case. Assumptions have to be made as to what the alternative situation would realistically be. The 'bottom line' is the *net effect* that takes into account the opportunities forgone by the presence or absence of the TNC. But the situation has become vastly more complex in today's highly interconnected global economy.

From a home economy's viewpoint, what would be the effect if the country's firms were not involved in overseas activities? Can a firm opt out of international operations in today's increasingly global production environment? Would jobs at home be lost anyway even if the firm did not invest overseas? Would the economy be better or worse off? Everything will depend, as we have seen, on whether the overseas investment is obligatory or discretionary, but even here differences of viewpoint will exist. From a host economy's viewpoint, could the particular item of technology, the particular level of employment, and so on, be created without the involvement of the TNC? In some cases, the answer will undoubtedly be 'yes': in others, the answer will just as undoubtedly be 'no'. What is quite clear, however, is that *TNCs tie national and local economies more closely into the global economy*.

NOTES

1 Hawkins (1972).
2 Sklair (1995: 167).
3 UNCTAD (1995: 142–3).
4 See Mytelka and Barclay (2004). The potential for development of R&D by TNCs is the major theme of UNCTAD's *World Development Report 2005*.
5 Rasiah (2004: 619).
6 Mytelka and Barclay (2004: 553).
7 Smarzynska and Wei (2001).
8 UNCTAD (2000; 2001).
9 Van Grunsven et al. (1995: 3).
10 Linden (2000: 210).
11 UNCTAD (2001: 135, emphasis added).
12 Belderbos and Capannelli (2001).
13 Driffield and Hughes (2003).
14 *The Economist* (17 December 2005).
15 UNCTAD (2001).
16 Wills (1998).
17 Wills (1998: 122).
18 Kay (2006: 25).

Seventeen
Making a Living in Developed Countries: Where Will the Jobs Come From?

Increasing affluence – but not everybody is a winner

At the global scale the developed countries are clearly 'winners'. They continue to contain a disproportionate share of the world's wealth, trade, investment and access to modern technologies (especially information technologies). But if we refocus our lens to look at what is happening *within* the developed world, we find wide variations in economic well-being, both between individual countries and – perhaps more surprisingly – within individual countries, even though most people in developed countries are significantly better off than in the past.

> It is remarkable that the extent of deprivation for particular groups in very rich countries can be comparable to that in the so-called third world. For example, in the United States, African Americans as a group have no higher – indeed have a lower – chance of reaching advanced ages than do people born in the immensely poorer economies of China or the Indian state of Kerala (or in Sri Lanka, Jamaica or Costa Rica).[1]

In this chapter we examine the position of people in developed countries in terms of their access to employment, and to the income that such employment brings, in the context of a globalizing world economy. The discussion is organized as follows:

- First, we look at general trends in employment, unemployment and incomes and at variations between social groups and between different parts of individual countries.
- Second, we explore the relationships between these trends and the processes we have been discussing throughout this book.
- Third, we look at some of the 'policy fixes' that have been proposed to alleviate the problems of 'Where will the jobs come from?'

What is happening to jobs and to incomes?

Employment

During the past 50 years, two particularly important developments have occurred in the employment structure of developed economies:

- the displacement of jobs in manufacturing industries by jobs in services
- the increasing participation of women in the labour market.

The shift from manufacturing to services

One of the most striking trends, since at least the 1960s but, especially since the 1970s, has been the disappearance of manufacturing jobs (Figure 17.1), leading to the view that developed economies have become *deindustrialized*[2] and that they are now effectively service economies. Today, around three-quarters of their labour forces are employed in service occupations – even higher in the United States. In recent years, virtually all the net employment growth in the developed economies has been in services.

A common criticism levelled at the new service jobs is that they are essentially poorly paid, low skilled, part-time and insecure – at least compared with the kinds of jobs in manufacturing that were characteristic of the developed countries until the 1960s. There is certainly some truth in this. Many of the new service jobs are, indeed, 'McJobs'. But that isn't the entire story. An OECD report suggested that

> most of the growth in new private services jobs in western industrialized countries is well-paid and skilled … the expansion in service employment brought faster growth in the 1990s in high-paid than low-paid work … 'There does not appear to be any simple trade-off between job quality and employment.' While the US has a higher proportion of its working-age population employed in low-paying jobs than in most other OECD countries, it also has a higher proportion in higher-paying jobs.[3]

Indeed, at least some of the shift from manufacturing to services may be more apparent than real. Many so-called 'service' jobs are performed within manufacturing firms as the nature of production circuits has become more complex.

Women's work

The shift in the balance of employment towards services has been closely associated with the increasing participation of women in the labour force.[4] In all developed economies, the changing roles of women, away from an automatically assumed domestic role, has gone hand-in-hand with the growth of service jobs. Although women are certainly employed in manufacturing industries, their relative importance is far higher in service industries. This is especially so where there are greater opportunities for part-time work, which allows women with families a degree of flexibility to combine a paid job with their traditional gender roles.

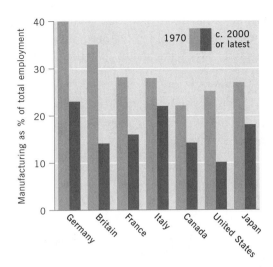

**Figure 17.1 The declining
share of manufacturing in total
employment**

Source: based on *The Economist*,
1 October 2005: Chart 1

Female participation in the labour market has increased in virtually all countries.
In the United States, for example, it was around 38 per cent in 1960; today it is
60 per cent. In the United Kingdom, a similar trend is evident: from 40 per cent
in 1971 to 55 per cent today. But, as Figure 17.2 shows, there is considerable variation between countries. The highest female participation rates are found in the
Scandinavian countries and the Netherlands; significantly lower rates are found in
Germany, France, Spain, Belgium and Italy.

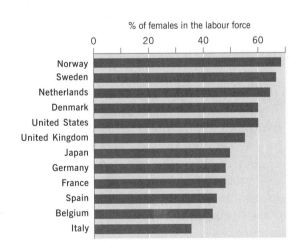

**Figure 17.2 Variations in
female participation in the
labour markets of
developed countries**

Source: based on ILO, 2001:
Table 2

Geographical contrasts in job creation

There are large differences between individual developed economies in their rates
of job creation. In terms of *total* jobs created, there is no doubt that, in recent
years, the United States has been in a different league from most of the other

industrialized countries. During the 1990s almost 22 million new jobs were created there.[5] Between the mid 1980s and mid 1990s, total employment in the United States grew almost four times faster than in the European Union and 50 per cent faster than in Japan. But this differential is now far less pronounced. As Figure 17.3 shows, rates of employment growth overall have been modest – and variable – for all the developed economies. Between 1996 and 2005, US employment grew by 1.5 per cent, EU employment grew by 1.3 per cent. Within the EU, the UK has had significantly higher employment growth rates in recent years.

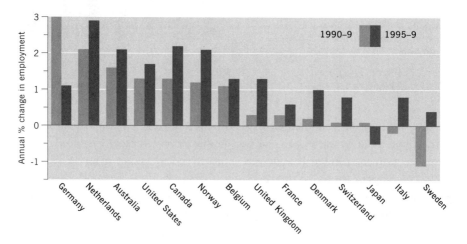

Figure 17.3   Differences in employment growth in selected developed economies

Source: based on ILO, 2001: Table 1.10

Unemployment

Levels of unemployment

The obverse of employment growth is, of course, *unemployment*. Compared with the 1960s and early 1970s – the so-called 'golden age of growth' – unemployment rates in the industrialized countries have increased dramatically (Figure 17.4). Between 1960 and 1973, unemployment rates were highest in the US and lowest in Japan, with the EC12 and the UK falling in between. But the post-1973 experience was very different. Whereas both Japan and the United States experienced an increase in the general level of their unemployment – more so in the case of the United States – the European experience was appalling, with average unemployment rates in the EC12 rising to above 12 per cent in the early 1990s compared with less than 4 per cent in Japan and a little over 5 per cent in the United States.

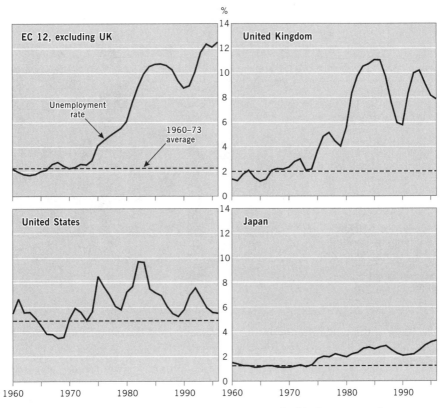

Figure 17.4   Unemployment rates 1960–96 compared with the average for 1960–73

Source: based on ILO, 1997: Figure 3.1

However, the late 1990s and early 2000s saw a considerable improvement in the unemployment situation in most developed countries, as Figure 17.5 indicates. In every case – with the exception of Japan and Germany – the unemployment rate in 2000–2 was below that of 1990–2. But such rates are very volatile. The sharp economic slowdown in the United States, coupled with the economic shock of the events of 9/11, was rapidly reflected in waves of corporate job reductions. The level of unemployment in Germany in the 1990s was heavily influenced by the effects of political reunification, which brought major problems of economic adjustment reflected in increased levels of unemployment, especially in the former East Germany.

The unemployment situation in Japan represents a particularly massive change. Historically, unemployment rates in Japan have been extremely low. A combination of a rapidly growing economy and a very strong orientation towards job security in the large-company sector of the economy sustained lower rates of unemployment than in any other industrialized country for almost 30 years. But

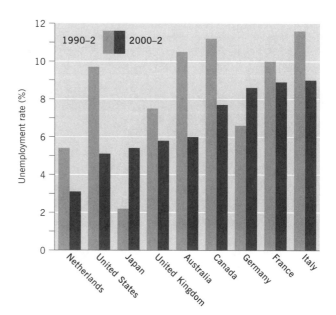

Figure 17.5   Rates of unemployment 1990–92 and 2000–2

Source: World Bank 2005a: Table 2.4

the burst of the bubble economy at the end of the 1980s and persistent domestic recession throughout the 1990s has changed this. The lifetime job system is crumbling: the days of the 'salaryman' appear to be numbered. As a result – and much to the dismay of Japanese society – Japanese unemployment rates are now comparable with those in the United States.

Selective impacts

Although the general level of unemployment, including long-term unemployment, remains very high in most industrialized countries, it is a *socially selective* process. For example, males aged between 25 and 54 years, with a good education and training, are far less likely to be unemployed, on average, than women, younger people, older workers and minorities. Most of these latter categories tend to be unskilled or semi-skilled workers. The vulnerability of women and young people to unemployment reflects two major features of the labour markets of the older industrialized countries. First, as we have seen, the increased participation of women in the labour force – particularly married women – has increased dramatically. A large proportion of these are employed as part-time workers in both manufacturing and services, especially the latter. Second, *youth unemployment* during the 1980s partly arose from the entry on to the labour market of vast numbers of 1960s 'baby boom' teenagers. In most industrialized countries, therefore, unemployment rates among the young (under 25 years) have been roughly twice as high as those for the over-25s. In some cases youth unemployment is three times higher than adult unemployment.

On the other hand, the demographic trends in the developed countries discussed in Chapter 15 mean that, over the medium and long term, the unemployment problem should ease. In fact, there is likely to be an increasing need for immigrant workers to fill the gaps created by the 'greying' of the populations in all the developed economies. This, of course, raises big social and political issues (see later in this chapter).

Unemployment tends to be especially high among *minority groups* within a population. In the United States, for example, unemployment among black youths can be 150 per cent higher than among white youths. Similarly, unemployment rates among Hispanic youths are at least 50 per cent higher than among white youths. In Europe the problem of minority group unemployment reflects the large-scale immigration of labour in the boom years of the 1960s.

Incomes

For most people, income is derived from their employment. During the long economic boom of the 1960s and early 1970s, most people became better off. This is no longer the case. Inequality is increasing even within the developed economies and this has huge implications for the overall well-being of societies.[6] Figure 17.6 shows the trends in the ratio of the 10 per cent highest-paid to the 10 per cent lowest-paid workers in leading industrialized countries between the mid 1980s and mid 1990s, the latest years for which comparable data are available. Most striking is the extent to which the United States and the United Kingdom stand out from the rest. In those two countries, the gap between the highest and lowest paid increased by more than one-third.

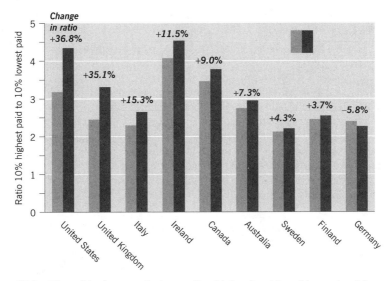

Figure 17.6   The widening gap between the highest-paid and lowest-paid workers in developed economies

Source: Based on ILO 2004b: Figure 16

> Even more striking has been the sharp increase in the share of the top 1 per cent of income earners in the United States, United Kingdom and Canada ... In the United States, the share of this group reached 17 per cent of gross income in 2000, a level last seen in the 1920s. This increased concentration in wealth has been the prime factor in the rise in income inequality in the United States; the declining share of the bottom decile of wage earners has been in reverse since 1995.[7]

The pattern is more varied across the other industrialized countries shown in Figure 17.6. For example, the same degree of increasing income dispersion within the labour force did not occur in many of the continental European countries. In the case of Germany the gap actually narrowed slightly. On the other hand, these countries experienced much higher levels of unemployment than the United States in particular and even the United Kingdom between the mid 1970s and mid 1990s (Figure 17.4). This suggests that labour market adjustments occurred in different ways in different countries. In the United States and the United Kingdom, adjustment has been primarily in the form of a relative lowering of wages at the bottom of the scale and a consequent *increase in income inequality*. In Western Europe, on the other hand, such wage levels may have been maintained at the expense of jobs with an *increase in unemployment*.

> Indeed, it is said – probably too simplistically – that the United States has accepted greater inequality, whereas Europe has accepted more unemployment.[8]

These differences of detail between individual countries reflect specific social policies, most notably the contrast between the neo-liberal market capitalist systems of the United States and the United Kingdom and the social market systems that still prevail (though perhaps less so than in the past) in continental Europe (see Chapter 6). But it is generally the case – as with employment – that it is the less skilled workers who have been most adversely affected.

> as compensation has fallen for the unskilled worker, it has increased mightily for highly educated workers ... in 1979 male workers with a college degree earned on average about 50 per cent more than unskilled workers; by 1993 that difference was nearly 90 per cent. To put the inequality problem in its starkest terms, between 1979 and 1994 the upper 5 per cent of American families captured *99 per cent* of the nation's per capita gains in gross domestic product! That is, with a mean family gain over this period of $4,419, $4,365 went to the upper 5 per cent.[9]

Uneven micro-geographies

Not only do employment creation, unemployment and income differentials vary between countries, but also there are huge differences *within* countries: there are distinct *micro-geographies*.[10] Within the older industrialized countries, three broad trends are apparent:

- *Broad interregional shifts* in employment opportunities, as exemplified by the relative shift of investment from 'Snowbelt' to 'Sunbelt' in the United States, and from north to south within the United Kingdom.
- *Relative decline of the large urban-metropolitan areas* as centres of manufacturing activity and the growth of new manufacturing investment in non-metropolitan and rural areas.
- *Hollowing out of the inner cities of the older industrialized countries*: in virtually every case, the inner urban cores have experienced massive employment loss as the focus of economic activity shifted first from central city to suburb and subsequently to less urbanized areas.

Thus, deindustrialization has been experienced most dramatically in the older industrial cities as well as in those broad regions in which the decline of specific industries (including agriculture) has been especially heavy. In many cases, the vacuum left by the decline of traditional manufacturing remains unfilled. The physical expression of these processes is the mile upon mile of industrial wasteland; the human expression is the despair of whole communities, families and individuals whose means of livelihood have disappeared. One outcome of these cataclysmic changes has been the growth of an *informal* or *hidden economy,* a world of interpersonal cash transactions or payments in kind for services rendered, a world much of which borders on the illegal and some of which is transparently criminal. Figures 17.7 to 17.9 illustrate the enormous geographical unevenness characteristic of the US, European and Japanese economies.

Within the older industrialized economies, some of the sharpest contrasts in levels of income, nature of employment, unemployment and minority participation in the labour force occur in the major cities, especially in the so-called 'global cities'.[11] Because of their particular functions in the global economy as the 'control points' of global financial markets and of transnational corporate activity, cities like New York and London contain both highly sophisticated economic activities, with their highly paid, cosmopolitan workforces, and also large supporting workforces in low- and medium-level services. The result is a high degree of social and spatial *polarization* within these cities.

Figure 17.10 shows the general ways in which such processes may operate. Although this description of the processes involved is rather over-generalized, and the precise form they take depends on the history and circumstances of each individual city, it provides a useful device for understanding a very complex situation. Empirical evidence certainly bears out its major features.[12]

- *Employment trends*: in both London and New York, the proportion of the labour force employed in manufacturing declined from over 20 per cent in 1977 to well under 10 per cent in the mid 1990s.
- *Income inequalities*: during the 1990s, income inequality in New York City grew much more sharply than in the United States as a whole. 'New York has the

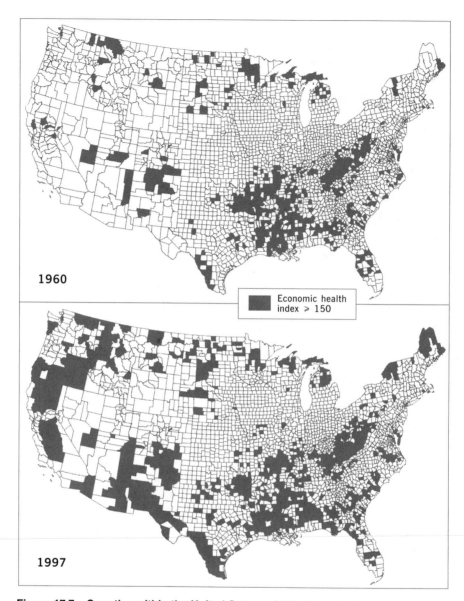

**Figure 17.7   Counties within the United States with high levels of
economic distress**

Source: based on Glasmeier, 2002: Figure 1

worst income inequality in the U.S'.[13] Similarly, earnings differentials within
London increased substantially between the mid 1980s and the late 1990s.

- *Casual and informal labour markets*: 'There has been a pronounced increase in
casual employment and in the informalization of work in both New York and
London ... In addition, jobs that were once full-time ones are now being

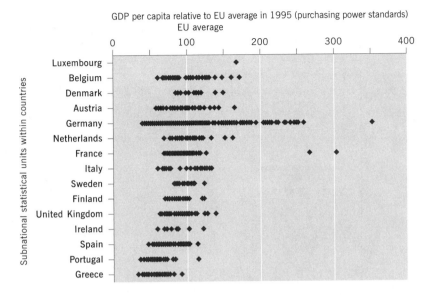

Figure 17.8 Regional inequalities in income in Europe

Source: based on Dunford and Smith, 2000: Figure 2

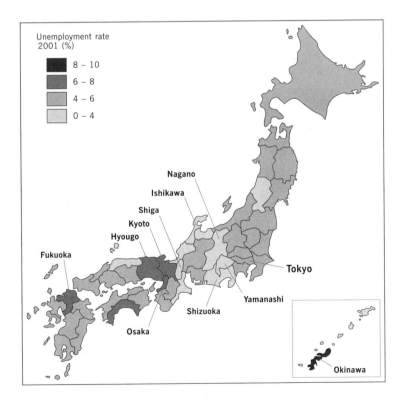

Figure 17.9 Regional unemployment in Japan

Source: Japanese Ministry of Labour

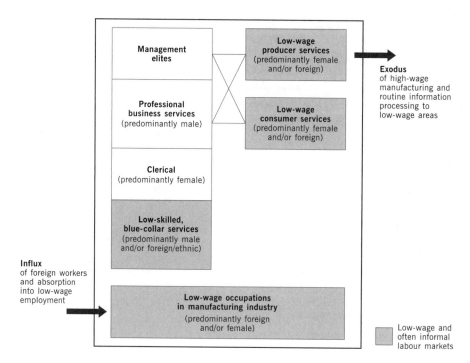

Figure 17.10   Processes of restructuring in global cities

Source: based on Friedmann, 1986: Figure 2

made into part-time or temporary jobs ... A majority are low-wage jobs'.[14] The precise forms of casualization vary between individual cities: in New York the emphasis has been on informal work, whereas in London most growth has been in part-time work.[15]

- *Race and nationality in the labour market*: in New York, 'Blacks and Hispanics increased their share of jobs, while whites lost share. Half of all resident workers in New York are now minority ... Blacks and Hispanics are far less likely than whites to hold the new high-income jobs and far more likely than whites to hold the new low-wage jobs'.[16] London is the major focus of Asian and Afro-Caribbean populations in the United Kingdom. 'In inner and outer London, 29 per cent and 22 per cent respectively of residents are from ethnic minorities in 2000'.[17]

Why is it happening?

Put this question to most politicians, journalists, quite a lot of academics and many ordinary people, and you are likely to get a simple answer: 'It's globalization, stupid.' In fact, it isn't as simple as that. These highly uneven trends in employment, unemployment and incomes in the industrialized economies cannot be explained simplistically in terms of a single set of causes. In so far as globalization contributes to these trends it does so precisely because globalizing processes are themselves highly complex and intrinsically uneven.

For example, the most general explanation of an overall high level of unemployment in the older industrialized countries between the early 1970s and mid 1980s and, again, in the 1990s is the effect of world recession. Recession, whatever its causes, drastically reduces levels of demand for goods and services. By this explanation, the bulk of unemployment in the older industrialized countries as a whole is cyclical: it is *demand-deficient* unemployment. But the general force of recession does not explain the *geographical variation* in unemployment between and within countries. In fact, a whole set of interconnected processes operates simultaneously to produce the changing map of employment, its reverse image, unemployment, and the increasingly uneven map of income.

Technological change

Technological developments in products and processes are widely regarded as being a major factor in changing both the number and the type of jobs available. In general, *product innovations* tend to increase employment opportunities overall as they create new demands. On the other hand, *process innovations* are generally introduced to reduce production costs and increase productive efficiency. They tend to be labour saving rather than job creating. Such process innovations are characteristic of the mature phase of product cycles (see Figure 3.10) and became a dominant phenomenon from the late 1960s onwards.

The general effect of process innovations, therefore, is to increase labour productivity: an increased volume of output from the same, or even a smaller, number of workers. Each of the case studies in Part Three demonstrated this tendency. But, again, the impact of such technological change on jobs tends to be uneven. In most cases, it has been the semi-skilled and unskilled workers who have been displaced in the largest numbers. Initially, it was manual workers rather than professional, technical and supervisory workers whose numbers were reduced most of all, although the spread of the new information technologies has changed this.

There is no doubt that changes in process technology have adversely affected the employment opportunities of less skilled members of the population. However, there is much disagreement about the overall contribution of technological change to unemployment. Some argue that the 'end of work' is nigh, and that much of this is due to the job-displacing effects of technological change. On the basis of his calculation that three out of four US jobs (including both white- and blue-collar jobs) could be automated, and that thousands of jobs have already disappeared from major US corporations, Rifkin predicted that hundreds of millions of workers would be without jobs by the middle of the century.[18] The explosive spread of new information and communication technologies (ICT) would seem to confirm such apocalyptic views.

But do they? Not in the opinion of the ILO,[19] which argues that:

- ICT does not destroy jobs.
- In the 'core' ICT sector, the jobs being lost in manufacturing are more than compensated for by rapid growth in the services segment of these industries (software, computer and data processing services).
- There is huge potential for high-cost economies to move up the value chain and to create higher-skill jobs.
- For most workers, employment stability remains the norm.
- However, these changes in employment tend to reinforce gender inequalities.

Although technological change may continue to create, in net terms, more jobs than it destroys, the problem is the actual *distribution* of such new jobs in relation to those destroyed. Although

> human labour in general has never yet become obsolete new technologies often displace particular workers and work hardships upon them, and upon whole industries and regions.[20]

Certainly, the ever-increasing pervasiveness of computerization in all areas of the economy is dramatically transforming existing divisions of labour and contributing towards a 'hollowing out' of labour market structures.[21] New technologies redefine the nature of the jobs performed, the skills required and the training and qualifications needed. They alter the balance of the labour force between different types of worker, involving processes of deskilling and reskilling. The geography of the employment effects of technological change is also extremely uneven. The 'anatomy of job creation' is rather different from the 'anatomy of job loss'; the terms 'sunrise' and 'sunset' industries imply (probably unconsciously) a geographical distinction (the sun does not rise and set in the same place).

The transnationalization of production

In Chapter 16 we explored the impact of TNCs on home and host economies. The development of complex transnational production networks in virtually all sectors of the economy, partly enabled by technological developments in transportation and communications, has a major impact on the geographical distribution of employment and incomes. In an increasingly volatile competitive environment, and driven by the imperatives of profit, TNCs are continuously reconfiguring their operations across, and within, national boundaries, as each of the case study chapters of Part III has shown. As a result, *which* jobs are created (or destroyed) and *where* are contingent upon the specific strategic behaviour of TNCs headquartered in different countries. The strong tendency to locate certain functions in particular kinds of place creates an internal geographical division of labour which, inevitably, is highly uneven by type of employment (and by income). As we have seen, TNCs are also increasingly connected into external networks of suppliers and collaborators and this creates indirect effects on employment patterns and more complex implications for local communities.

One way in which these externalized relationships impact on employment and incomes is through the currently highly controversial processes of outsourcing/offshoring of white-collar activities, especially in financial services (Chapter 13) but also in other sectors. Estimates of the scale and likely future trajectory of such offshoring vary enormously; scare stories abound.

> The McKinsey Global Institute extrapolating from a study of eight industrial sectors ... calculated that in 2003 there were 1.5m service jobs outsourced abroad from developed countries. By 2008, it reckons that number will have risen to 4.1m ... According to the OECD, close to 20% of total employment in the pre-expansion EU countries, America, Canada and Australia could 'potentially be affected' by the international sourcing of service activities.[22]

In other words, in addition to the long-established relocation of manufacturing jobs there is now an increasing tendency to relocate service jobs. On the one hand, the numbers involved are minuscule compared with the number of job changes that occur within individual countries all the time. For example, '"an average of 4.6m Americans started work with a new employer every month" in the year to March 2005'.[23] On the other hand, as always, the effects are experienced differentially, by different groups of people in different places. It is not so much the aggregate numbers affected by the reconfiguration of transnational production networks that matter but, rather, their distribution.

Trade competition from developing countries

One of the most widely accepted explanations for the employment and income problems facing workers in the older industrialized countries is the competition from imports of cheaper manufactured goods from developing countries, notably the NIEs. The rapid development of manufacturing production in a small number of NIEs, and their accelerating involvement in world trade, has been a major theme of this book. It is one of the most striking manifestations of global shifts in the world economy.

The basic question is: how far has the industrialization of these fast-growing economies – as expressed through *trade* – contributed towards the deindustrialization of the older industrialized countries, to the increased levels of unemployment, and to the pauperization of workers at the bottom end of the labour market? This has become an even more contentious issue with the recent emergence of China (and, to a lesser extent, India) as a major global economic force.[24]

> China is no longer a marginal supplier ... China's low production costs arise from and are coupled with growing industrial competence ... developments in these labour forces, when these economies are integrated into the global labour force, have the capacity to significantly affect global wage levels.

It is not just the wages of unskilled labour in the global economy which are being and will increasingly be undermined by the size of the labour reservoir in China (and India). One of the most striking features of the Chinese labour market is its growing level of education and skilling.[25]

There is a wide range of views on the relationship between developing country trade and employment and income changes in industrialized countries.[26] Wood, for example argues that trade with developing countries has had a considerable impact, especially in widening the gap between skilled and unskilled workers:

> Countries in the South have increased their production of labour-intensive goods (both for export and domestic use) and their imports of skill-intensive goods, raising the demand for unskilled but literate labour, relative to more skilled workers. In the North, the skill composition of labour demand has been twisted the other way. Production of skill-intensive goods for export has increased, while production of labour-intensive goods has been replaced by imports, reducing the demand for unskilled relative to skilled workers ... up to 1990 the changes in trade with the South had reduced the demand for unskilled relative to skilled labour in the North as a whole by something like 20 per cent ... Thus expansion of trade with the South was an important cause of the de-industrialization of employment in the North over the past few decades. However, it does not appear to have been the sole cause.[27]

The general conclusion of the ILO is that the results of the many economic studies of the relationship between trade and wage and income inequality in the older industrialized countries are 'inconclusive':

> Although international trade has contributed to income inequality trends to some extent, it has not played a major role in pushing down the relative wage of less-skilled workers ... [in the case of the United States] employment patterns in industries least affected by trade moved in the same direction as those in trade-affected manufacturing industry, increasing the share of high-wage employment. This pattern of change in the employment structure is not well explained by the argument relying on the trade effect.[28]

Again, however, the focus tends to be on the aggregate geographical picture. Given the particular ways in which the internal geographies of national economies have evolved, there will inevitably be a correspondingly uneven impact of trade on different parts of the same country. But such effects are very complex, as a recent study of US regions shows.[29]

> Many regions benefited from cheaper imports. The Southeast and South Central regions, however, both of which are dominated by low-wage, import-sensitive manufacturing industries, were made worse off by both cheaper imports and by greater orientation toward the production of import-competing goods. By contrast, the Great Lakes, a region with industries that are highly reliant on imported intermediate inputs, was helped by cheaper imports and a greater orientation toward the production of goods in import-competing

sectors. On the export side, cheaper exports hurt most regions, but helped states on the West Coast, a highly export-oriented region.[30]

Recent detailed empirical research into the Los Angeles labour market suggests that:

> An increase in foreign competition significantly reduces the wages of less-skilled workers in the Los Angeles CMSA. The wages of more highly educated workers are unaffected by imports and appear to rise with exports. Between 1990 and 2000, the negative impact of import competition moves up the skills ladder, suggesting that higher education may not insulate all workers from the pressures of the global economy over the long-run … the impact of trade on wage inequality eclipses the influence of technological change through the 1990s, at least in our study region.[31]

Searching for explanatory needles in messy haystacks

Which of these forces are responsible for changing employment and income levels and distribution in developed economies? Is one more important than the others? In fact, efforts to separate out individual influences, and to calculate their precise effects, have not been very successful. The basic problem in all of the individual factor explanations is that each of the factors is treated *independently* of the others. It is as though changes in one of the variables are unrelated to changes in the others. But this is clearly not the case.

For example, although the *direct* effects of trade may be relatively small, the *indirect* effects may be larger because of the ways in which firms respond to the threat of increased global competition. They may, for instance, invest in labour-saving technologies to raise labour productivity and to reduce costs. This would appear as a 'technology effect' whereas the underlying reason for such technological change may be quite different: a response to low-cost external competition. How do we separate out 'trade' effects from 'TNC' effects when so much of global trade either is intra-firm trade or is controlled and coordinated by TNCs? In some cases, a major driving force in import penetration has actually been the direct – or indirect – involvement of domestically owned TNCs. Is this a trade effect or a TNC effect?

In fact, the decline in overall manufacturing (and increasingly some service) employment in the older industrialized countries is primarily the result of increased productivity. But this has affected the labour force differentially, with the greatest relative losses of jobs and of income falling on the least skilled, least educated workers. The geographies of such effects are highly uneven, depending on the particular circumstances of individual regional and local economies.

In summary, Figure 17.11 sets out a rough balance sheet of the positive and negative effects of the globalizing processes on employment in developed economies.

Positive effects	Negative effects
Cheaper imports of relatively labour-intensive manufactures promote greater economic efficiency through the demand side while releasing labour for higher-productivity sectors	Particularly in relatively labour-intensive industries, the rising imports from developing countries, together with competition-driven changes in technology and other factors, lead to inevitable losses in employment and/or quality of jobs, including real wages. This increases inequality between skilled and unskilled workers, and causes extreme redeployment difficulties
Growth in developing countries through industry relocation and export-generated income leads to (a) increased demand for industrialized country exports and (b) shifts in production in industrialized countries from lower- to higher-valued consumer goods, to more capital- and/or skill-intensive manufacturing and services	Employment gains from rising industrialized country exports are unlikely to compensate fully for the job losses, especially if (a) industrialized country wages remain well above those of the NIEs and other emerging developing countries and (b) the rates of world economic growth are relatively low, and/or excessively concentrated in East and South East Asia
Employment growth and job quality improvement for skilled workers are likely to be significant in the short and medium term, even though in the long run the effects are unclear	The employment growth and job quality improvement for skilled workers will dwindle in the long run, as a result of relatively cheaper and more productive skilled labour in the NIEs
Relocation of production and/or imports causes negative short-term effects on workers but promotes labour market flexibility and efficiency through greater mobility of workers within countries (and, to a lesser extent, within regional economic spaces) to economic activities and areas with relative scarcity of labour	Increased trade will further reduce demand for unskilled labour. This exacerbates unemployment because, in a world of mobile capital, the industrialized countries no longer retain a capital-based comparative advantage

Figure 17.11   A balance sheet of effects of globalizing processes on employment in developed economies

Source: based on ILO, 1996: Table Int. 1

What is being done?

In reacting to the problems of job loss discussed in this chapter, developed country governments have utilized some of the policy measures outlined in Chapter 6 (Figures 6.4 to 6.8) and Chapter 7. But, as might be expected, there is considerable diversity in policy, depending on the country's social welfare perspective and its specific style of economic management.

Protecting against imports

For many policy makers, industry leaders and labour interest groups, an obvious response is to call for increased protection against what are often claimed to be 'unfair' imports. From time to time, especially when an election is imminent, the intensity of such calls increases. As we have seen, currently the primary focus of concern is China. The former CEO of General Motors was recently quoted as saying: 'Walk around Wal-Mart and it looks as if everything is made in China'.[32] Such statements create an atmosphere in which trade protectionist sentiments thrive (similar comments were made 20 years ago about Japanese imports and those from other East Asian NIEs). In 2005, for example, there were moves within the US Congress to impose substantial tariffs on China unless the Chinese government adjusted the exchange value of its currency. But there is also persistent trade friction between developed countries, notably between the United States and the EU.

A 'protectionist fix' is especially contentious not just in terms of its likely effects on the older industrialized countries themselves but also because of its implications for the economic well-being of developing countries. The crux of the issue is this: how far should those industries in which the older industrialized countries no longer have a comparative advantage be left to run down and be allowed to develop elsewhere? The neo-liberal answer is self-evident. The older industrialized countries should move out of what are, for them, obsolete activities, involving products that can be manufactured more cheaply in developing countries, and move into higher-technology products and into the more sophisticated service industries. But it is not as simple as this. The basic argument for protection against imports is that it is necessary to give domestic industry time to adjust: it provides a breathing space.

Such temporary measures in those sectors where international competition has intensified very rapidly are undoubtedly justified. However, if the breathing space is used, as it should be, to restore an industry's competitiveness or to shift into new activities, it will not necessarily preserve employment. As we have seen, new investment is likely to be labour saving: new activities may be located in different places. It may also be the case that protection against imports in the sectors most sharply affected by NIE growth will not prevent an inevitable decline in employment in the older industrialized countries.

There is also the broader question of the effects on consumers of protecting domestic industries. This is not just a problem for individual purchasers of consumer goods. Keeping domestic prices high through protection means that products used by other industries will also cost more – and this may itself lead to job reductions in those industries. Most of us are consumers as well as producers. There is also a likely tension between the position of large, especially transnational, firms and small domestic firms. Whereas the former are deeply engaged in sourcing from low-cost countries, the latter find it much harder to compete against cheap imports. The protectionist fix is far from being unproblematical.

Attracting inward investment; discouraging 'outsourcing'

As we saw in Chapter 8, rivalry for the investment favours of TNCs has become intense and is often pursued at the highest governmental and state levels. Prime ministers and presidents try to exert their influence, either overtly or covertly, to persuade TNCs to locate new, job-creating investment in their particular countries. Whether such a policy is beneficial in the long term is a matter of debate (see Chapter 16). From a political point of view, of course, the 'foreign investment fix' has the advantages of having a high profile and being relatively quick. Undoubtedly new jobs are created by inward investment, but whether there is a gain in *net* terms depends on its impact on existing firms and on the effect of foreign investment on the country's technological development. In any case, there is a limited amount of internationally mobile investment to go round and, since much of it will be in higher-technology industries anyway, the number of jobs created may well be far less than in the past.

The other side of the foreign investment coin is, of course, the relocation of activities from home countries, either directly as direct investment or indirectly through outsourcing/offshoring. It is this latter phenomenon that, as we have seen, has recently captured most attention. Quite apart from the wringing of hands at the apparent loss of jobs overseas, there is pressure, notably in the United States, for legislation to be enacted. In 2004, the Senate voted to forbid the outsourcing of work on contracts financed by federal funds, on the grounds that 'taxpayers' money should not be used to send jobs overseas'. However, despite concerns over the employment implications of outsourcing, not all developed countries have taken such a stance (the UK is an example).

Developing new technologies

As we have seen at several points in this book, governments have become universally involved in attempting to stimulate technological developments within their own national territories. For example, one of the objectives of EU policy, following the 2000 Lisbon summit, was for the EU to 'become the most competitive and dynamic knowledge-based economy in the world'. But does new technology create or destroy jobs? If it creates new jobs, are they different from the old jobs either in terms of skills required or in their geographical location? Historical experience seems to show that, *in aggregate,* new technologies create more jobs than they destroy, at least over the longer term. This seems to occur because they create new demands for goods and services, many of which could not have been foreseen. However, we should beware of too ready an adoption of the notion that investing heavily in new technology will automatically increase employment opportunities. Much of the growth associated with the new technologies may well be 'jobless growth'. What is certain, however, is that new technologies require a substantial and continuing process of education and (re-)training.

Promoting entrepreneurship and small firms

There has been a remarkable swing of the pendulum in attitudes towards firms of different sizes. In the 1960s the key to economic (and employment) growth was seen to be the very large firm. Only these, it was argued, could achieve the economies of scale needed to enhance competitiveness. Many governments encouraged mergers between enterprises towards this end. By the 1970s, in complete contrast, disillusionment with the large enterprise had set in and the employment panacea was seen to be the small firm, with its supposed dynamism and lack of rigidity. This view was strongly reinforced by claims that the vast majority of all net new jobs created in the United States during the 1980s had been in firms employing fewer than 20 workers.

However, important as they are, there are limits to the job-creating capabilities of small firms.[33] It would take many thousands of small firms to replace the jobs lost through the rationalization processes of even a few large firms, let alone to create additional jobs. There is no doubt that small firms are an important source of growth in an economy. But it should not be forgotten that the majority of small firms are far from dynamic, that most depend upon large firms for their markets, and that the failure rate of small firms is very high. Thus, the 'small-firm fix' is likely to be limited in its impact on unemployment:

> small, *per se*, is neither unusually bountiful nor especially beautiful, at least when it comes to job creation in the age of flexibility.[34]

Removing 'obstacles' to adjustment: 'flexibilizing' the labour force

In recent years, a new conventional wisdom has emerged whose essence is that of removing what are seen to be *rigidities* in the labour markets of the older industrialized countries: to make their labour markets more *flexible,* in tune with what are seen to be the dominant characteristics of a globalizing world economy. The 'flexibilization' of labour markets through deregulation involves greatly increased pressures and restrictions on labour organizations, the drastic cutting back of welfare provisions, and the move away from welfare towards *workfare*.[35]

The process has gone furthest in the United States. Its apparent success in continuing to create large numbers of jobs (albeit with the widening of income gaps) has 'been the most persuasive argument for neo-liberal policies'.[36] It certainly stimulated the United Kingdom government to move along the same path. As yet, the countries of continental Europe have not moved as far, or as fast, down the flexibilization path. Most European governments are concerned that the social costs of reducing unemployment using the US model may be politically unacceptable in a system in which the social dimension of the labour market is very strongly entrenched. But there are clear signs of change as governments become increasingly concerned about the financial costs of sustaining existing practices and the continuing loss of competitive edge. However, government attempts at some types of labour market reform invariably run into serious opposition, as the French student riots of early 2006 demonstrated.

As a result, a variety of labour market measures, employed in various combinations in different European countries, has emerged. These include:

- the use of more temporary and fixed-term contracts
- the introduction of different forms of flexible working time
- moves to encourage greater wage flexibility by getting the long-term unemployed and the young to take low-paid jobs
- increased vocational training to provide more transferable skills
- reforms in state employment services
- incentives to employers to take on workers
- measures to encourage workers to leave the labour market
- reductions in the non-wage labour cost burdens on employers
- specific schemes to target the long-term unemployed.

One of the biggest problems facing displaced workers is that there is frequently a geographical mismatch between the destruction of old jobs and the creation of new ones. In such circumstances,

> Education and training and employment subsidies will be for naught if workers are unable or unwilling to take new jobs where they exist. Studies to date show that high-skilled workers are much more willing than the unskilled to move to a new work location. This would indicate that removal of the obstacles that block unskilled workers from moving has an important role in labor policy.[37]

But there are also pronounced geographical differences in the propensity for labour to move. On the whole, labour is geographically less mobile in Europe than in North America, for reasons lying deep in social and economic histories and in cultural attitudes. There are many real obstacles to the geographical mobility of labour. Not only must deep community and family ties be broken but also the nature of housing markets may make relocation difficult. The whole question of the geographical mobility of labour in the older industrialized countries is more complex than is often supposed. The issue of labour mobility is particularly paradoxical in the context of the European Union, where all citizens of member states have the automatic right to work in any other member state. In fact, the actual level of labour mobility between states is minuscule. This leads us to what has become one of the most contentious issues: in-migration.

The thorny problem of migration

It may seem paradoxical to think of migration as helping to solve the adjustment problems of the older industrialized countries. After all – especially in Europe – we have been talking about not enough jobs to meet the demands of the existing populations. To add further to what appears to be an over-supplied labour market seems perverse to say the least. It is such considerations, together with fears of social unrest between indigenous and immigrant populations, that have made current immigration policies in most developed countries so rigid. But, as ever, things are not as simple as aggregate figures suggest. In addition to humanitarian concerns for refugees, asylum seekers, or people simply trying to improve their lives, there are two reasons why developed countries need to create a sensible policy toward in-migration. One is immediate, the other is longer term.

The immediate reason for asserting the need for more enlightened immigration policies is the fact that, in most developed countries, there is a *severe shortage of labour*. This applies as much in high-skill sectors such as IT and health care as in some low-skill service sectors. The other, longer-term, reason is that the populations of such countries are *getting older* (see Table 15.1). Their active populations are shrinking. There will not be enough people of working age to support future dependent populations. For both short- and longer-term reasons, then, there is a pressing need to rethink immigration policies. But, of course, there are major political obstacles to doing so.

Fears (sometimes justified, often not) of being squeezed out of jobs by incomers, or of local cultures and practices being diluted by 'foreign ways', generate powerful forces of opposition. Such fears are easily exploited by political groups of the extreme right, as can be seen today in many European countries, as well as in the United States. Labour force displacement does indeed happen. But not invariably so – and not on the scale so often imagined. One of the biggest obstacles to popular support for more liberal migration polices is that the size of the host country's immigrant population tends to be greatly over estimated.[38] As Figure 17.12 shows, foreign workers make up a very small percentage of the working age population in EU states. Of course, the distribution of immigrants tends to be highly uneven *within* individual countries and this is an important factor in people's perceptions. Specific transnational communities tend to develop specific geographies, some of which are more apparent than others, and that is where the greatest tension tends to develop.

Controls on immigration are now much tighter than in the past. Despite the fact that labour migration is an integral part of the European Union, the enlargement of 2005 to incorporate a further 10 countries, mostly from the former Soviet bloc, has led to 12 of the existing EU member states imposing 'transitional' restrictions on migration from Eastern Europe (the exceptions are the UK, Sweden and Ireland). This is despite pleas from the European Commission for an open door policy for new members and the fact that 'in most EU15 countries, workers from the new members make up less than 1% of the workforce'.[39] Similar problems exist in the case of Mexican migration into the United States. Around 400,000 Mexicans cross the border into the US illegally every year. According to a 2005 survey, proposals to build a more robust physical barrier along the border are supported by 60 per cent of the US population.[40]

Yet, many parts of the European and US economies – as well as many public services – simply could not operate without the employment of migrant workers. The need for an influx of new workers will not go away. On the contrary, given the demographic trends in all the developed countries, the need will increase. It

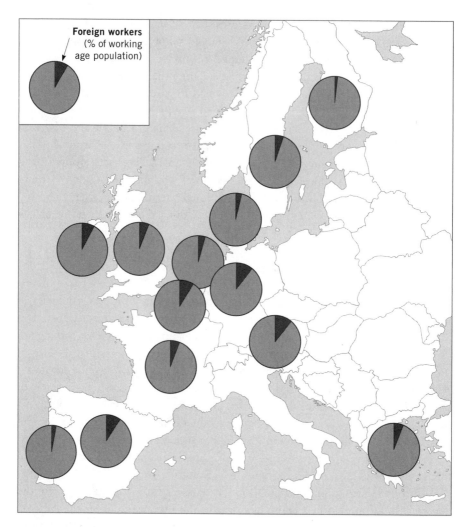

Figure 17.12   Foreign workers as a percentage of working age population in Europe

Source: European Commission data, cited in the *Financial Times*, 3 February 2006

can also be argued that not only does in-migration fill important needs – often performing tasks that, otherwise, will go unperformed – but also it need not have the negative effects claimed by opponents. A recent study of Europe claims that increased immigration leads to economic expansion rather than to job losses.[41]

According to a Home Office study of immigration in the UK:

> Native wages have *not* been depressed in the UK. Immigrants have tended to perform three types of job. They have worked in public services, especially health, where pay is determined by the government. Wages are well below market levels and the effect of newcomers is to reduce shortages ... At the other extreme, 'in relatively low paid and insecure sectors [such as] catering

and domestic services, unskilled natives are simply unwilling ... to take on the large number of available jobs ... if migrants do not fill these jobs they simply go unfilled or uncreated.' An estimated 70 per cent of catering jobs are filled by migrants.

There are also the highly skilled information technology workers. According to the Home Office study, the inflow of these technicians has enabled the IT sector to grow faster rather than to depress pay in it.[42]

Migrant workers into the UK from Eastern Europe since EU accession have not taken jobs from unemployed domestic workers.[43]

Conclusion

Although at a global scale the developed economies are clearly 'winners', that isn't true for everybody. There are undoubtedly 'losers' alongside the winners in terms of both jobs and incomes. While some – perhaps the majority – of developed country populations are better off than ever in material terms, others are relatively disadvantaged. There is both a social and a geographical bias to this. Certain groups, certain kinds of geographical area have fared substantially worse than others. The effects of globalizing processes within the developed countries are extremely uneven. In these respects, the older industrialized countries continue to face considerable difficulties in adjusting to the intensified competitive environment of a global economy. Preserving, let alone creating, jobs for their active populations has become infinitely more complex in today's highly interconnected world. The imperatives that drive nation-states to strive to enhance their international competitiveness will not always be job creating. However, the problems facing the affluent industrialized economies pale into insignificance when set beside the problems facing the world's developing countries. It is to these countries that we turn in the next chapter.

NOTES

1 Sen (1999: 21).
2 See Rowthorn and Ramaswamy (1997).
3 *Financial Times* (22 June 2001).
4 See ILO (2004b).
5 *The Economist* (15 January 2000).
6 Wilkinson (2005).
7 ILO (2004a: 42).
8 Kapstein (2000: 379).
9 Kapstein (1999: 101).
10 See OECD (2005b).

11 See Friedmann (1986), Sassen (2001).

12 Sassen (2001: Chapters 8 and 9) provides much detailed empirical data on London, New York and Tokyo.

13 Sassen (2001: 270).

14 Sassen (2001: 289, 290).

15 Sassen (2001: 294).

16 Sassen (2001: 306).

17 Sassen (2001: 309).

18 Rifkin (1995).

19 ILO (2001: 140–1).

20 Tobin (1984: 83).

21 Levy and Murname (2004).

22 *The Economist* (2 July 2005).

23 *The Economist* (2 July 2005).

24 For example, much of the special issue of *Der Spiegel* (2005: 7) is devoted to this.

25 Kaplinsky (2001: 56–7).

26 See, for example, Cline (1997), ILO (1997), Kaplinsky (2001), Kapstein (1999; 2000), Rigby and Breau (2006), Silva and Leichenko (2004), Wood (1994).

27 Wood (1994: 8, 11, 13).

28 ILO (1997: 71, 73).

29 Silva and Leichenko (2004).

30 Silva and Leichenko (2004: 283).

31 Rigby and Breau (2006: 18).

32 Quoted in *The Economist* (27 September 2003).

33 For a stimulating polemic against the small-firm policy fix see Harrison (1997).

34 Harrison (1997: 52).

35 Peck (2001: 10).

36 Faux and Mishel (2000: 101).

37 Kapstein (1999: 156-7).

38 Dustmann and Glitz (2005).

39 *The Economist* (11 February 2006).

40 *Financial Times* (15 December 2005).

41 Dustmann and Glitz (2005).

42 *Financial Times* (25 October 2001).

43 UK Department for Work and Pensions, cited in the *Financial Times* (1 March 2006).

Eighteen
Making a Living in Developing Countries: Sustaining Growth, Enhancing Equity, Ensuring Survival

Some winners – but mostly losers

In large part, though by no means entirely, the economic progress and material well-being of developing countries are linked to what happens in the developed economies. A continuation of buoyant economic conditions in those economies, stimulating a general expansion of demand for both primary and manufactured products, would undoubtedly help developing countries. But the notion that 'a rising tide will lift all boats', while containing a kernel of truth, ignores the enormous variations that exist between countries. The shape of the 'economic coastline' is highly irregular; some economies are beached and stranded way above the present water level.

For such countries there is no automatic guarantee that a rising tide of economic activity would, on its own, do very much to refloat them. The internal conditions of individual developing countries – their histories, cultures, political institutions, forms of civil society, resource base (both natural and human) – obviously influence their developmental prospects. However, despite the claims of the 'neo-environmental determinists', low levels of development cannot be explained simplistically in terms of the natural environment (for example, climatic conditions).[1] As always, it is a specific combination of external and internal conditions which determines the developmental trajectory of individual countries.

For the developing world as a whole, as we saw in Chapter 15, the basic problems are those of extreme poverty, continuing rapid population growth, and a lack of adequate employment opportunities. Apart from the yawning gap between developed and developing countries as a whole, however, there are enormous disparities within the developing world itself. It is these internal variations that we address in this chapter. The discussion is organized in the following way:

- First, we look at the heterogeneity of the developing world, focusing in particular on issues of income, employment and labour migration.
- Second, we look at the 'winners' among developing countries: the so-called newly industrializing economies (NIEs). Two related questions are addressed: how can economic growth be sustained, and how can this be achieved while sustaining some degree of equity?
- Finally, we focus on the clearest of the 'losers' – the poorest developing countries, whose populations exist on the very margins of survival.

Heterogeneity of the developing world

Income differentials

Using broad terms like 'developing countries' or the 'Third World' implies a degree of homogeneity, which simply does not exist within this group of countries. In this respect, the World Bank makes a useful distinction between three groups of developing countries based on per capita income level: upper-middle income, lower-middle income, and low-income countries.

Figure 18.1 shows the geographical distribution of each of these groups of developing countries. There are some striking features, most notably the very heavy concentration of the lowest-income countries in sub-Saharan Africa, which contains almost two-thirds of the world's low-income countries. The whole of the Indian subcontinent (with the exception of Sri Lanka, but including India itself) also falls into the low-income category. In contrast, most East Asian countries are in the middle-income categories. Despite its recent spectacular economic growth, China is still in the lower-middle-income category; Malaysia is in the upper-middle-income group; while Singapore, Taiwan and Korea now have sufficiently high per capita incomes to fall into the high-income category, along with the world's developed countries. Singapore, in particular, now has a per capita income level higher than several European countries. All countries in Latin America and the Caribbean are in the middle-income group (apart from Haiti and Nicaragua, which are in the low-income category).

The averages for three development indicators – infant mortality, life expectancy and adult literacy – give some impression of the heterogeneity within the developing world as well as a stark indication of the gap between these countries and the high-income countries of the world (Table 18.1). The income disparities are especially marked. The average for the upper-middle-income group is roughly 12 times greater than that of the low-income group. The income gradient is, not surprisingly, reflected in the data for infant mortality and life expectancy, although the range of values within each income category is enormous.

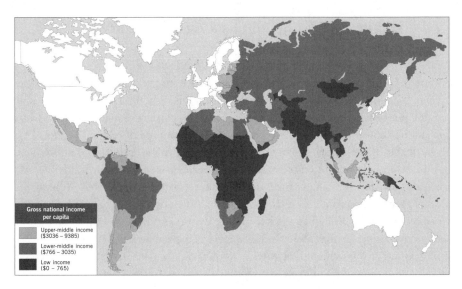

Figure 18.1 Distribution of upper-middle-, lower-middle-, and low-income countries

Source: calculated from World Bank, 2005b: Table 1

Table 18.1 Variations in income and other indicators within the developing world

Country group	Per capita income ($)	Under-5 mortality rate (per 1000)	Life expectancy at birth	Adult literacy rate (% over 15 years)
Low-income				
average	450	126	58	61
range	90–735	26–284	37–70	17–99
Lower-middle income:				
average	1,480	40	69	90
range	810–2,790	18–86	42–76	51–100
Upper-middle income:				
average	5,340	22	73	91
range	3,430–8,530	5–110	38–78	78–100
Reference group, high income:				
average	28,550	7	78	–
range	11,830–43,350			

Source: based on material in World Bank, 2005b: Table 1

Employment, unemployment, underemployment

Labour force growth exceeds the rate of job creation

Although the employment structure of developing countries has undergone marked change, most developing countries are predominantly agricultural economies. More than 50 per cent of the labour force in the lowest-income countries is employed in agriculture (compared with 4 per cent in the high-income developed economies). Even in the upper-middle-income group (in which most industrial development has occurred) agriculture employs almost 20 per cent of the labour force. In each category the relative importance of agriculture has declined even though, in absolute terms, the numbers employed in agriculture continue to grow. The balance of employment has shifted towards the other sectors in the economy: industry and services. However, as Figure 18.2 shows, there is a clear geography to these developing country employment structures – emphasizing, yet again, the marked heterogeneity of developing countries as a group.

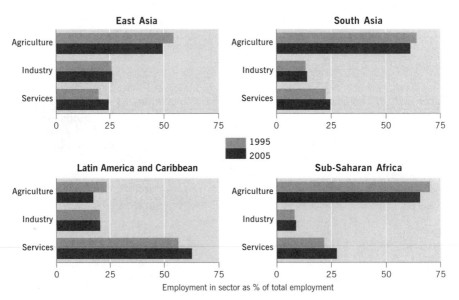

Figure 18.2   Geographical variations in employment structures

Source: ILO, 2006a: Table 5

These broad structural changes in employment in developing countries have to be seen within the broader context of growth in the overall size of the labour force. The contrast with the experience of the industrialized countries in the nineteenth century is especially sharp. During that earlier period, the European labour force increased by less than 1 per cent per year on average; in today's developing countries the labour force is growing at more than 2 per cent every year. Thus, the labour force in the developing world has doubled roughly every 30 years compared with the 90 years taken in the nineteenth entury for the European labour force to double. Hence, it is very much more difficult for developing countries to absorb the exceptionally rapid growth of the labour force into the economy. The problem is not likely to ease in the near future because labour force growth is determined mainly by past population growth with a lag of about 15 years. As we saw in Chapter 15, virtually all of the world's population growth since around 1950 – more than 90 per cent of it – has occurred in the developing countries.

There is, therefore, an enormous difference in labour force growth between the older industrialized countries on the one hand and the developing countries on the other. But the scale of the problem also differs markedly between different parts of the developing world itself. Overall,

> every year another 80 million or more people reach working age and join the 2.1 billion-strong workforce in the developing world. Mexico, Turkey, and the Philippines must create 500,000 to 1 million new jobs annually to accommodate the youths entering their workforce.[2]

The situation is especially acute in low-income Asian countries like Bangladesh, India and some in South East Asia. It is also a major problem even for fast-growing East Asian economies, which are not creating sufficient numbers of jobs for their burgeoning labour forces. In the case of China, for example, it is estimated that 15 million jobs need to be created every year.[3]

Of course, pressure on the labour market is lessened where lower population growth rates occur. Although this is undoubtedly occurring in many parts of the developing world (see Chapter 15), fertility rates are still sufficiently high to ensure that developing countries' labour forces will continue to increase for the next few decades at least. The basic dilemma facing most developing countries, therefore, is that the growth of the labour force vastly exceeds the growth in the number of employment opportunities available.

Formal and informal labour markets

It is extremely difficult to quantify the actual size of the unemployment problem in developing countries. Published figures tend to show a very low level of unemployment, in some cases lower than those recorded in the industrialized countries. But the two sets of figures are not comparable. One reason is the paucity of accurate statistics. But the other reason is that unemployment in developing countries is not the same as unemployment in industrial economies. To understand this we need to appreciate the strongly segmented nature of the labour market in developing countries, in particular its division into two distinctive, though closely linked, sectors: *formal* and *informal*.

- In the *formal sector,* employment is in the form of wage labour, where jobs are (relatively) secure and hours and conditions of work clearly established. It is the kind of employment characteristic of the majority of the workforce in the developed market economies. But in most developing countries the formal sector is not the dominant employer, even though it is the sector in which the modern forms of economic activity are found.
- *The informal sector* encompasses both legal and illegal activities, but it is not totally separate from the formal sector: the two are interrelated in a variety of complex ways. The informal sector is especially important in urban areas; some estimates suggest that between 40 and 70 per cent of the urban labour force may work in this sector. But measuring its size accurately is virtually impossible. By its very nature, the informal sector is a floating, kaleidoscopic phenomenon, continually changing in response to shifting circumstances and opportunities.

In a situation where only a minority of the population of working age are 'employed' in the sense of working for wages or salaries, defining unemployment is, thus, a very different issue from that in the developed economies, although even there an increasing informalization of the economy is apparent. The major problem in developing countries is *underemployment*, whereby people may be able to find work of varying kinds on a transitory basis, for example, in seasonal agriculture, or as casual labour in workshops or in services, but not permanent employment.

Positive and negative effects of globalizing processes on developing country employment

There is no question that the magnitude of the employment and unemployment problem in developing countries is infinitely greater than that facing the older industrialized countries. The high rate of labour force growth in many developing countries continues to exert enormous pressures on the labour markets of both rural and urban areas. Such pressures are unlikely to be alleviated very much by the development of manufacturing industry alone. Despite its considerable development in at least some developing countries, manufacturing industry has made barely a dent in the unemployment and underemployment problems of most developing countries.

Only in small, essentially urban, NIEs (like Hong Kong and Singapore) has manufacturing growth absorbed large numbers of people. Indeed, Singapore has a labour shortage and has had to resort to controlled in-migration, while Hong Kong firms have had to relocate most of their manufacturing production to southern China. In most other cases, the problem is not so much that large numbers of people have not been absorbed into employment – they have – but that the *rate of absorption cannot keep pace with the growth of the labour force*. Globalizing processes, whilst offering some considerable employment benefits to some developing countries, are again a double-edged sword, as Figure 18.3 shows.

Positive effects	Negative effects
Higher export-generated income promotes investment in productive capacity with a potentially positive local development impact, depending on intersectoral and inter-firm linkages, the ability to maintain competitiveness etc.	The increases in employment and/or earnings are (in contradiction to the supposed positive effects) unlikely to be sufficiently large and widespread to reduce inequality. On the contrary, in most countries, inequality is likely to grow because unequal controls over profits and earnings will cause profits to grow faster
Employment growth in relatively labour-intensive manufacturing of tradeable goods causes (a) an increase in overall employment and/or (b) a reduction of employment in lower-wage sectors Either of these outcomes tends to drive up wages, to a point which depends on the relative international mobility of each particular industry, labour supply–demand pressure and national wage-setting/bargaining practices	Relocations of relatively mobile, labour-intensive manufacturing from industrialized to developing countries, in some conditions, can have disruptive social effects if – in the absence of effective planning and negotiations between international companies and the government and/or companies of the host country – the relocated activity promotes urban-bound migration and its length of stay is short. Especially in cases of export assembly operations with very limited participation and development of local industry and limited improvement of skills, the short-term benefits of employment creation may not offset those negative social effects
These increases in employment and/or wages – if substantial and widespread – have the potential effect of reducing social inequality if the social structure, political institutions and social policies play a favourable role	
Exposure to new technology and, in some industries, a considerable absorption of technological capacity leads to improvements in skills and labour productivity, which facilitate the upgrading of industry into more value-added output, while either enabling further wage growth or relaxing the downward pressure	Pressures to create local employment, and international competition in bidding for it, often put international firms in a powerful position to impose or negotiate labour standards and labour management practices that are inferior to those of industrialized countries and, as in the case of some EPZs, even inferior to the prevailing ones in the host country

Figure 18.3   Positive and negative effects of globalizing processes on developing country employment

Source: ILO, 1996: Table Int. 1

Labour migration: a 'solution' to the jobs problem?

Not surprisingly, in the face of lack of employment opportunities at home, large numbers of potential workers have migrated to other countries. As we saw in Chapter 15, the number of people living outside their country of birth or citizenship runs into the tens of millions. Of course, some of these are political or religious refugees, fleeing persecution. But the majority are migrant workers. And remember, these are pretty conservative estimates: much migration is illegal and, therefore, undocumented. At one level, the decision to migrate abroad in search of work is an individual decision, made in the context of social and family circumstances. When successful, that is, when the migrant succeeds in obtaining work and building a life in a new environment, the benefits to the individual and his/her family are clear (although there may be problems of dislocation and emotional stress). There is invariably, as well, discrimination against migrant workers in host countries. In many cases, migrants are employed in very low-grade occupations, they may have few, if any, rights, and their employment security is often non-existent. They may also be subject to abuse and maltreatment.

But what are the effects of out-migration on the exporting country? From a *positive* perspective, out-migration helps to reduce pressures in local labour markets. In addition, the remittances sent back home by migrant workers make a huge contribution, not only to the individual recipients and their local communities but also to the home country's balance of payments position and to its foreign exchange situation. Indeed, migrants' remittances have reached epic proportions: almost $173 billion in 2003. Annual remittances to Latin America and the Caribbean ($54 billion in 2005) are greater than the combined flows of foreign direct investment and development aid.[4] In many cases, the value of foreign remittances is equivalent to a large share of the country's export earnings (Figure 18.4). In a few cases, remittances are worth between one and three times *more* than their total exports. More commonly, remittances can account for between one-fifth and one-half of exports. In absolute terms, the biggest recipients of migrant remittances are India ($17.4 billion), Mexico ($14.6 billion) and the Philippines ($7.9 billion).

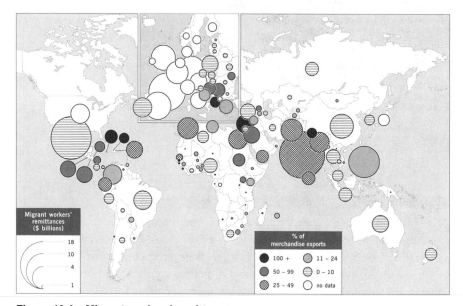

Figure 18.4   Migrant workers' remittances

Source: World Bank, 2005a: Table 6.13

Paradoxically, such remittances do not always help the poorest people back home, as recent Mexican research indicates.

> 'For some people, remittances allow them to buy a basic basket of essential goods,' says Rodolfo Tuiran, of Sedesol, Mexico's social development ministry. 'But overall, in terms of poverty, remittances do not have a significant impact. They do, however, have an important impact on inequality – they increase it. Of every $100 received, $75 goes to homes that aren't poor.' Anecdotal evidence supports this. In areas of high migration, the houses in good repair, with a satellite dish, are the ones that receive remittances.[5]

On the other hand, again in Mexico, there are schemes which capitalize on the fact that migrants from the same home town often tend to cluster together in their host country. As a result, there is now a network of Mexican 'home-town associations' across the United States. Collective remittances to a home town in several Mexican states are organized in a '"three-for-one" programme, where each dollar from the home-town association for a development project is matched by a dollar each from the municipal, state, and federal governments'.[6] However, most developed countries are now far less open to in-migration than they have been in the past.

The other side of the out-migration coin is less attractive for the labour exporting countries: there are important *negative* consequences. The migrants are often the young and most active members of the population. Further,

> returning migrants are rarely bearers of initiative and generators of employment. Only a small number acquire appropriate vocational training – most are trapped in dead-end jobs – and their prime interest on return is to enhance their social status. This they attempt to achieve by disdaining manual employment, by early retirement, by the construction of a new house, by the purchase of land, a car and other consumer durables, or by taking over a small service establishment like a bar or taxi business; there is also a tendency for formerly rural dwellers to settle in urban centres. There is thus a reinforcement of the very conditions that promoted emigration in the first place. It is ironic that those migrants who are potentially most valuable for stimulating development in their home area – the minority who have acquired valuable skills abroad – are the very ones who, because of successful adaptation abroad, are least likely to return. There are also problems of demographic imbalance stemming from the selective nature of emigration.[7]

Urban–rural contrasts

Although most developing countries are still predominantly rural, the trend, as we saw in Chapter 15, is inexorably towards the urban. Of course, cities are always the primary focus of economic growth – a point we have made several times in this book. But much of the growth of cities in the developing world is driven by a combination of high population fertility and the push of rural poverty, which drives millions of people towards what are seen to be the economic honeypots of the city. In these latter cases, therefore, what we have is a process of *over-urbanization*: circumstances where the basic physical, social and economic infrastructures are not commensurate with the sheer size and rate of growth. The sprawling shantytowns endemic throughout the developing world are the physical expression of this explosive growth.

In the developing countries, virtually all industrial growth is in the big cities. Stark polarization between rich and poor is one of the most striking features of developing country cities. United Nations data show that 80 per cent of the urban population of the 30 least developed countries live in slums.[8] Increasingly, very high levels of poverty tend to be concentrated in urban areas. Whereas rural dwellers may be able to feed themselves and their families from the land, such an option is not available in the cities. In addition, there is a whole syndrome of urban pathologies to contend with:

> About 220 million urban dwellers, 13 per cent of the world's urban population, do not have access to safe drinking water, and about twice this number lack even the simplest of latrines. Women suffer the most from these deficiencies … poverty also includes exposure to contaminated environments and being at risk of criminal victimization … Poverty is closely linked to the wide spread of preventable diseases and health risks in urban areas.[9]

Environmental problems also tend to reach their peak in the big cities of the developing world. This is one of the most serious negative externalities of rapid urban/industrial growth. In particular, extremely high levels of air pollution – the presence of excessive levels of solid particulate matter (SPM), sulphur dioxide, lead, carbon monoxide, nitrogen dioxide and other atmospheric impurities – are endemic. Such levels are especially high in Mexico City, Delhi, Calcutta, São Paulo, Beijing, Tianjin, Shanghai, Jakarta and Manila, to name just a few. They result, in particular, from very rapid industrialization and the accelerated increase in car ownership (often of old, inefficient vehicles) in the big cities of the developing world. One extremely serious byproduct is the high number of premature deaths (Figure 18.5).

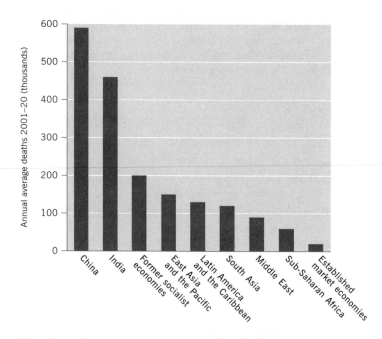

Figure 18.5   Premature deaths due to urban air pollution

Source: data from World Bank, WorldWatch Institute, cited in the *Financial Times*, 27 January 2006

Underemployment and a general lack of employment opportunities are widespread in both rural and urban areas in developing countries. There is a massive underemployment and poverty crisis in rural areas arising from the inability of the agricultural sector to provide an adequate livelihood for the rapidly growing population and from the very limited development of the formal sector in rural areas. Some industrial development has occurred in rural areas, notably in those countries with a well-developed transport network. Mostly this is subcontracting work to small workshops and households in industries such as clothing manufacture. But most modern industries are overwhelmingly concentrated in the major cities or in the export processing zones.

It is in the big cities that the locational needs of manufacturing firms are most easily satisfied. Yet despite the considerable growth of manufacturing and service industries in the cities, the supply of jobs in no way keeps pace with the growth of the urban labour force. Not only is natural population increase very high in the cities of many developing countries, but also migration from rural areas has reached gigantic dimensions. The pull of the city for rural dwellers is directly related to the fact that urban employment opportunities, scarce as they are, are much greater than those in rural areas. In the words of the (very) old song: 'How can they keep 'em down on the farm, now that they've seen Paris?' The bright lights of the big city (to echo Jay McInerney's novel) exert a very powerful attraction.

Sustaining growth and ensuring equity in newly industrializing economies

The spectacular economic growth of a relatively small number of developing countries – especially in East Asia – has been one of the most significant developments in the global economy in the past 50 years. In particular, the four East Asian tigers – Korea, Singapore, Taiwan and Hong Kong (although the last is now incorporated into China) – have industrialized at a rate 'unmatched in the 19th and 20th centuries by Western countries and, for that matter, by Latin American economies'.[10] The spread of the growth processes to encompass some other East Asian countries during the 1980s and 1990s – Malaysia, Thailand, the Philippines, Indonesia and, most of all, China – means that

> in little more than a generation, hundreds of millions of people have been lifted out of abject poverty, and many of these are now well on their way to enjoying the sort of prosperity that has been known in North America and Western Europe for some time ... While some groups have been obliged to bear the negative consequences of development far more than others ... it seems clear that the development project in the Asia–Pacific region holds out the promise of a scale of generalized prosperity unknown in human history.[11]

Certainly, there can be no doubting the remarkable economic progress of the NIEs, especially those in East Asia. But what of the future? What are the major problems facing the NIEs? Two are especially significant:

- sustaining economic growth
- ensuring that such growth is achieved with equity for the country's own people.

Sustaining economic growth

Measured in terms of increased per capita income, and larger shares of world production and trade, the East Asian NIEs, in particular, have been phenomenally successful. Far more so, in fact, than countries in other parts of the developing world, most notably in Latin America where, with one or two exceptions, economic growth has been more limited (see Chapter 2). But can such spectacular growth rates be maintained in the future? Although each of the four leading NIEs managed to sustain very high rates of growth for a very long period, there were signs in the mid 1990s that perhaps the 'miracle' was coming to an end. Press headlines such as 'Asia's precarious miracle', 'Is it over?', 'Roaring tiger is running out of breath' had begun to appear. A few academic observers had also begun to question the sustainability of East Asian growth.[12] But none had predicted the suddenness of the economic crisis that hit many parts of East Asia in 1997. Virtually overnight, it seemed, the East Asian miracle economies had become 'basket cases'. Millions of words were written on the causes of the 1997 East Asian crisis.[13] In fact, despite the gloom-laden predictions of many writers at the time of the crisis, most of the East Asian economies have recovered much (though by no means all) of their growth momentum. Whether that recovery is sustainable is the key issue.

Maintaining export growth

Economic growth in the East Asian NIEs has been based primarily upon an aggressive export-oriented strategy. It is no coincidence that the take-off of the first wave of East Asian NIEs occurred during the so-called 'golden age of growth' in the 1960s and early 1970s, or that it was made possible by the relative openness of industrialized country markets. The growth and openness of such markets is, therefore, vital for the continued economic growth and development of the NIEs. During the 1960s the conditions were indeed favourable; future prospects look far less propitious as the older industrialized countries have reduced their demands for NIE exports, partly through the deliberate operation of protectionist trade measures. From the NIEs' viewpoint, therefore, the macroeconomic expansion of the industrialized economies is vital. But this will be effective only if trade barriers – especially non-tariff barriers – are also removed or at least lowered. The present political climate in the older industrialized countries makes both possibilities somewhat remote.

Trade tensions – particularly between the United States, their biggest export market, and the leading Asian NIEs – are palpable and show little sign of disappearing. Countries like Korea and Taiwan – as well as more recent fast-growing economies like China – are regarded by the United States as being less open to industrialized country imports than they might be. Of course, this is changing as Korea has joined the OECD and China has joined the WTO. In both cases, liberalization of domestic markets is inevitably occurring.

> More than any other part of the world, East Asia needs the multilateral [trading] system and the rules and disciplines that underpin it. One reason is to keep the US and EU, its two biggest markets, open.[14]

One major problem facing the NIEs, therefore, is maintaining access to their major export markets, especially in the United States and Europe. In the light of less favourable external conditions for trade, the NIEs also need to develop further their own domestic markets. This poses a problem for the smaller NIEs, like Singapore, which have very limited domestic markets. However, it is significant that the regional market has become increasingly important.[15] The East Asian NIEs' share of global GDP has virtually doubled since 1976, to 25 per cent of the world total, whilst its share of world imports also doubled. Hence, not only are the first-tier NIEs big producers; they are also, together with their regional neighbours, increasingly major global markets.

Facing intensifying competition

A second major problem is that competition *between* NIEs has increased markedly, especially as labour costs in the first tier of NIEs have risen. The competition is obviously most severe in the lower-skill, labour-intensive activities on which NIE industrialization was originally based. Indeed, one of the major problems is that

> most East Asian economies have been following growth trajectories which involve ever-intensifying competition in external product markets … as the pace of globalisation speeds up … firms and economies constantly need to run faster just to stand still, fighting to raise productivity and product innovation faster than the decline in margins as competitive pressures intensify. What is scarce today – for example, the ability to fabricate semiconductor chips, or to assemble automobiles – will become common tomorrow. Thus the ability to sustain income growth depends upon the capability to respond flexibly to changing competitive circumstances … [however] the ability to physically transform inputs into outputs is increasingly widespread and in many sectors involves few barriers to entry … there are two possible responses to growing competitive pressures. The 'high road' is to upgrade production in such a way as to capture rents, to create barriers to entry, and hence to escape the competitive pressures. The 'low road' is to lower prices as a way of maintaining market share.[16]

It is in this context that the entry of China into the equation has to be seen.[17] China has become, or is becoming, the leading exporter of many of the manufactured products in which developing countries compete: 'The Chinese export structure is rapidly coming to resemble that of its neighbours'.[18] On the other hand, China is also a major *market* for the East Asian NIEs, both first and second generation.

> The [East Asia] region is the single largest trading partner for China, accounting for over twice the share of exports sold to the United States and over five times that sold to West Europe ... By 2000, China was importing more from every neighbour (apart from Hong Kong and Singapore) than it was exporting to it ... In direct trade, therefore, China acts *more as an engine of export growth than as a competitive threat to most of its neighbours* (Hong Kong excepted). It is difficult to predict if this will continue. While China will import more as it grows, how this will affect its regional trade balances depends on its neighbours' *patterns of specialization* with respect to China and their *competitiveness with respect to other exporters* to China.[19]

Prospects for NIEs outside Asia depend very much on their specific internal and external circumstances. One of the internal features differentiating the leading East Asian NIEs from many others is the big contrast in educational levels. Largely following the example of Japan, the East Asian countries have invested heavily, and from a very early stage, in education – 'and they have reaped as they have sown'.[20] In terms of external factors, regional context matters a great deal. In the case of Mexico, for example, the key issue is its ability to prosper within the NAFTA. Whilst its access to the US and Canadian markets is now guaranteed, its own economy is now feeling the full force of external competition within its own borders. Mexico is also threatened by Chinese competition, especially in its *maquiladora* industries. Similarly, the Southern European NIEs (especially Portugal and Greece), which have been members of the EU since the 1980s, now face increasing competition from the opening up of Eastern Europe.

Ensuring economic growth with equity

Sustaining economic growth is only one of the difficulties facing the NIEs (both existing and potential) in today's less favourable global environment. Sustaining growth *with equity* for their populations is also a major problem. Two aspects of this issue are especially important: income distribution and the socio-political climate within individual countries.

Income distribution

A widely voiced criticism of industrialization in developing countries has been that its material benefits have not been widely diffused to the majority of the population. There is indeed evidence of highly uneven *income distribution* within many developing countries, as Table 18.2 reveals. In countries such as Brazil, Chile and Mexico, for example, the share of total household income received by the top 20 per cent of households is very much higher than that in the industrial market economies. However, this pattern does not apply in all cases. For example, Korea has a household income distribution very similar to that of the industrial market economies. Of course, the question of income distribution is very much more complex than these simple figures suggest and is the subject of much disagreement among analysts. The fact remains, however, that in general the Asian NIEs have a more equitable income distribution than the Latin American countries, where there is a 'hugely unequal distribution of income and wealth. A disproportionately large number of Latin Americans are poor – some 222m or 43% of the total population'.[21]

Table 18.2   Distribution of income within selected developing and developed countries

Country (year)	Lowest 10%	Lowest 20%	Highest 20%	Highest 10%
Brazil (2001)	0.7	2.4	63.2	46.9
Chile (2000)	1.2	3.3	62.2	47.0
Mexico (2000)	1.0	3.1	59.1	43.1
Malaysia (1997)	1.7	4.4	54.3	38.4
Philippines (2000)	2.2	5.4	52.3	36.3
Korea (1998)	2.9	7.9	37.5	22.5
Singapore (1998)	1.9	5.0	49.0	32.8
China (2001)	1.8	4.7	50.0	33.1
India (1999–2000)	3.9	8.9	43.3	28.5
France (1995)	2.8	7.2	40.2	25.1
Germany (2000)	3.2	8.5	36.9	22.1
Sweden (2000)	3.6	9.1	36.6	22.2
United Kingdom (1999)	2.1	6.1	44.0	28.5
United States (2000)	1.9	5.4	45.8	29.9

Source: based on World Bank, 2005b: Table 2.7

Such differences reflect specific historical experiences, in particular, the different patterns of land ownership and reform. In Korea and in Taiwan, for example, post-war reform of land ownership had a massive effect, increasing individual incomes through greater agricultural productivity, expanding domestic demand, and contributing to political stability.[22]

> In 1953, while Taiwan was still recovering from World War II, the island had a level of income inequality that was about the level found in *present-day* Latin America. Ten years later it had dropped to the level *now found* in France. At the same time, growth rates in this period were of the order of 9 per cent per annum … this outcome was due primarily to improved income distribution in Taiwan's agriculture sector. This improvement, in turn, rested on a specific set of governmental policies, that focused in the first instance on agricultural reforms – especially land reform, infrastructure investment, and price reform – coupled with a rapid proliferation of educational opportunities for Taiwanese students at all levels. In terms of distributive measures, land reform was of greatest significance.[23]

However, especially in geographically extensive countries like Brazil, China or India, such aggregative income distribution data are misleading. As always, there are vast differences in income levels (and in other measures of well-being) between different parts of the same country. China, for example, faces massive internal problems. Its spectacular economic growth since its opening up in the early 1980s has created vast inequalities between different parts of the country (see Chapter 7).

Civil society

Income distribution is one aspect of the 'growth with equity' question. Another is the nature of civil society in NIEs: the broader social and political framework of *democratic institutions, civil rights and labour freedom*. Although the degree of repression and centralized control in NIEs may sometimes be exaggerated, the fact is that such conditions have existed in a number of cases.[24] The very strong state involvement in economic management in most NIEs has brought with it often draconian measures to control the labour force. Labour laws have tended to be extremely stringent and restrictive; in many instances strikes have been banned.

Although we have treated the questions of sustaining growth and of sustaining growth with equity as separate they are, in fact, closely related. It is an open question as to how far the various forms of the developmental state, which are manifested in different NIEs, can continue to provide the basis for future economic development. In this respect, it is significant that, from the late 1980s, both Korea and Taiwan began to democratize. It remains to be seen whether China can continue to develop as a market economy while still retaining an authoritarian communist political system. Certainly its 2005 policy statement on democracy in China makes it clear that the Communist Party remains the dominant force.

The conventional view in the West is that economic development and democracy must go together; that the first depends on the second. Of course, there is a view – expressed, for example, by the founder of the Singaporean state, Lee Kuan Yew, and by the former Malaysian Prime Minister, Mahathir Mohamad – that 'Asian' democracy is different from the Western model. In Lee's words, 'I do not believe that democracy necessarily leads to development. I believe what a country needs to develop is discipline more than democracy'.[25] His view is that Asian value systems are different from those of the West in their emphasis on collective responsibility rather than individualism and on the roles and responsibilities of the state, which is seen as essentially paternalistic.

Figure 18.6 sets out the major components of this concept of 'Asian values' which, in effect, 'recast "Asia" as a moral opposite of the West. Thus … the Asian penchant for hard work, frugality and love of the family are unproblematically figured as things the West lacks or has lost'.[26] Whether such opinions reflect the situation across the whole of East Asia (let alone across Asia as a whole) is unclear to say the least. This is, after all, a region of immense social, cultural and religious diversity. But in so far as they reflect at least some of the social and political characteristics of some successful East Asian economies, they form a considerable contrast with the situation in other parts of the world and help to explain some of the differences between East Asia's and, say, Latin America's economic development trajectories.

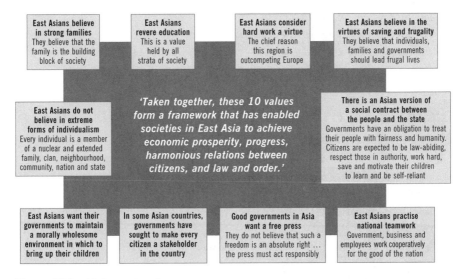

Figure 18.6 'Asian values'

Source: based on material in Koh, 1993

Environmental problems

Environmental degradation is also a major social problem and, although extensive environmental damage is certainly not confined to the NIEs,

> their single-minded pursuit of rapid economic growth has caused particularly severe environmental degradation ... Rising environmental costs in both urban and rural areas are materializing in poor health, physical damage, loss of amenities, and other problems that demand extensive remedial spending ... In order to stimulate rapid growth, the NIEs have used up significant environmental capital that can only be restored, if at all, at considerable cost to future generations.[27]

Similar comments can be made about the situation in the Mexico–United States border zone, where decades of unregulated pollution by the *maquiladora* factories has led to disastrous levels of toxicity in the water supplies and the atmosphere. It was for such reasons – and the United States' fear of 'environmental dumping' – that a specific environmental side agreement was incorporated into the NAFTA. Today, however, the environmental spotlight shines most of all on China.

The scale of China's environmental problems is difficult to exaggerate and puts an enormous strain on the country's social fabric. Headlong industrial growth has occurred along with massive atmospheric, water and ground pollution. In energy terms, China is heavily dependent on its huge reserves of coal transformed into usable energy using, in many cases, highly inefficient technologies. It is the second largest emitter of greenhouse gases in the world, after the United States. China has by far the highest number of premature deaths from urban air pollution (Figure 18.5). However, it is important to emphasize that 'unlike the US, the only country that produces more greenhouse gases than China, Beijing's leaders have put the environment at the heart of their rhetoric about economic development'.[28]

Ensuring survival and reducing poverty in the least developed countries

Although the NIEs certainly face problems, they are not of the same magnitude as those facing the least developed, lowest-income countries which are poor not just in terms of income but also in virtually every other aspect of material well-being. They are the countries of the deepest poverty, several of which face mass starvation. Figure 18.7 maps the 33 most impoverished countries in the world. These are the countries with per capita incomes less than the average for the World Bank's low-income group of $450. No fewer than 26 of the 33 (almost four in every five) are in Africa. Figure 18.7 also shows that a significant subset of these countries – the poorest of

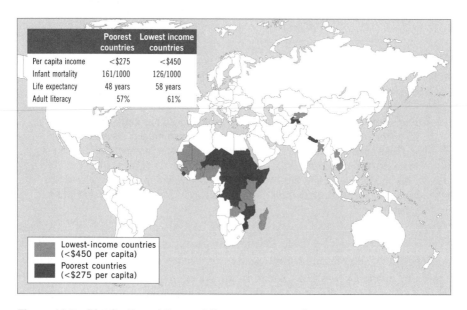

	Poorest countries	Lowest income countries
Per capita income	<$275	<$450
Infant mortality	161/1000	126/1000
Life expectancy	48 years	58 years
Adult literacy	57%	61%

Lowest-income countries (<$450 per capita)

Poorest countries (<$275 per capita)

Figure 18.7   Distribution of the world's poorest countries

Source: calculated from World Bank 2005a: Table 1

all – has an average per capita income of less than \$275. This subgroup also has significantly higher rates of infant mortality and lower rates of life expectancy and adult literacy.

Over-dependence on a narrow economic base

There is no single explanation for the deep poverty of low-income countries (or that of some of the lower-middle-income countries). There is no doubt, for example, that problems of inadequate internal governance (including corruption) play a major role in some cases. But in the context of the global economy, one factor is especially significant: an over-dependence on a very narrow economic base, together with the nature of the conditions of trade. We saw earlier (Figure 18.2) that the overwhelming majority of the labour force in low-income countries is employed in agriculture. This, together with the extraction of other primary products, forms the basis of these countries' involvement in the world economy. Two-thirds of developing countries have more than a 50 per cent dependence on commodity exports (including agricultural food and non-food products, ferrous metals, industrial raw materials and energy). In most sub-Saharan African countries, the level of dependence is around 80 per cent. Apart from the more successful of the East Asian NIEs, therefore,

> the exports of developing countries are still concentrated on the exploitation of natural resources or unskilled labour; these products generally lack dynamism in world markets.[29]

In the classical theories of international trade, based upon the comparative advantage of different factor endowments, it is totally logical for countries to specialize in the production of those goods for which they are well endowed by nature. Thus, it is argued, countries with an abundance of particular primary resources should concentrate on producing and exporting these, and should import those goods in which they have a comparative disadvantage. This was the rationale underlying the 'old' international division of labour in which the core countries produced and exported manufactured goods and the countries of the global periphery supplied the basic materials (Figure 2.1). According to traditional trade theory, *all* countries benefit from such an arrangement. But such a neat sharing of the benefits of trade presupposes:

- some degree of equality between trading partners
- some stability in the relative prices of traded goods
- an efficient mechanism – the market – which ensures that, over time, the benefits are indeed shared equitably.

In the real world – and especially in the trading relationships between the industrialized countries and the low-income, primary-producing countries – these conditions do not hold. In the first place, there is a long-term tendency for the composition of demand to change as incomes rise. Thus, the growth in demand for manufactured goods is greater than the growth in demand for primary products. This immediately builds a bias into trade relationships between the two groups of countries, favouring the industrialized countries at the expense of the primary producers.

Over time, these inequalities tend to be reinforced through the operation of the *cumulative* processes of economic growth. The prices of manufactured goods tend to increase more rapidly than those of primary products and, therefore, the *terms of trade* for manufactured and primary products tend to diverge. (The terms of trade are simply the ratio of export prices to import prices for any particular country or group of countries.) As the price of manufactured goods increases relative to the price of primary products, the terms of trade move against the primary producers and in favour of the industrial producers. For the primary producers it becomes necessary to export a larger quantity of goods in order to buy the same, or even a smaller, quantity of manufactured goods. In other words, they have to run faster just to stand still or to avoid going backwards. Although the terms of trade do indeed fluctuate over time, there is no doubt that they have generally, and systematically, deteriorated for the non-oil primary-producing countries over many years.

Table 18.3   Deteriorating terms of trade for primary-producing countries (ratio of export to import prices)

Country	1998	1999	2000	Main commodity	Exports (%)
Uganda	−5	−17	−34	Coffee	56
Zambia	−20	−26	−25	Copper	56
Mali	−11	−23	−28	Cotton	46
Rwanda	6	−11	−25	Coffee	45
Chad	−6	−15	−20	Cotton	42
Burkina Faso	−4	−16	−25	Cotton	39
Guyana	0	7	−14	Gold	16
Tanzania	1	−7	−13	Coffee	11

Source: based on material in *Financial Times,* 30 January 2002

Table 18.3 illustrates the seriousness of this problem for some African countries, each of which is heavily dependent on a single commodity for export earnings. There was a dramatic deterioration of the terms of trade for these countries at the end of the 1990s (although, as suggested above, this was nothing especially new). The figures show how the effects of the deepening slowdown in Western economies at the end of the 1990s and in the early 2000s were transmitted to these commodity producers.

Low levels of investment; high levels of debt

The world's poorest countries do not attract high levels of foreign direct investment. In some cases, they attract *hardly any* at all. As we saw in Chapter 2, the vast bulk of the world's FDI goes to developed economies, while the share going to developing countries is concentrated in a small number of NIEs. Neither do the poorest countries attract much portfolio investment because there are few attractive investment opportunities in their domestic economies. For the very poorest countries, therefore, the major external flow of income is through aid programmes. However, not only are these far below the necessary levels but they are also declining. For the group of low-income countries as a whole, aid amounted to just $12 per capita in 2002.

Although the world's poorest countries receive relatively little financial investment, many of them still face a crippling external *debt problem*. In fact, the problem of debt affects more than just the poorest countries: much of the economic growth of the NIEs was financed by overseas borrowing (as was that of the industrializing countries in the nineteenth century). A number of even the more successful NIEs have experienced periodic debt problems and financial crises (for example, several of the East Asian NIEs after 1997, Mexico in 1995, and Argentina in 2002). Hence, some of the highest absolute levels of debt are found in some of the NIEs rather than in the poorest countries, as Figure 18.8 reveals. However, absolute debt levels matter far less than *relative* levels – the scale of a country's debt relative to the size of its overall economy. Figure 18.8 shows the global distribution of debt as a proportion of developing countries' gross national income (GNI). Measured in this way, we can see clearly that most of the highly indebted countries are also among the poorest. In comparison, the large absolute debt of countries such as Brazil, China or Mexico looks more reasonable.

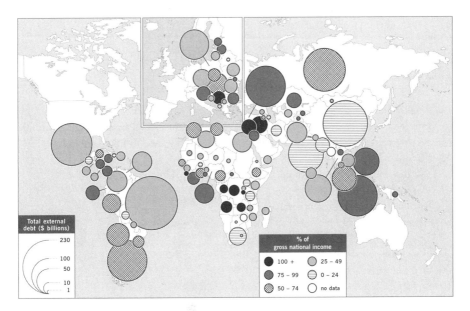

Figure 18.8   The map of developing country debt

Source: calculated from World Bank, 2005b: Table 4

Conclusion

In terms of winning and losing in the global economy, the position of developing countries is extremely variable. Overall, far too few jobs are being created to absorb rapidly growing populations. One result is a continuing high level of out-migration, a process with mixed results for the labour exporting country. A few countries – mostly, though not all, in East Asia – have undoubtedly prospered through globalizing processes. In some cases, the fruits of economic success have been distributed relatively equitably among their populations. In other cases, especially in Latin America, income distribution remains very polarized. Issues of civil society institutions and of environmental effects of economic growth remain important although, in the former case, we have to be careful not to impose Western-centric models of political structures.

Within the developing country group, there is a number of exceptionally poor countries which are impoverished even by the very low standards of the low-income group; four out of every five of these poorest countries are in Africa. Here the problems are manifold, and both internally and externally generated. In the latter case, the poor countries are overwhelmingly dependent on a very narrow range of primary activities (sometimes only one). These products operate in a global economy in which the terms of trade have generally been against the primary producers. Not surprisingly, then, such countries tend to be heavily dependent on aid; many of them are simply not attractive to foreign direct or portfolio investors. All of these issues beg the questions of what should, or can, be done. Such questions form the focus of the next chapter.

NOTES

1 One of the surprising developments of recent years has been the return to the old – and totally discredited – 'environmental' explanations of differences in economic development by the historian Landes (1998) and the economist Sachs (1997).
2 Martin and Widgren (1996: 10).
3 *The Economist* (14 January 2006), *Financial Times* (7 October 2005).
4 *Financial Times* (30 March 2006).
5 *Financial Times* (13 December 2005).
6 *Financial Times* (13 December 2005).
7 Jones (1990: 250).
8 UN Centre for Human Settlements (2003).
9 UN Centre for Human Settlements (2001: 14).
10 Ito (2001: 77).
11 Henderson (1998: 356–7).
12 Bello and Rosenfeld (1990) were strongly of the view that the first tier of NIEs was in serious crisis.

13 See, for example, Haggard and MacIntyre (1998), Stiglitz and Yusuf (2001).

14 De Jonquieres (2005: 24).

15 Dicken and Yeung (1999: 109–11).

16 Kaplinsky (1999: 4, 5).

17 Kaplinsky (1999; 2001), Lall and Albaladejo (2004).

18 Lall and Albaladejo (2004: 1447).

19 Lall and Albaladejo (2004: 1454, 1456).

20 Sen (1999: 41).

21 *The Economist* (17 September 2005).

22 Kapstein (1999: 118).

23 Kapstein (1999: 119).

24 See Bello and Rosenfeld (1990), Brohman (1996), Douglass (1994).

25 Lee (2000: 342).

26 Yao (1997: 238).

27 Brohman (1996: 126, 127).

28 McGregor and Harvey (2006).

29 UNCTAD (2002: 53).

Nineteen
Making the World a Better Place

'The best of all possible worlds'?

Voltaire, the eighteenth century French writer, wrote a wonderful satirical novel, *Candide*, in which the eponymous hero lives in a world of immense suffering and hardship yet whose tutor, Dr Pangloss, insists that Candide's world is 'the best of all possible worlds, where everything is connected and arranged for the best'.[1] This Panglossian view of the world is not far removed from those to whom an un-fettered capitalist market system – based on the unhindered flow of commodities, goods, services and investment capital – constitutes the 'best of all possible worlds'. Indeed, for some, it constitutes the very 'thread of history' (and of geography).

Without doubt, large numbers of people in the developed economies, and also in the rapidly growing economies of East Asia, have benefited from much-increased material affluence. There has been immense growth in the production and con-sumption of goods and services and, through international trade, a huge increase in the variety of goods available. But the evidence discussed in the preceding chapters of Part IV suggests a very different reality for a substantial proportion of the world's population, particularly in the poorest countries and regions but also among cer-tain sectors of the population in affluent countries, who have not benefited – or benefited very little – from the overall rise in material well-being. There remains vast inequality between the haves and the have-nots. If anything, that gap has been widening, despite the operation of precisely those globalizing processes that are supposed to create benefits for everybody.

What can or should be done about such large and pressing problems? How can the world be made a better place for all, including those at the bottom of the heap? There is, of course, no simple answer to such a disarmingly simple question. Much depends on one's political and ideological point of view. As we saw in Chapter 1, there are immense disagreements over the nature and impacts of 'globalization'. In terms of 'making the world a better place', one person's 'utopia' is another person's 'dystopia'.

The position taken throughout this book is that 'globalization' is a *multidimen-sional syndrome of processes* grounded in, and helping to create, *specific geographies*, involving *multiple actors* engaged in processes of both *conflict and collaboration*, and connected through asymmetrical *power relationships* (see Figure 1.3). It is far more complex than many of the highly polemicized debates suggest. For that reason, we would not expect there to be any simple, quick-fix solutions. There are many possible paths and, ultimately, we each have to decide for ourselves which paths we should take and act accordingly in our daily lives. Of course, choices are rarely – if ever – unconstrained. We are all deeply embedded in specific contexts, structures and places – and constrained by our knowledge and resources. As we have seen, the map of such constraints is immensely uneven; for many people, in many parts of the world, the exercise of choice is extremely limited.

The chapter is organized as follows:

- First, we outline some of the different perspectives on the problems created by, and the solutions advocated to deal with, the effects of globalization.
- Second, we examine existing global governance structures, particularly in the areas of finance and trade, together with the so far unsuccessful attempts to regulate transnational corporations.
- Third, we focus on two highly contentious problem areas: labour standards and environmental regulation.
- Finally, we ask what the future might be – and what it should be.

Globalization and its 'discontents': emergence of a global civil society?

In early December 1999, most people in Western countries were probably thinking about their Christmas shopping. Certainly, very few would have had the issues of 'globalization' at the front of their minds – or in their minds at all. Then, overnight, the TV screens were full of pictures of street protesters in Seattle, objecting to the attempts being made to initiate a new round of international trade negotiations within the WTO. The protests became the story. Suddenly, people in their sitting rooms at home became aware of an anti-globalization movement – of a broad coalition of 'discontents'.[2] Since then, similar protests (both peaceful and violent) have occurred at virtually every international meeting of government leaders and of bodies such as the IMF, the WTO, the World Bank, the European Union or the World Economic Forum.

In fact, the emergence of global civil society organizations (GCSOs) began long before Seattle, as Figure 19.1 shows. However, it may or may not be a coincidence that the spectacular take-off in their number more or less paralleled the increasing interconnectedness of the global economy (as we saw in Chapter 2). GCSOs come in a bewildering variety of forms. Figure 19.2 identifies six broad groups roughly in the chronological order of their appearance.

What is especially striking about the specifically anti-globalization movements is their sheer diversity. They include the long-established NGO pressure groups

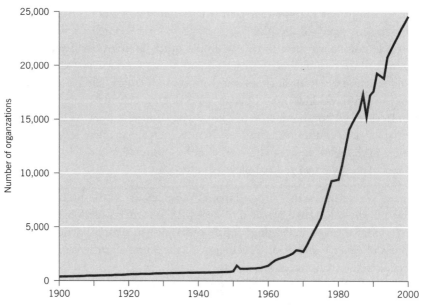

Figure 19.1   The growth of global civil society organizations

Source: based on Glasius et al., 2002: Figure 8.1

	'Old' social movements pre-1970	'New' social movements c. 1970s and 1980s	NGOs, think-tanks, commissions c. late 1980s and 1990s	Transnational civic networks c. late 1980s and 1990s	'New' nationalist and fundamentalist movements 1990s	'New' anti-capitalist movement c. late 1990s and 2000s
Issues	Redistribution, employment and welfare; self-determination and anti-colonialism	Human rights; peace; women; environment; Third World solidarity	Human rights; development and poverty reduction; humanitarianism; conflict resolution	Women; dams; land mines; International Criminal Court; global climate change	Identity politics	Solidarity with victims of globalization; abolition or reform of global institutions
Forms of organization	Vertical, hierarchical	Loose, horizontal coalitions	Ranges from bureaucratic and corporate to small scale and informal	Networks of NGOs, social movements and grassroots groups	Vertical and horizontal, charismatic leadership	Networks of NGOs, social movements and grassroots groups
Forms of action	Petition, demonstration, strike, lobbying	Use of media, direct action	Service provision; advocacy; expert knowledge; use of media	Parallel summits; use of media; use of local and expert knowledge; advocacy	Media, mass rallies, violence	Parallel summits; direct action; use of media; mobilization through Internet
Relation to power	Capturing state power	Changing state/society relations	Influencing civil society, the state and international institutions	Pressure on states and international institutions	Capturing state power	Confrontation with states, international institutions and transnational corporations

Figure 19.2   A typology of global civil society organizations

Source: based on Kaldor, 2003: Table 4.1

such as Oxfam, Greenpeace and Friends of the Earth; more recent ones like Jubilee 2000; organized labour unions like the AFL-CIO or the TUC; labour support organizations like Women Working Worldwide or the Maquila Solidarity Network; organizations focused primarily on TNCs and big corporations (like Corporate Watch or Global Exchange); right-wing nationalist/populist groups (exemplified by such figures as Pat Buchanan in the United States or Jean-Marie Le Pen's extreme right party in France); anti-capitalist groups (like ATTAC or the Socialist Workers Party); and various anarchist groups.

Within these groups' general focus on the costs of globalization there is huge variation both in their agendas and in how these agendas are pursued, from vociferous, often violent, confrontation through to more reformist movements. For the anti-capitalist groups, nothing less than the replacement of the capitalist system is advocated, although precisely what the alternative should be varies between groups. For some, it would be a democratically elected world government; to others, a structure in which the means of production were controlled by a nationally elected government. For others, it would be a system of locally self-sufficient communities in which long-distance trade would be minimized. This is the position, for example, of the 'deep green' environmental groups. For some, the focus is on 'fair' rather than 'free' trade – although who decides what is 'fair' varies a good deal. For the more nationalist-populist groups, and for some of the labour unions, the agenda is one of protecting domestic industries and jobs from external competition (especially from developing countries) and restricting immigration. For some, the objective is removing the burden of debt from the world's poorest countries or improving labour standards in the developing world (especially of child labour).

The problem is that, very often, these agendas are contradictory. There are some very unholy alliances involved. GCSOs have themselves attracted considerable criticism from some quarters, which question their legitimacy and, in some cases, their abilities to further economic and social development goals for the poor.[3] Despite the phenomenal recent growth of GCSOs, we certainly do not have, at present, a truly – or even a partial – global civil society. Although the proliferation of GCSOs has 'unquestionably projected the globalization debate into the popular political consciousness in important ways … the movements themselves have a severe democratic deficit: representing humanity ultimately requires legitimation through some sort of people's mandate'.[4]

Nevertheless, GCSOs undoubtedly force people – including politicians and business leaders – to recognize, and to engage with, the uncomfortable reality that both the benefits and the costs of globalization are very unevenly distributed and that there are severe and pressing problems that need resolution.

> The advocatory movements of global civil society are the originators, advocates and judges of global values and norms. The way they create and hone this everyday, local and global awareness of values is by sparking public outrage and generating global public indignation over spectacular norm violations. This they do by focusing on individual cases.[5]

Some response from the global institutions, such as the World Bank and the WTO, has been achieved. There is now more of a dialogue between the non-violent protest groups and the institutions. One potentially significant development was the emergence of the World Social Forum in 2001, originating in Brazil and consisting of many of the GCSOs, set up in response to the World Economic Forum (an annual meeting of the world's leading companies which, essentially, promotes a neo-liberal economic agenda).

Global governance structures

> While the world has become much more highly integrated economically, the mechanisms for managing the system in a stable, sustainable way have lagged behind.[6]

Virtually the entire world economy is now a *global capitalist market economy*. The collapse of the state socialist systems at the end of the 1980s and the headlong rush to embrace the market, together with the more controlled opening up of the Chinese economy since 1979, have created a very different global system from the one which emerged after World War II. The massive flows of goods, services and especially finance, in its increasingly bewildering variety, have created a world whose rules of governance have not kept pace with such changes. In particular, it is the sheer *speed,* especially of financial flows, that has so transformed economic-geographical and, therefore, power relationships. As a result, as Susan Strange[7] argues:

- Power has shifted upwards, from weak states to stronger states with global or regional reach beyond their frontiers.
- Power has moved sideways from states to markets and, hence, to non-state authorities which derive their power from their market shares.
- Some power has 'evaporated' in so far as no one exercises it.

As Figure 19.3 demonstrates, therefore, there is a veritable 'confusion' of global governance institutions – public, private and mixed – which operate at different but interconnected geographical scales, and in which

> institutions of global governance represent the outcome of a series of negotiations among corporations, states, and non-state actors.[8]

They reflect intricate bargaining relationships based upon asymmetrical power relationships between institutions, in which the exercise of 'soft' power predominates.[9] In what follows, we focus specifically on the *global* scale of governance in three areas, each of which is central to our concerns throughout this book:

- international finance
- international trade
- transnational corporations.

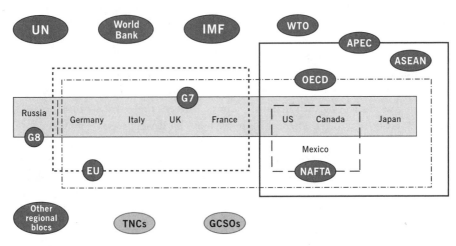

Figure 19.3   A 'confusion' of global governance structures and institutions

Source: based, in part, on Cable, 1999: Figure 3.1

Regulating the global financial system

The 'architecture' of the global financial system

The regulatory 'architecture' of the modern global financial system came into being formally at an international conference at Bretton Woods, New Hampshire, in 1944.[10] Two international financial institutions were created: the *International Monetary Fund* (IMF) and the *International Bank for Reconstruction and Development* (later renamed the *World Bank*).

The IMF's primary purpose was to encourage international monetary cooperation among nations through a set of rules for world payments and currencies. Each member nation contributes to the fund (a quota) and voting rights are proportional to the size of a nation's quota. A major function of the IMF has been to aid member states in temporary balance of payments difficulties. A country can obtain foreign exchange from the IMF in return for its own currency which is deposited with the IMF. A condition of such aid is IMF supervision or advice on the necessary corrective policies – the *conditionality* requirements. The World Bank's role is to facilitate development through capital investment. Its initial focus was Europe in the immediate post-war period. Subsequently, its attention shifted to the developing economies.

The primary objective of the Bretton Woods system was to stabilize and regulate international financial transactions *between nations* on the basis of fixed currency exchange rates, in which the US dollar played the central role. In this way it was hoped to provide the necessary financial 'lubricant' for a reconstructed world economy. However, through a whole series of developments (discussed in Chapter 13) the relatively stable basis of the Bretton Woods system was progressively undermined, particularly after the early 1970s. In effect, we have moved from a 'government-led international monetary system (G-IMS) of the Bretton Woods era to the market-led international monetary system (M-IMS) of today'.[11]

What we do not have, therefore, is a comprehensive and integrated global system of governance of the financial system. Instead, there are various areas of regulation performed by different bodies which are strongly *nationally* based:[12]

- The 'G' groups (G3, G7, G8) take an overall view of the monetary, fiscal and exchange rate relationships between themselves although, in practice, this has been confined primarily to attempts to determine the global money supply and to manage exchange rates. This has not resolved the basic problem of the continuing gap between an increasingly globalized financial system, the nationally based major central banks, and the mechanisms for regulating financial markets. The problem is that the 'G' groups have no real institutional base. They are informal arrangements structured around periodic summits of national leaders.
- The international payments system is operated through the national central banks rather than through an international central bank. 'While this central banking function remains unfulfilled at the international level the risks of default increase and disturbances threaten to become magnified across the whole system ... [there is] a growing network of cooperative and coordinative institutionalized mechanisms for monitoring, codifying and regulating such transactions (centred on the Bank for International Settlements and headed by the Group of Experts on Payment Systems).'
- The supervision of financial institutions themselves is carried out through the Bank for International Settlements (BIS), established in 1975 and subsequently based upon the 1988 Basel Committee's Capital Accord (which is currently being revised). 'Much like the monetary summits of the G3 to G7 the Basel Committee was initially designed as a forum for the exchange of ideas in an informal atmosphere with no set rules or procedures or decision-making powers. But, although it maintains this original informal atmosphere, its evolution has been toward much more involvement in hard-headed rule-making and implementation monitoring.'

Thus, there are several regulatory bodies in operation at the global level. Yet the fear remains that, in the absence of a more coordinated and institutionalized system, the global financial system could easily spiral out of control. Indeed, this is what appeared to be happening following the East Asian financial crisis of 1997 and its subsequent spillover effects on countries like Russia and Brazil. Suddenly, there was a chorus of calls for a new, or reformed, *financial architecture*. There was particular concern over the volatile nature of global capital flows in terms of their impact on both the financial system itself and the individual countries and their populations most seriously affected by volatile flows of 'hot money'.

There have been renewed calls for the introduction of a tax on international financial transactions, such as the 'Tobin tax' proposed in the early 1970s, whose purpose was to 'throw some sand in the wheels' of cross-border financial transactions by levying a small tax on each one. The idea was to discourage excessive flows of 'hot' money – short-term capital flows which can so easily destabilize financial systems, especially of weaker countries. So far, no such tax has been implemented, although in 2001 several European political leaders expressed renewed interest in the concept and the IMF acknowledged the need to look again at such proposals. Even so, a Tobin tax would address only a small – though important – part of the problem.

The other area of concern is the massive trade imbalances between countries (see Figures 2.13, 2.17, 2.20). In 2006, an agreement was reached giving the IMF a surveillance role over global trade imbalances with the hope of stimulating more balanced growth. Though seemingly a small measure, it could be seen as 'the biggest structural change since the break-up of the Bretton Woods system of fixed exchange rates in the early 1970s'.[13]

Developing countries in the global financial system

Developing countries are particularly vulnerable to the volatilities of global capital flows. One of the weaknesses of the various reforms to the global financial architecture since the breakdown of the Bretton Woods system is that they maintain a separation between the problems facing developed countries and those facing developing countries, instead of seeing them as inextricably linked together.

> Efforts in the past few years have focused on measures designed to discipline debtors and provide costly self-defence mechanisms. Countries have been urged to better manage risk by adopting strict financial standards, improving transparency, adopting appropriate rate regimes, carrying large amounts of reserves, and making voluntary arrangements with private creditors to involve them in crisis resolution. *While some of these reforms undoubtedly have their merits, they presume that the cause of crises rests primarily with policy and institutional weaknesses in the debtor countries and accordingly place the onus of responsibility for reform firmly on their shoulders. By contrast, little attention is given to the role played by institutions and policies in creditor countries in triggering international financial crises.*[14]

In fact, the IMF's and World Bank's conditionality 'medicine' may well make the patient worse rather than better. By imposing massive financial stringency – including raising domestic interest rates, insisting on increased openness of the domestic economy, reducing social spending, and the like – conditionality makes it extremely difficult for countries to help themselves out of debt. Even the former Chief Economist of the World Bank, Joseph Stiglitz, has strongly criticized such a policy.

> Conditionality, at least in the manner and extent to which it has been used by the IMF, is a bad idea; there is little evidence that it leads to improved economic policy, but it does have adverse political effects because countries resent having conditions imposed on them ... In some cases it even *reduced* the likelihood of repayment.[15]

Conditionality requirements are also frequently employed in the distribution of *aid* to developing countries and in attitudes towards *debt*. As we saw in Chapter 18, aid levels are very low whereas levels of debt are extremely high. At the 2002 UN Conference on Financing International Development held in Monterrey, Mexico, the United States belatedly offered to increase its aid to developing countries – but it still ranks among the lowest aid donors in the world. The UN target of aid equivalent to 0.7 per cent of developed countries' GDP is far from being fulfilled. Insistence on certain conditions has become more common as it has in the case of debt relief. In the view of one NGO representative at Monterrey:

> The Monterrey consensus (the statement of intent adopted by the conference) 'is just the Washington consensus wearing a sombrero'.[16]

On the other hand, there has been considerable progress in relieving the debt burden of the poorest countries. Not least because of the persistent lobbying of such pressure groups as Jubilee 2000, Oxfam and other aid agencies, there is now a programme to alleviate the debt burden of the highly indebted poor countries (HIPCs). At the G8 summit in 2005, it was agreed to write off debt for 18 of the world's poorest countries and that of a further nine countries within a year and a half. But there is still a very long way to go. Many more poor countries are in desperate need of debt relief.

In summary,

> the current global financial system is highly imperfect. More than any other markets, the global financial market is heavily dominated by financial interests in the industrialized countries. The governments of these countries, especially the economically strongest, determine the rules governing the market through their influence on the IFIs [international financial institutions]. These latter institutions in turn exercise great leverage over the macroeconomic and financial policies of developing countries. At the same time, the banks and financial houses from these same countries enjoy tremendous market power within the global financial system. The system is also characterized by severe market failures and is unstable. The upshot of all this is that most of the risks and the negative consequences of financial instability have been borne by the middle-income countries, currently the weakest players in the system ...

> Today there is a consensus on the need to reform the international financial architecture ... The bottom line is that the international financial system should support the integration of developing countries into the global economy in a manner that promoted development ...

> Progress towards attaining this goal has been slow and limited. So far, reform has been mainly focused on crisis prevention measures, such as greater disclosure of information, attempts to develop early warning systems, and the formulation of international standards and codes in financial sector supervision. While these initiatives are useful, their impact will be gradual and probably insufficient ...[17]

Of particular concern is that developing countries are not being adequately involved in the design of these new standards and codes.[17]

Regulating international trade: battles within the WTO

Compared with the international financial system, the governance of international trade is much clearer (though just as controversial).[18] In 1947, the General Agreement on Tariffs and Trade (GATT) was established as the third international institution formed in the aftermath of the Second World War, along with the IMF and the World Bank – completing what some have called the 'unholy trinity'.[19] Establishment of the GATT reflected the view that the 'beggar-my-neighbour' protectionist policies of the 1930s should not be allowed to recur. The objective was to be 'free' trade based upon the *principle of comparative advantage*, first introduced by David Ricardo in 1817. This states that a country (or any geographical area) should specialize in producing and exporting those products in which it has a comparative or relative cost advantage compared with other countries and should import those goods in which it has a comparative disadvantage. Out of such specialization, it is argued, will accrue greater benefit for all.

Whether or not there is such a thing as 'free' trade is a contentious issue; in order to work it needs some degree of equality between trading partners. As we have seen, at the global scale this does not exist. Nevertheless, the purpose of the GATT was to create a set of *multilateral rules* to facilitate free trade through the reduction of tariff barriers and other types of trade discrimination. The GATT was eventually replaced by the World Trade Organization (WTO) in 1995, an institutional change which greatly broadened the remit of the trade regulator. Today, around 95 per cent of world trade is covered by the WTO framework. Figure 19.4 traces its evolution.

Since 1947 there have been nine 'rounds' of multilateral trade negotiations, including the current Doha Round, initiated in 2001. Prior to the mid 1960s, the GATT was most concerned with trade of manufactures between developed nations. As a result, widespread dissatisfaction emerged among the developing countries: a particularly sensitive issue was the lack of access of developing country exports to developed country markets. Pressure led, in 1965, to the adoption

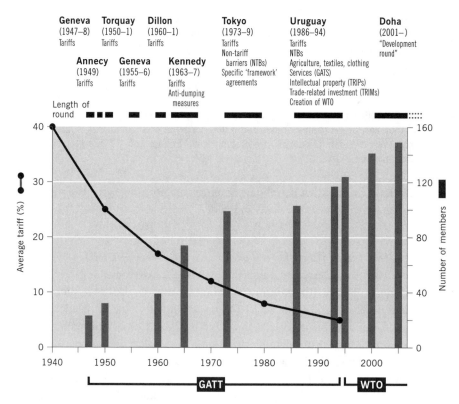

Figure 19.4   Evolution of the international trade regulatory framework: from the GATT to the WTO

within the GATT of a generalized system of preferences (GSP) under which exports of manufactured and semi-manufactured goods from developing countries were granted preferential access to developed country markets. In fact, there were a number of exclusions from the GSP, of which one of the most important was textiles and clothing (separately regulated under the MFA: see Chapter 9).

As Figure 19.4 shows, the first seven GATT rounds were both quite brief and also very limited in scope, although the Kennedy and Tokyo Rounds had slightly more extensive agendas. However, it was the Uruguay Round, started in 1986 and eventually concluded in 1994, which constituted the most ambitious and wide-ranging of all the GATT rounds to that point. For the first time, a number of additional trade issues was addressed. Notably, agriculture, textiles and clothing were brought into the GATT, and special agreements were concluded in services (GATS – the General Agreement on Trade in Services), intellectual property (TRIPS – Trade-Related Aspects of Intellectual Property Rights) and investment (TRIMS – Trade-Related Investment Measures). There was a further large reduction in overall tariff levels. The major organizational change was the creation of a new World Trade Organization.

The WTO, like the GATT, constitutes a *rule-oriented* approach to multilateral trade cooperation.

> Rule-oriented approaches focus not on outcomes, but on the rules of the game, and involve agreements on the level of trade barriers that are permitted as well as attempts to establish the general conditions of competition facing foreign producers in export markets.[20]

The fundamental basis is that of *non-discrimination*, based upon two components:

- The *most-favoured nation principle* (MFN) states that a trade concession negotiated between two countries must also apply to all other countries; that is, all must be treated in the same way. The MFN is 'one of the pillars of the GATT' and 'applies unconditionally, the only major exception being if a subset of Members form a free-trade area or a customs union or grant preferential access to developing countries'.[21]
- The *national treatment rule* requires that imported foreign goods are treated in the same way as domestic goods.

In November 1999, a WTO meeting was held in Seattle to try to initiate a new round of trade negotiations. The meeting was unsuccessful, not so much because of the now notorious anti-WTO/globalization protests but, as the UN Secretary-General Kofi Annan argued, because it failed to initiate a

> 'development round' that would at last deliver to the developing countries the benefits they have so often been promised from free trade ... instead [we] saw governments – particularly those of the world's leading economic powers – unable to agree on their priorities. As a result, no round was launched at all ... The [developed country] governments all favour free trade in principle, but too often they lack the political strength to confront those within their own countries who have come to rely on protectionist arrangements. They have not yet succeeded in putting across to their peoples the wider interest that we all share in having a global market from which everyone, not just the lucky few, can benefit.[22]

Three-quarters of the WTO's membership consists of developing countries. They face, as we have seen in previous chapters, immense economic and social problems. The Uruguay Round helped in some respects but created major difficulties in others. In particular,

> of the three big agreements coming out of the Uruguay Round – on investment measures (TRIMS), trade in services (GATS), and intellectual property rights (TRIPS) – the first two limit the authority of developing country governments to constrain the choices of companies operating in their territory, while the third requires the governments to enforce rigorous property rights of foreign (generally Western) firms. Together, the agreements make comprehensively illegal many of the industrial policy instruments used in the successful East Asian developers to nurture their own industrial and technological capacities.[23]

It was not until the end of 2001 that a new global trade round was announced at Doha in Qatar, with the official title of the 'Doha Development Agenda', to be concluded by 2004. However, such good intentions have not been fulfilled. The Doha Round has been possibly even more acrimonious than the Uruguay Round.[24] Deadlines have been missed with (un)impressive regularity. The Ministerial Meeting at Cancún in 2003 collapsed without producing any significant results. The most recent attempt to salvage the round was held in Geneva in June 2006, with little progress being made.

One of the biggest pressures to complete the Doha Round as quickly as possible has been that the US President's 'fast-track' authority to put trade deals to Congress in their entirety, rather than have to submit them 'line by line', expires in mid 2007. It may well not be renewed, especially in the current climate of increased protectionist feelings in the US (not least in relation to China, which joined the WTO in 2001).

> There are many reasons for the lack of agreement within Doha. One is that the WTO (like its predecessor the GATT) has been, by process and structure, a mercantilist institution that has worked on a principle of self-interested bargaining. The concept of a Development Round implies a fundamental departure from the system of mercantilism towards collectively agreed principles. However, there has been almost no discussion, let alone agreement, on what such principles might be.[25]

For this reason, some have argued that developing countries might be better off rejecting a deal that fails to deliver on their basic needs although not all agree in this. But lack of agreement not only reflects the division between developed and developing countries in the spheres of agricultural trade (see Chapter 12), industrial tariffs, access for services or intellectual property rights. It also reflects significant tensions between developed countries themselves (especially between the US and the EU over agricultural subsidies) and between developing countries (again, in agriculture, the interests of big global producers like Brazil are very different from those less well integrated into the global trading system).

The WTO continues to be subject to widespread criticism from many directions and from interest groups in both developed and developing countries.[26] For example, unilateralist groups within the United States regard the WTO as a basic infringement of the country's national sovereignty. Among developing countries, there is resentment over what is regarded as the bullying and unfair behaviour of the powerful industrialized countries. To the anti-globalization protesters, the WTO is an undemocratic, self-interested body acting primarily in the interests of global corporations.

There are certainly problems inherent in the workings of the WTO, but we should not forget that the WTO itself

> is an inter-governmental organization … [governments act] on behalf of their constituents in accordance with multilaterally agreed rules that have been adopted by consensus. *If the WTO is accused of acting against public interests, it is this collectivity of governments that is doing so …*

Nevertheless, there is certainly a problem relating to the systematic absence of many developing countries (particularly those that are small) from both informal and formal meetings in the WTO. Most small delegations from developing countries do not have the appropriate resources either in Geneva or at home to service the increasingly frequent, complex, and resource-intensive negotiation process at the WTO ...

However, knowledge and resources are not enough for all countries to be effective in WTO negotiations. *An important reality is that the WTO rules do not entirely remove the inequality in the power of nations. It remains the case that countries with big markets have a greater ability than countries with small markets to secure market access and to deter actions against their exporters.*[27]

(Not) regulating transnational corporations

In the case of foreign direct investment and transnational corporations there is no international body comparable to the WTO although, as we have seen, the Uruguay Round included a set of *trade-related investment measures* (TRIMS). Within this framework, some of the industrialized countries, led by the United States, wish to prohibit or restrict a number of the measures listed in Figure 6.5, notably local content rules, export performance requirements, and the like. TRIMS' advocates argue that such measures restrict or distort trade; its opponents see such measures as essential elements of their economic development strategies. They, in turn, wish to see a tightening of the regulations against the restrictive business practices of transnational corporations. Similarly, organized labour groups are generally opposed to measures that they feel will increase the ability of TNCs to affect workers' interests or to switch their operations from country to country.

In fact, there is a lengthy history of attempts to introduce an international framework relating to FDI and TNCs (apart from those agreed bilaterally or within the context of regional trade blocs).[28]

- OECD Guidelines for Multinational Enterprises (first introduced in 1976).
- ILO Tripartite Declaration of Principles Concerning Multinational Enterprises and Social Policy (1977).
- UN Code of Conduct for Transnational Corporations (initiated in 1982, abandoned in 1992).
- OECD Multilateral Agreement on Investment (MAI). The most recent attempt was launched in the mid 1990s. Its main provisions were:

  o Countries were to open up all sectors of the economy to foreign investment or ownership (the ability to exempt key sectors in the national interest was to be removed).
  o Foreign firms were to be treated in exactly the same manner as domestic firms.
  o Performance requirements (for example, local content requirements) were to be removed.

- ○   Capital movements (including profits) were to be unrestricted.
- ○   A dispute resolution process would enable foreign firms to be able to sue governments for damages if they felt that local rules violated MAI rules.
- ○   All states were to comply with the MAI.

Not surprisingly, the MAI generated a huge amount of opposition and was eventually blocked. Opposition was drawn from a very broad spectrum indeed, across both developed and developing countries.

> The choice of the OECD as the venue for the negotiations was a serious mistake because the OECD is a rich-country club and many LDCs were excluded from the discussions; why would LDCs accept an agreement that they had no part in formulating and that protected the interests of [T]NCs? Even many OECD countries objected to rules that would harm their own interests ... Labor and environmentalists objected that MAI would give [T]NCs license to disregard workers' interests and pollute the environment. Many critics charged that no protection was provided against the evils committed by [T]NCs. Even official American enthusiasm cooled when people realized that the MAI dispute mechanism could be used against the United States and its [T]NCs.[29]

The major dilemma in any attempt to establish a global regulatory framework for FDI and TNCs is the sharp conflict of interest inherent in the process involving TNCs, states, labour groups and CSOs. Should the focus be on regulating the *conduct* of TNCs (the viewpoint of most developing countries, some developed countries, labour and environmental groups) or should it be concerned with the *protection* of TNCs' interests? Both TRIMS and the aborted MAI were stacked in favour of TNCs.

But how far should TNCs in particular, and businesses in general, themselves explicitly pursue strategies of *corporate social responsibility* (CSR)?[30] In Milton Friedman's oft-quoted view, 'the business of business is business': the only thing that matters is maximizing profits and returns to shareholders. Although some business leaders may still share this view, there has been a rush to formulate *corporate responsibility* statements. Some of this may well be altruistic, in other cases mere self-interest. It does appear to vary between companies from different countries of origin.[31] In fact, as Figure 19.5 suggests, corporate social responsibilities span the entire spectrum of relationships between firms, states, civil society and markets.

It is difficult to avoid the conclusion that a major catalyst for CSR has been the increasing pressure on TNCs to recognize their social responsibilities and to conform to acceptable ethical standards.[32] Anti-globalization protesters routinely target high-profile TNCs like Nike, McDonald's, Shell, Starbucks or The Gap. In industries such as clothing and footwear, there have been sustained campaigns aimed at forcing TNCs to comply with decent labour standards (an issue we look at in more detail in the next section). As a result, a number of *voluntary codes of conduct* have appeared and most leading TNCs now produce an annual CSR report. For example, Nike included a list of the names and addresses of all its suppliers in

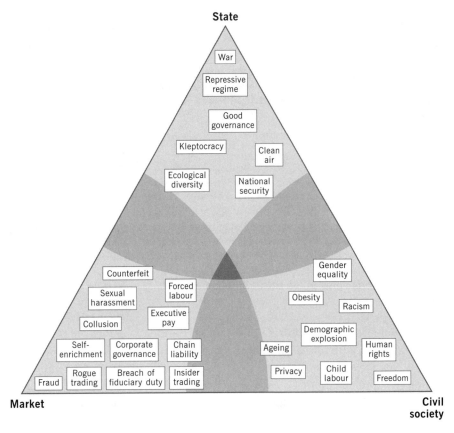

Figure 19.5 Primary issues of corporate social responsibility

Source: Based on van Tulder with van der Zwart, 2006: Figure 10.1

its 2005 CSR Report, an unprecedented step for a company which has always been highly secretive about its supply network (see Figure 5.11).

An example of a broader voluntary coalition is the Global Alliance for Workers and Communities, which involves Nike and The Gap together with the World Bank and the International Youth Foundation. Its mission is 'to improve the lives and future prospects of workers involved in global production and service supply chains'. A rather higher-profile voluntary code of good corporate conduct is the UN's *Global Compact*. This initially involved some 50 companies and 12 NGOs; today it has a membership of 3000, of which 2500 are companies spread across 90 countries. Figure 19.6 sets out its 10 principles.

Inevitably, there is a good deal of scepticism about such voluntary codes, whether at the individual firm or the collective level. This is not just because they are 'voluntary' but also because they are rather marginal in their scope and effect. In one sense, of course, anything which contributes to better conditions for people and communities should be welcomed. But without some degree of

Human rights	
Principle 1:	Support and respect the protection of international human rights within their sphere of influence
Principle 2:	Make sure their own corporations are not complicit in human rights abuses

Labour standards	
Principle 3:	The freedom of association and the effective recognition of the right to collective bargaining
Principle 4:	The elimination of all forms of forced and compulsory labour
Principle 5:	The effective abolition of child labour
Principle 6:	The elimination of discrimination in respect of employment and occupation

Environment	
Principle 7:	Support a precautionary approach to environmental challenges
Principle 8:	Undertake initiatives to promote greater environmental responsibility
Principle 9:	Encourage the development and diffusion of environmentally friendly technologies

Anti-corruption	
Principle 10:	Work against all forms of corruption, including extortion and bribery

Figure 19.6   Principles of the UN Global Compact

compulsion – and the monitoring of compliance – there is always the danger that such codes will amount to little more than a gesture or that companies will be able to influence how the process works. Nevertheless, they do represent a response to pressure from GCSOs as well as some governments. The two areas where there has been greatest activity in this regard are labour standards and the environment.

Two key concerns: labour standards and environmental regulation

How far do international differences in labour standards and regulations (such as the use of child labour, poor health and safety conditions, repression of labour unions and workers' rights) and in environmental standards and regulations (such as industrial pollution, the unsafe use of toxic materials in production processes) distort the trading system and create unfair advantages? In both cases, a common argument is that firms – as well as individual countries – may be able to undercut their competitors by capitalizing on cheap and exploited labour and lax environmental standards.

These two issues of labour and the environment were explicitly addressed in the negotiations for the NAFTA, in which the United States insisted on the signing of two side agreements to protect its domestic firms from low labour and environmental standards in Mexico. More recently, a group of countries led primarily by the United States, but also including some European countries, has made a concerted attempt formally to incorporate the issue of *labour standards* into the WTO. The attempt has failed, partly because not all industrialized countries supported it but also because the developing countries were vehemently opposed.

The argument of those opposed to its inclusion within the WTO's remit is that labour standards are the responsibility of the International Labour Organization (ILO). Indeed, all members of the ILO have agreed to the set of core principles shown in Figure 19.7. The counter-argument is that the ILO lacks any powers of enforcement. It is also notable that the United States, despite its current position on including labour standards in trade agreements, has not signed up to many of the ILO's core labour conventions, arguing that they do not comply with US law.

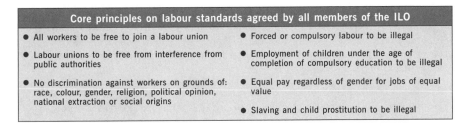

Core principles on labour standards agreed by all members of the ILO

- All workers to be free to join a labour union
- Labour unions to be free from interference from public authorities
- No discrimination against workers on grounds of: race, colour, gender, religion, political opinion, national extraction or social origins
- Forced or compulsory labour to be illegal
- Employment of children under the age of completion of compulsory education to be illegal
- Equal pay regardless of gender for jobs of equal value
- Slaving and child prostitution to be illegal

Figure 19.7   The core principles of the International Labour Organization

Labour standards

Stark differences exist in labour standards in different parts of the world. As we have seen at various points in this book, basic workers' rights are denied in many countries. Working conditions are often appalling. As far as child labour is concerned, the ILO calculates that around 246 million children are employed throughout the world, 'two-thirds of them engaged in hazardous forms of work'.[33] Most of these children work in agriculture or linked activities in rural areas and most are employed within the family rather than for outside employers. Even so, there is substantial evidence that, in many cases, young children are employed by manufacturers (whether in factories or as outworkers) in such industries as garments, footwear, toys, sports goods, artificial flowers, plastic products and the like. Their wages are a pittance and their working conditions often abysmal.

From the viewpoint of many developing countries, however, there is a strong feeling that the overall labour standards stance of many developed countries is merely another form of protectionism against their exports and, as such, an obstacle to their much-needed economic development. There is a suspicion, for example, that at least some of the developed country lobbies press for international agreements on a minimum wage in order to lessen the low-labour-cost advantages of developing countries.

But even where labour standard campaigns are based upon principles of human rights, the casualties may be the very people the campaigners are attempting to help. For example,

In 1993 American television broadcast pictures of Bangladeshi children making clothes for Wal-Mart. After a public outcry the store cancelled its contracts in Bangladesh and a senator proposed a bill to block imports of goods produced by child labour. Faced with losing one of its biggest markets, the Bangladesh garment manufacturers association announced it would eliminate the use of child labour within four months. Thousands of children were sacked, many of whom then found work in more dangerous industries. Few returned to school because their families simply could not afford it.

In the end the campaign against child labour in Bangladesh did have some positive results. Unicef, the United Nations' children's agency, negotiated a voluntary code of standards with the factories and provided money to send some of the children back to school.

Yet the incident illustrates another weakness of using boycotts of exports to raise labour standards in the third world: workers in the export sector usually enjoy better conditions than the vast majority of other third world workers who earn their livelihoods in the informal sector where there are no employment rights or health and safety provisions …

Keith Watkins of Oxfam says that a majority in the developing world are not working in the formal sector and they are not producing goods for export. 'There is no case you can point at where sanctions have improved labour standards. It's about a labour aristocracy in the north protecting its jobs and conditions.'[34]

There is clearly a basic dilemma for the international community. On the one hand, ethical considerations *must* be a basic component of international trade agreements. There is a clear moral responsibility to ensure that workers are not excessively exploited or ill-treated. On the other hand, there is a real danger of some threatened developed country interest groups using labour standards issues as a device to protect their own commercial interests. And even where this may not be the objective, there is always the possibility of unintended consequences. However, child labour is a symptom and consequence of poverty. One clear way of helping is for developed countries to lower their barriers to imports from developing countries. At the same time, efforts can continue to raise labour standards, for which there is both an ethical and an economic case.[35]

Encouragingly, a 2006 ILO report concluded that:

the number of child labourers globally fell by 11 per cent over the last four years whilst those engaged in hazardous work decreased by 26 per cent … Latin America and the Caribbean are making the greatest progress … Least progress has been made in sub-Saharan Africa where … child labour remain(s) alarmingly high. [36]

Globalization, trade and the environment

To what extent should variations in environmental standards be incorporated into international trade regulations? How far should developing countries be expected to forgo possibilities of development through industrialization because of the imposition of environmental regulations, which the industrialized countries did not have to face at a similar stage in their development? If they should, then who should foot the bill? Again, there is no doubting the existence of huge differences in the nature, scope and enforcement of environmental regulations across the world. There is no doubt, either, that the highest incidence of low environmental standards is in developing countries (see Chapter 18). The existence of such an *environmental gradient* certainly constitutes a stimulus for some firms at least to take advantage of low standards.

But there are also broader environmental issues associated with global production and trade. As we saw in Chapter 1, the production process is a system of materials flows and balances in which there are, inevitably, negative consequences for the natural environment (see Figure 1.9). Many of these negative environmental effects are not confined within national boundaries but spill over to affect other communities. Growing realization of the seriousness and pervasiveness of environmental problems associated with economic activity has led to attempts to devise global regulatory mechanisms.[37]

The most important agreements concluded in the 1990s included the

- Montreal Protocol on Substances that Deplete the Ozone Layer (1989–92)
- Convention on Biological Diversity (1992)
- Framework Convention on Climate Change (FCCC) (1992)
- Kyoto Protocol on Climate Change (1997).

Particular attention has focused on the climate change negotiations because of the concern over global warming. A key objective of the FCCC was to achieve

> the stabilization of greenhouse gas concentrations in the atmosphere at a level that would prevent dangerous anthropogenic interference with the climate system.[38]

The FCCC was based upon *voluntary* reduction of carbon dioxide levels. Its lack of success led to the signing of the Kyoto Protocol, which incorporates *binding* emissions targets for developed countries based on 1990 levels. A total of 186 countries (excluding, most significantly, the United States) signed the Protocol.

Since 2001, when the Kyoto Protocol was finally ratified, there has been much argument among the signatories, and with the United States, over the details of climate change regulation. In December 2005, the UN Conference in Montreal agreed the following measures:

- To strengthen the 'clean development mechanism', which allows developed countries to invest in sustainable development projects in developing countries whilst earning emission allowances.
- To launch the 'Joint Implementation' mechanism of the Kyoto agreement which allows developed countries to invest in other developed countries, especially the

transition economies of Eastern Europe. In doing so they can earn carbon allowances which can be used to meet their emission reduction commitments.

- To implement the Kyoto 'compliance regime' which ensures that countries have clear accountability for meeting their emission reduction targets.

Other problems remain, however, not least the extent to which developing countries can or should be expected to comply with emissions targets. It seems likely that greenhouse gas emissions from developing countries will reach the level of those of developed countries by between 2015 and 2020. Yet the costs to developing countries of introducing new clean technologies will be huge, even with assistance from developed countries. There is also the political issue – the feeling among developing countries that the imposition of carbon emission targets is another form of protectionism.

Hence, a closely related issue is the link between environmental standards and their regulation and international trade. At one level, the problem is exactly the same as that of labour standards. If a country allows lax environmental standards, it is argued, then it should not be able to use what is, in effect, a subsidy on firms located there to be able to sell its products more cheaply on the international market. The question then becomes one of whether the solution lies in using international trade regulations or in some other forms of sanction.

However, there is an even more extreme position adopted by some environmentalists: that the pursuit of ever-increasing international trade – which is clearly encouraged by a free trade regime like the WTO – should be totally abandoned, not merely regulated. The argument here is basically that *sustainable* development is incompatible with the pursuit of further economic growth and, especially, with an economic system based upon very high levels of geographical specialization, since such specialization inevitably depends upon, and generates, ever-increasing trade in materials and products.

One of the criticisms voiced by environmentalists is that the energy costs of transporting materials and goods across the world are not taken into account in setting the prices of traded goods and that, in effect, trade is being massively subsidized at a huge short-term and long-term environmental cost.

> More than half of all international trade involves the simultaneous import and export of essentially the same goods. For example, Americans import Danish sugar cookies, and Danes import American sugar cookies. Exchanging recipes would surely be more efficient.[39]

We noted in Chapter 12 the vast distances over which food products now travel.

But by no means all environmentalists agree with this kind of viewpoint, as David Pearce argues.

> Unquestionably, there are environmental problems inherent in the existing trading system. But there is also extensive confusion in the environmentalist critique of free trade ... Given the potentially large gains to be obtained from

free trade, adopting restrictions on trade for environmental purposes is a policy that needs to be approached with caution. Most importantly, all other approaches to reducing environmental damage should be exhausted before trade policy measures are contemplated ... the policy implication of a negative association between freer trade and environmental degradation is not that freer trade should be halted. What matters is the adoption of the most cost-effective policies to optimize the externality. Restricting trade is unlikely to be the most efficient way of controlling the problem ... The losses can best be minimized by firm *domestic* environmental policy design to uncouple the environmental impacts from economic activity ... The 'first' best approach to correcting externalities is to tackle them directly through implementation of the polluter pays principle (PPP), not through restrictions on the level of trade. Where the PPP is not feasible (e.g. if the exporter is a poor developing country), it is likely to be preferable to engage in cooperative policies, e.g. making clean technology transfers, assisting with clean-up policies etc., rather than adopting import restrictions.[40]

As in the case of labour standards, therefore, there is intense disagreement both over the impact of globalization processes and trade on the environment and also over what should (or can) be done.[41] There are, in other words, very different 'shades of green', ranging from the position that human ingenuity and new technologies will find the solutions without necessitating a change in lifestyles (the Panglossian view) through to the 'deep green' arguments that only a return to a totally different, small-scale, highly localized mode of existence will suffice. But such a path, rather than improving the position for the poor in the world economy, 'would condemn the vast majority of people to a miserable future, at best on the margins of the bare minimum of physical existence'.[42] It is not a socially acceptable policy:

> The alternative to an economic system that involves trade is not bucolic simplicity and hardy self-sufficiency, but extreme poverty. South Korea has plenty of problems, but not nearly so many as its neighbour to the north.[43]

This is to argue not for the status quo but, rather, for a policy towards the environment that recognizes people's needs and aspirations within an equitable context. In other words, the goal is

> development that meets the needs of the present without compromising the ability of future generations to meet their own needs.[44]

To be 'globalized' or not to be 'globalized': that is the question

> The main losers in today's very unequal world are not those who are too much exposed to globalization. They are those who have been left out.[45]

> Is the problem actually globalization or not-globalization? Is the difficulty being part of the system or not being part of it? How can globalization be the source of problems for those excluded from it?[46]

It is abundantly clear that the position of many of the world's poorest countries is highly marginal in terms of the global economy. The usual prescription of the IMF and World Bank 'doctors' is that they should open their economies more, for example, by positively encouraging exports and by liberalizing their regulatory structures.

> For policymakers around the world, the appeal of opening up to global markets is based on a simple but powerful promise: international economic integration will improve economic performance ... *The trouble is ... that there is no convincing evidence that openness, in the sense of low barriers to trade and capital flows, systematically produces these consequences. In practice, the links between openness and economic growth tend to be weak and contingent on the presence of complementary policies and institutions.*[47]

'Openness', then, is the name of the game. But this will work only if the playing field is relatively level – which it clearly is not. And it also has to work both ways – which clearly it does not. Tariffs imposed by the developed countries on imports of many developing country products remain very high. It is common for tariffs to increase with the degree of processing (so-called tariff escalation), so that higher-value products from developing countries are discriminated against. At the same time, agricultural subsidies make imports from developing countries uncompetitive. In other words, the odds are stacked against them.[48]

> The human costs of unfair trade are immense. If Africa, South Asia, and Latin America were each to increase their share of world exports by one per cent, the resulting gains in income could lift 128 million people out of poverty ...
>
> When developing countries export to rich-country markets, they face tariff barriers that are four times higher than those encountered by rich countries. Those barriers cost them $100bn a year – twice as much as they receive in aid.[49]

Simply opening up a developing economy on its own will almost certainly lead to further disaster. There is the danger of local businesses being wiped out by more efficient foreign competition before they can get a toehold in the wider world. Hence a prerequisite for positive and beneficial engagement with the global economy is the development of robust internal structures: 'The development of a national economy is more about internal integration than about external integration'.[50]

What might the future be?
What *should* the future be?

It is always tempting to look at recent trends and to extrapolate them into the future. There is, of course, some logic in this. After all, there is a strong element of *path dependency* in human affairs (see Chapter 1). But it isn't as simple as that. Path dependency does not mean *determinacy*. All paths have branching points: some go off in unexpected directions, others into dead-ends. This means that we need both a good map and a clear sense of the direction we wish to travel. The first requires a better understanding of how the world actually works (that's what this book has been about). The second requires an ethical and moral vision. It is about values.[51] It is about where we want to be.

The perils of prediction

We're not very good at making predictions. Every year (at least), new books or articles appear claiming to set out what the world will be like in X years' time. Most are soon forgotten — usually for the very good reason that what was predicted hasn't actually happened. It is very difficult indeed to identify which contemporary events and circumstances are likely to have long-lasting effects. For example, when the East Asian financial crisis broke with such suddenness in 1997, the literature was full of predictions of doom: the end of the East Asian 'miracle' had arrived. The future of the region was dire. Few would make those same predictions today.

Similarly, looking a little further back in time, who, from the standpoint of 1960, would have predicted that Japan would soon challenge the United States as an economic power — and, in some respects, overtake it to the extent that, in the 1980s, doomsayers in the US were lamenting the demise of the United States as the world's leading economy? Japan bashing became a national pastime (and not only in the US: there were outbreaks in Europe too, especially in France). Who would have predicted that the Japanese economy would suddenly find itself deep in economic recession lasting for more than a decade and a half?

Who would have predicted that South Korea would become one of the world's most dynamic economies within the space of 20 years or so? After all, in 1960 South Korea was one of the poorest countries in the world, with a per capita income comparable with that of Ghana. Which observer in the early 1970s would have predicted that China would open up its economy and become, in a very short time, the most dynamic economy in the world? Or that the command economies of the Soviet Union and Eastern Europe would, by the end of the 1980s, begin to be transformed into capitalist market economies, or that Germany would be reunited? Such examples should make us wary of prediction. But we don't learn. Today's big bets are on Chinese world economic dominance within the next few decades.

We are seduced, far too easily, by big numbers. We focus too eagerly on the quantitative, rather than the qualitative, dimensions and processes of change. This raises a much bigger question: will the tendency towards an increasingly highly interconnected and interdependent global economy intensify? Is 'globalization' an inexorable and unstoppable force? Not inevitably, as the period between 1919 and 1939 shows.

During that time, the unprecedented openness of the world economy that had come into being in the period between 1870 and 1913 was largely reversed through the actions of states responding to recession through increased protectionism. It took several decades to return to a similar degree of openness, by which time the world was a very different place.

Of course, the interconnections within the global economy are now much deeper – and faster – than in the past because of the ways in which the processes of production and distribution have been transformed. Development of the highly complex, geographically extensive, transnational production networks which have figured so prominently throughout this book epitomizes this. But such increased interdependence may, itself, be a source of vulnerability.[52] Unforeseen damage to one part of the system will inevitably have implications for the other parts. The sources of such potential damage are many and varied, ranging from natural phenomena like earthquakes to the human-made phenomena of geopolitical and religious conflicts.

But there are wider geopolitical problems, both directly and indirectly related to the economy. In the former case, there is undoubtedly the threat of a trade war between the United States and China, as the current raft of anti-China bills at various stages in the US Congress demonstrates. As we saw earlier in this chapter, the Doha Round of trade negotiations is in serious trouble and it is unlikely that anything other than a second-best agreement will be achieved. Not least, this is because of deep tensions that cut across the developed/developing country divide. In particular, there is continuing friction between the world's biggest trading areas: the United States and the European Union. Within parts of the EU, notably in France and Italy, as well as in some of the new Eastern European member states, there are renewed calls to protect national companies from foreign takeover (even from other EU firms). In the United States, the bid by the Dubai Port Authority to purchase P&O and its port facilities in the United States in 2006 was withdrawn in the face of intense US opposition, partly fuelled by security fears in a post 9/11 world.

These cases draw attention to the impossibility of separating out the geo-economic from the geopolitical. Three big geopolitical problems are especially relevant here. The first, and most obvious, is the set of issues that includes the post 9/11 US shift towards pre-emptive action and regime change, the Iraq War, the nuclear stand-off with Iran, the continuing impasse over an Israel–Palestine settlement, and the disturbing rise in religious fundamentalism (of all kinds) associated with these intractable problems. The implications are far-reaching and go beyond the human tragedies of those most closely affected by these conflicts.

From a global economic perspective, the most obvious is the effect on the continued supply (and price) of oil, the energy source which has underpinned the global economy for almost a century. Of course, this makes the need for energy efficiency and for the development of alternative, renewable energy resources even more urgent. In the short run, however, the effects could be serious. Not the least of the implications of the spectacular economic growth of China is its vastly increased need for energy and other natural resources.

The second major geopolitical problem is, again, not unconnected to the rise of China and, more broadly, to political developments in East Asia as whole. Ever since the end of the Second World War in 1945, the United States has been deeply involved in the Asia–Pacific for both security and economic reasons. Until recently, this was very much in the context of the Cold War. Indeed, the post-1945 economic revitalization of Japan, Korea, Taiwan and other parts of East Asia was strongly facilitated by US activities and financial aid. With the collapse of the Soviet empire, and the opening up of China economically, the position has changed. One symptom of this is the current Chinese dominance of much of the world's commodity trade.

Significant geopolitical problems remain in what is the world's most dynamic economic region. The US still sees China as a potential military threat (as well as an economic rival). The question of Taiwan is always there as a source of potential conflict, even though economic relations between China and Taiwan have improved markedly and there is huge Taiwanese investment in China. Relations between Japan and China remain extremely sensitive, not least because of Japan's reluctance to recognize some of the atrocities it perpetrated when it occupied China in the Second World War. More broadly, Japan's own future geopolitical intentions within East Asia are far from clear. Lastly, there is the intractable question of relationships between North Korea and South Korea, especially the nuclear issue.

The third big geopolitical problem is that of failed, dysfunctional or inadequate states.[53] Although many of the problems facing developing countries – especially the poorest – arise from their position in the global economy and their very weak power base in international negotiations, other problems are undoubtedly 'home grown'. There are substantial internal problems of governance, corruption and inhuman treatment of minority populations in some developing countries which cannot be ignored. The issue is how the world community deals with these problems to ensure the enhanced welfare of the populations of such failed states and to bring them fully, and effectively, into the world community and economy.

For all kinds of reasons, therefore, the future map of the global economy is far from clear. Nevertheless, the chances are that globalizing processes will continue to operate and that the world will become increasingly interconnected, although there may well be unanticipated upheavals. We should not simply extrapolate from past trends. Most of all, we need to think about the *kind* of world we, and our children, would want to live in. The key question is not so much what the world *might* be like in the future but what it *should* be like. There are choices to be made. What might these be? After all, globalization is not a force of nature; it is a social process.[54] What are the choices? In theory they are infinite; in practice they are not. There are three broad possibilities: to *reform* the present system; to *replace* the present system; and to *diversify* the present system.

A better world

Although the development of a highly interconnected global economy has generated unprecedented wealth and material opportunities for many, it has also helped to perpetuate – even to intensify – immense disparities between people in different parts of the world. But even increased affluence has not necessarily made people 'happier' in proportion to their increased wealth.[55] Research by the ILO[56] suggests that 'economic security', rather than wealth, 'promotes happiness'.

> The global distribution of economic security does not correspond to the global distribution of income ... South and South East Asia have greater shares of economic security than their share of the world's income ... By contrast, Latin American countries provide their citizens with much less economic security than could be expected from their relative income levels ... income security is a major determinant of other forms of labour-related security ... [and] income inequality worsens economic security in several ways ... *highly unequal societies are unlikely to achieve much by way of economic security or decent work.*[57]

In thinking about alternative futures in the context of globalization debates, there is a depressing tendency towards polarization of positions of the kind we identified in Chapter 1. The gung-ho, neo-liberal hyper-globalizers see the solution in yet more openness of markets, in unfettered flows of goods, services and capital. In other words, much, much more of the same. The deglobalizers argue for exactly the opposite. Whereas the first strategy would almost certainly create a world of even greater inequality (as well as environmental damage) the second embodies the dangers of reverting to a set of medieval subsistence economies. Although these may be caricatures of these positions, they do point to the tendency to see choices in terms of diametrical opposites.

In fact, there is far more potential diversity of economies[58] that offer different kinds of possibilities and which may occupy different positions in relation to the larger global economy. Many of these are, essentially, 'community economies'. Figure 19.8 summarizes the main ways in which such economies differ from the 'mainstream' economy: 'In some cases, these communities are geographically confined to the "local", while in others they span the "global".'[59] The fair trade networks we discussed in Chapter 12 are examples of the latter. The Mondragón Cooperative Corporation is an example of a complex of cooperatives spread across 14 countries but grounded in the Basque region of Spain and organized on worker-owned principles. Another is the LETS (local exchange trading system) which uses local 'currencies' to facilitate the exchange of services and self-produced or self-earned goods within a network of members in a local community.[60]

The diversity of economies that exists in the world offers significant possibilities for creating fulfilling and fair communities and, more generally, for reconsidering globalization as a transformable social process and not a force of nature. But they are, by definition, mostly small in scale and often highly local in scope. They

Mainstream economy	—	Community economy
Aspatial/global	—	Place attached
Specialized	—	Diversified
Singular	—	Multiple
Large scale	—	Small scale
Competitive	—	Cooperative
Centred	—	Decentred
Acultural	—	Culturally distinctive
Socially disembedded	—	Socially embedded
Non-local ownership	—	Local ownership
Agglomerative	—	Dispersed
Integrated	—	Autonomous
Export oriented	—	Oriented to local market
Privileges short-term return	—	Values long-term investment
Growth oriented	—	Vitality oriented
Outflow of extracted value	—	Recirculates value locally
Privately owned	—	Community owned
Management led	—	Community led
Controlled by private board	—	Community controlled
Private appropriation and distribution of surplus	—	Communal appropriation and distribution of surplus
Environmentally unsustainable	—	Environmentally sustainable
Fragmented	—	Whole
Amoral	—	Ethical
Crisis ridden	—	Harmonious
Participates in a spatial division of labour	—	Locally self-reliant

Figure 19.8   Contrasting characteristics of mainstream and community economies

Source: based on Gibson-Graham, 2006: Figure 23

raise important issues of how such economies connect into the bigger picture (unless they decide to opt out). And it is the 'bigger picture' that still demands our primary attention. The major global challenge facing us is to meet the material needs of the world community as a whole in ways that reduce, rather than increase, inequality and which do so without destroying the environment. That, of course, is far easier said than done. It requires the involvement of all major actors – business firms, states, international institutions, civil society organizations – in establishing mechanisms to capture the gains of globalization for the majority and not just for the powerful minority.

Such a system has to be built on world trading and financial systems that are *equitable*. This must involve reform of such institutions as the WTO, the World Bank and the IMF or, alternatively, their replacement by more effective, and more widely accountable, institutions. As we saw earlier, the exercise of developed country *power* through the various kinds of conditionality and trade-opening requirements imposed on poorer countries has seriously negative results. Without doubt, trade is one of the most effective ways of enhancing material well-being, but it has to be based upon a genuinely fairer basis than at present. The poorer countries must be allowed to open up their markets in a manner, and at a pace, appropriate to their needs and conditions. After all, that is precisely what the United States did during its phase of industrialization, as did Japan and the East Asian NIEs. At the same time, developed countries must operate a fairer system of access to their own markets for poor countries. Even the journal *Business Week* acknowledges this when it states that

> the flaws of trying to force every country into the same template have become clear ... it is time to forge a more enlightened consensus.[61]

Of course, this will cause problems for some people and communities in the developed countries and these should not be underestimated. They are reflected most clearly in the various lobbying groups that attempt to influence government trade policy. As we saw in Chapter 17, there are, indeed, many losers in the otherwise affluent economies. People who feel threatened by 'globalization' must be assisted effectively, whether this involves financial assistance or education and retraining. Governments must design and implement appropriate adjustment policies for such groups if trade policies helpful to developing countries are to be acceptable politically. Surveys of US workers have shown that 'less-educated, lower-income workers are much more likely to oppose policies aimed at freer trade and immigration'.[62] Equally, governments of developing countries must engage in their own internal reforms: to strengthen domestic institutions, enhance civil society, increase political participation, remove corruption, raise the quality of education, and reduce internal social polarization. Although difficult, such policies are not impossible if the social and political will is there.

The imperatives are both practical and moral. In practical terms, the continued existence across the world of vast numbers of people who are impoverished – but who can see the manifestations of immense wealth elsewhere through the electronic media – poses a serious threat to social and political stability. But the moral argument is, I believe, more powerful. It is utterly repellent that so many people live in such abject poverty and deprivation whilst, at the same time, others live in immense luxury. This is not an argument for levelling down but for raising up. The means for doing this are there. What matters is the *will* to do it – the real acceptance of, to use David Smith's words,

> the imperative of developing more caring relations with others, especially those most vulnerable, whoever and wherever they are, within a more egalitarian and environmentally sustainable way of life in which some of the traditional strengths of community can be realised and spatially extended.[63]

We all have a responsibility to ensure that the contours of the global economic map in the twenty-first century are not as steep as those of the twentieth century. We have a responsibility to treat others as equals. In a global context, this means being sensitive to the immense diversity that exists, to the world as a mosaic of people equally deserving of 'the good life'.

NOTES

1 Voltaire (1947: 8).
2 The term 'globalization and its discontents' has been used by several authors, notably Sassen (1998) and Stiglitz (2002). Recent discussions of global civil society include Anheier et al. (2005), Beck (2005: Chapter 6), Bircham and Charlton (2001), Brand and Wissen (2005), Kaldor (2003), van Tulder with van der Zwart (2006).
3 See Williamson (2005) for a review of this issue. Wolf (2004: Chapter 1) is an especially trenchant critic of anti-globalization groups.
4 Taylor et al. (2002: 15–16).
5 Beck (2005: 238).
6 Commission on Global Governance (1995: 135–6).
7 Strange (1996: 189).
8 Levy and Prakash (2003: 131).
9 Gritsch (2005).
10 For discussion of the global financial system see Braithwaite and Drahos (2000: Chapter 8), Eatwell and Taylor (2000), Eichengreen (1996), Hirst and Thompson (1999), Martin (1999), Peet et al. (2003: Chapters 2–4).
11 Hirst and Thompson (1999: 202).
12 Hirst and Thompson (1999: 202–4).
13 *The Guardian* (24 April 2006).
14 UNCTAD (2001: vi–vii, emphasis added).
15 Stiglitz (2002: 44).
16 *Financial Times* (25 March 2002).
17 ILO (2004b: 88, 89).
18 See Hoekman and Kostecki (1995), Peet et al. (2003: Chapter 5), Sampson (2001).
19 Peet et al. (2003).
20 Hoekman and Kostecki (1995: 24).
21 Hoekman and Kostecki (1995: 26).
22 Quoted in Sampson (2001: 19).
23 Wade (2003: 621).
24 Stiglitz and Charlton (2005) present a detailed critique of the Doha Round and show how 'fair' trade and development can go together.
25 Stiglitz and Charlton (2005: 67).
26 See, for example, Peet et al. (2003: Chapter 5), Singer (2004: Chapter 3).
27 Sampson (2001: 7–8, emphasis added).
28 On regulation of TNCs see Braithwaite and Drahos (2000: Chapter 10), Gilpin (2000: Chapter 6), Kolk and van Tulder (2005).
29 Gilpin (2000: 184–5).
30 Van Tulder with van der Zwart (2006: Parts II and III) provide a comprehensive analysis of CSR.

31 Maignan and Ralston (2002).
32 Gereffi et al. (2001).
33 ILO (2004a: 93).
34 *The Guardian* (16 February 2001).
35 Palley (2004) argues the economic case for international labour standards.
36 ILO (2006b: 1)
37 See Braithwaite and Drahos (2000: Chapter 12), Pearce (1995).
38 Quoted in Pearce (1995: 149).
39 Daly (1993: 25).
40 Pearce (1995: 74, 77, 78).
41 See Hudson (2001: 315–21), Turner et al. (1994: Chapter 2).
42 Hudson (2001: 315).
43 Elliott (2002). See also Elliott (2000).
44 UN World Commission on the Environment and Development (1987: 43).
45 Kofi Annan, UN Secretary-General, Speech to UNCTAD Meeting in Bangkok, 2000.
46 Mittelman (2000: 241).
47 Rodrik (1999: 136–7, emphasis added).
48 See UNCTAD (2002).
49 Oxfam (2002: 1).
50 Wade (2003: 635).
51 See Lee (2007), Lee and Smith (2004), Smith (2000).
52 For a US-centric analysis of this problem, see Lynn (2005).
53 Wolf (2004).
54 Massey (2000: 24).
55 Layard (2005), Oswald (1997).
56 ILO (2004a). See also Scheve and Slaughter (2004).
57 ILO press release: www.ilo.org
58 Gibson-Graham (2006).
59 Gibson-Graham (2006: 80).
60 Lee (1996).
61 *Business Week* (6 November 2000: 45).
62 Scheve and Slaughter (2001: 87).
63 Smith (2000: 208).

Bibliography

Abernathy, F., Dunlop, J., Hammond, J.H. and Weil, D. (1999) *A Stitch in Time: Lean Retailing and the Transformation of Manufacturing: Lessons from the Apparel and Textile Industry*. New York: Oxford University Press.

Abernathy, F.H., Dunlop, J.T., Hammond, J.H. and Weil, D. (2004) Globalization in the apparel and textile industries: what is new and what is not? In M. Kenney and R. Florida (eds), *Locating Global Advantage: Industry Dynamics in the International Economy*. Stanford, CA: Stanford University Press. Chapter 2.

ABN-AMRO (2000) *European Auto Components: Driven by Technology*. London: ABN-AMRO.

Abo, T. (ed.) (1994) *Hybrid Factory: The Japanese Production System in the United States*. New York: Oxford University Press.

Abo, T. (1996) The Japanese production system: the process of adaptation to national settings. In R. Boyer and D. Drache (eds), *States Against Markets: The Limits of Globalization*. London: Routledge. Chapter 5.

Abo, T. (2000) Spontaneous integration in Japan and East Asia: development, crisis, and beyond. In G.L. Clark, M.P. Feldman and M.S. Gertler (eds), *The Oxford Handbook of Economic Geography*. Oxford: Oxford University Press. Chapter 31.

Agnew, J. and Corbridge, S. (1995) *Mastering Space*. London: Routledge.

Alderson, A.S. and Beckfield, J. (2004) Power and position in the world city system. *American Journal of Sociology*, 109: 811–51.

Amin, A. (2000) The European Union as more than a triad market for national economic spaces. In G.L. Clark, M.P. Feldman and M.S. Gertler (eds), *The Oxford Handbook of Economic Geography*. Oxford: Oxford University Press. Chapter 33.

Amin, A. (2002) Spatialities of globalization. *Environment and Planning A*, 34: 385–99.

Amin, A. (2004) Regulating economic globalization. *Transactions of the Institute of British Geographers*, 29: 217–33.

Amin, A. and Robins, K. (1990) The re-emergence of regional economies? The mythical geography of flexible accumulation. *Environment and Planning D: Society and Space*, 8: 7–34.

Amin, A. and Thrift, N.J. (1992) Neo-Marshallian nodes in global networks. *International Journal of Urban and Regional Research*, 16: 571–87.

Amsden, A. (1989) *Asia's Next Giant: South Korea and Late Industrialization*. Oxford: Oxford University Press.

Andersen, P.H. and Christensen, P.R. (2004) Bridges over troubled water: suppliers as connective nodes in global supply networks. *Journal of Business Research*, 58: 1261–73.

Anderson, M. (1995) The role of collaborative integration in industrial organization: observations from the Canadian aerospace industry. *Economic Geography*, 71: 55–78.

Anderson, S. and Cavanagh, J. (2000) *Top 200: The Rise of Corporate Global Power*. Washington, DC: Institute for Policy Studies.

Angel, D. (1994) *Restructuring for Innovation: The Remaking of the US Semiconductor Industry*. New York: Guilford.

Anheier, H., Glasius, M. and Kaldor, M. (eds) (2005) *Global Civil Society, 2004/5*. Oxford: Oxford University Press.

Aoki, A. and Tachiki, D. (1992) Overseas Japanese business operations: the emerging role of regional headquarters. *RIM Pacific Business and Industries*, 1: 28–39.

Aoki, M. (ed.) (1984) *The Economic Analysis of the Japanese Firm*. Dordrecht: North Holland.

Archibugi, D., Howells, J. and Michie, J. (eds) (1999) *Innovation Policy in a Global Economy*. Cambridge: Cambridge University Press.

Archibugi, D. and Michie, J. (eds) (1997) *Technology, Globalization, and Economic Performance*. Cambridge: Cambridge University Press.

Baaij, M., van den Bosch, F. and Volberda, H. (2004) The international relocation of corporate centres: are corporate centres sticky? *European Management Journal*, 22: 141–9.

Badaracco, J.L. Jr (1991) The boundaries of the firm. In A. Etzioni and P.R. Lawrence (eds), *Socio-Economics: Towards a New Synthesis*. Armonk, NY: Sharpe. pp. 293–327.

Bair, J. (2002) Beyond the maquila model: NAFTA and the Mexican apparel industry. *Industry and Innovation*, 9: 203–25.

Bair, J. and Gereffi, G. (2001) Local clusters in global chains: the causes and consequences of export dynamism in Torreón's blue jeans industry. *World Development*, 29: 1885–903.

Bair, J. and Gereffi, G. (2002) NAFTA and the apparel commodity chain: corporate strategies, interfirm networks, and industrial upgrading. In G. Gereffi, D. Spener and J. Bair (eds), *Free Trade and Uneven Development: The North American Apparel Industry after NAFTA*. Philadelphia: Temple University Press. Chapter 2.

Bair, J. and Gereffi, G. (2003) Upgrading, uneven development, and jobs in the North American apparel industry. *Global Networks*, 3: 143–96.

Barrett, H.R., Ilbery, B.W., Browne, A.W. and Binns, T. (1997) Globalization and the changing networks of food supply: the importation of fresh horticultural produce from Kenya into the UK. *Transactions of the Institute of British Geographers*, 24: 159–74.

Bartlett, C.A. and Ghoshal, S. (1998) *Managing Across Borders: The Transnational Solution*, 2nd edn. New York: Random House.

Bartlett, D. and Seleny, A. (1998) The political enforcement of liberalism: bargaining, institutions, and auto multinationals in Hungary. *International Studies Quarterly*, 42: 319–38.

Bathelt, H. and Gertler, M.S. (2005) The German variety of capitalism: forces and dynamics of evolutionary change. *Economic Geography*, 81: 1–9.

Bathelt, H., Malmberg, A. and Maskell, P. (2004) Clusters and knowledge: local buzz, global pipelines and the process of knowledge creation. *Progress in Human Geography*, 28: 31–56.

Batty, M. and Barr, R. (1994) The electronic frontier: exploring and mapping cyberspace. *Futures*, 26: 699–712.

Baylin, F. (1996) *World Satellite Yearbook*, 4th edn. Boulder, CO: Baylin.

Bayly, C.A. (2004) *The Birth of the Modern World, 1780–1914*. Oxford: Blackwell.

Beattie, A. (2005) Top of the crops: Brazil's huge heartland is yielding farms that can feed the world. *Financial Times*, 23 June: 17.

Beck, U. (2005) *Power in the Global Age: A New Global Political Economy*. Cambridge: Polity.

Beechler, S.L. and Bird, A. (eds) (1999) *Japanese Multinationals Abroad: Individual and Organizational Learning*. New York: Oxford University Press.

Begg, B., Pickles, J. and Smith, A. (2003) Cutting it: European integration, trade regimes, and the configuration of East–Central European apparel production. *Environment and Planning A*, 35: 2191–207.

Belderbos, R. and Capannelli, G. (2001) Backward vertical linkages of foreign manufacturing affiliates: evidence from Japanese multinationals. *World Development*, 29: 189–208.

Bello, W. and Rosenfeld, S. (1990) *Dragons in Distress: Asian Miracle Economies in Crisis*. San Francisco: Institute for Food and Development Policy.

Benewick, R. and Wingrove, P. (eds) (1995) *China in the 1990s*. London: Macmillan.

Berger, A.N., Dai, Q., Ongena, S. and Smith, D.C. (2003) To what extent will the banking industry be globalized? A study of bank nationality and reach in 20 European nations. *Journal of Banking and Finance*, 27: 383–415.

Berger, S. (2005) *How We Compete: What Companies Around the World Are Doing To Make It In Today's Global Economy*. New York: Doubleday.

Berger, S. and Dore, R. (eds) (1996) *National Diversity and Global Capitalism*. Ithaca, NY: Cornell University Press.

Berggren, C. and Bengtsson, L. (2004) Rethinking outsourcing in manufacturing: a tale of two telecom firms. *European Management Journal*, 22: 211–23.

Bernard, M. and Ravenhill, J. (1995) Beyond product cycles and flying geese. *World Politics*, 47: 171–209.

Beukema, L. and Coenen, H. (1999) Global logistic chains: the increasing importance of local labour relations. In P. Leisenk (ed.), *Globalization and Labour Relations*. Cheltenham: Elgar. Chapter 7.

Bhagwati, J. (2004) *In Defense of Globalization*. New York: Oxford University Press.

Bircham, E. and Charlton, J. (eds) (2001) *Anti-Capitalism: A Guide to the Movement*. London: Bookmarks.

Birkinshaw, J. and Morrison, A.J. (1995) Configurations of strategy and structure in subsidiaries of multinational corporations. *Journal of International Business Studies*, 26: 729–55.

Blanc, H. and Sierra, C. (1999) The internationalisation of R&D by multinationals: a trade-off between external and internal proximity. *Cambridge Journal of Economics*, 23: 187–206.

Blowfield, M. (1999) Ethical trade: a review of developments and issues. *Third World Quarterly*, 20: 753–70.

Blythman, J. (2004) *Shopped: The Shocking Power of British Supermarkets*. London: HarperCollins.

Bonanno, A., Busch, L., Friedland, W. and Mingione, E. (eds) (1994) *From Columbus to ConAgra: The Globalization of Agriculture and Food*. Lawrence, KA: University Press of Kansas.

Borrus, M. (2000) The resurgence of US electronics: Asian production networks and the rise of Wintelism. In M. Borrus, D. Ernst and S. Haggard (eds), *International Production Networks in Asia: Rivalry or Riches?* London: Routledge. Chapter 3.

Borrus, M., Ernst, D. and Haggard, S. (eds) (2000) *International Production Networks in Asia: Rivalry or Riches?* London: Routledge.

Bowles, P. (2002) Asia's post-crisis regionalism: bringing the state back in, keeping the (United) States out. *Review of International Political Economy*, 9: 230–56.

Boyd, W. and Watts, M.J. (1997) Agro-industrial just-in-time: the chicken industry and postwar American capitalism. In D. Goodman and M.J. Watts (eds), *Globalizing Food: Agrarian Questions and Global Restructuring*. London: Routledge. Chapter 8.

Braithwaite, J. and Drahos, P. (2000) *Global Business Regulation*. Cambridge: Cambridge University Press.

Brand, U. and Wissen, M. (2005) Neoliberal globalization and the internationalization of protest: a European perspective. *Antipode*, 37: 9–17.

Brenner, N. (1998) Between fixity and motion: accumulation, territorial organization and the historical geography of spatial scales. *Environment and Planning D: Society and Space*, 16: 459–81.

Brohman, J. (1996) Postwar development in the Asian NICs: does the neoliberal model fit reality? *Economic Geography*, 72: 107–31.

Browne, A.W., Harris, P.J.C., Hofny-Collins, A.H., Pasiecznik, N. and Wallace, R.R. (2000) Organic production and ethical trade: definition, practice and links. *Food Policy*, 25: 69–89.

Brunn, S.D. and Leinbach, T.R. (1991) *Collapsing Space and Time: Geographic Aspects of Communication and Information*. New York: HarperCollins.

Buck, T. and Shahrim, A. (2005) The translation of corporate governance changes across national cultures: the case of Germany. *Journal of International Business Studies*, 36: 42–61.

Buckley, P.J. and Casson, M. (1976) *The Future of the Multinational Enterprise*. London: Macmillan.

Budd, L. (1999) Globalization and the crisis of territorial embeddedness of international financial markets. In R. Martin (ed.), *Money and the Space-Economy*. Chichester: Wiley. Chapter 6.

Bunnell, T.G. and Coe, N.M. (2001) Spaces and scales of innovation. *Progress in Human Geography*, 25: 569–89.

Cable, V. (1999) *Globalization and Global Governance*. London: Royal Institute of International Affairs.

Cable, V. and Henderson, D. (eds) (1994) *Trade Blocs? The Future of Regional Integration*. London: Royal Institute of International Affairs.

Cairncross, F. (1992) *Costing the Earth*. Boston, MA: Harvard Business School Press.

Cairncross, F. (1997) *The Death of Distance: How the Communications Revolution Will Change Our Lives*. London: Orion.

Cameron, A. and Palan, R. (2004) *The Imagined Economies of Globalization*. London: Sage.

Cantwell, J. (1997) The globalization of technology: what remains of the product cycle model? In D. Archibugi and J. Michie (eds), *Technology, Globalization, and Economic Performance*. Cambridge: Cambridge University Press. Chapter 8.

Cantwell, J. and Iammarino, S. (2000) Multinational corporations and the location of technological innovation in the UK regions. *Regional Studies*, 34: 317–32.

Carillo, J. (2004) Transnational strategies and regional development: the case of GM and Delphi in Mexico. *Industry and Innovation*, 11: 127–53.

Carroll, W.K. and Carson, C. (2003) Forging a new hegemony? The role of transnational policy groups in the network and discourses of global corporate governance. *Journal of World-Systems Research*, IX: 67–102.

Casson, M. (ed.) (1983) *The Growth of International Business*. London: Allen & Unwin.

Castells, M. (1996) *The Information Age: Economy, Society and Culture*, 3 vols. Oxford: Blackwell.

Castells, M. and Hall, P. (1994) *Technopoles of the World: The Making of 21st Century Industrial Complexes*. London: Routledge.

Castles, S. and Miller, M.J. (1993) *The Age of Migration: International Population Movements in the Modern World*. New York: Guilford.

Castree, N., Coe, N.M., Ward, K. and Samers, M. (2004) *Spaces of Work: Global Capitalism and Geographies of Labour*. London: Sage.

Cerny, P.G. (1991) The limits of deregulation: transnational interpenetrations and polic change. *European Journal of Political Research*, 19: 173–96.

Cerny, P.G. (1997) Paradoxes of the competition state: the dynamics of political globalization. *Government and Opposition*, 32: 251–74.

Chang, H.-J. (1998a) South Korea: the misunderstood crisis. In K.S. Jomo (ed.), *Tigers in Trouble: Financial Governance, Liberalization and Crises in East Asia*. London: Zed. Chapter 10.

Chang, H.-J. (1998b) Transnational corporations and strategic industrial policy. In R. Kozul-Wright and R. Rowthorn (eds), *Transnational Corporations and the Global Economy*. London: Macmillan. Chapter 7.

Clark, G.L. (2002) London in the European financial services industry: locational advantages and product complementarities. *Journal of Economic Geography*, 2: 433–53.

Clark, G.L. and O'Connor, K. (1997) The informational content of financial products and the spatial structure of the global finance industry. In K. Cox (ed.), *Spaces of Globalization: Reasserting the Power of the Local*. New York: Guilford. Chapter 4.

Cline, W.R. (1997) *Trade and Income Distribution*. Washington, DC: Institute for International Economics.

Coe, N.M. (2003a) Globalization, regionalization and 'scales of integration': US IT industry investment in South East Asia. In M. Miozzo and I. Miles (eds), *Internationalization, Technology and Services*. Cheltenham: Elgar. pp. 117–36.

Coe, N.M. (2003b) Information highways and digital divides: the evolving ICT landscapes of Southeast Asia. In L.S. Chia (ed.), *SouthEast Asia Transformed: A Geography of Change*. Singapore: ISEAS. Chapter 9.

Coe, N.M. and Hess, M. (2005) The internationalization of retailing: implications for supply network restructuring in East Asia and Eastern Europe. *Journal of Economic Geography*, 5: 449–73.

Coe, N.M., Hess, M., Yeung, H.W.-c., Dicken, P. and Henderson, J. (2004) 'Globalizing' regional development: a global production networks perspective. *Transactions of the Institute of British Geographers*, 29: 468–84.

Coe, N.M., Kelly, P.F. and Olds, K. (2003) Globalization, transnationalism and the Asia-Pacific. In J. Peck and H.W.-c. Yeung (eds), *Remaking the Global Economy: Economic-Geographical Perspectives*. London: Sage. Chapter 3.

Coe, N.M. and Lee, Y.-S. (2006) The strategic localization of transnational retailers: the case of Samsung-Tesco in South Korea. *Economic Geography*, 82: 61–88.

Cohen, B. (2004) Urban growth in developing countries: a review of current trends and a caution regarding existing forecasts. *World Development*, 32: 23–51.

Cohen, D. (1998) *The Wealth of the World and the Poverty of Nations*. Cambridge, MA: MIT Press.

Cohen, S.S. and Zysman, J. (1987) *Manufacturing Matters: The Myth of the Post-Industrial Economy*. New York: Basic.

Commission on Global Governance (1995) *Our Global Neighbourhood*. New York: Oxford University Press.

Crane, G.T. (1990) *The Political Economy of China's Special Economic Zones*. Armonk, NY: Sharpe.

Crook, C. (2001) Globalization and its critics: a survey of globalization. *The Economist*, 29 September.

Curry, J. and Kenney, M. (2004) The organizational and geographic configuration of the personal computer value chain. In M. Kenney and R. Florida (eds), *Locating Global Advantage: Industry Dynamics in the International Economy*. Stanford, CA: Stanford University Press. Chapter 5.

Daly, H.E. (1993) The perils of free trade. *Scientific American*, November: 24–9.

Dauvergne, P. (2005) Globalization and the environment. In J. Ravenhill (ed.), *Global Political Economy*. Oxford: Oxford University Press. Chapter 14.

D'Aveni, R.A. (1994) *Hypercompetition: Managing the Dynamics of Strategic Manoeuvring*. New York: Free.

Davis, E. and Smailes, C. (1989) The integration of European financial services. In E. Davis et al. (eds), *1992: Myths and Realities*. London: London Business School. Chapter 5.

De Jonquieres, G. (2005) Asia has the most to lose from a Doha collapse. *Financial Times*, 1 November: 24.

Department of International Development (2000) *Eliminating World Poverty: Making Globalization Work for the Poor*. London: DoID.

Der Spiegel (2005) *Globalization: The New World*. Special Issue, 7.

Deyo, F.C. (1992) The political economy of social policy formation: East Asia's newly industrialized countries. In R.P. Appelbaum and J. Henderson (eds), *States and Development in the Asian Pacific Rim*. London: Sage. Chapter 11.

Dicken, P. (2000) Places and flows: situating international investment. In G.L. Clark, M.P. Feldman and M.S. Gertler (eds), *The Oxford Handbook of Economic Geography*. Oxford: Oxford University Press. Chapter 14.

Dicken, P. (2003a) Global production networks in Europe and East Asia: the automobile components industries. University of Manchester, GPN Working Paper, 7 (www.sed.manchester.ac.uk/geography/research/gpn/gpnwp.htm).

Dicken, P. (2003b) 'Placing' firms: grounding the debate on the 'global' corporation. In J. Peck and H.W.-c. Yeung (eds), *Remaking the Global Economy: Economic-Geographical Perspectives*. London: Sage. Chapter 2.

Dicken, P. (2004) Geographers and 'globalization': (yet) another missed boat? *Transactions of the Institute of British Geographers*, 29: 5–26.

Dicken, P. (2005) Tangled webs: transnational production networks and regional integration. *Spaces* (www.uni-marburg.de/geographie/spaces/).

Dicken, P., Forsgren, M. and Malmberg, A. (1994) The local embeddedness of transnational corporations. In A. Amin and N.J. Thrift (eds), *Globalization, Institutions, and Regional Development in Europe*. Oxford: Oxford University Press. Chapter 2.

Dicken, P. and Hassler, M. (2000) Organizing the Indonesian clothing industry in the global economy: the role of business networks. *Environment and Planning A*, 32: 263–80.

Dicken, P. and Lloyd, P.E. (1990) *Location in Space: Theoretical Perspectives in Economic Geography*, 3rd edn. New York: Harper & Row.

Dicken, P. and Malmberg, A. (2001) Firms in territories: a relational perspective. *Economic Geography*, 77: 345–63.

Dicken, P. and Miyamachi, Y. (1998) 'From noodles to satellites': the changing geography of the Japanese *sogo shosha*. *Transactions of the Institute of British Geographers*, 23: 55–78.

Dicken, P. and Yeung, H.W.-c. (1999) Investing in the future: East and Southeast Asian firms in the global economy. In K. Olds, P. Dicken, P.F. Kelly, L. Kong and H.W.-c. Yeung (eds), *Globalization and the Asia–Pacific: Contested Territories*. London: Routledge. Chapter 7.

Dieter, H. and Higgott, R. (2003) Exploring alternative theories of economic regionalism: from trade to finance in Asian cooperation. *Review of International Political Economy*, 10: 430–54.

Dodge, M. and Kitchin, R. (2001) *Atlas of Cyberspace*. Reading, MA: Addison-Wesley.

Dolan, C. and Humphrey, J. (2002) Changing governance patterns in the trade in fresh vegetables between Africa and the United Kingdom. *Environment and Planning A*, 36: 491–509.

Donaghu, M.T. and Barff, R. (1990) Nike just did it: international subcontracting and flexibility in athletic footwear production. *Regional Studies*, 24: 537–52.

Dore, R. (1986) *Flexible Rigidities: Industrial Policy and Structural Adjustment in the Japanese Economy, 1970–1980*. Stanford, CA: Stanford University Press.

Doremus, P.N., Keller, W.W., Pauly, L.W. and Reich, S. (1998) *The Myth of the Global Corporation*. Princeton, NJ: Princeton University Press.

Dosi, G., Freeman, C., Nelson, R., Silverberg, G. and Soete, L. (eds) (1988) *Technical Change and Economic Theory*. London: Pinter.

Douglass, M. (1994) The 'developmental state' and the newly industrialized economies of Asia. *Environment and Planning A*, 26: 543–66.

Dow, S.C. (1999) The stages of banking development and the spatial evolution of financial systems. In R. Martin (ed.), *Money and the Space-Economy*. Chichester: Wiley. Chapter 2.

Doz, Y. (1986a) Government polices and global industries. In M.E. Porter (ed.), *Competition in Global Industries*. Boston: Harvard Business School. Chapter 7.

Doz, Y. (1986b) *Strategic Management in Multinational Companies*. Oxford: Pergamon.

Driffield, N. and Hughes, D. (2003) Foreign and domestic investment: regional development crowding out? *Regional Studies*, 37: 277–89.

Drucker, P. (1946) *The Concept of the Corporation*. New York: Day.

Dunford, M. and Kafkalas, G. (1992) The global–local interplay, corporate geographies and spatial development strategies in Europe. In M. Dunford and G. Kafkalas (eds), *Cities and Regions in the New Europe: The Global–Local Interplay and Spatial Development Strategies*. London: Belhaven. Chapter 1.

Dunford, M. and Smith, A. (2000) Catching up or falling behind? Economic performance and regional trajectories in the 'new Europe'. *Economic Geography*, 76: 169–95.

Dunning, J.H. (1979) Explaining changing patterns of international production: in defence of the eclectic theory. *Oxford Bulletin of Economics and Statistics*, 41: 269–96.

Dunning, J.H. (1980) Towards an eclectic theory of international production: some empirical tests. *Journal of International Business Studies*, 11: 9–31.

Dunning, J.H. (1992) The competitive advantages of countries and the activities of transnational corporations. *Transnational Corporations*, 1: 135–68.

Dunning, J.H. (1993) *Multinational Enterprises and the Global Economy*. Reading, MA: Addison-Wesley.

Dunning, J.H. (2000a) The eclectic paradigm as an envelope for economic and business theories of MNE activity. *International Business Review*, 9: 163–90.

Dunning, J.H. (ed.) (2000b) *Regions, Globalization and the Knowledge-Based Economy*. Oxford: Oxford University Press.

Dustmann, C. and Glitz, A. (2005) *Immigration, Jobs and Wages: Theory, Evidence and Opinion*. London: CEPR.

Eatwell, J. and Taylor, L. (2000) *Global Finance at Risk: The Case for International Regulation.* Cambridge: Polity.

Eden, L. and Molot, M.A. (1993) Insiders and outsiders: defining 'who is us' in the North American automobile industry. *Transnational Corporations*, 2 (3): 31–64.

Eden, L. and Monteils, A. (2000) Regional integration: NAFTA and the reconfiguration of North American industry. In J.H. Dunning (ed.), *Regions, Globalization and the Knowledge-Based Economy.* Oxford: Oxford University Press. Chapter 7.

Eichengreen, B. (1996) *Globalizing Capital: A History of the International Monetary System.* Princeton, NJ: Princeton University Press.

EIU (2000) *World Automotive Components: Market Prospects to 2005.* London: The Economist Intelligence Unit.

Elango, B. (2004) Geographic scope of operations by multinational companies: an exploratory study of regional and global strategies. *European Management Journal*, 22: 431–41.

Elliott, L. (2000) Free trade, no choice. In B. Gunnell and D. Timms (eds), *After Seattle: Globalization and its Discontents.* London: Catalyst. Chapter 2.

Elliott, L. (2002) Morals of the brothel. *The Guardian*, 15 April.

Ernst, D. and Kim, L. (2002) Global production networks, knowledge diffusion, and local capability formation. *Research Policy*, 1419: 1–13.

Faux, J. and Mishel, L. (2000) Inequality and the global economy. In W. Hutton and A. Giddens (eds), *On the Edge: Living with Global Capitalism.* London: Cape. pp. 93–111.

Feenstra, R.C. (1998) Integration of trade and disintegration of production in the global economy. *Journal of Economic Perspectives*, 12: 31–50.

Fields, G. (2004) *Territories of Profit: Communications, Capitalist Development, and the Innovative Enterprises of G.F. Swift and Dell Computers.* Stanford, CA: Stanford Business Books.

FIET (1996) *A Social Dimension to Globalization.* Geneva: FIET.

Frank, A.G. (1998) *ReORIENT: Global Economy in the Asian Age.* Berkeley, CA: University of California Press.

Fransman, M. (ed.) (2006) *Global Broadband Battles: Why the US and Europe Lag While Asia Leads.* Stanford: Stanford University Press.

Freeman, C. (1982) *The Economics of Industrial Innovation.* London: Pinter.

Freeman, C. (1987) The challenge of new technologies. In OECD (ed.), *Interdependence and Cooperation in Tomorrow's World.* Paris: OECD. pp. 123–56.

Freeman, C. (1988) Introduction. In G. Dosi, C. Freeman, R. Nelson, G. Silverberg and L. Soete (eds), *Technical Change and Economic Theory.* London: Pinter. Chapter 1.

Freeman, C. (1997) The 'national system of innovation' in historical perspective. In D. Archibugi and J. Michie (eds), *Technology, Globalization, and Economic Performance.* Cambridge: Cambridge University Press. Chapter 2.

Freeman, C., Clark, J. and Soete, L. (1982) *Unemployment and Technical Change.* London: Pinter.

Freeman, C. and Perez, C. (1988) Structural crises of adjustment, business cycles and investment behaviour. In G. Dosi, C. Freeman, R. Nelson, G. Silverberg and L. Soete (eds), *Technical Change and Economic Theory.* London: Pinter. Chapter 3.

Freidberg, S. (2004) *French Beans and Food Scares: Culture and Commerce in an Anxious Age.* Oxford: Oxford University Press.

French, S. and Leyshon, A. (2004) The new, new financial system? Towards a reconceptualization of financial reintermediation. *Review of International Political Economy*, 11: 263–88.

Freyssenet, M., Shimizu, K. and Volpato, G. (eds) (2003) *Globalization or Regionalization of the American and Asian Car Industry?* Basingstoke: Palgrave Macmillan.

Friedland, W. (1994) The new globalization: the case of fresh produce. In A. Bonanno, L. Busch, W. Friedland and E. Mingione (eds), *From Columbus to ConAgra: The Globalization of Agriculture and Food*. Lawrence, KA: University Press of Kansas. Chapter 10.

Friedman, T. (1999) *The Lexus and the Olive Tree*. New York: HarperCollins.

Friedman, T. (2005) *The World Is Flat: A Brief History of the Twenty-First Century*. London: Allen Lane.

Friedmann, H. (1993) The political economy of food. *New Left Review*, 197: 29–57.

Friedmann, J. (1986) The world city hypothesis. *Development and Change*, 17: 69–83.

Frigant, V. and Lung, Y. (2002) Geographical proximity and supplying relationships in modular production. *International Journal of Urban and Regional Research*, 26: 742–55.

Fröbel, F., Heinrichs, J. and Kreye, O. (1980) *The New International Division of Labour*. Cambridge: Cambridge University Press.

Fruin, W.M. (1992) *The Japanese Enterprise System*. Oxford: Clarendon.

Fujita, M. and Ishigaki, K. (1986) The internationalisation of commercial banking. In M.J. Taylor and N.J. Thrift (eds), *Multinationals and the Restructuring of the World Economy*. London: Croom Helm. Chapter 7.

Fukuyama, F. (1992) *The End of History and the Last Man*. London: Hamish Hamilton.

Fuller, D.B., Akinwande, A. and Sodini, C.G. (2003) Leading, following or cooked goose? Innovation, successes and failures in Taiwan's electronics industry. *Industry and Innovation*, 10: 179–96.

Gabriel, P. (1966) The investment in the LDC: asset with a fixed maturity. *Columbia Journal of World Business*, 1: 113–20.

Gabrielsson, M. and Kirpalani, V.H.M. (2004) Born globals: how to reach new business space rapidly. *International Business Review*, 13: 555–71.

Gamble, A. and Payne, A. (eds) (1996) *Regionalism and World Order*. London: Macmillan.

Garrett, G. (1998) Global markets and national politics: collision course or virtuous circle? *International Organization*, 52: 787–824.

Gereffi, G. (1990) Paths of industrialization: an overview. In G. Gereffi and D.L. Wyman (eds), *Manufacturing Miracles: Paths of Industrialization in Latin America and East Asia*. Princeton, NJ: Princeton University Press. Chapter 1.

Gereffi, G. (1994) The organization of buyer-driven global commodity chains: how US retailers shape overseas production networks. In G. Gereffi and M. Korzeniewicz (eds), *Commodity Chains and Global Capitalism*. Westport, CT: Praeger. Chapter 5.

Gereffi, G. (1996) Commodity chains and regional divisions of labor in East Asia. *Journal of Asian Business*, 12: 75–112.

Gereffi, G. (1999) International trade and industrial upgrading in the apparel commodity chain. *Journal of International Economics*, 48: 37–70.

Gereffi, G. (2001) Shifting governance structures in global commodity chains, with special reference to the Internet. *American Behavioral Scientist*, 44: 1616–37.

Gereffi, G. (2005) The global economy: organization, governance and development. In N. Smelser and R. Swedberg (eds), *The Handbook of Economic Sociology*, 2nd edn. Princeton, NJ: Princeton University Press. Chapter 8.

Gereffi, G., Garcia-Johnson, R. and Sasser, E. (2001) The NGO–industrial complex. *Foreign Policy*, July–August: 56–65.

Gereffi, G., Humphrey, J. and Sturgeon, T. (2005) The governance of global value chains. *Review of International Political Economy*, 12: 78–104.

Gereffi, G. and Korzeniewicz, M. (eds) (1994) *Commodity Chains and Global Capitalism.* Westport, CT: Praeger.

Gereffi, G. and Memedovic, O. (2004) The global apparel value chain: what prospects for upgrading by developing countries? In J. O'Loughlin, L. Staeheli and E. Greenberg (eds), *Globalization and its Outcomes.* New York: Guilford. Chapter 4.

Gereffi, G., Spener, D. and Bair, J. (eds) (2002) *Free Trade and Uneven Development: The North American Apparel Industry after NAFTA.* Philadelphia: Temple University Press.

Gereffi, G. and Wyman, D.L. (eds) (1990) *Manufacturing Miracles: Paths of Industrialization in Latin America and East Asia.* Princeton, NJ: Princeton University Press.

Gerlach, M. (1992) *Alliance Capitalism: The Social Organization of Japanese Business.* Berkeley, CA: University of California Press.

Gertler, M.S. (1995) 'Being there': proximity, organization and culture in the development and adoption of advanced manufacturing technologies. *Economic Geography,* 71: 1–26.

Gibb, R. and Michalak, W. (eds) (1994) *Continental Trading Blocs: The Growth of Regionalism in the World Economy.* Chichester: Wiley.

Gibson-Graham, J.K. (2006) *Post-Capitalist Politics.* Minneapolis, MN: Minnesota University Press.

Gilpin, R. (1987) *The Political Economy of International Relations.* Princeton, NJ: Princeton University Press.

Gilpin, R. (2000) *The Challenge of Global Capitalism: The World Economy in the 21st Century.* Princeton, NJ: Princeton University Press.

Gilpin, R. (2001) *Global Political Economy: Understanding the International Economic Order.* Princeton, NJ: Princeton University Press.

Glasius, M., Kaldor, M. and Anheier, H. (eds) (2002) *Global Civil Society, 2002.* Oxford: Oxford University Press.

Glasmeier, A. (2002) One nation pulling apart: the basis of persistent poverty in the United States. *Progress in Human Geography,* 26: 155–74.

Glasmeier, A., Thompson, J.W. and Kays, A.J. (1993) The geography of trade policy: trade regimes and location decisions in the textile and apparel complex. *Transactions of the Institute of British Geographers,* 18: 19–35.

Glassner, M.I. (1993) *Political Geography.* New York: Wiley.

Gomes-Casseres, B. (1996) *The Alliance Revolution: The New Shape of Business Rivalry.* Cambridge, MA: Harvard University Press.

Goodman, D., Sorj, B. and Wilkinson, J. (1987) *From Farming to Biotechnology: A Theory of Agro-Industrial Development.* Oxford: Blackwell.

Gordon, D.M. (1988) The global economy: new edifice or crumbling foundations? *New Left Review,* 168: 24–64.

Govindarajan, V. and Gupta, A. (2000) Analysis of the emerging global arena. *European Management Journal,* 18: 274–84.

Graham, S. and Marvin, S. (1996) *Telecommunications and the City: Electronic Spaces, Urban Places.* London: Routledge.

Granovetter, M. (1985) Economic action and social structure: the problem of embeddedness. *American Journal of Sociology,* 91: 481–510.

Granovetter, M. and Swedberg, R. (eds) (1992) *The Sociology of Economic Life.* Boulder, CO: Westview.

Gretschmann, K. (1994) Germany in the global economy of the 1990s: from player to pawn? In R. Stubbs and G.R.D. Underhill (eds), *Political Economy and the Changing Global Order*. London: Macmillan. Chapter 29.

Grigg, D. (1993) *The World Food Problem*, 2nd edn. Oxford: Blackwell.

Gritsch, M. (2005) The nation-state and economic globalization: soft geo-politics and increased state autonomy? *Review of International Political Economy*, 12: 1–25.

Grugel, J. (1996) Latin America and the remaking of the Americas. In A. Gamble and A. Payne (eds), *Regionalism and World Order*. London: Macmillan. Chapter 5.

Guiher, G. and Lecler, Y. (2000) Japanese car manufacturers' component makers in the ASEAN region: a case of expatriation under duress – or a strategy of regionally integrated production? In J. Humphrey, Y. Lecler and M.S. Salerno (eds), *Global Strategies and Local Realities: The Auto Industry in Emerging Markets*. London: Macmillan. Chapter 9.

Gwynne, R. (1994) Regional integration in Latin America: the revival of a concept? In R. Gibb and W. Michalak (eds), *Continental Trading Blocs: The Growth of Regionalism in the World Economy*. Chichester: Wiley. Chapter 7.

Haggard, S. (1995) *Developing Nations and the Politics of Global Integration*. Washington, DC: Brookings Institution.

Haggard, S. and MacIntyre, A. (1998) The political economy of the Asian economic crisis. *Review of International Political Economy*, 5: 381–92.

Hall, P. and Preston, P. (1988) *The Carrier Wave: New Information Technology and the Geography of Innovation, 1846–2003*. London: Unwin Hyman.

Hall, P.A. and Soskice, D. (2001) *Varieties of Capitalism: The Institutional Foundations of Comparative Advantage*. Oxford: Oxford University Press.

Ham, R.M., Linden, G. and Appleyard, M.M. (1998) The evolving role of semiconductor consortia in the United States and Japan. *California Management Review*, 41 (1): 137–63.

Hamilton, G.G. and Feenstra, R.C. (1998) Varieties of hierarchies and markets: an introduction. In G. Dosi, D.J. Teece and J. Chytry (eds), *Technology, Organization and Competitiveness*. Oxford: Oxford University Press. pp. 105–46.

Hamilton-Hart, N. (2003) Asia's new regionalism: government capacity and cooperation in the Western Pacific. *Review of International Political Economy*, 10: 222–45.

Harrison, B. (1997) *Lean and Mean: The Changing Landscape of Corporate Power in the Age of Flexibility*. New York: Guilford.

Harvey, N. (ed.) (1993) *Mexico: Dilemmas of Transition*. London: Institute of Latin American Studies.

Harzing, A.-W. (2000) An empirical analysis and extension of the Bartlett and Ghoshal typology of multinational companies. *Journal of International Business Studies*, 31: 101–20.

Havas, A. (2000) Changing patterns of inter- and intra-regional division of labour: Central Europe's long and winding road. In J. Humphrey, Y. Lecler and M.S. Salerno (eds), *Global Strategies and Local Realities: The Auto Industry in Emerging Markets*. London: Macmillan. Chapter 10.

Hawkins, R.G. (1972) Job displacement and the multinational firm: a methodological review. Center for Multinational Studies, Occasional Paper 3, Washington, DC.

Hedlund, G. (1986) The hypermodern MNC – a heterarchy. *Human Resource Management*, 25: 9–35.

Heenan, D.A. and Perlmutter, H. (1979) *Multinational Organizational Development: A Social Architecture Perspective*. Reading, MA: Addison-Wesley.

Held, D., McGrew, A., Goldblatt, D. and Perraton, J. (1999) *Global Transformations: Politics, Economics and Culture*. Cambridge: Polity.

Helou, A. (1991) The nature and competitiveness of Japan's *keiretsu*. *Journal of World Trade*, 25: 99–131.

Henderson, J. (1998) Danger and opportunity in the Asia–Pacific. In G. Thompson (ed.), *Economic Dynamism in the Asia–Pacific*. London: Routledge. Chapter 14.

Henderson, J. and Castells, M. (eds) (1987) *Global Restructuring and Territorial Development*. London: Sage.

Henderson, J., Dicken, P., Hess, M., Coe, N. and Yeung, H. W.-c. (2002) Global production networks and the analysis of economic development. *Review of International Political Economy*, 9: 436–64.

Herod, A. (1997) From a geography of labor to a labor geography: rethinking conceptions of labor in economic geography. *Antipode*, 29: 1–31.

Herod, A. (2001) *Labor Geographies: Workers and the Landscape of Capitalism*. New York: Guilford.

Hess, M. (2004) 'Spatial' relationships? Towards a re-conceptualization of embeddedness. *Progress in Human Geography*, 28: 165–86.

Higgott, R. (1999) The political economy of globalization in East Asia: the salience of 'region building'. In K. Olds, P. Dicken, P.F. Kelly, L. Kong and H.W.-c. Yeung (eds), *Globalization and the Asia–Pacific: Contested Territories*. London: Routledge. Chapter 6.

Hirsch, F. (1977) *Social Limits to Growth*. London: Routledge and Kegan Paul.

Hirsch, S. (1967) *Location of Industry and International Competitiveness*. Oxford: Clarendon.

Hirst, P. and Thompson, G. (1992) The problem of 'globalization': international economic relations, national economic management and the formation of trading blocs. *Economy and Society*, 21: 357–96.

Hirst, P. and Thompson, G. (1999) *Globalization in Question: The International Economy and the Possibilities of Governance*, 2nd edn. Cambridge: Polity.

Hoekman, B. and Kostecki, M. (1995) *The Political Economy of the World Trading System: From GATT to WTO*. Oxford: Oxford University Press.

Hofheinz, R. Jr and Calder, K.E. (1982) *The East Asia Edge*. New York: Basic.

Hofstede, G. (1980) *Culture's Consequences*. London: Sage.

Hofstede, G. (1983) The cultural relativity of organizational practices and theories. *Journal of International Business Studies*, Fall: 75–89.

Hollingsworth, J.R. (1997) Continuities and changes in social systems of production: the cases of Japan, Germany, and the United States. In J.R. Hollingsworth and R. Boyer (eds), *Contemporary Capitalism: The Embeddedness of Institutions*. Cambridge: Cambridge University Press. Chapter 9.

Hollingsworth, J.R. and Boyer, R. (eds) (1997) *Contemporary Capitalism: The Embeddedness of Institutions*. Cambridge: Cambridge University Press.

Holmes, J. (1992) The continental integration of the North American automobile industry: from the Auto Pact to the NAFTA. *Environment and Planning A*, 24: 95–120.

Holmes, J. (2000) Regional economic integration in North America. In G.L. Clark, M.P. Feldman and M.S. Gertler (eds), *The Oxford Handbook of Economic Geography*. Oxford: Oxford University Press. Chapter 32.

Hood, N. and Young, S. (2000) Globalization, corporate strategies, and business services. In N. Hood and S. Young (eds), *The Globalization of Multinational Enterprise Activity and Economic Development*. London: Macmillan. Chapter 4.

Horaguchi, H. and Shimokawa, K. (eds) (2002) *Japanese Foreign Direct Investment and the East Asia Industrial System*. Berlin: Springer.

Hotz-Hart, B. (2000) Innovation networks, regions and globalization. In G.L. Clark, M.P. Feldman and M.S. Gertles (eds), *The Oxford Handbook of Economic Geography*. Oxford: Oxford University Press. Chapter 22.

Howells, J. (2000) Knowledge, innovation and location. In J.R. Bryson, P.W. Daniels, N. Henry and J. Pollard (eds), *Knowledge, Space, Economy*. London: Routledge. Chapter 4.

Hu, Y.-S. (1992) Global firms are national firms with international operations. *California Management Review*, 34: 107–26.

Huang, S.W. (2004) *Global Trade Patterns in Fruits and Vegetables*. US Department of Agriculture, Agriculture and Trade Report WRS-04-06.

Huang, Y. (2002) Between two coordination failures: automotive industrial policy in China with a comparison to Korea. *Review of International Political Economy*, 9: 538–73.

Hudson, A. (2000) Offshoreness, globalization and sovereignty: a postmodern geo-political economy? *Transactions of the Institute of British Geographers*, 25: 269–83.

Hudson, R. (2001) *Producing Places*. New York: Guilford.

Hudson, R. (2002) Changing industrial production systems and regional development in the New Europe. *Transactions of the Institute of British Geographers*, 27: 262–81.

Hudson, R. (2004) Conceptualizing economies and their geographies: spaces, flows and circuits. *Progress in Human Geography*, 28: 447–71.

Hudson, R. (2005) *Economic Geographies: Circuits, Flows and Spaces*. London: Sage.

Hughes, A. (2001) Global commodity networks, ethical trade and governmentality: organizing business responsibility in the Kenyan cut flower industry. *Transactions of the Institute of British Geographers*, 26: 390–417.

Humbert, M. (1994) Strategic industrial policies in a global industrial system. *Review of International Political Economy*, 1: 445–64.

Humphrey, J. and Oeter, A. (2000) Motor industry policies in emerging markets: globalization and promotion of domestic industry. In J. Humphrey, Y. Lecler and M.S. Salerno (eds), *Global Strategies and Local Realities: The Auto Industry in Emerging Markets*. London: Macmillan. Chapter 3.

Hymer, S.H. (1976) *The International Operations of National Firms: A Study of Direct Foreign Investment*. Cambridge, MA: MIT Press.

ILO (1996) *Globalization of the Footwear, Textiles, and Clothing Industries*. Geneva: ILO.

ILO (1997) *World Employment, 1996/97: National Policies in a Global Context*. Geneva: ILO.

ILO (1998) *Labour and Social Issues Relating to Export Processing Zones*. Geneva: ILO.

ILO (2001) *World Employment Report 2001: Life at Work in the Information Economy*. Geneva: ILO.

ILO (2003) *Employment and Social Policy in Respect of Export Processing Zones (EPZs)*. Geneva: ILO.

ILO (2004a) *Economic Security for a Better World*. Geneva: ILO.

ILO (2004b) *A Fair Globalization: Creating Opportunities for All*. Geneva: ILO.

ILO (2005) *Promoting Fair Globalization in Textiles and Clothing in a Post-MFA Environment*. Geneva: ILO.

ILO (2006a) *Global Employment Trends Brief, January 2006*. Geneva: ILO.

ILO (2006b) *The End of Child Labour: Within Reach*. Geneva: ILO.

International Organization for Migration (2005) *World Migration 2005: Costs and Benefits of International Migration*. London: IOM.

Ito, T. (2001) Growth, crisis, and the future of economic recovery in East Asia. In J.E. Stiglitz and S. Yusuf (eds), *Rethinking the East Asia Miracle*. New York: Oxford University Press. Chapter 2.

Jarillo, J.C. (1993) *Strategic Networks: Creating the Borderless Organization*. London: Butterworth-Heinemann.

Jessop, B. (1994) Post-Fordism and the state. In A. Amin (ed.), *Post-Fordism: A Reader*. Oxford: Blackwell. Chapter 8.

Jessop, R. (2002) *The Future of the Capitalist State*. Cambridge: Polity.

Johnson, C. (1985) The institutional foundations of Japanese industrial policy. *California Management Review*, XXVII: 59–69.

Jones, H.R. (1990) *A Population Geography*, 2nd edn. London: Chapman.

Kaldor, M. (2003) *Global Civil Society: An Answer to War*. Cambridge: Polity.

Kaltenthaler, K. and Mora, F.O. (2002) Explaining Latin American economic integration: the case of Mercosur. *Review of International Political Economy*, 9: 72–97.

Kang, N.-H. and Sakai, K. (2000) International strategic alliances: their role in industrial globalization. OECD STI Working Paper 2000/5.

Kao, J. (1993) The worldwide web of Chinese business. *Harvard Business Review*, March–April: 24–36.

Kaplinsky, R. (1999) 'If you want to get somewhere else, you must run at least twice as fast as that!': the roots of the East Asian crisis. *Competition and Change*, 4: 1–30.

Kaplinsky, R. (2001) Is globalization all it is cracked up to be? *Review of International Political Economy*, 8: 45–65.

Kapstein, E.B. (1999) *Sharing the Wealth: Workers and the World Economy*. New York: Norton.

Kapstein, E. (2000) Winners and losers in the global economy. *International Organization*, 54: 359–84.

Kay, J. (2006) Scrutiny of foreign takeovers is prudent not protectionist. *Financial Times*, 7 February: 25.

Kelly, P.F. (1999) The geographies and politics of globalization. *Progress in Human Geography*, 23: 379–400.

Kelly, R. (1995) Derivatives: a growing threat to the international financial system. In J. Michie and J. Grieve-Smith (eds), *Managing the Global Economy*. Oxford: Oxford University Press. Chapter 9.

Kennedy, P. (2002) Global challenges at the beginning of the twenty-first century. In P. Kennedy, D. Messner and F. Nuscheler (eds), *Global Trends and Global Governance*. London: Pluto.

Kenney, M. (ed.) (2000) *Understanding Silicon Valley: The Anatomy of an Entrepreneurial Region*. Stanford, CA: Stanford University Press.

Kenney, M. and Curry, J. (2001) Beyond transaction costs: e-commerce and the power of the Internet dataspace. In T.R. Leinbach and S.D. Brunn (eds), *Worlds of e-Commerce: Economic, Geographical, and Social Dimensions*. Chichester: Wiley. Chapter 3.

Kessler, J.A. (1999) The North American Free Trade Agreement, emerging apparel production networks and industrial upgrading: the southern California/Mexico connection. *Review of International Political Economy*, 6: 565–608.

Khanna, S.R. (1993) Structural changes in Asian textiles and clothing industries: the second migration of production. *Textile Outlook International*, September: 11–32.

Kim, H.Y. and Lee, S.-H. (1994) Commodity chains and the Korean automobile industry. In G. Gereffi and M. Korzeniewicz (eds), *Commodity Chains and Global Capitalism*. Westport, CT: Praeger. Chapter 14.

Kim, L. (1997) The dynamics of Samsung's technological learning in semiconductors. *California Management Review*, 39 (3): 86–100.

Kindleberger, C.P. (1969) *American Business Abroad*. New Haven, CT: Yale University Press.

King, A.D. (1983) The world economy is everywhere: urban history and the world-system. *Urban History Yearbook*: 7–18.

Klein, N. (2000) *No Logo*. London: Flamingo.

Kobrin, S.J. (1987) Testing the bargaining hypothesis in the manufacturing sector in developing countries. *International Organization*, 41: 609–38.

Koh, T. (1993) 10 values that help East Asia's economic progress and prosperity. *Singapore Straits Times*, 14 December.

Kolk, A. and van Tulder, R. (2005) Setting new global rules? TNCs and codes of conduct, *Transnational Corporations*, 14: 1–27.

Koo, H. and Kim, E.M. (1992) The developmental state and capital accumulation in South Korea. In R.P. Appelbaum and J. Henderson (eds), *States and Development in the Asian Pacific Rim*. London: Sage. Chapter 5.

Kozul-Wright, R. (1995) Transnational corporations and the nation-state. In J. Michie and J. Grieve-Smith (eds), *Managing the Global Economy*. Oxford: Oxford University Press. Chapter 6.

Kozul-Wright, R. and Rowthorn, R. (1998) Spoilt for choice? Multinational corporations and the geography of international production. *Oxford Review of Economic Policy*, 14: 74–92.

Krugman, P. (ed.) (1986) *Strategic Trade Policy and the New International Economics*. Cambridge, MA: MIT Press.

Krugman, P. (1990) *Rethinking International Trade*. Cambridge, MA: MIT Press.

Krugman, P. (1994) Competitiveness: a dangerous obsession. *Foreign Affairs*, March-April: 28–44.

Krugman, P. (1998) What's new about the new economic geography? *Oxford Review of Economic Policy*, 14: 7–17.

Lall, S. (1994) Industrial policy: the role of government in promoting industrial and technological development. *UNCTAD Review*: 65–90.

Lall, S. and Albaladejo, M. (2004) China's competitive performance: a threat to East Asian manufactured exports? *World Development*, 32: 1441–66.

Landes, D.S. (1998) *The Wealth and Poverty of Nations*. New York: Norton.

Langdale, J.V. (2000) Telecommunications and 24-hour trading in the international securities industry. In M.I. Wilson and K.E. Corey (eds), *Information Tectonics: Space, Place, and Technology in an Electronic Age*. Chichester: Wiley. Chapter 6.

Lasserre, P. (1996) Regional headquarters: the spearhead for Asia–Pacific markets. *Long Range Planning*, 29: 30–7.

Law, J. (1986) On the methods of long-distance control: vessels, navigation and the Portuguese route to India. *Sociological Review Monograph*, 32: 234–63.

Lawrence, R.Z. (1996) *Regionalism, Multilateralism, and Deeper Integration*. Washington, DC: Brookings Institution.

Layard, R. (2005) *Happiness: Lessons from a New Science*. London: Penguin Allen Lane.

Leachman, R.C. and Leachman, C.H. (2004) Globalization of semiconductors: do real men have fabs, or virtual fabs? In M. Kenney and R. Florida (eds), *Locating Global Advantage: Industry Dynamics in the International Economy*. Stanford, CA: Stanford University Press. Chapter 8.

League of Nations (1945) *Industrialization and Foreign Trade*. New York: League of Nations.

Leclair, M.S. (2002) Fighting the tide: alternative trade organizations in the era of global free trade. *World Development*, 30: 949–58.

Lee, K.-Y. (2000) *From Third World to First. The Singapore Story: 1965–2000*. Singapore: Times Publishing Group.

Lee, N. and Cason, J. (1994) Automobile commodity chains in the NICs: a comparison of South Korea, Mexico, and Brazil. In G. Gereffi and M. Korzeniewicz (eds), *Commodity Chains and Global Capitalism*. Westport, CT: Praeger. Chapter 11.

Lee, R. (1996) Moral money? LETS and the social construction of local economic geographies in southeast England. *Environment and Planning A*, 28: 1377–94.

Lee, R. (2007) The ordinary economy: tangled up in values and geography. *Transactions of the Institute of British Geographers,* 32:

Lee, R. and Smith, D.M. (eds) (2004) *Geographies and Moralities: International Perspectives on Development, Justice and Place*. Oxford: Blackwell.

Leinbach, T.R. (2001) Emergence of the digital economy and e-commerce. In T.R. Leinbach and S.D. Brunn (eds), *Worlds of e-Commerce: Economic, Geographical, and Social Dimensions*. Chichester: Wiley. Chapter 1.

Levinson, M. (2006) *The Box: How the Shipping Container Made the World Smaller and the World Economy Bigger*. Princeton, NJ: Princeton University Press.

Levy, D.L. and Prakash, A. (2003) Bargains old and new: multinational corporations in global governance. *Business and Politics*, 5: 131–20.

Levy, F. and Murname, R. (2004) *The New Division of Labour: How Computers are Creating the Next Job Market*. Princeton, NJ: Princeton University Press.

Lewis, M.K. and Davis, J.T. (1987) *Domestic and International Banking*. Oxford: Allan.

Leyshon, A. (1992) The transformation of regulatory order: regulating the global economy and environment. *Geoforum*, 23: 249–67.

Liao, S. (1997) ASEAN model in international economic cooperation. In H. Soesastro (ed.), *One South East Asia in a New Regional and International Setting*. Jakarta: Centre for Strategic and International Studies.

Linden, G. (2000) Japan and the United States in the Malaysian electronics sector. In M. Borrus, D. Ernst and S. Haggard (eds), *International Production Networks in Asia: Rivalry or Riches?* London: Routledge. Chapter 8.

Linden, G., Brown, C. and Appleyard, M.M. (2004) The net world order's influence on global leadership in the semiconductor industry. In M. Kenney and R. Florida (eds), *Locating Global Advantage: Industry Dynamics in the International Economy*. Stanford, CA: Stanford University Press. Chapter 9.

Linden, G. and Somaya, D. (2003) System-on-a-chip integration in the semiconductor industry: industry structure and firm strategies. *Industrial and Corporate Change*, 12: 545–76.

Liu, W. (2006) The changing geography of the automobile industry in China: a general picture. Presented to the Association of American Geographers Conference, Chicago.

Liu, W. and Dicken, P. (2006) Transnational corporations and 'obligated embeddedness': foreign direct investment in China's automobile industry, *Environment and Planning A,* 38: 1229–47.

Lovering, J. (1990) Fordism's unknown successor: a comment on Scott's theory of flexible accumulation and the re-emergence of regional economies. *International Journal of Urban and Regional Research*, 14: 159–74.

Lowes, P., Celner, A. and Gentle, C. (2004) A global shift: how offshoring is changing the financial services business model. *Financial Technology International*, June: 24–34.

Lundvall, B.-Å. (ed.) (1992) *National Systems of Innovation*. London: Pinter.

Lundvall, B.-Å. and Maskell, P. (2000) Nation-states and economic development: from national systems of production to national systems of knowledge creation and learning. In G.L. Clark, M.P. Feldman and M.S. Gertler (eds), *The Oxford Handbook of Economic Geography*. Oxford: Oxford University Press. Chapter 18.

Lüthje, B.-Å. (2002) Electronics contract manufacturing: global production and the international division of labor in the age of the Internet. *Industry and Innovation*, 9: 227–47.

Luttwak, E. (1999) *Turbo-Capitalism: Winners and Losers in the Global Economy*. London: Orion.

Lynn, B.C. (2005) *End of the Line: The Rise and Coming Fall of the Global Corporation*. New York: Doubleday.

Lyons, D. and Salmon, S. (1995) World cities, multinational corporations, and urban hierarchy: the case of the United States. In P.L. Knox and P.J. Taylor (eds), *World Cities in a World-System*. Cambridge: Cambridge Unviersity Press. Chapter 6.

Macher, J.T., Mowery, D.C. and Hodges, D.A. (1998) Reversal of fortune? The recovery of the U.S. semiconductor industry. *California Management Review*, 41 (1): 107–36.

Macher, J.T., Mowery, D.C. and Simcoe, T.S. (2002) E-business and disintegration of the semiconductor industry value chain. *Industry and Innovation*, 9: 155–81.

Maddison, A. (2001) *The World Economy: A Millennial Perspective*. Paris: OECD.

Magaziner, I.C. and Hout, T.M. (1980) *Japanese Industrial Policy*. London: Policy Studies Association.

Magretta, J. (1998) Fast, global, and entrepreneurial: supply chain management, Hong Kong style. An interview with Victor Fung. *Harvard Business Review*, September–October: 103–14.

Maignan, I. and Ralston, D.D. (2002) Corporate social responsibility in Europe and the US: insights from businesses' self-presentations. *Journal of International Business Studies*, 33: 497–514.

Malecki, E.J. (2002) The economic geography of the Internet's infrastructure. *Economic Geography*, 78: 399–424.

Malecki, E.J. and Hu, W. (2006) A wired world: the evolving geography of submarine cables and the shift to Asia. Presented at the Association of American Geographers Annual Conference, Chicago.

Malmberg, A. (1999) The elusive concept of agglomeration economies: theoretical principles and empirical paradoxes. SCASSS Seminar Paper.

Malnight, T.W. (1996) The transition from decentralized to network-based MNC structures: an evolutionary perspective. *Journal of International Business Studies*, 27: 43–63.

Mansfield, E. and Milner, H.V. (1999) The new wave of regionalism. *International Organization*, 53: 589–627.

Markusen, A. (1996) Sticky places in slippery space: a typology of industrial districts. *Economic Geography*, 72: 293–313.

Marsden, T.K. (1997) Creating space for food: the distinctiveness of recent agrarian development. In D. Goodman and M.J. Watts (eds), *Globalizing Food: Agrarian Questions and Global Restructuring*. London: Routledge. Chapter 7.

Marsden, T.K. and Cavalcanti, J.S.B. (2001) Globalization, sustainability and the new agrarian regions: food, labour and environmental values. *Cadernos de Ciência & Technologia, Brasilia*, 18: 39–68.

Martin, P. and Widgren, J. (1996) International migration: a global challenge. *Population Bulletin*, 51.

Martin, R. (1999) The new economic geography of money. In R. Martin (ed.), *Money and the Space-Economy*. Chichester: Wiley. Chapter 1.

Martin, R. and Sunley, P. (2003) Deconstructing clusters: chaotic concept or policy panacea? *Journal of Economic Geography* 3: 5–36.

Maslow, A.H. (1970) *Motivation and Personality*, 2nd edn. New York: Harper & Row.

Mason, M. (1994) Historical perspectives on Japanese direct investment in Europe. In M. Mason and D. Encarnation (eds), *Does Ownership Matter? Japanese Multinationals in Europe*. Oxford: Clarendon. Chapter 1.

Massey, D. (2000) The geography of power. In B. Gunnell and D. Timms (eds), *After Seattle: Globalization and its Discontents*. London: Catalyst.

Mathews, J.A. (1997) A Silicon Valley of the East: creating Taiwan's semiconductor industry. *California Management Review*, 39 (4): 26–54.

Mathews, J.A. (1999) A Silicon Island of the East: creating a semiconductor industry in Singapore. *California Management Review*, 41 (2): 55–78.

Mathews, J.A. and Cho, D.–S. (2000) *Tiger Technology: The Creation of a Semiconductor Industry in East Asia*. Cambridge: Cambridge University Press.

McConnell, J. and Macpherson, A. (1994) The North American Free Trade Agreement: an overview of issues and prospects. In R. Gibb and W. Michalak (eds), *Continental Trading Blocs: The Growth of Regionalism in the World Economy*. Chichester: Wiley. Chapter 6.

McGregor, R. and Harvey, F. (2006) The polluter pays: how environmental disaster is straining China's social fabric. *Financial Times*, 27 January: 17.

McHale, J. (1969) *The Future of the Future*. New York: Braziller.

McMichael, P. (1997) Rethinking globalization: the agrarian question revisited. *Review of International Political Economy*, 4: 630–62.

McNeill, J. (2000) *Something New Under the Sun: An Environmental History of the Twentieth Century*. London: Allen Lane.

Metcalfe, J.S. and Dilisio, N. (1996) Innovation, capabilities and knowledge: the epistemic connection. In J. de la Mothe and G. Paquet (eds), *Evolutionary Economics and the New International Political Economy*. London: Pinter. Chapter 3.

Micklethwait, J. and Wooldridge, A. (2000) *A Future Perfect: The Challenge and Hidden Promise of Globalization*. New York: Crown Business.

Miles, R. and Snow, C.C. (1986) Organizations: new concepts for new forms, *California Management Review*, XXVIII: 62–73.

Miles, R., Snow, C.C., Mathews, J.A. and Miles, G. (1999) Cellular-network organizations. In W.E. Halal and K.B. Taylor (eds), *Twenty-First Century Economics: Perspectives of Socioeconomics for a Changing World*. New York: St Martin's. Chapter 7.

Miller, D. (1995) Consumption as the vanguard of history. In D. Miller (ed.), *Acknowledging Consumption: A Review of New Studies*. London: Routledge. Chapter 1.

Millstone, E. and Lang, T. (2003) *The Atlas of Food*. London: Earthscan.

Min, S. and Keebler, J.S. (2001) The role of logistics in the supply chain. In J.T. Mentzer (ed.), *Supply Chain Management*. Thousand Oaks, CA: Sage. Chapter 10.

Mirza, H. (2000) The globalization of business and East Asian developing-country multinationals. In N. Hood and S. Young (eds), *The Globalization of Multinational Enterprise Activity and Economic Development*. London: Macmillan. Chapter 9.

Mitchell, K. (2000) Networks of ethnicity. In E. Sheppard and T. Barnes (eds), *A Companion to Economic Geography*. Oxford: Blackwell.

Mitchell, W. (1995) *City of Bits: Space, Place and the Infobahn*. Cambridge, MA: MIT Press.

Mittelman, J.H. (2000) *The Globalization Syndrome: Transformation and Resistance*. Princeton, NJ: Princeton University Press.

Mockler, R.J. (2000) *Multinational Strategic Alliances*. Chichester: Wiley.

Mol, M.J., van Tulder, R.J.M. and Beije, P.R. (2005) Antecedents and performance consequences of international outsourcing. *International Business Review*, 14: 599–617.

Morgan, G., Kristensen, P.H. and Whitley, R. (eds) (2004) *The Multinational Firm: Organizing Across Institutional and National Divides*. Oxford: Oxford University Press.

Morrison, A.J. and Roth, K. (1992) The regional solution: an alternative to globalization. *Transnational Corporations*, 1: 37–55.

Mortimore, M. (1998) Mexico's TNC–centric industrialization process. In R. Kozul-Wright and R. Rowthorn (eds), *Transnational Corporations and the Global Economy*. London: Macmillan. Chapter 13.

Mortimore, M. and Vergara, S. (2004) Targeting winners: can foreign direct investment policy help developing countries industrialise? *The European Journal of Development Research*, 16: 499–530.

Muller, A.R. (2004) *The Rise of Regionalism: Core Company Strategies under the Second Wave of Integration*. Rotterdam: Erasmus University.

Myrdal, G. (1958) *Rich Lands and Poor*. New York: Harper & Row.

Mytelka, L.K. and Barclay, L.A. (2004) Using foreign investment strategically for innovation. *The European Journal of Development Research,* 16: 531–60.

Nelson, R.R. (ed.) (1993) *National Innovation Systems: A Comparative Study*. New York: Oxford University Press.

Nike Inc. (2005) *Second Corporate Responsibility Report*. Beaverton, OR: Nike Inc. (www.nikeresponsibility.com/reports).

Nixson, F. (1988) The political economy of bargaining with transnational corporations: some preliminary observations. *Manchester Papers in Development*, IV: 377–90.

Nolan, P. (2001) *China and the Global Business Revolution*. London: Palgrave.

Nordås, H.K. (2004) The global textile and clothing industry post the Agreement on Textiles and Clothing. WTO Discussion Paper, 5.

Nye, J.S. Jr (2002) *The Paradox of American Power: Why the World's Only Superpower Can't Go It Alone*. Oxford: Oxford University Press.

O'Brien, K.L. and Leichenko, R.M. (2000) Double exposure: assessing the impacts of climate change within the context of economic globalization. *Global Environmental Change*, 10: 221–32.

O'Brien, R. (1992) *Global Financial Integration: The End of Geography*. London: Royal Institute of International Affairs.

OECD (2004) *A New World Map in Textiles and Clothing: Adjusting to Change*. Paris: OECD.

OECD (2005a) *Economic Survey of China*. Paris: OECD.

OECD (2005b) *OECD Regions at a Glance*. Paris: OECD.

Offe, C. (1996) *Varieties of Transition: The East European and East German Experience.* Cambridge: Polity.

Ohmae, K. (1985) *Triad Power: The Coming Shape of Global Competition.* New York: Free Press.

Ohmae, K. (1990) *The Borderless World: Power and Strategy in the Interlinked Economy.* New York: Free.

Ohmae, K. (ed.) (1995) *The Evolving Global Economy: Making Sense of the New World Order.* Boston, MA: Harvard Business Review Press.

O'Rourke, K.H. and Williamson, J.G. (1999) *Globalization and History: The Evolution of a Nineteenth-Century Atlantic Economy.* Cambridge, MA: MIT Press.

Orrù, M., Biggart, N.W. and Hamilton, G.G. (1997) *The Economic Organization of East Asian Capitalism.* Thousand Oaks, CA: Sage.

O'Shaughnessy, J. (1995) *Competitive Marketing: A Strategic Approach*, 3rd edn. London: Routledge.

Oshima, H.T. (1983) On the coming Pacific Century: perspectives and prospects. *The Singapore Economic Review*, 28: 6–21.

Oswald, A.J. (1997) Happiness and economic performance. *Economic Journal*, 107: 1815–31.

Oviatt, B.M. and McDougall, P.P. (2005) Toward a theory of international new ventures. *Journal of International Business Studies*, 36: 29–41.

Owen, G. (1999) *From Empire to Europe: The Decline and Revival of British Industry Since the Second World War.* London: HarperCollins.

Oxfam (2002) *Rigged Rules and Double Standards: Trade, Globalization, and the Fight against Poverty.* London: Oxfam.

Oxfam (2004) *Like Machines in the Fields: Workers without Rights in American Agriculture.* Boston: Oxfam America.

Page, B. (2000) Agriculture. In E. Sheppard and T.J. Barnes (eds), *A Companion to Economic Geography.* Oxford: Blackwell. Chapter 15.

Palan, R. (2003) *The Offshore World: Sovereign Markets, Virtual Places, and Nomad Millionaires.* Ithaca, NY: Cornell University Press.

Palley, T.I. (2004) The economic case for international labour standards. *Cambridge Journal of Economics*, 28: 21–36.

Palloix, C. (1975) The internationalization of capital and the circuit of social capital. In G. Radice (ed.) *International Firms and Modern Imperialism.* Harmondsworth: Penguin. Chapter 3.

Palloix, C. (1977) The self–expansion of capital on a world scale. *Review of Radical Political Economics*, 9: 1–28.

Palpacuer, F., Gibbon, P. and Thomsen, L. (2005) New challenges for developing country suppliers in global clothing chains: a comparative European perspective. *World Development,* 33: 409–30.

Patel, P. (1995) Localized production of technology for global markets. *Cambridge Journal of Economics*, 19: 141–53.

Patel, P. and Pavitt, K. (1998) Uneven (and divergent) technological accumulation among advanced countries: evidence and a framework of explanation. In G. Dosi, D.J. Teece and J. Chytry (eds), *Technology, Organization and Competitiveness.* Oxford: Oxford University Press. pp. 289–317.

Paul, H. and Steinbrecher, R. (2003) *Hungry Corporations: Transnational Biotech Companies Colonise the Food Chain.* London: Zed.

Pauly, L.W. (2000) Capital mobility and the new global order. In R. Stubbs and G.R.D. Underhill (eds), *Political Economy and the Changing Global Order*. Oxford: Oxford University Press. pp. 119–29.

Pauly, L.W. and Reich, S. (1997) National structures and multinational corporate behavior: enduring differences in the age of globalization. *International Organization*, 51: 1–30.

Pearce, D.E. (1995) *Capturing Environmental Value*. London: Earthscan.

Peck, J.A. (1996) *Work-Place: The Social Regulation of Labour Markets*. New York: Guilford.

Peck, J.A. (2001) *Workfare States*. New York: Guilford.

Peck, J.A. (2005) Economic sociologies in space. *Economic Geography*, 81: 129–75.

Peet, R. et al. (2003) *Unholy Trinity: The IMF, World Bank and WTO*. New York: Zed.

Perez, C. (1985) Microelectronics, long waves and world structural change. *World Development*, 13: 441–63.

Phillips, D.R. and Yeh, A.G.O. (1990) Foreign investment and trade: impact on spatial structure of the economy. In T. Cannon and A. Jenkins (eds), *The Geography of Contemporary China: The Impact of Deng Xiaoping's Decade*. London: Routledge. Chapter 9.

Picciotto, S. (1991) The internationalisation of the state. *Capital & Class*, 43: 43–63.

Pitelis, C. (1991) Beyond the nation-state? The transnational firm and the nation-state. *Capital and Class*, 43: 131–52.

Pitelis, C. and Sugden, R. (eds) (1991) *The Nature of the Transnational Firm*, 2nd edn. London: Routledge.

Ponte, S. (2002) The 'latte' revolution? Regulation, markets and consumption in the global coffee chain. *World Development*, 30: 1099–122.

Poon, J.P. (1997) The cosmopolization of trade regions: global trends and implications, 1965–1990. *Economic Geography*, 73: 390–404.

Poon, J.P.H., Thompson, E.R. and Kelly, P.F. (2000) Myth of the triad? The geography of trade and investment 'blocs'. *Transactions of the Institute of British Geographers*, 25: 427–44.

Poon, J.P.H. and Thompson, E.R. (2004) Convergence or differentiation? American and Japanese corporations in the Asia–Pacific. *Geoforum*, 35: 111–25.

Population Reference Bureau (1999) *World Population Data Sheet*. Washington, DC: Population Reference Bureau.

Porteous, D. (1999) The development of financial centres: locations, information externalities and path dependence. In R. Martin (ed.), *Money and the Space-Economy*. Chichester: Wiley. Chapter 5.

Porter, M.E. (1990) *The Competitive Advantage of Nations*. London: Macmillan.

Porter, M.E. (1998) Clusters and the new economics of competition. *Harvard Business Review,* November–December: 77–90.

Porter, M.E. (2000) Locations, clusters and company strategy. In G.L. Clark, M.P. Feldman and M.S. Gertler (eds), *The Oxford Handbook of Economic Geography*. Oxford: Oxford University Press. Chapter 13.

Porter, M.E., Takeuchi, H. and Sakakibara, M. (2000) *Can Japan Compete?* London: Macmillan.

Prahalad, C.K. and Doz, Y. (1987) *The Multinational Mission*. New York: Free Press.

Pritchard, B. (2000a) Geographies of the firm and transnational agro-food corporations in East Asia. *Singapore Journal of Tropical Geography*, 21: 246–62.

Pritchard, B. (2000b) The tangible and intangible spaces of agro-food capital. Paper presented at the International Rural Sociology Association World Congress X, Rio de Janeiro.

Pritchard, B. (2000c) The transnational corporate networks of breakfast cereals in Asia. *Environment and Planning A*, 32: 789–804.

Rabach, E. and Kim, E.M. (1994) Where is the chain in commodity chains? The service sector nexus. In G. Gereffi and M. Korzeniewicz (eds), *Commodity Chains and Global Capitalism*. Westport, CT: Praeger. Chapter 6.

Ramesh, M. (1995) Economic globalization and policy choices: Singapore. *Governance: An International Journal of Policy and Administration*, 8: 243–60.

Ramesh, R. (2005) Chindia, where the world's workshop meets its office. *The Guardian*, 30 September.

Rasiah, R. (2004) Exports and technological capabilities: a study of foreign and local firms in the electronics industry in Malaysia, the Philippines and Thailand. *The European Journal of Development Research*, 16: 587–623.

Ravoli, P. (2005) *The Travels of a T-Shirt in the Global Economy*. New York: Wiley.

Raynolds, L.T. (2004) The globalization of organic agro-food networks. *World Development*, 32: 725–43.

Redding, S.G. (1991) Weak organizations and strong linkages: managerial ideology and Chinese family business networks. In G. Hamilton (ed.), *Business Networks and Economic Development in East and South East Asia*. Hong Kong: Centre for Asian Studies, University of Hong Kong. Chapter 3.

Redding, S.G. (2005) The thick description and comparison of societal systems of capitalism. *Journal of International Business Studies*, 36: 123–55.

Reich, S. (1989) Roads to follow: regulating direct foreign investment. *International Organization*, 43: 543–84.

Reid, H.C. (1989) Financial centre hegemony, interest rates, and the global political economy. In Y.S. Park and N. Essayyad (eds), *International Banking and Financial Centres*. Boston: Kluwer. Chapter 16.

Rennstich, J. (2002) The new economy, the leadership long cycle and the nineteenth K-wave. *Review of International Political Economy*, 9: 150–82.

Richardson, J.D. (1990) The political economy of strategic trade policy. *International Organization*, 44: 107–35.

Rifkin, J. (1995) *The End of Work: The Decline of the Global Labour Force and the Dawn of the Post-Market Era*. New York: Putnam.

Rigby, D.L. and Breau, S. (2006) Impacts of trade on wage inequality in Los Angeles: analysis using matched employer–employee data. UCLA Center for Economic Studies Discussion Paper.

Roberts, S. (1994) Fictitious capital, fictitious spaces: the geography of offshore financial flows. In S. Corbridge, R. Martin and N.J. Thrift (eds), *Money, Power and Space*. Oxford: Blackwell. Chapter 5.

Roche, E.M. and Blaine, M.J. (2000) Telecommunications and governance in multinational enterprises. In M.I. Wilson and K.E. Corey (eds), *Information Tectonics: Space, Place, and Technology in an Electronic Age*. Chichester: Wiley. Chapter 5.

Rodan, G. (1991) *The Political Economy of Singapore's Industrialization*. Petalang Jaya: Forum.

Rodrik, D. (1999) *The New Global Economy and Developing Countries: Making Openness Work*. Washington, DC: Overseas Development Council.

Ross, R.J.S. (2002) The new sweatshops in the United States: how new, how real, how many and why? In G. Gereffi, D. Spener and J. Bair (eds), *Free Trade and Uneven*

Development: The North American Apparel Industry after NAFTA. Philadelphia: Temple University Press. Chapter 5.

Rowthorn, R. and Ramaswamy, R. (1997) De-industrialization: causes and implications. IMF Working Paper.

Rugman, A.M. and Brain, C. (2003) Multinational enterprises are regional, not global. *Multinational Business Review*, 11: 3–12.

Rugman, A.M. and Collinson, S. (2004) The regional nature of the world's automotive industry. *European Management Journal*, 22: 471–82.

Sachs, J. (1997) Nature, nurture, and growth. *The Economist*, 14 June: 19–22.

Sampson, G.P. (ed.) (2001) *The Role of the World Trade Organization in Global Governance.* Tokyo: United Nations University Press.

Sanderson, S.E. (1986) The emergence of the 'world steer': international and foreign domination in Latin American cattle production. In F.L. Tullis and W. Ladd Hollist (eds), *Food, the State, and International Political Economy: Dilemmas of Developing Countries.* Lincoln, NB: University of Nebraska Press. pp. 123–48.

Sassen, S. (1998) *Globalization and its Discontents.* New York: New Press.

Sassen, S. (2001) *The Global City: New York, London, Tokyo.* 2nd edn. Princeton, NJ: Princeton University Press.

Saul, J.R. (2005) *The Collapse of Globalism and the Reinvention of the World.* London: Atlantic.

Saxenian, A. (1994) *Regional Advantage: Culture and Competition in Silicon Valley and Route 128.* Cambridge, MA: Harvard University Press.

Saxenian, A. (2002) Transnational communities and the evolution of global production networks: the cases of Taiwan, China and India. *Industry and Innovation*, 9: 183–202.

Sayer, A. (1986) New developments in manufacturing: the just-in-time system. *Capital & Class,* 30: 43–72.

Schary, P.B. and Skjøtt-Larsen, T. (2001) *Managing the Global Supply Chain*, 2nd edn. Copenhagen: Copenhagen Business School Press.

Scheve, K. and Slaughter, M.J. (2001) *Globalization and the Perceptions of American Workers.* Washington, DC: Institute for International Economics.

Scheve, K. and Slaughter, M.J. (2004) Economic insecurity and the globalization of production. *American Journal of Political Science*, 48: 662–74.

Schiff, M. and Winters, L.A. (2003) *Regional Integration and Development.* Washington, DC: World Bank.

Schoenberger, E. (1997) *The Cultural Crisis of the Firm.* Oxford: Blackwell.

Schoenberger, E. (1999) The firm in the region and the region in the firm. In T. Barnes and M. Gertler (eds), *The New Industrial Geography: Regions, Regulation and Institutions.* London: Routledge. Chapter 9.

Schoenberger, E. (2000) The management of time and space. In G.L. Clark, M.P. Feldman and M. Gertler (eds), *The Oxford Handbook of Economic Geography.* Oxford: Oxford University Press. Chapter 16.

Schumpeter, J. (1943) *Capitalism, Socialism and Democracy.* London: Allen & Unwin.

Scott, A.J. (1988) Flexible production systems and regional development. *International Journal of Urban and Regional Research*, 12: 171–85.

Scott, A.J. (1998) *Regions and the World Economy: The Coming Shape of Global Production, Competition and Political Order.* Oxford: Oxford University Press.

Scott-Quinn, B. (1990) US investment banks as multinationals. In G. Jones (ed.), *Banks as Multinationals*. London: Routledge. Chapter 5.

Seidler, E. (1976) *Let's Call it Fiesta*. London: Stephens.

Sen, A. (1999) *Development as Freedom*. Oxford: Oxford University Press.

Silva, J.A. and Leichenko, R.M. (2004) Regional income inequality and international trade. *Economic Geography*, 80: 261–86.

Simonis, U.D. and Brühl, T. (2002) World ecology – structures and trends. In P. Kennedy, D. Messner and F. Nuscheler (eds), *Global Trends and Global Governance*. London: Pluto. Chapter 4.

Singer, P. (2004) *One World: The Ethics of Globalization*, 2nd edn. New Haven, CT: Yale University Press.

Sklair, L. (ed.) (1995) *Sociology of the Global System*, 2nd edn. Hemel Hempstead: Prentice Hall/Harvester Wheatsheaf.

Sklair, L. (2001) *The Transnational Capitalist Class*. Oxford: Blackwell.

Smarzynska, B.K. and Wei, S.-J. (2001) Pollution havens and foreign direct investment: dirty secret or popular myth? NBER Working Paper 8465.

Smelser, N. and Swedberg, R. (eds) (2005) *The Handbook of Economic Sociology*, 2nd edn. Princeton, NJ: Princeton University Press.

Smith, A., Pickles, J., Begg, B., Roukova, P. and Bucek, M. (2005) Outward processing, EU enlargement and regional relocation in the European textiles and clothing industry. *European Urban and Regional Studies*, 12: 83–91.

Smith, D.M. (1981) *Industrial Location: An Industrial-Geographical Analysis*, 2nd edn. New York: Wiley.

Smith, D.M. (2000) *Moral Geographies: Ethics in a World of Difference*. Edinburgh: Edinburgh University Press.

Sonnino, R. and Marsden, T. (2006) Beyond the divide: rethinking relationships between alternative and conventional food networks in Europe. *Journal of Economic Geography*, 6: 181–99.

Sorj, B. and Wilkinson, J. (1994) Biotechnologies, multinationals, and the agro-food systems of developing countries. In A. Bonanno, L. Busch, W. Friedland and E. Mingione (eds), *From Columbus to ConAgra: The Globalization of Agriculture and Food*. Lawrence, KA: University Press of Kansas. Chapter 4.

Stalk, G., Evans, P. and Shulman, L.E. (1998) Competing on capabilities: the new rules of corporate strategy. *Harvard Business Review*, March–April: 57–69.

Stalk, G. Jr and Hout, T.M. (1990) *Competing against Time: How Time-Based Competition is Reshaping Global Markets*. New York: Free.

Stallings, B. (ed.) (1995) *Global Change, Regional Response: The New International Context of Development*. Cambridge: Cambridge University Press.

Stephens, P. (2005) Why nostalgia is futile at the beginning of history. *Financial Times*, 27 May: 26.

Stiglitz, J.E. (2002) *Globalization and its Discontents*. London: Allen Lane.

Stiglitz, J.E. and Charlton, A. (2005) *Fair Trade for All: How Trade Can Promote Development*. Oxford: Oxford University Press.

Stiglitz, J.E. and Yusuf, S. (eds) (2001) *Rethinking the East Asia Miracle*. New York: Oxford University Press.

Stopford, J.M. and Strange, S. (1991) *Rival States, Rival Firms: Competition for World Market Shares*. Cambridge: Cambridge University Press.

Storper, M. (1992) The limits to globalization: technology districts and international trade. *Economic Geography*, 68: 60–93.

Storper, M. (1995) The resurgence of regional economies, ten years later: the region as a nexus of untraded interdependencies. *European Urban and Regional Studies*, 2: 191–221.

Storper, M. (1997) *The Regional World: Territorial Development in a Global Economy*. New York: Guilford.

Storper, M. and Walker, R. (1984) The spatial division of labour: labour and the location of industries. In L. Sawers and W.K. Tabb (eds), *Sunbelt/Snowbelt: Urban Development and Regional Restructuring*. New York: Oxford University Press. Chapter 2.

Strange, S. (1986) *Casino Capitalism*. Oxford: Blackwell

Strange, S. (1995) The limits of politics. *Government and Opposition*, 30: 291–311.

Strange, S. (1996) *The Retreat of the State: The Diffusion of Power in the World Economy*. Cambridge: Cambridge University Press.

Strange, S. (1998) *Mad Money*. Manchester: Manchester University Press.

Sturgeon, T.J. (2002) Modular production networks: a new American model of industrial organization. *Industrial and Corporate Change*, 11: 451–96.

Sturgeon, T.J. (2003) What really goes on in Silicon Valley? Spatial clustering and dispersal in modular production networks. *Journal of Economic Geography*, 3: 199–225.

Sturgeon, T.J. and Florida, R. (2004) Globalization, deverticalization, and employment in the motor vehicle industry. In M. Kenney and R. Florida (eds), *Locating Global Advantage: Industry Dynamics in the International Economy*. Stanford, CA: Stanford University Press. Chapter 3.

Sturgeon, T.J. and Lester, R.K. (2002) Upgrading East Asian industries: new challenges for local suppliers. Industrial Performance Center, MIT.

Swyngedouw, E. (2000) Elite power, global forces, and the political economy of 'glocal' development. In G.L. Clark, M.P. Feldman and M.S. Gertler (eds), *The Oxford Handbook of Economic Geography*. Oxford: Oxford University Press. Chapter 27.

Taplin, I.M. (1994) Strategic reorientations of US apparel firms. In G. Gereffi and M. Korzeniewicz (eds), *Commodity Chains and Global Capitalism*. Westport, CT: Praeger. Chapter 10.

Taylor, P.J. (1994) The state as container: territoriality in the modern world-system. *Progress in Human Geography*, 18: 151–62.

Taylor, P.J. (2001) Commentary: being economical with the geography. *Environment and Planning A*, 33: 949–54.

Taylor, P.J. (2004) *World City Network: A Global Urban Analysis*. London: Routledge.

Taylor, P.J., Watts, M.J. and Johnston, R.J. (2002) Geography/globalization. In R.J. Johnston, P.J. Taylor and M.J. Watts (eds), *Geographies of Global Change: Remapping the World*. Oxford: Blackwell. Chapter 1.

Terpstra, V. and David, K. (1991) *The Cultural Environment of International Business*. Cincinnati, OH: South-Western.

The Economist (2004) Perpetual motion: a survey of the car industry. 4 September.

The Economist (2005a) Open wider: a survey of international banking. 21 May.

The Economist (2005b) The sun also rises: a survey of Japan. 8 October.

Thoburn, J. and Howell, J. (1995) Trade and development: the political economy of China's open policy. In R. Benewick and P. Wingrove (eds), *China in the 1990s*. London: Macmillan. Chapter 14.

Thrift, N.J. (1990) Doing regional geography in a global system: the new international financial system, the City of London and the South East of England, 1984–1987. In

R.J. Johnston, J. Hauer and G.A. Hoekveld (eds), *Regional Geography: Current Developments and Future Prospects*. London: Routledge. pp. 180–207.

Thrift, N.J. (1994) On the social and cultural determinants of international financial centres: the case of the City of London. In S. Corbridge, R. Martin and N.J. Thrift (eds), *Money, Power and Space*. Oxford: Blackwell. Chapter 14.

Thrift, N.J. (1996) *Spatial Formations*. London: Sage.

Tickell, A. and Peck, J.A. (2003) Making global rules: globalization or neo-liberalization? In J.A. Peck and H.W.-c. Yeung (eds), *Remaking the Global Economy: Economic–Geographical Perspectives*. London: Sage. Chapter 10.

Tobin, J. (1984) Unemployment in the 1980s: macroeconomic diagnosis and prescription. In A.J. Pierre (ed.), *Unemployment and Growth in the Western Economies*. New York: Council for Foreign Relations. pp. 79–112.

Tudor, G. (2000) *Rollercoaster: The Incredible Story of the Emerging Markets*. London: Pearson.

Turner, A. (2001) *Just Capital: The Liberal Economy*. London: Macmillan.

Turner, R.K., Pearce, D. and Bateman, I. (1994) *Environmental Economics: An Elementary Introduction*. Hemel Hempstead: Harvester Wheatsheaf.

Turok, I. (1993) Inward investment and local linkages: how deeply embedded is 'Silicon Glen'? *Regional Studies*, 27: 401–17.

UN Centre for Human Settlements (Habitat) (2001) *Global Report on Human Settlements, 2001*. New York: United Nations.

UN Centre for Human Settlements (Habitat) (2003) *Global Report on Human Settlements, 2003*. New York: United Nations.

UNCTAD (1994) *World Investment Report, 1994: Transnational Corporations, Employment and the Workplace*. New York: United Nations.

UNCTAD (1995) *World Investment Report 1995: Transnational Corporations and Competitiveness*. New York: United Nations.

UNCTAD (2000) *The Competitiveness Challenge: Transnational Corporations and Industrial Restructuring in Developing Countries*. New York: United Nations.

UNCTAD (2001) *World Investment Report 2001: Promoting Linkages*. New York: United Nations.

UNCTAD (2002) *Trade and Development Report 2002: Developing Countries in World Trade*. New York: United Nations.

UNCTAD (2004) *World Investment Report 2004: The Shift towards Services*. New York: United Nations.

UNCTAD (2005) *World Investment Report 2005: Transnational Corporations and the Internationalization of R&D*. New York: United Nations.

UNDP (1999) *Human Development Report 1999*. New York: Oxford University Press.

UNDP (2005) *Human Development Report 2005*. New York: United Nations.

UNIDO (2003) *The Global Automotive Industry Value Chain: What Prospects for Upgrading by Developing Countries?* Vienna: UNIDO.

UN Population Division (2001) *World Population Prospects: The 2000 Revision*. New York: United Nations.

UN Population Division (2005) *World Population Prospects: The 2004 Revision*. New York: United Nations.

UN World Commission on the Environment and Development (1987) *Our Common Future*. Oxford: Oxford University Press.

US Department of Commerce (2000) *Digital Economy 2000.* Washington, DC: US Department of Commerce.

Van Grunsven, L., van Egeraat, C. and Meijsen, S. (1995) New manufacturing establishments and regional economy in Johor: production linkages, employment and labour fields. Department of Geography of Developing Countries, University of Utrecht Report Series.

Van Tulder, R. with van der Zwart, A. (2006) *International Business–Society Management: Linking Corporate Responsibility and Globalization.* London: Routledge.

Veloso, F. and Kumar (2002) The automotive supply chain: global trends and Asian perspectives. Asian Development Bank ERD Working Paper 3.

Vernon, R. (1966) International investment and international trade in the product cycle. *Quarterly Journal of Economics,* 80: 190–207.

Vernon, R. (1979) The product cycle hypothesis in a new international environment. *Oxford Bulletin of Economics and Statistics,* 41: 255–68.

Vernon, R. (1998) *In the Hurricane's Eye: The Troubled Prospects of Multinational Enterprises.* Cambridge, MA: Harvard University Press.

Vitols, S. (ed.) (2004) *The Transformation of the German Model?* Special issue of *Competition and Change,* 8 (4).

Voltaire (Arouet, F.-M.) (1947) *Candide,* trans. J. Butt. London: Penguin.

Wade, R. (1990) Industrial policy in East Asia: does it lead or follow the market? In G. Gereffi and D.L. Wyman (eds), *Manufacturing Miracles: Paths of Industrialization in Latin America and East Asia.* Princeton, NJ: Princeton University Press. Chapter 9.

Wade, R.H. (1996) Globalization and its limits: reports of the death of the national economy are greatly exaggerated. In S. Berger and R. Dore (eds), *National Diversity and Global Capitalism.* Ithaca, NY: Cornell University Press. Chapter 2.

Wade, R.H. (2003) What strategies are viable for developing countries today? The World Trade Organization and the shrinking of 'development space'. *Review of International Political Economy,* 10: 621–44.

Wade, R.H. (2004a) *Governing the Market: Economic Theory and the Role of Government in East Asian Industrialization,* 2nd edn. Princeton, NJ: Princeton University Press.

Wade, R.H. (2004b) Is globalization reducing poverty and inequality? *World Development,* 32: 567–89.

Wallerstein, I. (1979) *The Capitalist World-Economy.* Cambridge: Cambridge University Press.

Ward, N. and Almås, R. (1997) Explaining change in the international agro-food system. *Review of International Political Economy,* 4: 611–29.

Warf, B. (1989) Telecommunications and the globalization of financial services. *Professional Geographer,* 41: 257–71.

Warf, B. (2006) International competition between satellite and fiber optic carriers: a geographic perspective. *Professional Geographer,* 58: 1–11.

Warf, B. and Purcell, D. (2001) The currency of currency: speed, sovereignty, and electronic finance. In T.R. Leinbach and S.D. Brunn (eds), *Worlds of e-Commerce: Economic, Geographical and Social Dimensions.* Chichester: Wiley. Chapter 12.

Watts, M.J. and Goodman, D. (1997) Agrarian questions. Global appetite, local metabolism: nature, culture and industry in *fin-de-siècle* agro-food systems. In D. Goodman and M.J. Watts (eds), *Globalizing Food: Agrarian Questions and Global Restructuring.* London: Routledge. Chapter 1.

Webber, M.J. and Rigby, D.L. (1996) *The Golden Age Illusion: Rethinking Postwar Capitalism.* New York: Guilford.

Weidenbaum, M. and Hughes, S. (1996) *The Bamboo Network: How Expatriate Chinese Entrepreneurs Are Creating a New Economic Superpower in Asia.* New York: Free Press.

Weiss, L. (1998) *The Myth of the Powerless State: Governing the Economy in a Global Era.* Cambridge: Polity.

Weiss, L. (ed.) (2003) *States in the Global Economy: Bringing Domestic Institutions Back In.* Cambridge: Cambridge University Press.

Wells, L.T. Jr (ed.) (1972) *The Product Life Cycle and International Trade.* Boston: Harvard Business School Press.

Whatmore, S. (2002) *Hybrid Geographies: Natures, Cultures, Spaces.* London: Sage.

Whatmore, S., Stassart, P. and Renting, H. (2003) What's alternative about alternative food networks? *Environment and Planning A,* 35: 389–91.

Whatmore, S. and Thorne, L. (1997) Nourishing networks: alternative geographies of food. In D. Goodman and M.J. Watts (eds), *Globalizing Food: Agrarian Questions and Global Restructuring.* London: Routledge. Chapter 11.

Whitley, R. (1992) *Business Systems in East Asia: Firms, Markets and Societies.* London: Sage.

Whitley, R. (1999) *Divergent Capitalisms: The Social Structuring and Change of Business Systems.* Oxford: Oxford University Press.

Whitley, R. (2004) How and why are international firms different? The consequences of cross-border managerial coordination for firm characteristics and behaviour. In G. Morgan, P.H. Kristensen and R. Whitley (eds), *The Multinational Firm: Organizing across Institutional and National Divides.* Oxford: Oxford University Press. Chapter 2.

Wilkinson, J. (2002) The final foods industry and the changing face of the global agro-food system. *Sociologia Ruralis,* 42: 329–46.

Wilkinson, R.G. (2005) *The Impact of Inequality: How to Make Sick Societies Healthier.* London: Routledge.

Williamson, H. (2005) Under attack. *Development and Cooperation,* April.

Wills, J. (1998) Taking on the CosmoCorps? Experiments in transnational labour organization. *Economic Geography,* 74: 111–30.

Wolf, M. (2002) Countries still rule the world. *Financial Times,* 6 February.

Wolf, M. (2004) *Why Globalization Works.* New Haven, CT: Yale University Press.

Womack, J.R., Jones, D.T. and Roos, D. (1990) *The Machine that Changed the World.* New York: Rawson Associates.

Wong, K.Y. and Chu, D.K.Y. (eds) (1995) *Coordination in China: The Case of the Shenzhen Special Economic Zone.* Hong Kong: Oxford University Press.

Wood, A. (1994) *North–South Trade, Employment, and Inequality.* Oxford: Oxford University Press.

World Bank (2001) *World Development Report, 2001: Attacking Poverty.* New York: Oxford University Press.

World Bank (2005a) *World Development Indicators, 2005.* Washington, DC: World Bank.

World Bank (2005b) *World Development Report, 2005.* Washington, DC: World Bank.

Wrigley, N. (2000) The globalization of retail capital: themes for economic geography. In G.L. Clark, M.P. Feldman and M.S. Gertler (eds), *The Oxford Handbook of Economic Geography.* Oxford: Oxford University Press. Chapter 15.

WTO (2005) *The Future of the WTO: Addressing Institutional Changes in the New Millennium.* Geneva: World Trade Organization.

Yao, S. (1997) The romance of Asian capitalism: geography, desire and Chinese business. In M.T. Berger and D.A. Borer (eds), *The Rise of East Asia: Critical Visions of the Pacific Century*. London: Routledge. Chapter 9.

Yeung, H.W.-c. (1998) The political economy of transnational corporations: a study of the regionalization of Singaporean firms. *Political Geography*, 17: 389–416.

Yeung, H.W.-c. (1999) Regulating investment abroad: the political economy of the regionalization of Singaporean firms. *Antipode*, 31: 245–73.

Yeung, H.W.-c. (2000) The dynamics of Asian business systems in a globalizing era. *Review of International Political Economy,* 7: 399–433.

Yeung, H.W.-c. (2001) Organising regional production networks in South East Asia: implications for production fragmentation, trade, and rules of origin. *Journal of Economic Geography*, 1: 299–321.

Yeung, H.W.-c. (2002) *Entrepreneurship and the Internationalization of Asian Firms.* Cheltenham: Elgar.

Yeung, H.W.-c. (2004) *Chinese Capitalism in a Global Era: Towards Hybrid Capitalism.* London: Routledge.

Yeung, H.W.-c., Liu, W. and Dicken, P. (2006) Transnational corporations and network effects of a local manufacturing cluster in mobile telecommunications equipment in China. *World Development*, 34: 520–40.

Yeung, H.W.-c. and Olds, K. (eds) (2000) *Globalization of Chinese Business Firms*. London: Macmillan.

Yeung, H.W.-c., Poon, J. and Perry, M. (2001) Towards a regional strategy: the role of regional headquarters of foreign firms in Singapore. *Urban Studies*, 38: 157–83.

Yeung, Y.-M. and Lo, F.-C. (1996) Global restructuring and emerging urban corridors in Pacific Asia. In Y.-M. Yeung and F.-C. Lo (eds), *Emerging World Cities in Pacific Asia*. Tokyo: United Nations University Press. Chapter 2.

Yoffie, D.B. (ed.) (1993) *Beyond Free Trade: Firms, Governments, and Global Competition*. Boston: Harvard Business School Press.

Yoffie, D.B. and Milner, H.V. (1989) An alternative to free trade or protectionism: why corporations seek strategic trade policy. *California Management Review*, 31: 111–31.

Young, D., Goold, M. et al. (2000) *Corporate Headquarters: An International Analysis of their Roles and Staffing*. London: Pearson.

Zanfei, A. (2000) Transnational firms and the changing organization of innovative activities. *Cambridge Journal of Economics*, 24: 515–42.

Ziltener, P. (2004) The economic effects of the European Single Market Project: projections, simulations – and the reality. *Review of International Political Economy*, 11: 953–79.

Zook, M.A. (2001) Old hierarchies or new networks of centrality: the global geography of the internet content market. *American Behavioral Scientist*, 44: 1679–96.

Zook, M.A. (2005) *The Geography of the Internet Industry*. Oxford: Blackwell.

Index

About the Author

Peter Dicken is Emeritus Professor of Economic Geography in the School of Environment and Development at the University of Manchester, UK. He has held visiting academic appointments at universities and research institutes in Australia, Canada, China, Hong Kong, Mexico, Singapore, Sweden, and the United States, and lectured in many other countries throughout Europe and Asia. He is an Academician of the Social Sciences, is a recipient of the Victoria Medal of the Royal Geographical Society (with the Institute of British Geographers), and holds an Honorary Doctorate from the University of Uppsala, Sweden.